# The Direct Use of
# COAL
## PROSPECTS AND PROBLEMS OF PRODUCTION AND COMBUSTION

# The Direct Use of
# COAL

## PROSPECTS AND PROBLEMS OF PRODUCTION AND COMBUSTION

**GRAND RIVER BOOKS**
1249 Washington Blvd. • Detroit, Mich. 48226

**Bibliographic Note**

Originally published by the Office of Technology
Assessment, Congress of the United States.
Washington, D.C., 1979.

**Library of Congress Cataloging in Publication Data**

United States. Congress. Office of Technology
    Assessment.
    The direct use of coal.

    Includes bibliographical references and index.
    1. Coal--United States.  I. Title.
TN800.U54   1980      333.8'22'0973      80-39563
ISBN 0-8103-1022-8

# Contents

# Foreword

Coal production in the United States will more than double by the year 2000, to as much as 2 billion tons per year. The investments in facilities, manpower, and technologies for the mining, transportation, and use of such vast quantities of coal will be immense.

This report assesses the benefits and risks of such a massive shift to coal and away from other fuels. It deals with the social, economic, physical, and biological impacts of such a shift. It examines the complete coal system, from extraction to combustion, including the key steps and institutions that policy can influence. The environmental impacts and possible effects on the public health are documented. Occupational and community impacts are examined. Technologies under development that might reduce costs, increase convenience of use, or mitigate negative impacts are reviewed.

This assessment was requested by the House Committee on Science and Technology. It is another in a series of assessments that are being provided to Congress to assist in the development of national energy policy.

The Project Director for this assessment was Mr. Alan T. Crane, who was supported by the Energy Group staff and Dr. Richard E. Rowberg, Group Manager. We are grateful for the assistance provided in this assessment by the Coal Advisory Panel, chaired by Mr. Harry Perry, and the overall guidance provided by OTA's Energy Advisory Committee, chaired by Professor Milton Katz.

Daniel De Simone
*Acting Director*

# OTA Energy Advisory Committee

Milton Katz, *Chairman*
Director, International Legal Studies, Harvard Law School

Thomas C. Ayers
*President and Chairman of the Board*
*Commonwealth Edison Company*

Kenneth E. Boulding
*Professor of Economics*
*Institute of Behavioral Science*
*University of Colorado*

Eugene G. Fubini
*Fubini Consultants, Ltd.*

Levi (J.M.) Leathers
*Executive Vice President*
*Dow Chemical USA*

Wassily Leontief
*Department of Economics*
*New York University*

George E. Mueller
*President and Chairman of the Board*
*System Development Corporation*

Gerard Piel
*Publisher*
*Scientific American*

John F. Redmond, *Retired*
*Shell Oil Company*

John C. Sawhill
*President*
*New York University*

Chauncey Starr
*Electric Power Research Institute*

# Direct Use of Coal Advisory Panel

Harry Perry, *Chairman*
*Resources for the Future, Inc.*

Mike Clark
*Highlander Research Center*

David Comey*
*Citizens for a Better Environment*

A. W. Deurbrouck
*Department of Energy*

Michael Enzi
*Mayor, City of Gillette*

Don Gasper
*Consolidation Coal Company*

W. L. Johns
*E. I. du Pont de Nemours & Co.*

Lorin Kerr
*United Mine Workers of America*

George Land
*AMAX Coal Corporation*

Ed Light
*West Virginia Citizens Action Group*

Robert Lundberg
*Commonwealth Edison*

Paul Martinka
*American Electric Power (Retired)*

David Mastbaum
*Environmental Defense Fund*

Ken Mills
*Tug Valley Recovery Center*

Michael Rieber
*University of Arizona*

Steve Shapiro
*United Mine Workers of America, LU 6025*

Ronald Surdam
*University of Wyoming*

Joana Underwood
*INFORM*

Ralph Perhac
*Electric Power Research Institute*

Joseph Yancik
*National Coal Association*

The Advisory Panel provided advice and constructive criticism throughout this project. The panel does not, however, necessarily approve, disapprove, or endorse this report. OTA assumes full responsibility for the report and the accuracy of its contents.

---

*Died January 1979. The OTA Coal Project staff expresses their sincere appreciation for the assistance and support that David Comey had contributed throughout this assessment.*

# Direct Use of Coal Project Staff

Lionel S. Johns, *Assistant Director*
*Energy, Materials, and Global Security Division*

Richard E. Rowberg, *Energy Group Manager*

Alan T. Crane, *Project Director*

Marvin Ott, *Policy*
Steven Plotkin, *Environment and Public Health*
A. Jenifer Robison, *Regulatory Issues*
Curtis Seltzer, *Mining Implications and Community Impacts*
Joanne Sedor, *Research Associate*

Lisa Jacobson    Lillian Quigg    Rosaleen Sutton

*Supplements to Staff*

Robert Jimeson
Barbara G. Levi
W. Gene Tucker
L. William Richardson

*Contributors*

Lynda L. Brothers
J. Bradford Hollomon

# OTA Publishing Staff

John C. Holmes, *Publishing Officer*
Kathie S. Boss    Joanne Heming

# Summary

Coal can be America's most productive domestic energy source well into the 21st century. No other fuel, however, evokes such memories of environmental damage and social disruption. Although new programs, practices, and legislation have addressed many of these problems, their adequacy and effectiveness are uncertain, and new concerns are emerging. Thus, coal's future is a mixture of promise and risk.

This report analyzes the role that coal combustion may play in the Nation's energy future, the factors that will shape that role, and the impacts that may result.

## SUPPLY AND DEMAND

A tripling of coal production and use by 2000 appears to be possible without either substantial regulatory relaxation or major technological innovations. Such rapid growth may be in the national interest even though total energy demand will probably grow more slowly than it has in the past. However, several factors might inhibit this rapid rate of development. Primary among these is the likelihood that demand for coal will not rise this quickly. Coal is not normally the fuel of choice unless it offers a large cost advantage. Many users will prefer oil and natural gas as long as they are available, despite their higher prices, because of their greater convenience and lower capital investment requirements. Recent regulations have increased the cost of mining and burning coal, thus reducing its attractiveness. The 1978 National Energy Act discourages large new facilities from burning gas or oil, but additional legislation or other incentives may be required if coal's share of the energy system is to be further increased. If demand does triple, several potential supply constraints could still interfere with achieving a tripling of coal supply. Mine-to-market transportation systems will need upgrading and expansion. Federal coal leasing may have to resume. Conflicts between mine labor and management could escalate to the point where national production is compromised. Some communities, particularly in the East, may need assistance to accommodate the required development. Public concerns about the siting of mines, powerplants, and other facilities may delay expansion.

In addition, questions have been raised about the reliability of available control technologies and their ability to meet environmental regulations. Significant problems with these technologies could constrain future coal use. The technological viability of flue-gas desulfurization (FGD) to meet proposed air pollution emission standards is a focus of this controversy. A growing body of evidence, however, indicates that this technology will be adequate though expensive. Safe disposal of FGD sludge also appears to be feasible technologically, but expensive. Finally, requirements of the Surface Mine Control and Reclamation Act will raise the cost of coal but should not pose technical barriers to meeting demand.

## ENVIRONMENT AND PUBLIC HEALTH

This report does not identify any significant violations of existing environmental standards that inevitably would result from a substantial rise in coal use. However, some adverse environmental impacts are likely to occur with increased coal mining and combustion because of conscious economic tradeoffs, lack of efficient control devices or insufficient enforcement of existing standards. For example, some underground mines will cause acid mine drainage or subsidence problems long after their abandonment; permanent controls for these

ix

problems are not available for all mining situations. Some emissions may prove to be inadequately regulated because of scientific uncertainty concerning the cause and significance of their effects. Emissions of sulfur dioxide, nitrogen dioxide, carbon dioxide ($CO_2$) and fine particulates will all increase significantly by the end of the century under a high coal scenario and present or pending regulations. Increases in coal-related air pollution may cause or aggravate three potential environmental problems whose magnitude and importance cannot yet be evaluated adequately. Additional evidence of these problems could force a reappraisal of our environmental control strategy or even our commitment to increased coal development.

First, fossil fuel combustion (along with the clearing of forest land) appears to be causing an atmospheric $CO_2$ buildup that may lead to significant changes in global climate. $CO_2$ is not likely to become a serious problem before the next century, but if it does, a dramatic worldwide reduction in both fossil fuel combustion and deforestation may be the only way to halt or reduce climate change. Coal is the fossil fuel of greatest concern because of its large reserves and high carbon content; presumably, the faster coal use increases the sooner a critical point will be reached and the more difficult it will be to switch to nonfossil fuels.

Second, there is a possibility that present ambient air standards are inadequate to protect the public health from problems resulting from long-term exposure to low levels of pollutants. Some controversial analyses suggest that current exposure levels may be responsible for tens of thousands of premature deaths annually. Increases in emissions of coal combustion-related pollutants could aggravate existing problems.

Third, acid rain, which has been linked to extensive damage to aquatic ecosystems and may also cause terrestrial damage, may increase as a result of increased emissions of sulfur and nitrogen oxides from coal combustion.

If additional evidence confirms current concerns, further reduction in air emissions may be desirable to reduce the health and environmental risks associated with increasing U.S. coal use. Existing coal-burning facilities would be primary targets for emission regulations, since many of these will not be stringently controlled under State air quality enforcement plans. The technical and economic problems of retrofitting capital-intensive control systems on existing plants can be severe. Thus, emission reductions may be most efficiently realized by supplying these facilities with cleaned coal. The various techniques for coal cleaning (physical desulfurization, solvent refined coal, and synthetic fluids) may become the primary component of a national strategy for achieving the dual objective of increased use of coal and an improved environment.

## OCCUPATIONAL SAFETY AND HEALTH

Considerable improvement has been recorded in reducing mine fatalities since the enactment of the 1969 Federal Coal Mine Health and Safety Act, but there has been no reduction in the rate of disabling injuries, and the total number of injuries has been increasing. In 1977, 139 coal workers were killed and about 15,000 suffered a disabling injury—each of which cost an average of almost two months lost time. A tripling of production, under the assumptions of this report, will result in approximately 370 fatalities and 42,000 disabling injuries per year.

The prevalence of coal workers' pneumoconiosis should be lower in 2000 than it was in 1970 as a result of the Federal respirable dust standard, but thousands of workers will nevertheless be disabled by various respiratory diseases (collectively called black lung disease). Furthermore, important questions have been raised concerning the effectiveness of enforcement of the existing dust standard and the adequacy of the research on which it is based.

## COMMUNITY IMPACTS

Coal development brings both beneficial and adverse social and economic consequences. Communities with diversified economies are better equipped to deal with rapid development and are more likely to benefit. Where existing services are already under severe stress, coal development is likely to bring additional social and economic problems. Public services and facilities will not be able to expand sufficiently to cope with increased demand. Tax revenues will not increase fast enough to finance needed improvements.

Some communities in Appalachia will be particularly vulnerable to these adverse impacts. Stagnant coal demand and excess production capacity during the 1920-70 period led to depressed economic conditions in the mining industry and communities. Inadequate services and facilities are a legacy of these conditions. Increased coal production without substantial new revenues and technical aid from the Federal or State governments is likely to exacerbate current inadequacies in housing, transportation, health care, and other public and private services.

Some western communities may suffer serious social and economic dislocations if mining expands too rapidly. States with underground reserves are especially vulnerable. Development of these reserves will require many new workers and will bring substantial population growth to sparsely populated areas. Mining of Indian coal may create significant social problems on reservations. However, with some exceptions, the tax revenues that could be generated by coal development can provide significant opportunities for economic benefit. To minimize economic dislocations, provision will have to be made to overcome the initial shortfalls of funds that follow the onset of development. Also, mismatches between areas absorbing the impacts and those reaping the economic benefits must be resolved.

Finally, the diversified economies of many Midwestern coal counties will allow them to benefit from coal development. These counties combine the population and adequate service base needed to absorb development without severe stress.

The following section describes some of the related policy issues that Congress may consider. A detailed summary of the entire report is found in the overview of chapter I.

## ISSUES AND OPTIONS

Legislative activity bearing on coal will fall into three categories: further amelioration of negative impacts; promotion of coal to reduce the use of other fuels; and oversight to ensure that existing legislation is being appropriately implemented. This section lists the major steps discussed in this report that Congress may consider. The options are not recommendations. Some, in fact, may be contradictory. The options are intended only to show how this analysis can be used to reach a given end, the desirability of which depends in part on the perspective of the policymaker. Cross references to the relevant sections of the main report are included.

### Control of Coal-Related Air Pollution

Coal combustion is accompanied by a risk of health damage (possible increased incidence of illness and premature death) and ecological disturbances (e.g., by acid rainfall) from air pollution.

Although existing pollution sources are rapidly being controlled by State implementation plan (SIP) regulations and new sources will be subject to new source performance standards and other regulations, ambient concentrations of fine particulates and nitrates could rise rapidly (and those of sulfates more slowly) during the next few decades if coal development continues at the expected rate. The reasons for these increases include: failure of particulate control requirements to distinguish among different particle sizes, allowing the use of controls that are less effective in controlling the smaller (and more dangerous) particles; unavailability of efficient $NO_x$ controls; and basing of SIP emission control requirements on local impacts only, thus allowing certain sources to escape tight controls because their pollution is widely dispersed.

The tradeoff between emission requirements and costs of control is severely complicated by the scientific controversy surrounding the effects associated with these pollutants. Scientists still do not agree on the existence or importance of some of the effects or their precise cause.

Options

- Increase Federal research emphasis on the impacts and mechanisms of long-range pollutant transport. Increase Federal effort on development of more effective, less expensive controls for $NO_x$, $SO_x$, and fine particulates. Emphasize clean fuels and low cost technological controls suitable for retrofit.

If the public health and ecological risks are seen as sufficiently dangerous that further limitations should be applied before the risks are better defined by further research, the following actions should be considered:

- Establish emission and ambient standards for fine particulates.
- Revise the Clean Air Act to allow direct Federal control of existing sources of pollution. Actions could range from requiring recently built facilities to comply with NSPS requirements, to selective emission reductions on large facilities when their State Implementation Plan requirements are lenient or when they are located in areas that are the source of problems associated with long range transport of pollutants.
- Revise the Clean Air Act to allow regional and/or Federal control of siting of major facilities when they might contribute to pre-existing long range transport problems.
- Give maximum priority to development of $NO_x$ controls.

## The Role of Cost Benefit Analysis in Environmental Decisionmaking

The need to weigh costs against benefits has recently become a critical topic of debate in regulatory decisionmaking. An array of environmental decisions affecting coal development will be affected by the outcome of this debate. For example, the Environmental Protection Agency has recently proposed new source performance standards (NSPS) for sulfur dioxide emissions from coal-fired utility boilers. The additional cost of installing and operating the equipment (flue gas desulfurization) to meet these standards is of great concern to utilities. Alternative strategies have been proposed that could grant utilities somewhat relaxed standards that could be met at lower cost. No one, however, has produced a convincing analysis of the relative value of the different strategies in ameliorating adverse impacts—public health risks, crop, ecological and material damage, etc.—to compare with the costs.

The NSPS debate shares a common problem with other environmental issues: the costs of controls are more easily computed than their environmental benefits. Thus, a fair balancing of costs and benefits in setting environmental standards is always difficult and frequently impossible.

Text discussions:

## Option A:

Maintain the status quo. The present regulatory emphasis is on the use of "Best Available (pollution) Control Technology," although the agencies have used legislative language requiring consideration of the cost of achieving the standards as a restraint on how expensive the required controls can be. (The degree of their "restraint'is of course a subject of some disagreement).

## Option B:

Require regulatory agencies to develop and publish "impact statements" that state their estimates of the costs and benefits of proposed standards and regulations.

## Option C:

Establish a mechanism to decide whether a forthcoming standard must be set by an explicit balancing of costs and benefits. The basis for this decision would be the state of the art of impact assessment for the pollutant in question.

## Occupational Health

Despite a strict standard for respirable coal mine dust mandated by Congress in 1969, miners continue to be exposed to harmful dusts and fumes. The current standard should greatly lower the number of miners disabled by coal workers' pneumoconiosis if it is complied with regularly. Exposure to nonrespirable dust (associated with other respiratory diseases) and other toxic substances is neither measured nor regulated. Hearing loss is another major occupational hazard. The present dust and noise sampling programs are of questionable reliability. Some mines may not be in compliance even when their sample data say they are. Even with current standards and diligent compliance, it is likely that thousands of miners will show evidence of work-related lung disease in the future.

Options

- Reassess the inherent safeness of the current respirable dust standard.
- Consider alternatives to current dust sampling, including continuous in-mine monitoring, and possibly more effective ways of carrying out sampling, such as miner- or MSHA-control of the program.
- Encourage the establishment of health standards for other harmful substances— nonrespirable dust, trace elements, fumes, etc. —that are now unregulated.
- Consider lowering the Federal noise standard for mining.
- Promote occupational health training for miners.

## Occupational Safety

Coal mining involves higher probabilities of occupational death and disabling injury than almost any other trade. Although coal's fatality rate has improved considerably with the implementation of the 1969 Federal Coal Mine Health and Safety Act, the frequency and number of disabling injuries—about 15,000 in 1977—has not improved. Both fatalities and injuries will rise in proportion to increases in production unless accident frequency rates are lowered.

Options

- Consider the feasibility of requiring or encouraging conversion to the "safest available mining equipment" (adjusted to individual mine characteristics), consistent with the intent of the 1969 and 1977 Federal mine safety laws.

- Establish Federal safety standards for mine equipment.
- Require 90-day apprentice training before a miner is allowed to operate an unfamiliar piece of mobile mine equipment.
- Clarify the right of individual miners to withdraw from conditions of imminent danger under Federal law.
- Establish Federal limits on fatality and injury frequency for different kinds of mines. Substantial penalties could be levied against mine operators exceeding these limits.
- Establish performance standards for the Mine Safety and Health Administration.

## Community Impacts

In communities where coal mining has dominated local life, rapid coal growth tends to produce a broad range of socio-economic problems. The chronic underdevelopment of community services and facilities associated with earlier coal development has left a substantial legacy of needs. Where mining has been a negligible factor in local affairs as in the West, rapid growth can easily overwhelm existing institutions. The level of current need and ability to manage growth depends on site-specific conditions, coal company policies and the local tax system, among other factors. Expanding mining communities suffer from a lack of housing, planning capability, experience with Federal programs, and money to develop services and facilities concurrently with mine development. Such needs can be expected to be met at the state and local level after mining growth stabilizes if the local economy is healthy. Chronically underdeveloped and temporarily overwhelmed communities, however, would benefit from Federal assistance if appropriate funding channels can be devised.

Text discussion: Eastern . . . . . . Ch. VI
              Western . . . . . Ch. VI

Options

- Enact a national severance tax on coal (or on other fuels to avoid an economic disin-
centive to use coal) to help finance needed improvement in impacted communities.
- Provide loans or subsidies for programs to provide better public services in these places, possibly through a public, non-profit coalfield development bank.
- Develop programs that would encourage economic diversification in coal-dominated regions.
- Require operators to submit a "community impact statement" to local and Federal officials before mining begins.
- Promote access to land in communities where it is presently unavailable for housing.
- Improve flood control measures in Appalachian coalfields as a way of encouraging investment in coal and noncoal development and housing.
- Maintain roads by enforcing coal truck weight limits and institute a highway coal haulage tax to fund road reconstruction.
- Institute a rural health care system for coalfields.

## Labor-Management Relations

The history of coal's labor-management relations is a chronicle of turmoil caused by uncertain markets, hostile attitudes, and inequities. Although lost time from strike activity increased in the 1970s over the 1950s and 1960s, no enduring coal shortage has been experienced, even during the 3½ month UMWA strike in 1977-78. Strikes disrupt normal production patterns, but experienced coal purchasers take adequate precautions by accumulating large stockpiles. As more and more coal is likely to be mined in non-UMWA operations, strikes may decline in national importance. Nevertheless, the potential for intermittent disruption will exist as coal assumes a more central role in the Nation's energy system. Government intervention has generally been resisted by both unions and operators, and it has not been effective. Harsh measures aimed at either party may do substantial long-term damage to labor-management relations even when they work in the short term.

Text discussion:

## Options

- Create contingency plans for emergency coal shortages.
- Promote programs to improve social conditions in the coalfields, thereby removing sources of conflict.
- Plan for government seizure of the mines in the event of a crippling strike.

## Increasing the Use of Coal

This report assumes coal use will grow substantially. Such growth is consistent with present expectations of energy needs, supply of alternative fuels and potential constraints on coal supply. National policy could seek to further increase coal's share of the total energy supply, perhaps to reduce dependency on foreign oil or nuclear power. Measures to stimulate demand would then be considered. The mining industry is expected to be able to meet any plausible demand for at least the next decade, but thereafter, constraints on production could become serious, especially if demand rises very rapidly.

## Options

- Increase demand for coal by strengthening penalties and incentives in the National Energy Act. Industrial users will be the most likely target for such initiatives (chapter II).
- Improve technology to make coal combustion more attractive. Clean coal and synthetic fuels are essentially the only ways of using coal in small furnaces and boilers while meeting clean air standards. Fluidized bed combustion may be an asset to industrial use of coal (III-G and IV-L)
- Facilitate supply by ensuring an early resumption of Federal leasing (IV-A), and

improving the transportation systems (IV-J).
- Reduce the costs and constraints of mining and combustion by easing regulatory standards. Although this study has not found any fundamental incompatability between increased coal use and existing environmental regulations, reclamation and emission controls are expensive. A less drastic step would be to institute more flexible regulations designed to be less stringent where the potential for damage is less.

## Oversight

Legislation has addressed many major impacts created in the past by coal production and use. Some issues and programs may require particularly intensive oversight to assure that congressional intent is being carried out. The list below identifies the most important such programs discussed in this report.

- Development of long-range air-pollution transport models and emplacement of ambient air-quality monitoring to determine the effect of siting and energy policy decisions. Epidemiological and ecological studies to determine the impacts of air quality changes (V.B.).
- Enforcement of control of surface mine runoff during operation and reclamation, especially for small mines. Development of ways to control acid mine drainage and subsidence from abandoned underground mines (V.C. and D.).
- Enforcement of adequate dust control in mines through sampling or continuous monitoring (VI.A.).
- Resumption of leasing of Federal coal with appropriate safeguards (IV.A. and VII.A.2.).
- Development of Federal strategies to assist the upgrading of the coal transportation system (IV.J.).
- Enforcement of current mine safety and health standards through inspections and penalties.
- Conversion of oil and gas burning facilities to coal under the National Energy Act (II and VII.A.8.).

Chapter I

# INTRODUCTION AND OVERVIEW

# Chapter I.—INTRODUCTION AND OVERVIEW

# Chapter I
# INTRODUCTION AND OVERVIEW

## INTRODUCTION

The increased use of coal has become the focus of a national debate over the need for energy and the desire to maintain a healthy physical and social environment.

U.S. reserves of coal are vast, and the technologies of coal production and combustion are well developed. Thus coal offers a major alternative to U.S. dependence on foreign supplies of oil. However, coal development has a history of exploitation and turmoil in the coalfields of Appalachia, of cities laden with soot and noxious fumes, and of destruction of land and water resources. Although an extensive regulatory system and other pressures have ameliorated many of coal's historic problems, it still retains potential for damage. Many believe that rapid development of U.S. coal resources cannot be accomplished without unacceptable damage both to the natural environment and to the communities where coal mining and combustion occur.

Federal policymakers face difficult choices in dealing with coal's adverse impacts. On the one hand, strict controls almost always result in higher costs of mining and using coal. These costs may inhibit demand for coal and slow its pace of development. Also, efforts to control one adverse impact sometimes cause another. For example, the scrubbers used to reduce sulfur oxide ($SO_x$) pollution produce large quantities of sludge that represent a land use and potential water pollution problem. Finally, gaps in our knowledge of the importance and causes of certain environmental effects create doubt as to whether all required control measures are necessary or effective. Failure to adequately control an environmental or health impact, however, may result in adverse consequences that outweigh the costs of control. These consequences may not be fully measurable in dollars and may take years to occur. In addition, the public opposition to new coal mines and powerplants aroused by a failure to control environmental and health impacts may be a more significant deterrent to development than are high prices. On the whole, Federal policy has tried both to encourage increased production and use of coal and to lower its environmental, occupational, and social costs. The attempt to write legislation and regulations that balance these two goals has been marked by controversy, with extensive litigation between environmental and coal development interests as one result.

This report presents the results of a broad study of the mining and direct combustion of coal. There are three major themes. The first is to determine coal's potential contribution to future U.S. energy needs. Next is an assessment of the environmental and social impacts that may result from rapid coal development. Finally, there is a guide for policy initiatives that Congress may consider in addressing a variety of coal-related issues. In addition, the report provides information on impacts and on new technologies that should be useful for legislative oversight of Federal energy and environmental R&D programs. It identifies those impacts that have escaped the regulatory system and those that have been only partially controlled. In many cases, it also describes the means available to eliminate or mitigate these impacts.

Assessment of adverse environmental impacts is particularly controversial because information is lacking in critical areas. Many important issues will not be resolved conclusively for many years. Nevertheless, decisions on economic, environmental, and social policy to be made over the next few years will be based on this incomplete information. Hence, some of this report's conclusions are tentative, based on an evaluation of the existing evidence. For those areas where the evidence is particularly controversial, alternative arguments are discussed.

quences of coal use with either the use of other fuels or with energy conservation. No alternative to coal development is free of adverse consequences; these must be taken into account before coal's "place" in the overall U.S. energy picture can be determined. Second, the report does not provide a complete and definitive accounting of the costs and benefits associated with coal mining and combustion. Not all costs and benefits are known or knowable. Often, they cannot be measured in the same terms. Therefore, the report attempts to identify the uncertainties surrounding the impacts described. The existence of high levels of uncertainty, as well as the strong role that personal values play in gauging the importance of impacts, clearly has profound implications for policymaking.

The report is organized as follows:

- Chapter II surveys U.S. energy patterns and alternative forecasts of national energy demand and coal supply.
- Chapter III describes the coal resource base and mining and combustion technologies. It also describes technologies avail-

able to control air pollution from coal combustion.

- Chapter IV analyzes the factors that will affect coal production and use, including leasing, industry structure, labor relations, regulatory restrictions, physical and economic constraints, and public attitudes.
- Chapter V describes the environmental and public health effects associated with mining and combustion under the current regulatory system.
- Chapter VI describes the occupational health and safety aspects of mining and the socioeconomic effects on communities that can be expected from increased coal use.
- Chapter VII reviews current legislation regulating coal production and use.
- Chapter VIII analyzes policy dilemmas that accompany coal development and outlines Federal policy options in the areas of environment and health, community impacts, labor-management relations, occupational health and safety, and coal leasing.

# OVERVIEW

This overview focuses on the elements of the report most useful to policymakers. It discusses the reasons for expecting coal use to grow rapidly and the manner in which that growth may occur. It then describes the expected environmental and social impacts. These impacts have been addressed by a number of Federal policy initiatives. Because such Federal actions will influence future coal development, they are discussed in the context of all other factors that will affect coal production and use. Finally, policy options are outlined that might be considered either to encourage coal use or to curb remaining adverse impacts.

## Energy and the Role of Coal

The importance of coal must be gauged in the context of the energy system of which it is

a part. Both the demand for energy in general and the supply of other fuels will affect the use of coal. Energy demand depends on the efficiency of use and the level of demand for goods and services that consume energy. Rising population and real gross national product will increase demand, while increases in efficiency of use in response to higher prices and Government policies will reduce it. Most estimates of energy demand in 2000 fall between 100 and 150 quadrillion Btu (Quads), compared with 73.1 Quads used in 1975. The upper end of this range is becoming increasingly unlikely because of a leveling off of population, rapidly increasing efficiency of use, and indications that many of the goods and services that consume energy are reaching saturation levels.

Coal's share of future energy demand will depend on its availability, cost, and attractiveness as compared to alternative fuels. Total U.S. petroleum and natural gas production cannot expand much and may well decline by 2000 unless ways are found to tap presently uneconomical resources such as geopressurized gas. Thus oil and gas are likely to continue to be discouraged as fuels for powerplants, leaving coal and nuclear energy as the only two commercial options for electric utilities. Nuclear power will expand rapidly as reactors on order are completed, but further growth faces uncertainties in public acceptance, costs, schedules, and (possibly) uranium supplies. Specific site characteristics such as proximity to coal and environmental factors will affect future selections. As coal and nuclear are generally competitive, policy decisions affecting one may well influence the other.

Oil, gas, and electricity will be coal's major competitors in the industrial and residential/commercial sectors. Solar energy and other new fuels have great potential, but economics and the logistics of expansion may delay extensive use until the next century. Coal use by industry will be increased by legal restrictions on future oil and gas use. Requiring smaller users to shift to coal will incur problems of high costs and less effective pollution control. New technologies such as fluidized-bed combustion (FBC), thoroughly cleaned coal, and synthetic fuels may be necessary if small facilities are to increase greatly their use of coal. The residential/commercial sector will be especially deterred by the inconvenience of delivery, storage, and ash disposal. Without rapid growth in central heating plants, the residential sector will only slightly increase coal consumption as long as other fuels are available.

Achieving a 150-Quad energy supply by 2000 would call for a very successful oil and gas discovery rate, substantial imports, an upsurge in orders for nuclear powerplants, and an all-out expansion of coal. Problems with any of these would preclude such a rapid expansion of supply.

Most energy scenarios assume a rapid expansion of coal. By 2000, production could be two to three times the present level. The regional distribution of production (in millions of tons) for this range is estimated as follows:

| | 1977 | Year 1985 | 2000 |
|---|---|---|---|
| **Surface mines** | | | |
| Appalachia .. | 185 | 130-155 | 130-175 |
| Midwest...... | 91 | 75-95 | 95-135 |
| West ........ | 141 | 415-495 | 700-1,005 |
| Total ...... | 417 | 620-745 | 925-1,315 |
| **Underground mines** | | | |
| Appalachia .. | 205 | 225-260 | 380-505 |
| Midwest..... | 54 | 60-80 | 120-180 |
| West ....... | 13 | 50-60 | 80-110 |
| Total ...... | 272 | 355-400 | 580-795 |
| **Total** | | | |
| Appalachia .. | 390 | 355-415 | 510-680 |
| Midwest..... | 145 | 135-175 | 215-315 |
| West ....... | 154 | 460-510 | 780-1,115 |
| Grand total | 689 | 955-1,145 | 1,505-2,110 |

The sharpest increases in coal consumption will be in the West-Central and Mountain regions. All regions east of the Mississippi (except New England) will sustain a more gradual growth in their already comparatively high consumption. By 2000, consumption will be much more evenly distributed than now with all regions except New England and the Pacific States burning large quantities.

The actual level and distribution of production and combustion will depend largely on the costs of production, transportation, and combustion. Surface mining is generally cheaper than underground mining. Large mechanized mines exploiting thick seams enjoy substantial cost advantages over smaller mines working thin seams. For example, new, western surface mines might produce coal at $5/ton while an Appalachian underground mine could have production costs of $30. The increasing share of Western coal will keep average coal costs down, but transportation costs can be significant if the shipping distance is more than a few hundred miles. Thus, Western coal is not necessarily competitive with Eastern coal for a particular site. Combustion costs depend on coal characteristics and the technology to accommodate them. For example, boilers burning high-sulfur or high-ash coals require more expensive control equipment. Coal characteristics vary widely even within a seam, but can be characterized roughly by region. In general, Eastern coal has a high heat content but often has a high concentration of sulfur. Midwestern coal is slightly lower in heat value and higher in sulfur, while Western coal is lowest in heat value and sulfur. These factors are weighed by potential users. Policy initiatives affecting these costs (e.g., limits on sulfur emissions) influence decisions of whether to use coal and where to purchase it.

## Environmental Impacts and Controls

If coal does indeed stage the comeback that has been forecast, it will return to prominence in a manner vastly different from the way it dominated national energy use in the past. The availability of pollution controls, better combustion techniques, and new mining methods, coupled with enforcement of a wide range of environmental protection requirements, should prevent a repetition of much of the environmental degradation—soot-laden cities, scarred landscapes, ruined and discolored streams—that accompanied coal development in the past. However, despite the laws and new equipment and techniques, large-scale coal development may still be accompanied by substantial environmental impacts. Some of these impacts could result from inadequacies in the enforcement of the laws or in the environmental controls. Other impacts may result from the failure to regulate a damaging pollutant or to specify an adequate level of protection from a regulated pollutant. These kinds of failures usually result from inadequate knowledge: the inability to recognize a subtle but important impact, to connect a known impact to its correct source, or to determine properly the quantitative relationship between impact and source.

This very real deficiency in our knowledge of environmental processes makes it difficult to determine whether current plans for coal development could cause unacceptable environmental impacts. Some of the more spectacular impacts that have been attributed to coal development—for example, the warming of the Earth's atmosphere by increasing levels of carbon dioxide ($CO_2$) (a possible long-term effect), or the thousands of premature deaths attributed to the particulate sulfate products of sulfur dioxide ($SO_2$) emissions (an effect said

to be occurring now)—represent risks rather than certainties. Scientists disagree sharply on the extent of the risks, greatly increasing the difficulty of developing environmental policies. Part of this disagreement involves sharply differing opinions about the quality of data and the validity of analytical methodology. Part also involves more basic philosophical differences about the nature of "proof." Because many environmental relationships are drawn from circumstantial and statistical evidence, considerable judgment must be used in determining when a "postulated" relationship turns into a "probable" one, and finally, into a "proven" one. The long fight to conclusively prove a relationship between smoking and cancer is a classic example; many environmental cause-and-effect relationships follow the same lines.

Finally, one additional problem with predicting environmental impacts is that their magnitude depends on the effectiveness of pollution control systems, and this can change. The regulatory systems often will respond to newly perceived environmental threats by requiring more stringent controls. Because it cannot be predicted how well the future environment will be monitored, the extent to which ongoing research will discover new evidence linking particular pollutants to specific impacts, or how policymakers will respond to such evidence, the discussion below focuses on the effects of coal development under the current regulatory system and attempts to place the postulated, probable, and proven environmental impacts into perspective. However, virtually all of the most severe of the impacts described below are capable of being mitigated or eliminated by controls that are available today or are under active development.

### Impacts on Air Quality

Although the mining and transportation of coal can cause local air pollution problems from fugitive dust (from mining operations, storage piles, and coal hauling) and noxious fumes (from smoldering mine fires), the major, national air quality impacts from large-scale coal development will come from the combustion portion of the fuel cycle—from coal-fired powerplants, industrial boilers and furnaces, and commercial space-heating plants. The air pollutants released in large quantities by combustion units include the oxides of sulfur, nitrogen, and carbon, as well as particles of ash that become entrained in the hot flue gases. Smaller quantities of trace inorganic elements, radionuclides, and hydrocarbons are also emitted; these are often adsorbed on the surface of the ash particles.

Of the major pollutants, $SO_x$ and nitrogen oxides ($NO_x$)—in the form of $SO_2$ and nitrogen dioxide ($NO_2$)—and particulate matter are controlled directly by Federal law. The present strategy for regulating these pollutants under the Clean Air Act combines a series of "local" control levels—incorporated in State implementation plan (SIP) requirements to meet National Ambient Air Quality Standards (NAAQS) and Prevention of Significant Deterioration (PSD) requirements—with New Source Performance Standards (NSPS) that set nationwide emissions limits on large new pollution sources. In general, the regulations are directed at the chemical form of the pollutants as they are emitted from their source, and are focused on maintaining air quality requirements within a local air quality control region. Unfortunately, the pollutants are neither chemically nor physically static within the atmosphere. Most are subject to complex chains of chemical transformations, with the importance of each reaction depending on the presence of catalysts, intensity of sunlight, degree of humidity, and other factors. Physical processes are also at work, mixing the pollutants and at times carrying them far from their source until they are removed from the atmosphere by settling, by colliding with terrestrial surfaces, or by being "washed out" by rainfall. Examples of long-range transport include instances of "hazy blobs" of urban pollutants crossing the boundaries of several States and persisting for periods of a week or more, and elevated levels of sulfate particulates in Pennsylvania that originate with $SO_2$ emissions from a cluster of coal-fired powerplants in the Ohio River Basin that are alined with the prevailing winds.

Two key problems are created by this long-range transport and transformation:

1. Areas can experience episodes of poor air quality caused by pollution sources that are outside their jurisdictional boundaries and thus essentially beyond their control under the current regulatory system.
2. The pollutants that may be present in the air and that may be doing the most damage are not necessarily the same as the pollutants being regulated. Controls on the "primary" emitted pollutants may not be effective on the "secondary," transformed pollutants.

A major reason why Federal regulatory strategy does not explicitly take these factors into account is that the analytical ability to trace pollutants from source to "receptor," through both chemical transformation and long-distance physical transport, is inadequate. Moreover, the identification and measurement of the physical impacts of these secondary pollutants are often disputed. To identify specific problems, each of the major pollutants, its transformation products, and its impacts is examined in turn.

**Sulfur oxides** have received more attention than any other emission from coal combustion, primarily because of the large quantity of emissions, the diversity and controversy surrounding its impacts (human health effects, acid rain, crop damage, etc.), and the great expense involved in $SO_x$ controls. Coal combustion is now, and is likely to continue to be, the major source of $SO_x$ emissions in the United States. Although existing coal combustion sources will be coming into compliance with local regulations in the next few years and new sources will be subject to strict NSPS, the expected expansion in coal combustion should prevent total $SO_x$ emissions from continuing to decline significantly in the next decade and should cause them to rise (although slowly) thereafter.

The $SO_2$ that emerges from the stack of a coal-fired boiler is either removed (primarily by impaction in an area close to its source) from the atmosphere or transformed into sulfuric acid or to some other form of sulfate ion. These sulfates tend to be transported over a wide area and are removed predominantly by rainfall. Transport of sulfates over distances of hundreds of miles is not uncommon.

The major effect of the locally deposited $SO_2$ is to damage crops and forests during meteorological conditions that cause the "plume" (column of exhaust gases from the stack) to touch the ground. Lower plant species may suffer severe losses downwind of coal-burning facilities, while a wide variety of crops are damaged when other pollutants react with $SO_2$. However, these effects should not increase greatly in severity on a nationwide basis because of the relative stability of $SO_2$ emissions for the next few decades and because new powerplants will operate with $SO_2$ scrubbers.

The major effects of the sulfate transformation products are continuation and possible aggravation of the effects of acid rain (discussed below) and the possibility of major human health effects. Both of these are excellent examples of research inadequacies; scientists as yet cannot credibly describe the relationship between emissions and the ambient concentrations that may occur far from the pollution source, and do not agree on the magnitude of the impacts. Improvement and verification of some existing long-range air pollution models may soon allow acceptable predictions of ambient concentrations caused by distant sources. This would still leave unresolved the prediction of the environmental and health impacts from these concentrations.

The estimation of human health effects from sulfates or from any air pollutant is extraordinarily difficult. Short-term tests on cultures and tests on animals can employ high levels of pollutants and thus can measure easily identifiable acute effects; however, the relevance of these effects to human beings is uncertain. Clinical studies utilize human subjects but only at low levels of concentration; the impacts that are measured are not easily translatable into estimates of more dangerous

effects at high concentration levels or at low levels over long periods of time. Human epidemiological studies have tended to suffer from problems with heterogeneous populations, poorly measured pollution exposures, and multiple pollutants.

A series of epidemiologic analyses of the relationship between mortality rates and air pollution in several American cities has linked current levels of sulfate (and particulate) concentrations to tens of thousands of premature deaths yearly in this country. These analyses suffer from the general problems associated above with most epidemiological studies as well as inadequate data on those population characteristics that might affect the death rate, and they have been rejected on these grounds by many health scientists. However, the arguments advanced by these scientists are inconclusive and have not invalidated the analyses. It remains a possibility that existing levels of air pollution are causing significant numbers of deaths. (Sulfate should be considered as a pollution indicator rather than as necessarily the prime cause, although presumably the causative pollutants are produced in association with the sulfate precursor, $SO_2$.) If the relationship between current levels of air pollution and large numbers of premature deaths were proven, or if it were perceived as proven by a politically significant portion of the population, future coal development could be substantially affected by demands for restrictions on development, deliberate shifts to alternative energy sources, or increases in pollution control requirements and cost.

Although coal combustion is now secondary to automobiles as a source of **Nitrogen oxides**, the present inability of coal-fired boilers to reduce their uncontrolled emissions by more than about half will yield a substantial increase—perhaps 20 percent by 1990—in total U.S. $NO_2$ emissions (unless more efficient controls are developed and used) despite further cleanup of automobile exhausts.

The threshold for almost all observed effects on ecosystems is well above the $NO_2$ levels in most U.S. cities, and the forecast increase in emissions should not cause significant new damage from $NO_2$. However, $NO_2$ is a precursor of photochemical oxidants such as ozone and peroxyacyl nitrates (PAN). Oxidants are the most damaging air pollutants affecting agriculture and forestry in the United States. Elevated ozone concentrations caused by long-range transport from urban areas have become a regional problem throughout the United States, causing widespread damage to crops on both coasts. However, the relationship between $NO_2$ emissions and oxidant formation is not well understood; thus, although increased $NO_2$ emissions may be expected to cause some increased oxidant formation and subsequent ecosystem damages, the severity of this outcome is highly uncertain.

The transformation of $SO_2$ and $NO_2$ into acid sulfate and nitrate and the long-range transport of these acid products can damage the environment by producing acid rain. The transformation products of $SO_2$ and $NO_2$ may be the major contributors to the current acidity of rainfall over most of the Eastern United States. The resulting increases in the acidity of lakes have seriously degraded some aquatic ecosystems. Although acid rain has been shown to affect individual components of the terrestrial ecosystem, it has not been proven to affect that ecosystem to the same extent that it clearly affects aquatic systems. However, postulated associations of acid rain with some ominous trends (such as declining forest growth) and potential impacts (such as leaching of toxic metals and damage to nitrogen-fixing bacteria from increasing soil acidity) suggest an urgent need for more research on acid rain effects. The problem is especially important in the Northeastern United States. The Southwest, which will absorb substantial amounts of acid sulfate and nitrate, will not be as troubled because its soil and water are predominantly alkaline.

Because of the widespread use of electrostatic precipitators (ESPs) that remove large particles with high efficiency but are less efficient in controlling smaller particles, any ecosystem impact of particulate emissions from coal-fired powerplants will be caused almost exclusively by **fine particulates** and the **associated trace elements and hydrocarbons** that are adsorbed on their surfaces. Although total particulate emissions will be substantially reduced in the future as a result of strict controls on new plants and progress in obtaining conformance with State regulations for existing plants, emissions of fine particulates may increase unless controls are installed that are equally effective for all particle size ranges.

The physical characteristics of fine particulates make them candidates for potential ecosystem and health damages, but virtually no data exist to verify such effects. Speculation about ecosystem damages is based on the potential for alteration of normal soil processes associated with nutrient recycling and with soil micro-organisms, as well as a small amount of evidence of trace hydrocarbon damage to aquatic systems. Similarly, there is a speculative potential for a human health impact based on the ability of fine particulates to penetrate the lung's defenses, and the coating of toxic materials on the particles. The same types of controversial epidemiological studies that have implicated sulfate as being associated with premature deaths have also implicated particulates.

In recognition of the possible health implications of increasing fine-particulate emissions, the 1977 Clean Air Act amendments require the Environmental Protection Agency (EPA) to study the health effects of fine particulates, associated trace elements, and polycyclic organic matter (POM) and to establish regulatory controls for these pollutants if necessary.

Both the fine particulates directly emitted from coal combustion facilities and the particulate sulfate that are the transformation products of $SO_2$ emissions are effective in scattering light and thus causing a degradation of visibility. The Southwest in particular would appear to be vulnerable to this damage, because regional shifts in power generation will add considerably to its pollution burden, and its vistas are an important resource. Although Federal PSD restrictions are designed in part to protect western visibility, the state of the art in predicting the visibility impacts of new coal-fired powerplants — or of any light-scattering particles — is not well advanced. Although visibility impacts may occur hundreds of miles from a large source, EPA has specified 50 km as the maximum distance over which current air quality models are credible for evaluating the effects of new pollution sources — and most modelers would consider this an optimistic assessment of the state of the art. Thus, the restrictions on visibility reductions will be difficult to enforce, until acceptable long-range transport and visibility models are available.

Most of the impacts discussed above would originate from large coal-fired powerplants or industrial boilers. Few predictions of energy growth expect major increases in coal combustion by small residential and commercial boilers or furnaces. Small boilers or furnaces, if poorly maintained, can experience incomplete combustion and generate elevated levels of hydrocarbon emissions. A major worry in such a situation would be increased generation of POM, some species of which are carcinogenic. Because these emissions would tend to come from low stacks, possibly in densely populated areas, special attention must be paid to these units if larger than expected growth of direct coal use occurs in the residential and commercial sectors.

Despite the serious uncertainties involved in identifying air pollution impacts and their causes, all but one of the major pollutants from coal combustion have sufficiently recognized impacts and means of control to have warranted Federal regulation of their emissions. The exception is **Carbon dioxide**, which at current and expected ambient levels displays no direct or immediate adverse impacts on human health or on the biota but may conceivably represent the greatest long-term danger from an increase in the use of coal or other fossil fuels. Fossil fuel combustion over the past century appears to be a major cause of increasing concentrations of $CO_2$ in the Earth's atmosphere (deforestation may be another major cause); $CO_2$ levels have increased 5 percent since 1958 alone. Some predictions show $CO_2$ concentrations as doubling by the middle of the next century. This could present a substantial risk of significant climatic change, because $CO_2$ in the Earth's atmosphere has a "greenhouse effect," allowing incoming sunlight to warm the Earth's surface but trapping outgoing heat radiation. Effects of such a climate change, if it occurred, could include massive shifts in the productivity of farmlands as well as partial melting of the polar icecaps and flooding of coastal cities. Current gaps in our understanding of how climate is regulated and how $CO_2$ is cycled between its sources and reservoirs leave this issue surrounded by considerable uncertainty. Although the problem is widely perceived by the scientific community as potentially serious, some scientists believe that any effect would be overwhelmed by the natural climatic cycle that may be moving the Earth to a cooler future climate. A further critical aspect of the problem is the apparent lack of a practical $CO_2$ emission control technology for fossil combustion. Should control of $CO_2$ be judged necessary, the available options are to reduce worldwide fossil fuel combustion (by energy conservation and switching to alternative energy sources) and to stop ecosystem changes (especially deforestation) that might be aggravating the problem.

As discussed above, each of the pollutants generated by coal combustion will have proven or postulated environmental and/or health impacts, even with the level of pollution control currently required by Federal, State, and local regulations. Our inability to assess accurately the level of impacts and their economic and social values leads to extreme difficulty in selecting appropriate levels of control. These inadequacies partially explain the Federal strategy of choosing the "best available control technology" instead of using an approach that would select control levels more oriented to weighing costs and benefits. Whichever approach is chosen, the selection of controls clearly requires an understanding of the control options and the associated costs and difficulties.

Feasible control options are available for three of the four major pollutants generated by coal combustion — particulates, $SO_2$, and $NO_2$. The fourth, $CO_2$, can theoretically be scrubbed out of the flue gases, but the quantity of absorbent required and volume of waste products generated appear to be too large for serious consideration.

Control of $SO_x$ emissions from large coal combustion sources can be accomplished by a variety of measures, including selection of low-sulfur coals, coal cleaning, flue-gas desulfurization (FGD), and several new combustion technologies. EPA's proposed NSPS for coalfired electric utilities would guarantee that virtually all new powerplants would use FGD (scrubbers) to reduce $SO_x$ emissions. However, stringent control of new plants and achievement of full compliance with SIP requirements for existing plants will not substantially reduce the 30 million or so tons of $SO_x$ emitted annually in the United States, because the SIP requirements are often not very severe. Powerplants, which have tall stacks, can in many circumstances emit very large amounts of pollutants without seriously affecting local air quality. In some instances, plants are burning relatively high-sulfur coal without controls even though they are in full compliance with their SIPs. Thus, there is considerable room for further reduction of $SO_x$ emissions if such a reduction were dictated by national policy.

The scrubbers required on all new powerplants are expensive to build and operate. The scrubber system on a 500-MW powerplant might cost $50 million to $75 million out of a total plant cost of $400 million, and add 0.4 to 0.8 cents to each kilowatthour of electricity the plant produces. Furthermore, lime/limestone scrubbers, the systems that currently dominate the utility market, are throwaway processes and therefore generate wastes that are themselves considered a significant environmental problem. This problem is discussed in more detail in the next section.

Scrubber installations in this country have been beset by many significant operating problems, including extensive scaling of surfaces, failure of stack linings, corrosion and erosion of critical components, and plugging of orifices. The utility industry generally considers scrubbers to be an unreliable technology and an inefficient means of achieving control of $SO_x$. There is available evidence, however, that U.S. utility operators can comply with proposed NSPS requirements for $SO_2$ removal efficiency and reliability. For instance, Japanese scrubber experience with medium-sulfur coal (the energy equivalent of 3 percent sulfur U.S. coal) and operating conditions similar to those in U.S. plants has been extremely successful, achieving reliabilities and control efficiencies in excess of 90 percent.

Alternative means of controlling $SO_x$ are available or under development for use on new and existing coal-fired powerplants. Some of the alternatives offer reduction in $NO_x$ and/or particulates in addition to $SO_x$ control.

Further control of existing powerplants can be accomplished by using low-sulfur coal or coal that has been physically or chemically cleaned. Options that may become suitable for new plants include solvent-refined coal (SRC-1, the solid form, which is the only one considered in this report), combined-cycle powerplants using liquid and gaseous fuels from coal, and fluidized-bed combustion (FBC). SRC-1 and FBC may also become useful in promoting environmentally sound coal use in smaller industrial and commercial units. Advanced electrical generation systems are undergoing R&D but do not appear to have the potential to make a serious impact on energy use for at least the next several decades.

The extent of application of these alternatives depends on their costs. Except for low-sulfur coals, state-of-the-art FGD systems, and mechanical cleaning systems now in use, the costs of the alternatives are speculative.

Control of $NO_x$ is currently accomplished by design modification and adjustment of operating conditions rather than using "add-on" controls. Effective techniques for minimizing $NO_x$ emissions are staged combustion, which reduces excess air in the boiler, and burner designs which delay mixing of fuel and air. Combinations of these and other strategies have succeeded in lowering $NO_x$ emissions from large utility boilers by 40 to 50 percent.

Although gas-cleaning systems for $NO_x$ control are not currently used in this country, a number of processes are under development. The Japanese are the most advanced in this field. These processes included injection of ammonia or various solids into the combustion chamber, and scrubbing of the flue gases using absorbents such as magnesium oxide. Capital costs of these systems are said to be of the same order of magnitude as FGD.

EPA is developing a low-$NO_x$ burner for coal-fired boilers that it hopes will be inexpensive and effective. (EPA hopes for an 85-percent emission reduction relative to an uncontrolled boiler.) If successful, the burner could, by the late 1980's, cut back sharply on predicted rises in $NO_x$ emissions.

High efficiency control of particulates has been a long-term practice in the electric utility industry, the dominant user of coal. Although mechanical or cyclone collectors were extensively used in the past, ESPs are the collection technology on most coal-fired utility boilers today. ESPs collect particulates by charging the individual particles and collecting them on plates to which a powerful opposite charge has been applied. They often can attain collection efficiencies of over 99 percent.

Requirements for control of $SO_x$ emissions from coal combustion and resulting increased use of low-sulfur coal have created problems for ESPs. The ash particles from low-sulfur coal are usually of high resistivity and are not easily charged and collected by the ESP. Degradation of performance from a shift to low-sulfur coal can be extremely severe, with particulate emissions increasing tenfold or more. Designers of new powerplants have attempted to solve this problem by increasing the size of the ESPs or installing them at a hotter (lower resistivity) part of the exhaust cycle, or by using controls that depend on the mechanical rather than electrical properties of the particles. These "mechanical" controls include wet scrubbers and baghouses (fabric filters). Although these options may be available to some existing plants, many plants have space limitations or want to avoid large capital expenditures because of their limited remaining operating life. Flue gas conditioning, which involves the injection of a chemical into the flue gas to coat the particles and reduce their resistivity, may be an attractive option for such plants (although questions have been raised

about increased gaseous emissions caused by these systems).

The proposed Federal requirements for particulate control are said to be too stringent to be universally met by ESPs. Compliance with the proposed requirements (a particulate NSPS of 0.03 lbs/million Btu of heat input) for utility boilers burning low-sulfur coal may require the use of baghouses. However, baghouses are a relatively untried technology for large coal-fired boilers. A few systems have been installed or ordered for such plants. The experience of these plants will be influential in determining the future direction of particulate control in the utility industry.

The choice of technology is particularly important in terms of controlling fine particulates and trace elements. ESPs are not efficient collectors of "respirable size" particles, and a continuation of current particulate-control technology could lead to substantial (and potentially dangerous) increases in fine-particulate emissions over the next several decades. Baghouses, on the other hand, can be designed to collect fine particles at greater than 99-percent efficiency.

## Impacts on Land and Water

In contrast to the impacts of air pollution, most impacts of water pollution, solid waste disposal, and land disturbance are more concentrated geographically. They are controlled less by technological devices than by adjustments in operating procedures, and the degree of control obtained is often extremely dependent on local conditions. Thus, appropriate enforcement and careful monitoring are especially important to the success of controlling these impacts.

In the past, coal development in general, and mining in particular, were often devastating to both land and water ecosystems. The major damage from mining was caused by the acid drainage from both underground and surface mines, the lack of adequate restoration of surface-mined land, and the subsidence of lands overlying underground mines. Ecological damage also resulted from the heating of surface waters by powerplant cooling systems.

All of these impacts are now addressed by Federal legislation. As a result, some problems—in particular, acid mine drainage from large active mines, and powerplant thermal pollution—have been virtually eliminated as significant problems for future development. All of the others have been reduced, although substantive problems of enforcement and/or availability of effective controls remain. Also, some new problems may result from the regional shift of coal production to areas where little experience can guide new operations, and from the generation of waste products from air pollution control measures.

**Mining.**—Approximately 60 percent of national coal production comes from surface mines, and the proportion will not rise much. The use of new mining methods that integrate reclamation into the mining process and enforcement of the Surface Mine Control and Reclamation Act (SMCRA) should reduce the importance of reclamation as a critical national issue. However, concern remains that the combination of development pressures and inadequate knowledge may lead to damage in particularly vulnerable areas—alluvial valley floors in the West, prime farmland in the Midwest, and hardwood forests, steep slope areas, and flood-prone basins in Appalachia. Although most of these areas are afforded special protection under SMCRA, the extent of any damage will depend on the adequacy of enforcement of the new strip-mining legislation. Assuming strong enforcement of SMCRA, no major problems with acid mine drainage from active surface and underground mines should result from increased coal development. However, inactive mines may still present some technical control problems. Although a very small percentage of inactive surface mines may suffer from acid seepage, problems with underground mines should be the primary problem. Despite a long history of Federal and State efforts aimed at controlling acid drainage from inactive underground mines, some mining situations do not allow adequate permanent control once active mining and water treatment cease. A significant percentage of the mines that are active at present or that will be opened in this century will present acid drainage problems on

closure. This problem may taper off as shallower reserves are exhausted and new mines begin to exploit coal seams that are deeper than the water table. Many of these mines will be flooded, allowing the seams to be shut off from the oxidation that creates the acid drainage.

Another impact of underground mining that will not be fully controlled is subsidence of the land above the mine workings. Unfortunately, there are no credible estimates of potential subsidence damage from future underground mining. Subsidence, like acid drainage, is a long-term problem. However, SMCRA does not hold the developers responsible for sufficient time periods to ensure elimination of the problem, nor does it specifically hold the developer responsible to the surface owner for subsidence damage. The major "control" for subsidence is to leave a large part of the coal resources—up to 50 percent or more—in place to act as a roof support. There is obviously a conflict between subsidence prevention and removal of the maximum amount of coal. Moreover, the supports can erode and the roof collapse over a long period of time. The resulting intermittent subsidence can destroy the value of the land for development. A second "control" technique—longwall mining—actually promotes subsidence, but in a swifter and more uniform fashion. Longwall mining is widely practiced in Europe but is in limited use in the United States. It is not suitable for all situations.

Although all types of mining have the potential to severely impact ground water quantity and quality by physical disruption of aquifers and by leaching or seepage into them, this problem is imperfectly understood. The shift of production to the West, where ground water is a particularly critical resource, will focus increased attention on this impact. As with other sensitive areas, SMCRA affords special protection to ground water resources, but the adequacy of this protection depends on the state of knowledge about the problem and on the level of enforcement.

**Impacts of Combustion and Waste Disposal.**—The major impact of coal-fired combustion sources on the land and water stems from the secondary effects of environmental controls— the effects of cooling tower blowdown and water consumption, and those of the waste products collected by air pollution controls.

Most powerplants built in the future will use closed-cycle cooling, so that thermal damage from once-through cooling systems will not be an impact of future coal development. However, the concentrated salts, or "blowdown," from these closed-cycle systems are discharged into the Nation's waterways. The dissolved solids discharged by coal-fired electric utilities, of which blowdown is the predominant source, are nearly 20 percent of the total national dissolved solids discharge. Although effluent limitations have been established under the Clean Water Act, increases in utility coal burning will play an important role in the expected substantial growth in the discharge of dissolved solids in the South Atlantic, Midwest, North-Central, and Central regions.

An additional impact of closed-cycle cooling systems results from their water consumption, which is approximately double that attributed to once-through cooling. Although the magnitude of consumption for a particular facility varies with location, a 3,000-MW powerplant can be expected to consume between 20,000 and 30,000 acre-feet per year (mostly for cooling). If a number of these plants are built in the arid portions of the West, their water requirements could exacerbate existing water problems in several river basins—for example, in the Upper Colorado and Yellowstone. Reduced flows can interfere with the rivers' assimilative capacity and in some extreme cases damage or destroy ecosystems from sheer lack of water. This problem is as much institutional as physical, because much of the water consumption in the West results from existing water allocation and pricing procedures that undervalue the water; the absence of a comprehensive Federal water policy hinders the resolution of these problems. Furthermore, technological means (dry or wet-dry cooling systems) are available to sharply reduce powerplant water consumption, although the systems are expensive and lower plant efficiency.

The impacts caused by the disposal of powerplant waste heat into the environment are relatively independent of the fuel source, although impacts from the current generation of nuclear powerplants will be somewhat higher per kilowatt than those of similarly sized fossil-fired plants because of the nuclear plants' lower efficiency. Thus, the effects of blowdown and water consumption should be attributed to electricity demand rather than specifically to coal use.

Both the particulate control devices and $SO_2$ scrubbers on coal combustion facilities produce massive quantities of wastes (projected to be approximately 80 million tons of ash and slag and 20 million tons of sludge by 1985, or fully half of the Nation's total noncombustible solid waste and industrial sludge), which can cause land use problems and environmental damage unless properly managed.

A 1,000-MW powerplant may require a disposal area of 500 acres or larger for a 30-year period, even assuming that the sludge is dried and the wastes are 20 feet thick. However, the land use problem posed by this requirement is eased by the likelihood that most new powerplants will be built outside of densely populated areas, where more land is available.

The major environmental problem associated with waste disposal is the contamination of surface and ground waters by leaching of trace elements from the ash and sludge. Although the ash contains by far the greater amount of trace elements, the fluid that is trapped in the sludge also presents a significant leaching problem for years after disposal.

The actual environmental damage caused by disposal of these wastes will depend on the form of the regulations and the firmness of enforcement of the Resource Conservation and Recovery Act (RCRA). Methods exist for reducing or controlling the potentially severe impact on ground water, but they are expensive and are likely to be applied only sporadically unless they are required by law; some may be difficult to monitor and enforce. These methods range from lining of disposal ponds and landfills, chemical stabilization of the scrubber sludge, and alteration of the chemistry of the scrubber (to create a more manageable waste), to utilizing regenerable scrubber systems and virtually eliminating the sludge portion of the problem altogether. Designation of ash and sludge as hazardous materials under RCRA would force use of these controls, but, according to the Department of Energy (DOE) estimates, could almost double disposal costs over present practices. (Industry thinks these estimates are too low.) A side effect could be to eliminate some present uses of ash, such as its use in roadbeds. Twenty percent of the ash produced by the utility industry is constructively used rather than being disposed. Even without a "hazardous" designation, however, a rigorously enforced RCRA will force substantial changes in present disposal practices to protect ground water.

## Workplace and Community Impacts

### Coal Worker Health and Safety

Coal mining has always been a hazardous occupation. The 1969 Federal Coal Mine Health and Safety Act addressed some work-related health and safety hazards, explosions and dust control in particular. But coal workers are likely to continue to suffer from occupational disease, injury, and death at a rate well above other occupations, and the total magnitude of these impacts will grow along with the growth in coal production.

The mine-worker health issue that has received most Congressional attention is black lung disease, the nonclinical name for a variety of respiratory illnesses of which coal workers' pneumoconiosis (CWP) is the most prominent. More than 420,000 Federal compensation awards were made between 1970 and the end of 1977, costing the Government more than $5.5 billion. The industry will pay a greater share of the compensation costs in the future as a result of the 1977 black lung legislation. Ten percent or more of working coal miners today show X-ray evidence of CWP, and perhaps twice that number show other black lung illnesses—including bronchitis, emphy-

sema, and other impairments—some of which are caused, or worsened, by cigarette smoking. The prevalence of respiratory disease is probably less today than it was 10 years ago because about 50,000 older—and often disabled—miners have retired and about 150,000 new workers have been hired.

To prevent CWP from disabling miners in the future, Congress mandated a 2-mg/m³ standard for respirable dust (the small particles that cause pneumoconiosis). This standard was based on British research done in the 1960's and is lower than any other country's. However, critics now question the inherent safeness of this standard and the soundness of the British methodology. The safeness of the standard, which was based on mathematical probabilities, must still be confirmed by long-term epidemiological evidence. The Federal dust-sampling program, which is intended to measure compliance with the standard, is not a reliable indicator of daily dust exposure. Many opportunities for intentional and inadvertent sampling errors exist. Sampling is so infrequent and the timelag in reporting the results back to the mine so great that sampling has limited relevance to actual dust control. A monitoring program based on continuous sampling and immediate correction may prove to be a more useful approach. Because black lung compensation is costly and because several hundred thousand more miners will be employed in the next 25 years, it may prove useful to reassess the safeness of the dust standard and to evaluate alternative dust-sampling procedures.

If respirable dust is controlled effectively, CWP will be reduced and a major source of black lung disease will have been addressed. However, other sources and potential sources of miners' respiratory disability may require attention. Other coal mine dust constituents—the large dust particles (that affect the upper respiratory tract) and trace elements—deserve additional research. Toxic fumes from mine equipment fires and diesel emissions are hazards that may deserve regulation. Cigarette smoking increases breathing difficulties in diseased miners and probably should be discouraged in this work force. Under the most optimistic assumptions, it is estimated that 11,000 to 18,000 working miners will show X-ray evidence of CWP in 2000 and at least an equal number will exhibit other respiratory impairment. Even more retired miners will have CWP, and many will be disabled by it. If dust control is not effective or adequate, disease prevalence will be higher and more cases of CWP and black lung will be found.

Mine safety—as distinct from mine health— has shown a mixed record of improvement since the 1969 Act was passed. The frequency of mining fatalities has decreased for both surface and underground mines, but no consistent improvement has been seen in the frequency of disabling injuries. Coal worker fatalities numbered 139 in 1977, and disabling injuries approached 15,000. Each disabling injury resulted in an average of 2 months or more of lost time. The number of disabling injuries has been increasing as more workers are drawn to mining and accident frequency remains constant.

On the whole, surface mining is several times safer than underground mining. But some underground mines show safety records equal to or better than some surface mines. Generally, western surface mines are safer than eastern surface mines. As western surface-mine production assumes increasing prominance, accident frequency industrywide is likely to decline when expressed as accidents per ton of output. But this statistical trend may conceal no improvement or even a worsening of safety in deep mines.

Post-1969 mine safety performance is related to several factors. High-fatality mine disasters have been reduced substantially. Operator compliance with Federal safety standards and frequency of inspection appear to be correlated with improved safety. Big, new mines that have been opened since 1970 tend to be safer than older, smaller mines. Greatly improved coal profitability in 1974 and 1975 coincided with the industry's two best years in reducing injuries. The emphasis on safe work by the United Mine Workers of America (UMWA) and Federal agencies were other important factors.

Labor-saving technology or different work processes may or may not improve safety and may conflict with improved health goals. The introduction of continuous miners in the 1950's raised productivity and lowered fatalities and injuries because the work force was reduced by 70 percent. But the new machines produced higher dust levels, which caused higher rates of black lung. Longwall mining systems appear to be more productive than other units and safer in terms of fatalities, but not in injuries. About half of the longwalls surveyed recently did not meet the respirable dust standard. Speeding up the pace of work to increase output is likely to increase accident frequency. Improved workplace relations and less absenteeism are likely to increase both productivity and safety.

Mine-safety analysis is made difficult by weaknesses in the recorded Federal data. Reporting practices are not uniform throughout the industry. Some companies have adopted light-duty policies (with respect to injured workers) that allow them to report fewer disabling injuries. Some companies probably do not report all injuries and illnesses. The Mine Safety and Health Administration (MSHA) has tightened its reporting requirements recently, and better reporting should result.

As more coal is mined and more miners are hired, the number of coal worker fatalities and injuries is likely to increase even if accident frequency improves somewhat. Depending on future coal production levels, between 157 and 187 coal workers are likely to be killed and between 17,400 and 20,800 injured in 1985. That represents a 13- to 35-percent increase in fatalities over 1977 and a 17- to 39-percent increase in injuries. In 2000, between 259 and 371 coal worker fatalities are forecast and between 29,200 and 41,800 injuries. These estimates represent an 86- to 167-percent increase in fatalities over 1977 and a 95- to 180-percent increase in injuries. These calculations assume no underreporting and undercounting. The 25-year total (1976-2000) of mine fatalities may exceed 5,000 and injuries may exceed 500,000.

Reducing these numbers calls for action in several areas that may conflict with increased productivity. Federal policy should address three main issues:

1. reducing the number and frequency of disabling injuries,
2. assessing the safety costs of increased production and productivity strategies, and
3. reevaluating the effectiveness of existing Federal enforcement programs.

Equipment can be designed with improved safety as a factor. Broader Federal equipment standards could help. Western and midwestern surface mining could be encouraged by Federal policy, but this emphasis would probably have severe employment and price impacts in the East. Predictable growth in coal demand and steady profitability might be encouraged to further enable companies to devote time and money to safety and health. Federal inspections—the factor apparently most directly correlated with good safety—could be increased. More severe sanctions against chronic high-accident operations might be considered. More effective safety and job training, coupled with increased participation of coal miners in safety and health efforts, may also prove fruitful.

## Community Impacts

Coal mining brings many diverse benefits and costs to coalfield communities. The private sector—mine operators, coal workers, and local business—benefits from steady coal output through profits, wages, and spendable income, respectively. Private production costs are paid by the mine operator. But when costs are externalized, mine workers and local private interests often pay them. These costs can be measured in dollars, human health, and environmental quality. Coal production also entails public costs and benefits. Local communities will benefit from population growth (at a manageable pace), more tax revenues, increased employment, and better services from wisely planned coal development. On the other hand, if growth is too rapid, communities may be unable to expand services fast enough. Even without much coal growth, the public costs of past mining in Appalachia are substantial and deserve redress. Chronic community underdevelopment exists throughout this region and may impede rapid coal growth in several dozen counties.

Coal development will occur in three different kinds of communities. Many Appalachian communities are hampered by the legacy of historical underdevelopment. Coal-based growth will continue the pattern of a "one-crop" economy with all of the costs and benefits of that mode of economic development. In other Appalachian communities and in the Midwest, coal will be mined in communities with diversified economies. The social costs of coal development are likely to be the least substantial there. In the West, coal development will create boomtown growth in some counties and towns. The extent of coal's social costs (both public and private) depends on the rapidity of development, the adequacy of existing public services, the ability of local communities to manage growth, the level of local public participation in coal development, the kinds of mines that are planned, and the attitudes of mine operators and the local citizens toward development.

**East.**—The Appalachian coalfields have produced more than 90 percent of all coal ever mined in the United States. Because they never experienced a sustained period of growth and profitability, operators were forced to cut costs simply to stay in business. Local jurisdictions were usually unable to raise sufficient tax revenue to provide adequate public services because taxes on coal might have handicapped the ability of local operations to compete. Roads, schools, water and sewage facilities, recreation, health care, and local public administration suffered from this unwillingness or inability to tax adequately coal production and undeveloped coal reserves. The cyclical nature of coal demand also made rational, public financial planning difficult. The lack of diversified economies made spendable income and public revenue in coal towns almost totally dependent on the whims of demand for local coal. For most of the 1920-70 period, overall coal demand stagnated, which meant that State and local tax revenues did not increase greatly even during the short-lived booms. The company-town system, which was the main form of community socialization throughout most of the Appalachian coalfields, intensified community dependence on local coal sales. Many of today's coal towns began as company towns. They have yet to compensate for the public deprivations characteristic of their earlier underdevelopment. On the other hand, whatever public infrastructure exists today in these communities stems from coal development. Had coal mining not begun 80 or 90 years ago, these counties might have even fewer public services than they do today. They would have small populations and little industry. But they would also have few of the social costs of coal development.

No significant net increase in Eastern coal production is anticipated before 1985, although substantial production gains should occur thereafter. This interval can be used to plan how coal development can be managed best. Yet severe community overloads are already apparent in much of Appalachian coalfields, where the combination of existing inadequacies and thousands of new miners has overwhelmed the ability of communities to provide services.

Appalachia will experience two contrasting patterns in the early 1980's. Some communities where metallurgical or high-sulfur coal is mined are likely to continue to stagnate because of weak demand. Production will not increase, although the coal-related population may grow. Historical patterns of underdevelopment will continue. Two to three dozen counties may fit this pattern. For more than a year, 10,000 to 15,000 Appalachian miners have been working irregularly because of slack demand. Little has been done to encourage coal demand from these hard-hit areas or to plan the economic diversification of their economies (an even sounder long-term approach).

The opposite case will find boom-like conditions imposed on underdeveloped communities. Production and population will increase rapidly. Spendable income and demand in consumer goods will rise. Demand for housing and public services will increase sharply and may not be met. Local tax resources may not provide needed public services. Production and productivity may be slowed because of these shortcomings. In communities where underdevelopment is least, boom conditions are likely to be most readily accommodated. Several growth-related problems are worth identifying. Housing is crucial to expanding coal production. Both the supply and adequacy of coalfield housing are deficient. Most coalfield land is owned by coal producers or land-owning companies, which generally refuse to sell land suitable for housing. It is often uneconomic for them to sell the surface rights to coal-bearing land, and future liability for subsidence is also a constraint. Consequently, coalfield housing supply is deficient and many areas have experienced severe congestion and inflated land and housing prices. Private builders and mortgage money are in short supply. The quick-fix "solution" to the lack of land and housing has been the mobile home. This is widely seen as less desirable than single-family construction. The lack of housing has made for long commutes between home and workplace, which may contribute to absenteeism. To increase housing supply, land and mortgage money will have to be made available. In many places, a housing-construction industry must be assembled almost from scratch. Increased flood-control measures are necessary to permit building on valley floors.

Roads in the Appalachian coalfields are generally in poor condition. Most were badly constructed initially. The shortage of tax revenues and heavy coal-truck traffic have left them in a constant state of disrepair. Illegal overloading of coal trucks is widespread and contributes heavily to roadway destruction. To upgrade the 6,880 miles of coal-haul roads judged inadequate to meet the current volume of coal traffic would cost an estimated $4.1 to $4.9 billion, with another $600 million to $700 million to replace inadequate bridges. Small fractions of these sums are now spent on maintenance, and even less on reconstruction and rehabilitation. Strict enforcement of maximum load standards is necessary to take full advantage of dollars used for road repair. Alternatives to coal-truck haulage such as overland trams, conveyor belts, and slurry pipelines would limit future road damage and inconvenience. Increased appropriations for road repair (whether raised by general taxes or user taxes) are likely to be a less cost-effective—but a necessary—option.

Other kinds of services and facilities are often inadequate. Many towns have limited water and sewage treatment systems—if they have any at all. This constrains the development of private housing. Health care services fall short of national standards. Recent refinancing of the UMWA medical plan appears to have weakened coalfield health delivery systems. Public administration is competent in some places, but planning, familiarity with assistance programs and solid local development strategies are often lacking. Opportunities for recreation and education are often limited.

While some Appalachian coal communities will be able to handle rapid coal growth with little difficulty, many will not. It is difficult to quantify the implications for Eastern coal production of community underdevelopment. Common sense, as well as several recent studies, suggests that the ability of the East to produce more coal rests in part on controlling the future social costs of increased production and dealing with present inadequacies. To internalize fully the social and environmental costs of coal mining, coal will have to be priced to reflect its true costs of production.

Federal policymakers face difficult questions with respect to Appalachian coal development. The first is who should pay to rectify coal's accumulated social deficit. The general public? Current coal consumers? Coal operators out of their profits? State and local governments through coal taxes and general revenues? Second is the question of how to establish new patterns of economic development that will bring local social costs and benefits into balance. Policymakers may wish to examine alternative ways of accomplishing this end, such as economic diversification plans financed by revenues from coal development, public regulation of coal development, and better planning. No Federal policy now addresses these complex issues, although the Appalachian Regional Commission and the President's Commission on Coal are examining them. If serious breakdowns develop at the local level in the next few years, Congress may wish to consider this matter systematically.

**Midwest.**—Coal counties in the Midwest are likely to be in a better position to benefit from increased coal development than either Appalachia or the West. These counties generally have diversified economies, which have made them less vulnerable to coal's booms and busts. Because much of their tonnage will come from surface mines, the rate of population growth is not likely to be excessive. Because local tax revenues here were higher than in Appalachian counties, community services are generally better. As a rule of thumb, it has been found that where coal mining has been the single, dominant economic activity, communities are least prepared to manage rapid coal development. Counties where mining has been less prominent and economics more diversified appear to be more able to benefit from rapid development. Most of the Midwest fits into this category.

**West.**—Significant Western coal production is a recent phenomenon. Today, about 9,000 miners produce more than 20 percent of national production. Future growth will be rapid. Between 34,000 and 42,000 miners will be working in the West by 1985. Six States—Colorado, Montana, New Mexico, North Dakota, Texas, and Utah—are each expected to add 55 million to 61 million tons more annual capacity by 1987, according to optimistic forecasts. Wyoming may be able to produce more than 270 million additional tons by that year, although 147 million tons is a more plausible estimate. Depending on how much actual production occurs, from 6 to 11 western counties will experience population growth rates exceeding 5 percent annually. Four of these counties are in Utah, where underground mining (with its comparatively high labor requirements) will take place. Mercer County, N. Dak., and Campbell County, Wyo., are also vulnerable. If coal production matches optimistic industry predictions, two Colorado counties and three Montana counties will also show 5-percent growth rates. Towns in other counties will also experience boom conditions.

Western towns have had varied degrees of success coping with rapid coal growth. Some, like Rock Springs, Wyo., reflect a range of social problems and general community breakdown. Others, like Gillette, Wyo., have muddled through the initial boom problems and are in the process of managing growth with some success. Western towns often find themselves short of housing, water and sewage treatment systems, health care facilities, and front-end revenues. With some exceptions, local tax revenues appear to be able to provide needed services in the long run, although revenue shortages may be felt during the first 5 years or more of rapid growth.

Indian coal development presents special issues. Although considerable tonnage may be mined from Indian reservations, relatively few new miners—perhaps several thousand—are likely to be employed. Most are expected to be Indians already living on the reservations. Western tribes are now insisting on Indian-preference in mine employment in their reservations along with higher tonnage royalties. Boomtowns are not expected to occur on the reservations, although several reservation towns may show high population growth.

Boomtown conditions are not likely to constrain Western coal production goals. But Federal policymakers must face questions concerning equity of sacrifice, responsibility for controlling the social costs of private energy development, and responding to local demands for assistance. As in much of Appalachia, many western towns will need to deal with one-crop economies, rapid population growth, and revenue shortages.

### Social Impacts of Transportation

Coal transportation can be barely noticeable, or it can create a major disturbance—depending on means used and the route chosen. About two-thirds of all coal is shipped by railroad. More fatalities result from coal transport than from mining. New routes may have to be selected to avoid populated areas. Grade crossings will have to be improved to prevent accident and disruption. Trucks are used for short hauls, especially in Appalachia. They contribute greatly to road deterioration, dust, and highway safety problems. Barges and slurry pipelines are probably the least disruptive modes but can be used only under special conditions. An alternative to coal haulage is electric transmission. Its major liabilities are the health and safety concerns of high-voltage electric fields (as yet unquantified) and the visual impact of towers and rights-of-way. The health issue is particularly controversial and requires considerable research.

### Impacts on Coal-Using Communities

Coal mining can easily dominate local communities, but combustion facilities tend to be a much smaller—and more stable—part of community economic activity. When a coal-fired powerplant is built, several impacts occur—increased wage income and employment, side by side with more air pollution and disruption from transport. As the literature on the social impacts of powerplant operation is limited, this study surveyed public attitudes of residents near three large powerplants. In all cases, the respondents living within 3 miles of the plants found them objectionable. This attitude diminished rapidly with distance. The most widely perceived disadvantage was air pollution. The major perceived advantages were employment and the availability of energy. In general, slightly more than half the respondents found the plants to be reasonably acceptable neighbors.

## Factors Affecting Coal Production and Use

All of coal's major supply factors—reserves, labor, capital, and industrial infrastructure—must be available to support greatly increased production. From time to time, each may become a short-term bottleneck, as in the case of a national strike. But none, either singly or together, is expected to hamper the mining of as much coal as can be sold over the next 25 years. However, attainment of the highest coal scenarios could be precluded by any of several factors. Many potential constraints will be alleviated in the normal course of events. Others may require special attention. These factors are discussed below.

## Coal Availability and Leasing

Unlike other domestic fossil fuels, coal is still plentiful. In the East, most coal reserves are owned by producers or landowning corporations. About 65 percent of all coal reserves in the West, however, is owned by the Federal Government and can be mined by private companies only under Federal lease. The rest is owned by States, Indian tribes, or corporations. About 17 billion tons, or 5 percent federally owned coal resources have been competitively leased. This is far more than will be mined over the next decade in the West, but the characteristics of the tracts leased and the needs of the mining industry are such that further leasing may be required soon to meet production projections for 2000.

A leasing moratorium is now in effect while procedures are being devised to remedy past abuses. The Carter administration plans to end the moratorium in the early 1980's. In the interim, only leases necessary to maintain an existing mining operation or to meet existing contracts are issued. The new leasing program will emphasize public participation and environmental acceptability in the context of multiple-use land management. In addition, leases may be reorganized to reflect logical mining units, and they may be required to meet criteria of diligent development and continued operation. The terms of all existing leases will not be modified under the revised leasing program. This exclusion has drawn criticism from environmentalists, Indians, and others. With the resumption of leasing, coal operators will be able to plan with a degree of certainty that has been lacking in the recent past. Until Federal leasing policy is finalized, it is not possible to determine the extent to which the overall policy will encourage or constrain the development of coal reserves on public lands.

## Industry Structure

The structure of the bituminous coal industry is complicated and dynamic. Although there are more than 6,000 operating mines, it has been estimated that there are only about 600 independent producers or producer groups. The 15 top coal operators mined about 40 percent of all domestic production in 1977,

and between 40 to 45 percent of all utility consumption. Five of the top 15 producers are captives, owned by steel companies or utilities. The others are owned by horizontally integrated energy companies or conglomerates. Only two are independent coal companies. In the last 10 years, three trends have emerged and are likely to continue: 1) increasing concentration among companies selling in regional and specialized markets; 2) increasing utility-owned production (known as vertical integration); and 3) growing ownership of coal production by horizontally integrated energy companies.

Competition, price, and coal supply are matters directly affected by industry structure. Industry advocates and critics disagree over whether or to what extent these factors have been affected by the changes in coal ownership. Although a number of recent studies have examined the competitive structure of the coal industry, none has focused the quantitative analysis properly or developed the case study information that would justify definitive conclusions. As coal production from utility-owned mines and energy-company mines is likely to increase faster than coal production nationally, the competitive implications of such ownership patterns deserves close analysis and monitoring.

## Labor Profile

Today's 237,000 coal workers are a diverse group. More than 140,000 work in underground mines, while 65,000 work as surface miners. Others work on new mine construction projects, and in preparation plants, mine-related shops, or mine offices. About 2,000 are women. More than 10,000 are Black, Hispanic, or Indian. More than 90 percent live East of the Mississippi River, and most of those work in the Appalachian fields. Although about half of all coal production in 1985 and 2000 will come from west of the Mississippi, more than 80 percent of the labor force will work east of it. Most of this group will work in underground mines and in Appalachia. The work force is young: the median age for underground miners is about 33; for surface miners, about 37. As the current work force ages, job experience should increase. This should improve safety and productivity.

Between 65 and 70 percent of all coal workers are members of UMWA. They account for about 50 percent of total national production. The average annual income for most underground miners exceeds $17,000; for surface miners, more than $20,000. (Some miners who log a great deal of overtime can earn $35,000 annually, but they are exceptions.) Because of mining's high wage rates and certain social factors, the turnover rate industrywide is probably below average. However, where employment is not steady or where community conditions are perceived to be harsh, individual mines have experienced high turnover.

It is estimated that very little—if any—increase in net mine employment will occur between now and 1985. Although several thousand miners will retire in these years, a sufficient supply of younger workers appears to be available. By 2000, however, mine employment is estimated to increase by 45 percent to 110 percent over 1977 levels. Most of these additional workers will be needed in the East and should be available from indigenous coalfield populations. Labor supply does not seem to be a problem in the boom areas of the West because of high wage rates.

### Labor-Management Relations and Collective Bargaining

Labor-management relations in the coal industry have never been very good for very long. This stormy relationship has been shaped by the structure of the industry, the level of coal demand, the level of competition among coal producers, the nature of the underground workplace, the social experience of the coal camp, and the history of the effort to unionize the work force.

One measure of the stability—if not the quality—of day-to-day relations in the workplace is the level of wildcat strike activity. Miners engage in unauthorized work stoppages more than any other group of industrial workers; underground coal miners strike more than any other group. Over the last 40 years, coal miners have participated in wildcat

strikes more frequently when coal demand was firm and employment security was high.

Wildcat strikes occurred with unprecedented frequency between 1973 and 1978 when coal prices more than doubled and demand was growing. These strikes usually begin at a single mine. Sometimes they spread quickly to many others when a disputed condition exists across the coalfields. Most wildcats are limited to one mine and arise over a disputed work condition, interpretation of the contract, or the miners' perception of harassment. Several were precipitated by widespread disgruntlement over broader issues—compulsory shift-rotation, pending black lung legislation, controversial school textbooks, gasoline rationing, the right to strike, the use of Federal injunctions against wildcat strikers, and cutbacks in medical benefits. Since the last contract strike, wildcat strikes have been much less frequent. Poor market conditions, improvement in the grievance procedure, and depleted savings are the probable reasons for the improvement. UMWA and mine management appear to be improving their ability to resolve disputes, which should lead to less strike activity in the future. Many miners are as opposed to the recent level of wildcat strike activity as are their employers because of the loss of income these shutdowns entail. But these strikes have always been a part of coal mining and may never be eliminated. There appears to be little that Federal policy can—and possibly should—do to stabilize this area of private enterprise. Harsh legal penalties against strikers do not seem to deter wildcat strikes in this industry.

Collective bargaining in the coal industry has usually been characterized by acrimony, strikes, cataclysmic rhetoric, and reluctant Federal involvement. The most recent contract impasse came in the winter of 1977-78, when for 109 days, UMWA miners struck their employers, their Government and, it can be said, their own negotiators.

The 1977-78 strike lasted almost 4 months because the Bituminous Coal Operator's Association (BCOA) insisted on a set of dramatic changes in the old contract. These changes—

involving health and pension benefits, the future of the UMWA Funds, whether management would have the right to fire and discharge strikers and absentee workers—were objectionable to rank-and-file miners although they were accepted by UMWA negotiators. BCOA members felt they were necessary to achieve what they called "labor stability." The contract that was finally ratified incorporated many of the changes BCOA demanded. The ramifications of this contract—the attempted recall of UMWA President Arnold Miller, grassroot discontent, and health care changes—are still working themselves out. It is likely that UMWA members will demand major revisions in the current agreement in 1981.

The executive branch has had little success in mediating contract negotiations in this industry. Both the union and the operators balk at Federal intervention, and miners have little respect for Taft-Hartley injunctions. Short of nationalization of the industry or draconian labor controls, there is probably little that Congress can do to improve the negotiating/contract ratification process. It is essentially a private relationship in the private sector. Rising coal demand and continued prosperity for mine operators should improve the climate of negotiations. As more and more production comes from non-UMWA mines, the national impact of UMWA contract strikes should decline. This was clearly evident last year when, after the 3½-month strike, no power shortages were recorded and only 25,000 workers were laid off.

The dynamics of collective bargaining and labor-management relations are very much in flux. One factor behind the development of western mines is the more tranquil labor situation in the West. Members of BCOA, which has negotiated industrywide contracts with UMWA since 1950, may jointly or separately decide to begin negotiating regionally or on a company-by-company basis. Observers disagree about whether denationalization of collective bargaining will reduce net lost time from contract strikes. High-income western miners may become more inclined to unionize in the future as working conditions and job security supplant wages as their main job concerns. Western miners have had a tradition of militant unionism, and it may be premature to conclude that western mines will necessarily be nonunion operations. UMWA contract strikes are likely to continue to affect most metallurgical and export production, and roughly half of Eastern steam coal output.

Productivity

Although productivity has declined since 1969, the evidence suggests that it has not affected the ability of the industry to produce as much coal as can be sold. Much of the explanation for the rise in productivity in the 1950's and 1960's has to do with mechanization (which cut the work force by 70 percent by 1969), the absence of workplace health and safety regulation, the lack of surface mining controls, the high level of job experience among miners, and the ability of the industry to externalize a variety of production costs. The hidden costs of rising productivity in these decades were unemployment, community stagnation, black lung disease, unreclaimed surface mines, and other environmental problems. Several factors have combined to reduce productivity since 1970. A principal one is that 44,500 UMWA miners—about half of the 1969 UMWA working membership, which represented 74 percent of all miners—retired in the 1970's. These retireees had key production jobs, for the most part, and their productivity was high. Overall, about 60 percent of today's coal workers have been hired in the 1970's. State surface mining regulations cut productivity, as did Federal environmental controls. The 1974 UMWA contract added an estimated 5,000 additional workers to the work force for training and safety reasons. The 1969 Act slowed down certain mining cycles to control methane, dust, explosions, and ground conditions. High prices in the mid-1970's encouraged more than 1,000 small, low-productivity mines to open. In addition, railcar shortages, absenteeism, and poor labor-management relations contributed to low productivity.

The decline in productivity seems to have bottomed out. As more and more production is mined by western surface mines, larger mines, and newer mines (both surface and underground), workers' productivity should increase gradually. Improved labor-management relations and more appealing community conditions should reduce absenteeism, thus raising productivity. An increasingly experienced work force should boost efficiency. Stable prices should discourage inefficient producers and prune inefficient mines. If the industry can lower its injury experience, a contributing source of poor productivity would be addressed. No technological developments appear to be on the horizon that will quickly and dramatically raise productivity, although several innovations, such as continuous face-to-portal haulage and continuous roof support systems, should help. The most promising area for productivity improvement is better job training and restructuring work relations.

### Other Factors

Potential constraints on coal production include supplies of water, capital, and equipment, and transportation facilities. Although the United States appears to have an adequate national water supply to meet projected energy requirements in the near-term to mid-term future, severe local and regional shortages as well as institutional constraints on water availability could become problems. Although these problems could be alleviated through water conservation practices in a variety of sectors (e.g., energy, municipal, and agricultural uses), present water pricing and water rights allocation systems do not encourage these practices.

The availability of capital and equipment for coal development are market factors that should not become constraints if the industry remains economically sound. However, the availability of transportation facilities could impose local constraints on coal supply unless track conditions are improved and the supply of railroad cars is increased. Similar local transportation constraints are posed by inadequate coal haul roads and waterway systems.

## Federal Coal Policy

Federal coal policy emphasizes two goals: more coal should be mined and burned, but more attention should be paid to controlling the environmental, health, and safety costs of doing so. Congress has enacted legislation to meet both objectives. Sometimes the objective of one legislative measure conflicts with the objective of another. Compromises and trade-offs are often made, either as part of the regulatory process or in litigation.

Much of the discussion about Federal coal policy quickly narrows to a discussion of regulatory restrictions on supply and demand. Lost in this process is the considerable effort given to helping industry produce coal and burn it. Federal research money for coal amounted to about $73 million for the Bureau of Mines and about $669 million for the DOE fossil fuel program in FY 1979. Loan money has been made available to "small" underground operators. Federal tax policy on depreciation and depletion encourages investment and reduces tax burdens. Millions of Federal tax dollars are spent annually on highway construction, inland waterways, and railroads that benefit mine-to-market transportation. The National Energy Act of 1978 promotes coal as a primary energy source. The Act significantly strengthens Federal authority to order combustion facilities to convert to coal by prohibiting the use of oil or natural gas in new facilities as well as the use of gas in existing facilities after 1990. However, the impact of the prohibitions may be difficult to ascertain because most utilities are not planning new oil or gas units. Where the prohibitions could have a major impact — on smaller industries — the amount of coal involved is not as great and exemptions are more easily obtained.

Significant Federal regulation of coal production and combustion is relatively recent, beginning in 1969 with the Federal Coal Mine Health and Safety Act. Although mine safety legislation had been enacted as early as 1910, it appears to have had little impact on fatalities and injuries. The 1969 legislation improved some mine conditions significantly, resulting in fewer high-fatality disasters and lower dust levels. It did not, however, specifically address injury-prevention and little improvement has been recorded since 1970 in this area. The 1977 amendments required training programs and tightened other aspects of the 1969 Act. To control respirable dust, methane, poor roof conditions, explosive conditions, and other hazards, the 1969 Act required mine operators to take extra safety measures and hire more safety-related personnel, which contributed to declining productivity. However, this is a very complex issue and it has not been demonstrated that the 1969 Act is the principal cause of the decline, although it has been a factor.

The Surface Mine Control and Reclamation Act of 1977 is designed to change coal mining practices that generate severe social and environmental costs and to prohibit mining in areas that cannot be reclaimed. The Act sets performance standards intended to prevent adverse environmental impacts, such as ground and surface water contamination and degradation of agricultural land quality. Operators must demonstrate, as a prerequisite to obtaining a mining permit, that the land can be restored to a postmining land use equal to or better than the premining use. In addition, significant constraints are placed on coal mining in the prime farmlands of the Midwest and in the alluvial valleys of the arid and semiarid regions of the West. Enforcement of these standards will play a critical role in determining the effect of the Act on coal production.

Much controversy exists over whether and to what degree the 1977 Act will impede coal production. Certainly, it will increase the costs of surface-mined coal. It may also hit small operators harder than larger companies. Productivity may be affected as more worker time and equipment are devoted to preventing environmental damage. It will be several years, however, before an accurate assessment of the impact of this Act can be made.

The major concerns in managing the environmental impacts of coal combustion are its effects on air, water, and land quality. General environmental management is regulated under the National Environmental Policy Act of 1969 (NEPA), which requires that all Federal agencies include a detailed environmental impact statement (EIS) in every recommendation or report on legislative proposals and other major

Federal actions significantly affecting the quality of the human environment. Most major Federal coal-related programs (such as leasing) and federally permitted activities (such as construction of a powerplant) are subject to the EIS requirement. Although the EIS process has increased institutional awareness of the need to minimize adverse environmental impacts, it has been criticized for its alleged attention to procedure over substance and for the time it adds to the beginning of a project. In 1978 the Council on Environmental Quality promulgated new EIS regulations that are designed to reduce paperwork and delays, improve EIS quality, and better integrate the EIS into agency decisionmaking. These new procedures should remove most of the objections to NEPA and result in better decisions and greater environmental protection.

Federal policy toward air quality is implemented under the Clean Air Act Amendments of 1977, which will speed achievement of ambient standards and allow greater growth in the future since each new facility will limit its emissions as strictly as possible. However, the amendments do not reflect a consistent unified approach to some of the fundamental problems, and the overall effect of the Act is difficult to assess. Exemptions for smaller sources and the Amendments' failure to deal with pollutant transformation and transport may undermine the air quality protection intended to accommodate new large coal-fired sources. Coal-fired facilities may also require greater expenditures for pollution control equipment or be subject to stricter siting and other preconstruction review procedures; thus they may be at a competitive disadvantage relative to cleaner fuels. In addition to these increased costs, coal combustion may be prohibited near areas where air quality-related values are important, such as national parks and wilderness areas.

Similarly, water quality impacts of coal mining and combustion are regulated under the Clean Water Act, which requires coal mine and combustion facility operators to meet effluent limitations designed to achieve Federal water quality goals. The techniques for meeting these limitations are available, but may require mine operators to institute waste water treatment and may increase the costs of both mining and combustion.

The Resource Conservation and Recovery Act regulates the disposal of solid wastes from coal-related activities, including ash and scrubber sludge and mine wastes. If these wastes are listed as hazardous, RCRA will impose strict disposal and recordkeeping requirements that will significantly increase coal combustion costs. However, the environmental and health benefits from preventing the open dumping of these wastes are also substantial.

## Implementation

Numerous departments and agencies are responsible for implementing Federal policies that affect the production and use of coal. Those with substantial roles include EPA, which regulates the byproducts of coal use through administration of the Clean Air and Water Acts and RCRA; the Department of Labor, which is responsible for miners' health and safety; the Department of the Interior, which administers SMCRA; DOE, which has the authority to order coal conversions and, in cooperation with Interior, administers the Federal coal leasing program; and the Council on Environmental Quality, which has primary oversight responsibility for NEPA. In addition, numerous interagency consultation and coordination requirements involve a variety of other departments and agencies in policy implementation.

Some critics of Federal coal policy argue that energy development is overregulated. Others contend that more regulation is needed, either because of the way agencies have interpreted their mandates or because conflicts or gaps among those mandates preclude the existence of either a coherent national coal policy or a coherent environmental protection policy. The major factors affecting implementation of such policies, in addition to those mentioned above, include the lack of comprehensive Federal programs for leasing, land use, and water resource management; the absence of workable mechanisms for resolving

interstate or interregional pollution problems; and the focus on immediate problems to the detriment of long-range planning. These could hamper increased coal use in the short-term and midterm future.

In general, Federal policies that affect coal-related activities are not expected to constrain increased coal use in the long term. The requirements of the Clean Air Act impose the most significant constraints in the short term, both by limiting the number of available sites for combustion facilities and by substantially increasing the costs of combustion. However, the air quality benefits that will result should facilitate coal combustion in the long term. The cumulative effect of all other existing regulations may delay the construction of new facilities or the opening of new mines and will make both mining and combustion more expensive.

## Future Policy

Each of the national energy supply and demand scenarios in this report involves a substantial increase in coal use over the next two decades. There is no doubt that the resource is physically present and accessible to sustain a high level of use over that period. It is also clear from an engineering standpoint that coal can be extracted, processed, and burned at a cost that will make it very competitive with other fuels. What is not clear is how the external costs, institutional and social constraints, and other nonmarket factors associated with coal use will affect the validity of the economic and technological analysis. At one extreme, increased coal use might pose such serious external costs to the environment and public health that strict limits on its use would be required. At a minimum, the process of reducing external costs—by increasing, for example, pollution controls and coping with internal constraints (such as labor-management conflicts)—will moderately increase the economic costs of coal utilization. Given the central place of coal in future U.S. energy planning and projections, the stakes involved in formulating a national coal policy are substantial.

The task of policy analysis in this area is to identify the potential problems and constraints and to examine the range of governmental policies that offer some promise of amelioration. There are three basic types of criteria for choosing among these Government policies: 1) national objectives concerning the timetable for and level of coal production and use, 2) political and normative values, and 3) pragmatic calculations concerning the absolute and relative efficacy of policies in stimulating production and use and/or minimizing adverse impacts.

National objectives concerning the magnitude and timing of coal use set the context within which coal policy is formulated. The mining industry should be able to double or triple its production by 2000 if current conditions continue. Existing and pending environmental, health and safety, leasing, and other legislative and regulatory requirements may be costly but otherwise appear to be compatible with greatly increased coal production.

Nevertheless, there are actions that will provide an additional margin of safety against the possibility that these supply projections are overly optimistic or that it becomes necessary to raise coal's fraction of U.S. total energy supply above the levels posited in this report. Many of these measures have merit independent of their potential effect on coal supply. The list includes efforts to: 1) mitigate the adverse community impacts that might constrain coal development, 2) address the causes of labor-management disputes, 3) anticipate and avert potential coal transportation bottlenecks by upgrading existing modes (e.g., railroads) and facilitating the creation of new ones (e.g., slurry pipelines), 4) expedite the formation of a leasing policy and the designation of eligible tracts, 5) streamline the permitting process for new mines, and (6) develop procedures for anticipating and accommodating potential objections to new coal facilities in order to avoid extensive litigation and delay.

Demand is more likely to be a constraint on coal development over the next two decades than is supply. While demand will probably be adequate to sustain all but very high energy scenarios, this is far from certain. Several broad policy options are available to strengthen the future market for coal. These include: 1) tax pressures and incentives to induce utility and industrial conversion to coal, 2) RD&D support for technologies (e.g., FBC and SRC) that can help make coal an acceptable fuel for small users, 3) RD&D support for improved, less expensive emission control technologies, 4) RD&D support for coal gasification and liquefaction technologies, and 5) higher prices for natural gas and fuel oil through deregulation or surcharges. It is most likely, however, that no significant policy initiatives will be required to reach plausible supply and demand projections. In sum, the targets discussed in this report regarding coal production and use for the remainder of the century do not emerge as a critical basis for sorting among legislative and regulatory options; both rapid growth and the current controls can be accommodated.

The choice between conflicting courses of action will often require subjective judgments concerning what is desirable. With regard to coal policy, the most important value conflict involves the relative priorities to be assigned to increasing production and to reducing adverse impacts. The existing legislation and regulations define a rough but discernible balance between these two value sets. Future policy may maintain that balance or shift it in favor of either production or impact amelioration. This choice lies at the heart of national coal policy. Specific dimensions of the choice include tradeoffs between: coal extraction and environmental quality, coal combustion and environmental quality and public health, coal extraction and the well-being of coalfield communities, and coal extraction and workplace health and safety. A second value conflict arises over whether increased Government regulation is the appropriate way to achieve national energy policy goals, or whether the emphasis should be on broad guidelines (performance standards), negotiation, and mediation. Finally, if regulation is determined to be necessary, value conflicts will result from the allocation of decisionmaking authority among the various levels of government: Federal, State, local, and tribal. Value judgments and priorities, then, play an important role in shaping coal policy choices.

Judging policy options also involves an assessment of the utility of different policies in solving the specific production, utilization, and impact problems associated with coal. Five major areas of policy concern have been identified, each with a potential for significant influence on efforts to expand the production and use of coal. They are: 1) environmental impacts, 2) community and social impacts, 3) labor-management relations, 4) workplace health and safety, and 5) leasing of Federal coal reserves.

Environmental Impacts

Environmental considerations are an important potential constraint upon a substantial increase in coal production and combustion. The era of unregulated environmental impacts is clearly past for coal, as for other fuels. An elaborate, though still incomplete framework of legislation, regulation, and implementing institutions is in place. It constitutes a national policy system for managing the environmental impacts of increased coal use. The relevant control technologies are at various stages in their evolution from conception to maturity, but most have at least reached the point where a first-generation technology can actually be applied. Control technologies for combustion emissions are particularly important, and although existing technologies are far from optimal, the outlook is promising. In short, after the investment of substantial economic, technological, and human resources over recent years, the ingredients of a viable environmental policy for coal now exist.

Under these circumstances the paramount task of policy analysis is to identify ways that the existing policy system might be upgraded with regard to:

* gaps in present knowledge about the nature and magnitude of the risks to the environment associated with coal utilization,

* the performance and future prospects of specific control technologies and the means of stimulating improvements,

* omissions, inconsistencies, or disutilities in existing environmental laws and regulations,

* implementation of laws and regulations, and

* promising new policy innovations or instruments.

Community and Social Impacts

A comprehensive national energy program also may include policies designed to alleviate the adverse community impacts associated with coal development. These policies, if they are to be effective, must take into account two basic characteristics of the present situation. First, there is considerable uncertainty over the nature of future coal development impacts and the balance of benefits and costs that will accompany them. Second, value disagreements over what impacts are beneficial and what are adverse occur even where the nature of these effects is understood. Whether economic growth itself is to be viewed as a positive or negative phenomenon in particular localities is itself the subject of dispute. Nevertheless, a number of potential adverse community impacts can be identified including overloaded public services, hyperinflation, and various symptoms of social stress and instability. A number of general and specific policy measures designed to cope with these problems can be identified, ranging from Federal grants to coalfield communities for public works construction to studies of ways to limit the corrosive impact of energy development on Indian tribal culture. Given the uncertainties and value disagreements regarding community impacts, there is a need for policies that seek to deal with the concerns of interested parties in a context where compromise is encouraged. This will occur if all parties affected by increased coal use participate in decisions about the location, timing, and scale of coal developments that directly affect them and if Federal policies are designed to distribute the risks, costs, and benefits of increased use equitably among all affected parties.

## Labor/Management Relations

Although recent instability in labor-management relations in the coal industry has been significant mainly at the local and subregional level, future instability could have significant implications for the national economy. Consequently, Federal policymakers may try to actively influence the situation. If so, their task is twofold: deciding how to ameliorate the causes of destructive labor-management conflict and lay the groundwork for a more constructive long-term relationship, and planning ways to cope with future strikes, should they occur.

Under present legislation the Federal Government can do little to alter directly the terms of labor-management relations. The principal causes of wildcat and contract strikes are the conditions of work and terms of employment. They are essentially privately determined matters. However, some policies can alter the context in which the unions and operators interact. These include measures designed to ensure steady growth in the demand for coal in major sectors of the economy. The recently passed National Energy Act contains a number of these provisions. A healthy market should ease the historic economic insecurity—of both operators and miners—that has been such a large factor in the industry's labor problems. Other promising actions are equally indirect and relate to the basic social, economic, and environmental ills that contribute to the miners' discontent. They include such diverse measures as improving dust monitoring and control within underground mines to restoring Appalachian trout streams.

In the event of another major coal strike the Federal Government would have an interest in promoting a settlement that is prompt and noninflationary, and one that establishes the basis for long-term labor-management stability. In pursuit of these objectives, five major strategies will be available: 1) reliance on collective bargaining with limited Government intervention, 2 collective bargaining with strong Government involvement, 3) use of Taft-Hartley with limited efforts at enforcement, 4) Taft-Hartley with vigorous enforcement, and 5) Government seizure of the mines. Each of these options has opportunities and liabilities that vary with the particular circumstances of a strike; however, as noted previously, past Federal intervention in coalfield labor-management disputes has been counter-productive.

## Workplace Health and Safety

Mining, particularly underground, is a hazardous occupation as measured by the record of work-related accidents and diseases. Any substantial increase in coal production inevitably will mean thousands of diseased and injured miners. The question facing policymakers is what modifications in or additions to existing standards and enforcement might minimize that number? Efforts to answer this question will focus on improved dust control, research related to the appropriate dust standards, and research on the effects of new pollutants and the synergistic impacts of multiple pollutants in terms of mine health. The major need in mine safety is to develop an accident-reduction strategy to limit the rising number of disabling injuries, which now amount to about 15,000 a year and result in 2 months or more lost time each. Improved safety and job training, education in different work practices, changes in management and worker attitudes, and new Federal safety standards for mine equipment are some approaches to this problem.

## Leasing of Federal Coal Reserves

Western coal comprises roughly half of the Nation's coal reserves and the Federal Government owns 65 percent and indirectly controls another 20 percent of that resource. Consequently, Federal policy concerning the leasing of those reserves can have a substantial influence on the future of coal use. The most basic policy question concerns whether additional leasing of Federal coal lands will be required to meet projected increases in demand, and, if so, when and how much. The answer remains unclear; the 1977 amendments to the

Clean Air Act may have reduced the attractiveness of Western coal for utilities, and legal uncertainties surround the status of pending lease applications. In any case, recent litigation means that no new leasing will be possible until the early 1980's at best. Assuming that it will become necessary at some point to resume leasing, the Federal Government will have to decide to what extent private industry should be allowed to determine which lands will be made available. One option would permit operators to nominate those Federal lands they desire to mine and the Government to approve or disapprove the proposed lease. Under a second option the Government would identify areas eligible for leasing and industry would nominate specific tracts. Still another option would leave to the Government selection of both the areas and specific tracts for leasing. The comparative attractiveness of these options will depend on whether the principal selection criteria are reducing planning and administrative costs, minimizing environmental damage, increasing tonnage mined, forestalling litigation, or controlling adverse socioeconomic impacts. Whichever approach ultimately is selected, a number of specific issues will have to be clarified. They include definition of logical mining units, status of preference-right lease applications, requirements of diligent development and continued operation, estimated recoverable reserves, advance royalty payments, and the exchange of environmentally sensitive leased lands for other unleased Federal land. Some important institutional issues center on the division of leasing responsibility between the Departments of the Interior and Energy. The Department of the Interior has the overall responsibility for the leasing program, but DOE controls economic leasing terms and conditions.

Chapter II

# ENERGY AND THE ROLE OF COAL

# Chapter II.—ENERGY AND THE ROLE OF COAL

## TABLES

## FIGURES

# Chapter II
# ENERGY AND THE ROLE OF COAL

Coal is expected to rapidly become more important in the Nation's energy system. Its abundance relative to domestic reserves of oil and gas should preclude the steep price increases that may be in store for oil and gas as their production becomes more expensive, and most observers conclude that our present heavy dependence on foreign oil is a grave political and economic liability.

Many of the functions served by the energy of oil and gas could also be served by coal, as they were in the past. Others, however, such as residential heating and transportation, seem less appropriate for coal to serve unless the coal is first converted to a more convenient form. It is necessary to examine energy demand in general, and the various sectors of energy consumption in particular, to determine the degree to which coal can rejoin the energy system and how. This chapter describes the factors that determine energy and coal demand. Three energy scenarios are presented to demonstrate a range of possible coal growth requirements. These scenarios reflect the range of most current predictions, though recent projections have tended to be toward the lower end of this range. Thus the high development scenario serves to indicate the probable upper limit on the amount of coal that will be needed. If the actual level is lower than the low development scenario, both the challenge of meeting demand and the resulting impacts will be reduced. The coal production and combustion elements of the scenarios are described in detail to set the stage for the following chapters.

## KEY FACTORS AFFECTING ENERGY DEMAND

Three primary factors are ultimately responsible for determining future energy use: population, economic activity as indicated by the gross national product (GNP), and efficiency of energy use. Each factor is a complicated function of subfactors, which are often interrelated. Appendix I in volume II amplifies this abbreviated analysis.

People are the final consumers of the goods and services that use energy. The higher the population, the more energy will be needed. Changes in population growth are determined by fertility, death, and immigration rates. Estimates of the former range from 1.7 to 2.1 births per woman over her lifetime (1.8 in 1975), but the effects on energy consumption will not be great before 2000. The death rate is not expected to change significantly by 2000. Hence immigration is the least certain factor. Population is expected to be between 246 million and 260 million by 2000, an increase of 13 to 20 percent from the present 217 million. Of more immediate concern are the demographic shifts within the total population. The labor force will grow considerably faster than the population as a whole. The shift in the median age that this implies suggests not only that GNP will increase as described below, but that households and drivers, both major energy-consuming groups, will increase faster than population. Changes in tastes, lifestyles, and habits, perhaps engendered by rising prices, can also affect energy demand, but such shifts cannot be confidently predicted.

GNP is a measure of overall economic activity, most of which consumes energy. A close relationship has been observed in the past between GNP and energy consumption. The recent changes in this relationship are discussed in the next paragraph. GNP obviously depends in part on the population size and the labor force in particular. The number of persons in the labor force age group (16 to 65) can be predicted quite accurately to the year 2000. The number of persons actually working depends on the labor participation rate and the unemployment rate. The former is expected to continue its upward trend, reflecting the increased participation of women. Unemployment is expected to drop below 5 percent for most of the rest of the century. The positive effect of a large population and a higher rate of participation in the labor force is partially offset by the expected continuation of the long-term decline in work hours; average hours worked per week dropped from 40.0 in 1948 to 37.1 in 1973. The final element in estimating future GNP is labor productivity, the measure of output per worker, per hour worked. The replacement of manpower with capital, materials, knowledge, and energy has been the historic means of increasing productivity. It now appears that industry is finding a more attractive return on its capital when it restructures this equation to reduce the use of energy. This is one of several factors that have led to a long-term decline in the rate of growth of labor productivity. If this rate continues to decline, a very low-growth economy with lower energy needs than this report assumes will emerge. Recent concern has led to tax law changes, indicating that a national commitment exists to reverse the decline, which is the assumption of this analysis. All these factors combine to yield estimates of GNP in 2000 of $3,300 billion to $3,600 billion (constant dollars) compared to $1,516 billion in 1975. This increase of 120 to 140 percent will about double real, per capita income.

Energy efficiency relates the performance of a given task or process to the quantity of energy required. Efficiency (or the conservation measures implemented to enhance it) is not to be confused with constraint, which implies less consumption of the goods or services involved. Energy efficiency rises largely in response to economic pressures—fuel prices in particular—but also to tax benefits and other policies. History provides little help in estimating the response to energy price increases, as the cost of fuel was stable or slowly declining in real terms throughout the century. Until 1973 there was little incentive to design for energy efficiency and almost none to change an existing practice. The situation is quite different now, and decisions based on cost estimates will result in rising energy efficiency. There will be exceptions. As resources become scarcer, more energy will be required to produce them, and as energy conversions such as electricity and synthetic fuels assume a larger role, efficiency will be adversely affected. Nevertheless, recent economic and energy data imply that energy and GNP have been largely decoupled and a substantially different ratio will be established. Measuring this change is much more difficult than measuring the previous factors. Conservation will be a function of fuel prices, which will depend in part on factors such as domestic oil and gas reserves, foreign and domestic policy decisions, and technology developments. Thus the uncertainty in projecting future energy efficiencies largely accounts for the wide range in the scenarios to follow.

## DEMAND SCENARIOS

Forecasting energy demand is a highly uncertain art. As described in the previous section, there are too many important variables that can only be speculatively quantified. Depending on assumptions, modelers can produce scenarios for 2000 predicting anywhere from 60 to 190 Quads[1] (73.1 Quads in 1975). Both extremes are highly improbable. Most forecasts fall between 100 and 150 Quads.

[1] A Quad is short for quadrillion Btu.

Rather than selecting existing scenarios or creating more, the following analysis simply assumes energy consumption levels in 2000 of 100, 125, and 150 Quads and then determines the circumstances that would be consistent with arriving at each level and the patterns in which these aggregate levels would be distributed.

The objective of these energy demand scenarios is to determine the impact of a given level of aggregate energy demand on coal consumption. Aggregate demand is the sum of residential/commercial, industrial, and transportation demand. Most energy used in residences and commercial establishments is for heating and cooling. Oil and gas are the primary fuels for direct heating and electricity for cooling. The use of electricity for heating is increasing rapidly. The transportation sector is a major consumer of liquid fossil fuels. Industry and electric utilities, which use large quantities of energy to produce steam, are the prime candidates for the substitution of coal for oil and gas.

The 100-Quad demand scenario is a slow-growth scenario. An assumed fertility rate of 1.7 births per woman marks the leveling off of a long-term decline in U.S. fertility, and with moderate immigration results in a projected population of 246 million in the year 2000. This modest increase in population is accompanied by an equally modest average rate of GNP growth: 3.8 percent between 1975-85, and 2.8 percent between 1985-2000.

A key assumption in the 100-Quad scenario is a major increase in the price of oil and gas relative to coal because of a disappointing discovery rate. The price of oil is expected to increase to $25/bbl in 1975 dollars by the year 2000, while natural gas increases to $4.30/1,000 ft³ at the wellhead. These dramatic price increases reflect increased scarcity and deregulation of oil and gas. Long-term contracts and a competitive industry will prevent coal prices from rising as rapidly: from $17.50/ton in 1975 to $28.88/ton in 2000. Accordingly, oil prices increase by a factor of 2.4, gas prices by a factor of 10, and coal prices by a factor of 1.65. The price of energy to the consumer will not increase as much because other determinants of retail energy prices (refining costs, distribution costs, etc.) will not rise at the same rate as fuel costs. As a result, electric power increases from 2.7 cents/kWh to only 4.5 cents/kWh in 2000. The basis for these increases is discussed in the *Supply Alternatives* section of this chapter and in appendix I of volume II.

These increases in absolute energy prices lead to major efforts to implement energy-sav-

ing technology, resulting in increased energy efficiency. In industry a 30-percent increase in energy efficiency is assumed. Industrial use of petrochemical feedstocks is also forecast to grow at a much slower rate as a result of high prices. In the transportation sector auto efficiency increases from 14 miles per gallon (mpg) in 1975 to 27 mpg in 2000, small trucks and vans increase in efficiency from 11 to 18 mpg: heavy trucks, planes, and ships experience a 20-percent increase in efficiency. In the residential/commercial sector it is assumed that 2 percent of old homes (pre-1975) and commercial structures are retrofitted with insulation each year, and 10 percent are fitted with heat pumps by 2000. All new homes and commercial structures are insulated, and 25 percent of these are equipped with heat pumps.

The breakdown of sectoral energy demand resulting from these assumptions is shown in table 1. The increase in household/commercial energy demand greatly exceeds population growth because of demographic shifts and a substantial increase in per capita use; however, it is much less than the assumed increase in the number of households and in commercial footage, implying increased efficiency. The significant increase in industrial energy use is partially explained by the corresponding growth in GNP, 120 percent between 1975 and 2000. Energy consumption grows slower than GNP for two reasons: first, the achievement of increased energy efficiency already cited; and second, a substantial shift in the composition of industrial production away from petrochemical products. Transportation energy demand rises slightly faster than population, reflecting the increased gas mileage of automobiles and offsetting increases in mileage driven per capita, air travel, etc.

The 125-Quad scenario implies a more liberal supply system but a less successful conservation effort. Thus energy prices are the same as in the 100-Quad scenario. The key demographic and economic assumptions in the 125-Quad scenario are essentially the same as in the 100-Quad case, but the Nation has not been as successful in implementing energy-efficient technology, nor has there been a sub-

### Table 1.—Energy Demand Scenarios, Year 2000, and Their Determinants

| Demand scenarios/primary assumptions | Residential/commercial | | | Industry | | | Transportation | | |
|---|---|---|---|---|---|---|---|---|---|
| | Quads | % change | Statistics | Quads | % change | Statistics | Quads | % change | Statistics |
| **1975**<br>**73.1 Quads** | 25.7 | | # of homes: 72 x 10⁶<br>Commercial ft. ft²: 25.2 x 10⁹ | 28.5 | | High energy consumption in petrochemicals | 18.8 | | Autos: 104 x 10⁶<br>MPG: 14<br>MPG (trucks): 11 |
| **2000**<br>**Scenario A**<br>**100 Quads**<br>Fertility rate: 1.7<br>Hours worked: 198 x 10⁹<br>Labor productivity growth:<br>1970-80: 1.8% annual<br>1980-85: 2.2% annual<br>1985-2000: 2.6% annual<br>GNP growth per year:<br>1975-85: 3.8%<br>1985-2000: 2.8% | 33.5 | 30% | # of homes: 102 x 10⁶ (+42%)<br>Old (pre-'75): 43.4 x 10⁶<br>New:           58.6 x 10⁶<br>Commercial ft.: ft² 94.6 x 10⁹ (+275%)<br>Old:           15.2 x 10⁹<br>New:           29.4 x 10⁹<br>Retrofit rate:<br>Old:    2% insulated/year<br>          10% heat pumps<br>New:    all insulated<br>          25% heat pumps | 44.3 | 70.4% | Shift composition away from petrochemicals<br><br>30% increase in efficiency | 22.2 | 18.1% | Autos: 120 x 10⁶ (15%)<br>MPG: 27<br>MPG (trucks): 18<br><br>20% increase in efficiency |
| **2000**<br>**Scenario B**<br>**125 Quads**<br>Fertility rate: 1.9<br>Hours worked: 200 x 10⁹<br>Labor productivity growth:<br>(Same as Scenario A)<br>GNP growth per year:<br>(Same as Scenario A) | 44.4 | 72% | # of homes: same as Scenario A<br>Commercial ft.: same as Scenario A<br>Retrofit rate:<br>Old:    1% insulated/year<br>New:    10% heat pumps | 52.5 | 101% | Shift composition away from petrochemicals<br><br>20% increase in efficiency | 28.1 | 49.5% | Autos: 152 x 10⁶ (46%)<br>MPG: 27<br>MPG (trucks): 18<br><br>10% increase in efficiency |
| **2000**<br>**Scenario C**<br>**150 Quads**<br>Fertility Rate: 1.9<br>Immigration: +750,000<br>Hours worked: 203 x 10⁹<br>Labor productivity growth:<br>1975-80: 2.0% annual<br>1980-85: 2.4% annual<br>1985-2000: 2.8% annual<br>GNP growth per year:<br>1975-85: 4%<br>1985-2000: 3% | 54.2 | 111% | No change in efficiency factors: increased consumption of energy is the result of lowered energy prices occasioned by increased availability of liquid fuels and natural gas. | 64.3 | 147% | | 31.5 | 67.5% | Autos: 170 x 10⁶ (63%) |

stantial shift in industrial production away from petrochemicals.[2] The difference in energy consumption in the two cases is primarily attributable to the differences assumed in energy efficiency. In industry the increase in energy efficiency is 20 percent instead of 30 percent. In transportation, the efficiency increase for heavy trucks, planes, and ships is 10 percent instead of 20 percent. In the residential/commercial sector the retrofit rate of insulation into old homes is 1 percent instead of 2 percent, and only 10 percent of new homes and commercial structures have heat pumps. The only other significant departure from the 100-Quad scenario is the number of automobiles—152 million instead of 120 million.

The 150-Quad scenario may be characterized as a high-growth, cheap-oil case. It assumes no improvements in energy efficiency. Energy fuel prices are much lower than in the other two cases. Oil is only $12.48/bbl, natural gas is only $2.25/1,000 ft³, and the price of coal

and electricity is unchanged from 1975 levels of $17.50/ton and 2.7 cents/kWh respectively. Oil and gas are priced much more favorably in relation to coal and electricity than in the low- and medium-demand scenarios. The automobile population in this high-growth scenario is 170 million, much greater than in the previous scenarios.

## Implications

The greatest growth in these scenarios is in the use of electricity by the industrial and residential/commercial sectors. All these incorporate lower growth in electricity than the historical average, but this growth rate has been declining since about 1966 to the present 3.4 percent that is envisioned for the 100-Quad scenario. The direct use of energy by industry has grown slowly over the last three decades. Industrial consumption totaled 18.8 Quads in 1977, only 47 percent higher than in 1947 and actually lower than in 1968.[3] A resumption of

---

[2]The population assumptions are slightly higher for this scenario—an estimated 254 million based on a fertility rate of 1.9; however, no affect is assumed on GNP, which is the same as in the 100-Quad case.

[3]*Annual Report to Congress, 1977*, Department of Energy, Energy Information Administration.

rapid growth by industry is unlikely. Opportunities for the substitution of coal for oil and gas are limited and is discussed later in this chapter under the section, *Projections of Coal Production and Use.*

In the commercial sector, total demand has increased in all three scenarios, but the 100-Quad scenario actually lowers direct use of energy. Electric power and synthetic fuels are clearly substitutable for oil and gas in this sector. The extent to which direct combustion of coal is substitutable hinges upon a number of factors, which are examined subsequently. The transportation sector, for the most part, depends on liquid fuels, hence the direct combustion of coal and electric power is limited as a substitute. Demand growth in this sector means demand growth for oil, natural or synthetic. The greater number of automobiles in the higher scenarios implies a continuing development of outlying suburbs, less mass transit, and increased driving for recreation.

A word should be said about how these demand scenarios compare with other studies. The Department of Energy (DOE) conducted its Market Oriented Program Planning Study (MOPPS) in 1977. Total primary energy demand estimated by MOPPS in the year 2000 is 117.25 Quads, which is bracketed by the 100- and 150-Quad scenarios. The breakdown of the MOPPS estimate by sector is 36 Quads for the household/commercial sector, 20.8 Quads for transportation, 56.9 Quads for industry, and 3.5 Quads for metallurgical coal exports.

Comparing MOPPS to the 100-Quad scenario, the most significant difference is in the industrial sector, where the MOPPS estimate exceeds the 100-Quad scenario by 12.5 Quads. The MOPPS scenario projects increased use of gas and oil as petrochemical feedstocks, and the overall increase in energy efficiency is presumably lower than that assumed for the 100-Quad case. The MOPPS industrial energy demand more closely approximates the 52.5 Quads estimated in the 125-Quad scenario, where overall industrial energy efficiency increases are one-third less than in the 100-Quad case, and where petrochemical feedstocks continue at pre-1975 rates of use. The significant source of difference between the 125-

Quad scenario and MOPPS is in the household/commercial sector, 44.4 Quads compared to 36 Quads in MOPPS. This relatively large difference is accounted for by greater implementation of energy-saving technology in the MOPPS scenario.

The Energy Information Administration of DOE, in its Annual Report to Congress for 1977, projects a domestic consumption of about 100 to 110 Quads in 1990, with a sectoral distribution similar to this report's 100-Quad scenario. The Electric Power Research Institute (EPRI), in their report "Supply 77" of May 1978 uses a reference case for 2000 of 159 Quads and compares it to scenarios of 146 and 196 Quads. EPRI clearly expects a more favorable fuel supply situation, less public and governmental intervention, and lower price elasticities than this report considers probable. At the other extreme, the Committee on Nuclear and Alternative Energy Systems considered levels of energy consumption as low as 58 Quads and up to 180 Quads in 2010.[4]

In creating the three scenarios, various assumptions have been stipulated regarding key factors and relationships. Little has been said about the implications of a given scenario for lifestyles. The way people live and their attitudes have a great deal to do with their willingness to accept constraints on their behavior that might be implied by energy resource constraints and high prices. Whether consumers would resist conservation policies is a matter of speculation.

Americans have traditionally duly complied with requirements associated with emergency situations, but indefinite compliance is a different matter. Most people are economically rational, however, and if it is apparent that price increases are not contrived and are being applied fairly, they will adjust their patterns of consumption appropriately.

The historical growth rate in per capita energy consumption since 1950 has been 1.4 percent annually. The scenarios discussed here result from 0.9-, 1.7-, and 2.3-percent increases.

---

[4]"U.S. Energy Demand: Some Low Energy Futures," *Science,* Apr. 14, 1978.

Thus even 125 Quads implies an acceleration of per capita energy consumption growth. Given expectations regarding increased energy efficiency and higher fuel prices, there is no reason to assume that rational Americans cannot adjust to a reduced rate of increase in energy use. There is no basis for concluding that the lower per capita growth in energy use will result in deprivation. Hence low energy-growth scenarios cannot automatically be rejected as contrary to the demands of the American people. On the other hand, if 150 Quads are available at sufficiently low prices, ways will be found to use them, and it is incumbent on low energy-growth advocates to suggest how consumers will be spending their doubled real income, if not for energy-consuming goods and services, or whether in fact economic growth should be dropped as a national goal.

## SUPPLY ALTERNATIVES

Historically, energy supply has risen to meet demand without major increases in prices, largely because of the Nation's vast resources and the increasing efficiency of production. The situation is quite different now. Few if any analysts expect the Nation's production rates of oil and gas to double at any price. Total exhaustion is not a near-term concern, but resource depletion is sufficiently advanced that major new discoveries are noticeably harder to make, and new production is significantly more expensive. As oil and gas currently supply 75 percent of national energy demand, an indefinite continuation of past trends is not an option for the future. These expectations are reflected in the price assumptions for the 100- and 125-Quad scenarios. The 150-Quad scenario incorporates much lower prices for natural gas and oil on the assumption of major new discoveries.

The energy sources considered here are oil, gas, coal, nuclear, solar, and geothermal. Hydropower is not expected to increase significantly because of the lack of economic sites not subject to significant environmental degradation. Each energy source presents a different set of characteristics that are valued differently by the various users. The most important characteristics are price (including transportation), convenience, facility cost, cleanliness, and reliability of supply.

### 100-Quad Scenario

The essence of the 100-Quad scenario is the sharply higher price of oil and gas caused by resource constraints. The same high price structure could result from a high fuel tax policy, but such a policy is improbable unless impending resource constraints are widely perceived to be real. The higher prices keep U.S. oil and gas production close to current levels through enhanced recovery and exploitation of less attractive sites, such as Alaskan, offshore, and marginal fields. Oil imports are expected to drop in response to policies aimed at that end. Coal production is expected to more than double and actually become the most important fuel. Nuclear power increases rapidly but more modestly than most recent projections. Solar energy increases at a very rapid rate, but the time frame is too short to do much more than lay a significant base for the 21st century. New technologies, such as the production of liquid fuels from shale and biomass, are not expected to make major contributions by 2000.

The economic advantages of coal and nuclear power become overwhelming for electric utilities. Industry's energy growth is derived entirely from coal and electricity. Transportation's relatively small growth comes mostly from oil. The residential/commercial sector grows mostly with electricity. The actual breakdown is shown in table 2 for all scenarios.

### 125-Quad Scenario

The same general price structure is assumed to prevail as in the previous scenario, but the less elastic demand leads to higher consumption. It is assumed that national policies result

Table 2.—Energy Supply and Demand Scenarios

| | Type | | | Coal | | | Oil | | | Gas | | | Nuclear | Solar and other | | |
|---|---|---|---|---|---|---|---|---|---|---|---|---|---|---|---|---|
| | Total | Electric* | Direct | Total | Electric | Direct | Total | Electric | Direct | Total | Electric | Direct | Total (elec.) | Total | Electric | Direct |
| 1975 (actual)**......... | 73.1 | 20.2 | 52.7 | 15.3 | 8.8 | 6.5 | 32.8 | 3.2 | 29.4 | 19.9 | 3.2 | 16.7 | 1.8 | 3.2 | 3.2 | 0.0 |
| Transportation......... | 18.8 | .1 | 18.7 | | | 0 | | | 18.0 | | | | | | | |
| Residential/commercial. | 25.7 | 12.0 | 13.6 | | | .3 | | | 5.8 | | | 7.6 | | | | |
| Industrial ............. | 23.8 | 8.0 | 15.7 | | | 1.5 | | | 5.6 | | | 8.5 | | | | |
| Metallurgical .......... | 2.2 | | 2.2 | | | 2.2 | | | | | | | | | | |
| Exports............... | 1.8 | | 1.8 | | | 1.8 | | | | | | | | | | |
| Stock changes......... | .7 | | .7 | | | .7 | | | | | | | | | | |
| 100 Quads ............ | 100.0 | 42.5 | 57.5 | 31.1 | 20.2 | 10.9 | 30.1 | 1.7 | 28.4 | 17.2 | 1.0 | 16.2 | 14.5 | 7.1 | 5.1 | 2.0 |
| Transportation......... | 22.2 | .4 | 21.8 | | | 0 | | | 21.2 | | | .6 | | | | 0 |
| Residential/commercial. | 33.6 | 21.1 | 12.5 | | | .5 | | | 3.8 | | | 7.2 | | | | 1.0 |
| Industrial ............. | 36.9 | 21.0 | 15.9 | | | 3.1 | | | 3.4 | | | 8.4 | | | | 1.0 |
| Metallurgical .......... | 2.7 | | 2.7 | | | 2.7 | | | | | | | | | | |
| Exports............... | 4.3 | | 4.3 | | | 4.3 | | | | | | | | | | |
| Stock changes......... | .3 | | .3 | | | .3 | | | | | | | | | | |
| 125 Quads ............ | 125.0 | 53.8 | 71.2 | 37.8 | 26.4 | 11.4 | 40.4 | 2.2 | 38.2 | 19.1 | 1.5 | 17.6 | 18.6 | 9.1 | 5.1 | 4.0 |
| Transportation......... | 28.1 | .4 | 27.7 | | | 0 | | | 27.0 | | | .7 | | | | 0 |
| Residential/commercial. | 44.4 | 29.4 | 15.0 | | | .5 | | | 4.6 | | | 8.0 | | | | 2.0 |
| Industrial ............. | 44.9 | 24.0 | 20.9 | | | 3.4 | | | 6.6 | | | 8.9 | | | | 2.0 |
| Metallurgical .......... | 2.8 | | 2.8 | | | 2.8 | | | | | | | | | | |
| Exports............... | 4.4 | | 4.4 | | | 4.4 | | | | | | | | | | |
| Stock changes......... | .4 | | .4 | | | .4 | | | | | | | | | | |
| 150 Quads ............ | 150.0 | 67.8 | 82.2 | 41.3 | 28.9 | 12.4 | 48.4 | 4.0 | 44.4 | 26.4 | 3.0 | 23.4 | 26.8 | 7.1 | 5.1 | 2.0 |
| Transportation......... | 31.5 | .6 | 30.9 | | | 0 | | | 30.0 | | | .9 | | | | 0 |
| Residential/commercial. | 54.2 | 35.0 | 19.2 | | | 0.5 | | | 6.4 | | | 11.4 | | | | 1.0 |
| Industrial ............. | 56.4 | 32.2 | 24.2 | | | 4.0 | | | 8.0 | | | 11.1 | | | | 1.0 |
| Metallurgical .......... | 3.0 | | 3.0 | | | 3.0 | | | | | | | | | | |
| Exports............... | 4.5 | | 4.5 | | | 4.5 | | | | | | | | | | |
| Stock changes......... | .4 | | .7 | | | .4 | | | | | | | | | | |

*All values in columns labeled "electric" are for the heat produced at the powerplant. The values for hydroelectric power (under solar and other) represent the heat that would have been required at a typical thermal generating station to produce the same electrical power.
**Derived from Bureau of Mines, Department of the Interior, press release of Mar. 14, 1977.

in a significant increase in the availability of imported oil, and a vigorous expansion in electric power generation. These steps are required by the assumed increase in the consumption of petroleum in the transportation sector and by a greater demand for electric power. In providing for the increased electric power, coal- and nuclear-generated power are assumed to share equally in the expansion, and geothermal energy is starting to become significant. Both the industrial and residential/commercial sectors are turning to the direct use of solar energy.

Under this scenario, coal does not surpass oil as the leading fuel, largely because of increased imports.

## 150-Quad Scenario

In addition to the increased availability of oil and gas, nuclear power grows at a more optimistic rate. The economic advantages of these fuels limit the growth of coal to little more than that of the 125-Quad case, and actually makes solar energy less attractive than for slower growth scenarios.

## Interpretation

An infinite variety of scenarios can be drawn to meet any given demand scenario. The actual mix will depend upon relative prices, fuel availability, and the other characteristics listed above. Federal policy will have a considerable influence on the relative level of each fuel as well as on the level of demand. Much has been said here of the importance of the availability of oil and gas, but little about their actual availability. A detailed analysis of the subject is beyond the scope of this study, but the values here are consistent with other recent studies. The higher estimates could be met only with a substantial fraction of imported oil. If for political reasons it is necessary to strictly limit imports, other fuels could probably not be expanded much beyond the levels of the 150-Quad scenario to meet the deficit. Nuclear energy has in the past been projected for much higher levels than the highest in table 2 (500 plants of 1,000 MWe). Only about 200,000 MWe have been ordered so far, and many of these projects are inactive. In order to exceed 500,000 MWe by 2000, all these plants will have to be completed and more than 30 ordered each year throughout the 1980's. This is about the maximum rate of orders ever and might be accommodated by the industry, but utilities are showing no signs of resuming ordering at such a level. A utility ordering a reactor now faces considerable uncertainty over regulatory delays, expensive redesigns and retrofits, public opposition, and perhaps eventually, fuel shortages unless unproven uranium reserves are discovered or the breeder reactor is commercialized. Hence the levels assumed here are unlikely to be exceeded.

# PROJECTIONS OF COAL PRODUCTION AND USE

The scenarios of the previous section all show coal to be the fuel of greatest total growth. This growth can occur in a variety of ways and places. The primary distinction in use is between electricity generation and direct heat applications. At present, about 70 percent of the coal mined in the United States is burned for electricity, and the bulk of coal's growth will be in that sector. Direct use by industry is the only other large-scale use considered here. The historical distribution by end-use sector is shown in figure 1. The virtual disappearance of the transportation and retail (essentially the same as residential/commercial) markets is apparent, as is the growing dominance of the utility sector. Also included in figure 1 are the projections to meet the scenarios of the previous sections. Table 3 summarizes the coal consumption of the three scenarios.

Coal-based synthetic fuels are another frequently mentioned possibility. Projections of several million barrels per day equivalent by 2000 have been made, but these are becoming increasingly improbable. In order to build up the industry to meet these levels, a major commitment would have to be made—soon. Coal is an economically rational source of energy when electricity is desired, but for liquid or gaseous fuels, as long as natural sources are available at one-third to one-half the cost of synthetics, there is little incentive for any individual user to turn to the latter. A future OTA report will address the fundamental questions raised by the previous sentence: the logic of electricity vs. synthetics when natural fuels are limited, what these limits are, and the rate at which a synthetic fuels industry can be developed when needed. To the extent that any coal-based synthetic fuel production does oc-

cur, this analysis assumes it would come at the expense of electric generation or oil imports. The former is unlikely to occasion any significant change in the impacts discussed in this report. The latter possibility is accounted for by a slight increase in the coal production levels of the previous section.

As discussed in the previous section, coal as a fuel presents a set of characteristics that potential users will compare to alternative fuels. The balance, and hence the eventual level of coal consumption may be shifted by changes in several factors in addition to growth in demand for electricity and industrial processes that can use coal or coal-derived electricity. These factors, which include technological improvements and regulatory restrictions, are discussed in detail in chapter IV. This section analyzes the demand of the users consistent with the scenarios, and the patterns of supply needed to meet the demand.

## Electric Power Generation

Utilities currently produce about 44 percent of their electricity with coal. The major alternative for the rest of this century is nuclear power. Most new utility demand for coal will be from new powerplants, as conversion of existing plants to coal will be quite difficult and expensive.

The status of coal-fired electric power capacity in megawatts as of September 1978 is as follows:

| Existing | Planned or under construction | Total |
|---|---|---|
| 220,583 | 152,521 | 373,104 |

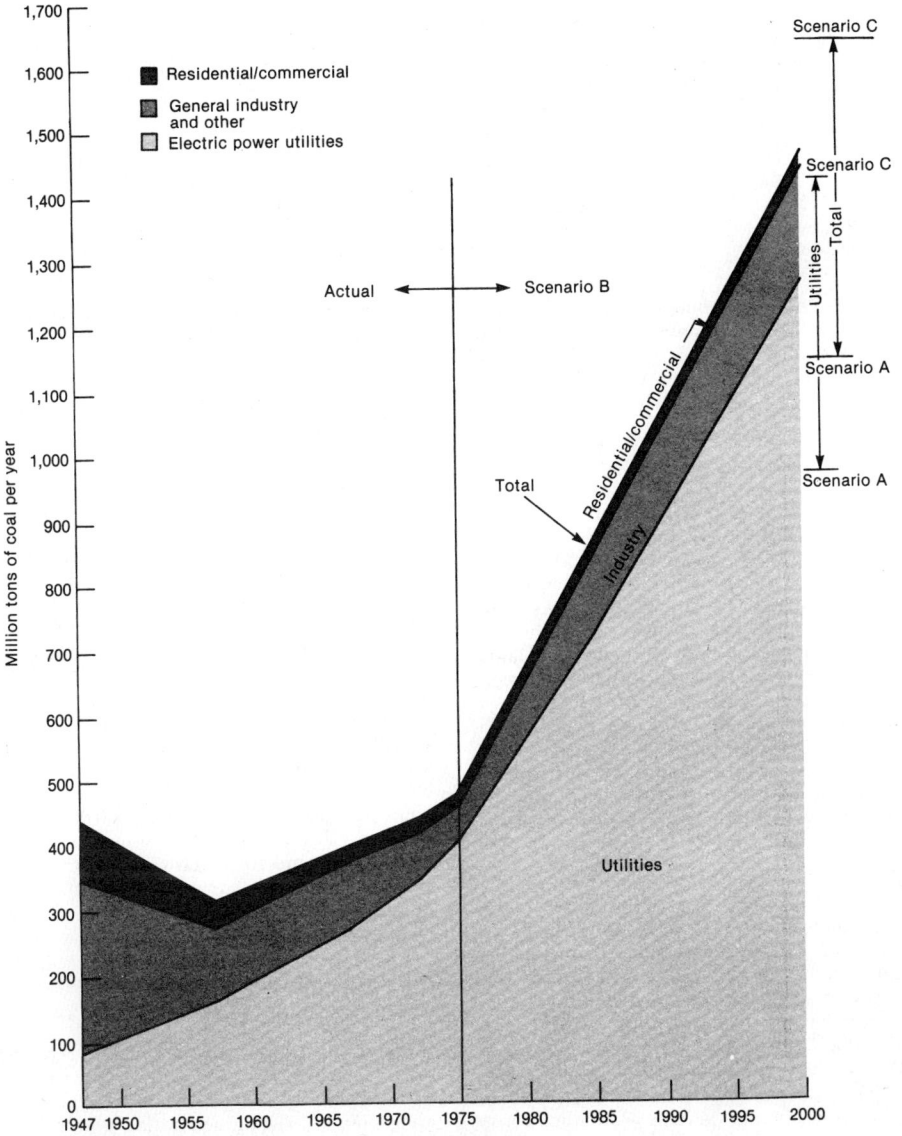

Figure 1.—Combustion of Coal and Lignite by End-Use Sector

SOURCES: *Annual Report to Congress*, Energy Information Administration, Department of Energy, vol. III, 1977; and the Office of Technology Assessment.

**Table 3.—Coal Consumption Projections**
**(millions of tons)**

| Scenario | Total production | Utilities | Industry | Residential/ commercial | Total domestic combustion | Potential synfuels | Nonenergy | Exports | Additions to stocks |
|---|---|---|---|---|---|---|---|---|---|
| 1975 | 655 | 403 | 62 | 12 | 477 | 0 | 83 | 67 | 32 |
| 1985 | | | | | | | | | |
| A | 955 | 675 | 90 | 15 | 780 | 0 | 95 | 70 | 10 |
| B | 1050 | 725 | 120 | 15 | 860 | 0 | 95 | 80 | 15 |
| C | 1145 | 775 | 150 | 15 | 940 | 0 | 100 | 90 | 15 |
| 2000 | | | | | | | | | |
| A | 1505 | 965 | 150 | 25 | 1140 | 65 | 100 | 185 | 15 |
| B | 1845 | 1275 | 160 | 25 | 1460 | 70 | 105 | 190 | 20 |
| C | 2110 | 1410 | 200 | 25 | 1635 | 145 | 110 | 200 | 20 |

Because of lengthening construction schedules, it is unlikely that any plants not included in these figures would be operating by 1985; the above total of 373,104 MW is thus an upper limit for 1985 capacity estimates.

If recent history is a guide, some of the units planned but not yet under construction will be dropped, and some construction schedules will slip. Some older units will also be retired, though probably only a few thousand megawatts worth over the next decade. Other old units will be downgraded from baseload to intermittent operation. This procedure was more pronounced in the past because new plant efficiencies were rising, making the newer plants cheaper to operate, but this trend has leveled off since the early 1960's. In fact, plants with scrubbers are less efficient and more expensive to operate. Utilities may favor the use of older plants when they do not need all their capacity, a factor that will tend to make attainment of clean air standards more difficult than expected.

The National Electric Reliability Council is predicting 309,476 MWe capacity (of units greater than 25 MWe) by 1985, consuming 824 million tons of coal. The above concerns indicate this is an optimistic schedule. If all plants with prevention of significant deterioration permits are completed on schedule, and if 3,000 MWe are retired, there will be 301,000 MWe, but a more conservative estimate might be 295,000 MWe. This would correspond to coal consumption of perhaps 775 million tons. This is obviously still a huge jump over the 1977 utility use of 475 million tons (estimated).

If major deferrals result from a continuation of the very low growth of the past few years, utility demand may be as much as 100 million tons lower.

Beyond 1985, projections are highly uncertain. A growth rate of 5 percent from 295,000 MWe in 1985 for the remainder of the century would result in about 450,000 MWe of coal-fired powerplants by 2000. A growth rate nearer historical levels would lead to 600,000 MWe. These would lead to coal demands of 1,400 million and 2,000 million tons annually. (Increases in coal are not proportional to increases in capacity, because the shift to lower heat value Western coal requires more tons for the same electrical output.) Utility forecasts are changing rapidly, mostly towards lower estimates. Historical growth patterns appear to be unlikely to continue, as pointed out in the previous sections. The year 2000 coal-fired electric generation of tables 2 and 3 is equivalent to about 330,000 to 450,000 MWe. The exact level will depend primarily on the level of demand for electricity and the competitiveness of nuclear power; it is therefore quite sensitive to policy decisions affecting nuclear power and the economics of coal.

## Decentralized Electric Power Generation: An Alternative Approach?

The projections have implicitly assumed that the utility industry would continue the trend toward larger (600 to 1,200 MWe) generating plants. Existing generating plants in 44 States average less than 100 MWe each, but only three States have any planned facilities at that low level, and planned additions in 20 States average greater than 500 MWe. Every State that has planned capacity additions (only Vermont and the District of Columbia have none) will significantly increase the average capacity per plant. This trend for generating units may be leveling off, but stations, which can include several units, are clearly still getting larger as sites become scarcer. Economics of powerplant construction and operation have been the primary cause of the growth in size. The physical size of the plant, and the materials and labor to construct it, increase less than proportionally with its size. Operating costs (excluding fuel) are only slightly higher for a 500 MWe plant than for a 50 MWe plant.

Considerable interest has recently been expressed in some quarters in small, dispersed plants. ("Small" is relative: a 50 MWe plant would serve an average community of 20,000 people.) This interest has been prompted partly by the downtime maintenance experience with large coal-fired powerplants and partly by the environmental and social impacts of large plants and the resulting issues of public acceptability. Appendix II of volume II discusses the issue in detail. Smaller plants can in theory be placed closer to the communities they serve because their siting criteria are so much more modest. This dispersal would substantially reduce the need for long-distance, high-voltage transmission, though the distribution system would remain the same. Such a siting policy also raises the possibility of using the waste heat from the plants (about 60 percent of the total heat of combustion) for district heating or industrial processes. The latter in particular, usually called cogeneration, has been espoused as a major element in energy conservation policy. There are no technical barriers to either district heating or industrial cogeneration. Both have been in operation for many years. The major impediments are economics and a variety of institutional problems.

In order to assure continuity of supply, a spinning reserve (virtually instantly available) equivalent to the biggest single unit online must be maintained. The smaller all the units are, the easier it is to assure the same degree of

reliability. It should be noted, however, that most utilities are part of large grids. The entire grid shares the spinning reserve, so the excess capacity for each utility is not large.

The present high interest rates throw the economies of scale into question. Interest on capital costs is one of the larger items in the final bill for a large plant that may take 6 or more years to construct. Small plants may take one-third to one-half the time to construct, thus providing a much faster return on their share of the capital. This time factor is also a great advantage to utilities as they plan in this era of uncertainty in load forecasting. Another economic advantage is that the smaller size makes possible the factory fabrication and shipment of components that now must be field fabricated. Factory labor is often cheaper and more productive than construction labor. Community impacts during construction will clearly be less severe for smaller plants. Remote siting of large plants needing more than 1,000 construction workers can provoke serious strains in the nearby small communities that must support them temporarily. Construction of small plants near bigger cities would be relatively inconspicuous to the community infrastructure.

One of the major arguments advanced in favor of decentralized power systems is enhancement of local control, particularly if the plants are owned by a local government or a cooperative. This argument is very difficult to analyze. Until recently, few consumers were concerned with the other end of their powerline as long as the power was there when they flipped the switch, and their bills weren't too high. Obviously many persons feel the latter criterion is not now being met, but it is questionable whether decentralization or local control would provide much relief. The advantages would be more subjective: a sense of involvement and control over factors affecting one's life. It is as easy to name examples showing indifference to involvement (e.g., the difficulty of getting neighborhood committees to do much) as it is to list advantages. Further exploration of this issue is beyond the scope of this report.

There would probably not be a great deal of difference in total environmental impacts between a centralized and decentralized system of the same capacity. Total emissions could be about the same, and though they would be more dispersed for the small-scale system they would probably also directly affect more people. Lower stack heights could lead to a different mix of local and long-range transport pollutants, as discussed in chapter V. Insofar as any system includes district heating or cogeneration, thermal pollution and the combustion of other fuels would be substantially reduced.

Small plants and dispersed siting have their disadvantages, of course. Construction clearly requires more materials and possibly more labor for the same output. Environmental control measures are often easier and cheaper to implement in large plants. Flue-gas desulfurization, for example, may prove prohibitively expensive for small units. Hence even if decentralization proves advantageous otherwise, full realization may have to await commercialization of developments such as fluidized-bed combustion or highly cleaned coal, which are discussed in the next chapter. It is also likely that public heath impacts of dispersed facilities would be greater even for the same efficiency of control, simply because such plants would be in more densely populated regions.

Fuel delivery and waste removal can be major drawbacks to dispersed units. Unit trains have cut transportation costs for big plants to about half that of conventional trains, but these and slurry lines would not be practical for small plants. Insofar as the plants are located in more densely populated areas, transport would be more obtrusive, especially if trucks are used.

Some of the problems and expenses facing a utility trying to get a plant constructed and online are not markedly different for a small plant. Hence the total effort for several small plants will be greater than for one large plant. Operating costs have already been mentioned. Licensing would be another. Opposition to particular sites may not be much less intense, and if a close-in site is selected, opposition could be much more intense than for a remote site. This factor is further discussed in chapter IV.

This largely qualitative discussion is ambiguous. The economic and environmental trade-offs are uncertain, and the social impacts depend on values and expectations. Thus neither approach appears to have an overwhelming advantage—a striking observation, as the trend has been so strong to centralization. The economies of scale discussed above are not the whole cause of the trend. Regulatory constraints are another factor as suggested by the controversial Avech-Johnson[5] effect:

Utilities subject to a constraint on their rate of return have incentives to expand the size of their rate base to unjustifiable levels.

Thus while large central stations are clearly in the utilities' best interest they are not necessarily in society's. Reversal of the trend may require drastic changes in utility operations and ownership, and in regulatory practices. Nevertheless, the subject seems worthy of more detailed analysis. Particular attention should be directed to the States with utility systems most resembling the decentralized concept. Michigan, Nebraska, Wisconsin, and Iowa, for example, have many small generating utilities and facilities and access to coal.

## Industrial Use of Coal

The outlook for coal use by industry is quite different from that for utilities. Industry burned 60 million tons in 1976, representing only about 10 percent of all energy purchased by industry. Both figures have been dropping steadily for 30 years, and there appears to be no major effort on industry's part to reverse the trend. These factors have left industry as the sector of greatest opportunity for policy actions for conversion from oil and gas to coal. A recent report by the Congressional Budget Office[6] extensively analyzed the use of fuel by manufacturers and the prospects for increasing their use of coal. This report found that coal use could be raised by about 90 million

[5]H. Avech and L. Johnson, "Behavior of the Firm Under Regulatory Constraint," American Economic Review, 211 (December 1962), pp. 1059-69.

[6]"Replacing Oil and Natural Gas With Coal: Prospects in the Manufacturing Industries," Congress of the United States, Congressional Budget Office, August 1978.

tons above present levels by 1985 under various tax policies.

Industry uses most of its energy for process steam, electric power generation, and direct heat applications. Most of these functions could in principle be provided by coal, but conversion of existing oil- and gas-fired facilities to coal will be quite difficult. There is little experience with large coal-fired furnaces for direct heat rather than steam generation, so most applications will be for steam-raising boilers. Boilers or furnaces not designed to burn coal must be essentially replaced to accommodate coal, and entirely new storage and handling equipment must be added. Coal requires twice the storage volume for the same heat content as oil, and the handling equipment is larger and more expensive. Ash removal equipment and disposal must also be included. Pollution control equipment is generally necessary, even when not required for oil or gas. If industry is to be held to emission limitations similar to those being promulgated for utilities, only the very largest facilities will be able to consider coal. New technologies such as fluidized-bed combustion or synthetic fuels would be necessary to meet both coal use and environmental goals. Strong financial incentives above the fuel cost savings must be considered if it is desired to force these conversions. New facilities are more favorable targets. Energy consumption by industry, however, is expected to grow much more slowly than consumption by utilities. Hence industry cannot adopt coal as a major fuel without wide-scale replacement of existing units.

Guiding industry toward coal will be difficult but not impossible. Small powerplants and process steam generators are used throughout industry. Coal boilers are now available to cover a wide range of needs. These units can be installed much faster than utility powerplants. Some of the small units (up to about 1 ton of coal burned per hour or equivalent to about 2 MWe) can be manufactured and delivered as a package. Thus if the economics and the less tangible factors, such as reliability of supply, prove favorable, industrial coal use could rise rapidly. Industrial process heaters must be designed with a specific purpose in mind. The cement industry uses several million tons of coal annually for direct heat and could expand this relative to other fuels. The glass and metals industries may also find coal use advantageous.[7] The 1962 Census of Manufactures by the Bureau of Census[8] reported that the chemical, paper, and food industries were also major users of coal, with most industrial use in the East North Central, Middle Atlantc, and South Atlantic regions.

Coal use by industry will almost certainly reverse its long-term decline because of the fuel cost advantage and policy initiatives. Hence a low estimate for 1985 for industrial use of steam coal is 90 million tons. On the higher side, perhaps 150 million tons might be feasible. By 2000, industrial coal will have risen even without further policies to encourage it. A minimum figure for energy use might be 150 million tons. An upper limit is almost arbitrary as the outcome is sensitive to so many factors, but for the purposes of this report 200 million tons is used. Oil and gas will be the preferred alternatives as long as they are available at competitive prices. Solar energy could prove uniquely advantageous for industry if the costs (including reliable backup) prove competitive, as it could be used to avoid both fuel and emission restrictions.

## Residential and Commercial Use of Coal

Coal has nearly disappeared as a fuel for homes and commercial facilities over the last three decades. In 1948 coal provided 50 percent of the energy used in this sector, but this has declined to less than 2 percent.[9] The reasons are obvious: for the small user in particular, coal required dirty and noisy truck deliveries, messy storage in the house, and daily

[7]Frank H. Boon, "Industrial Consumers May Loom Very Large in Coal's Future," *Coal Age*, February 1976.
[8]Richard L. Gordon, reprinted in *Coal in the U.S. Energy Market*, Lexington Books, 1978.
[9]E. N. Cart, Jr., M. M. Farmer, C. E. Johnig, M. Lieberman, and F. M. Spooner, "Evaluation of the Feasibility for Widespread Introduction of Coal Into the Residential and Commercial Sector," Government Research Laboratories of EXXON Research and Engineering Company for the Council on Environmental Quality, April 1977.

stoking and ash removal, and it left the air laden with smoke and fumes. Most Americans turned to oil and gas, and the coal retail market (essentially the same as residential and commercial use) nearly followed the transportation market into oblivion as shown in figure 1. Recently, however, concerns over the price and availability of oil and gas have sparked a flurry of inquiries, though not yet substantial orders, to manufacturers of coal-fired equipment. Although a return to coal might at first appear unlikely, the scenario warrants examination because of the large energy consumption of the residential and commercial sectors (21 percent of the U.S. energy budget). Appendix III analyzes this potential in greater detail.

The retail market now is about 12 million tons per year, including anthracite. Most of this is consumed in the Appalachian coal-producing States, plus New York, Illinois, and Michigan. There is no national or even regional market structure for coal or furnaces. Dealers purchase coal from the mines (usually small ones) and have it delivered by train or truck. As the purchases are usually made on a one-time basis (spot market) in small quantities, and delivery is made without the economies of scale utilities enjoy, the retail price of coal is startlingly high, perhaps $80 to $100 per ton. These prices are probably enough in themselves to preclude a resurgence of coal use, but they could drop if the retail market develops into a more stable operation. Many mines sell coal by the pickup truck load at about $30 per ton, and retail markets in the mining areas carry it at $40 to $45 per ton, but this will benefit only those living near coalfields.

Although improvements in equipment, such as automatic stokers, and improvements in combustion technology could increase residential use of coal, the scenarios used here do not anticipate much growth. Coal-fired district heating plants (such as the U.S. Capitol Power Plant) may be the only way that coal could experience a resurgence in the residential market, as such large plants make stringent environmental controls practical and relieve the end-use consumer of the inconvenience. Large commercial customers may, however, find it economical to use coal in either conventional boilers or fluidized-bed combustion units.

Given the marginal economic advantages, the choice of coal as a heating fuel for the residential/commercial sector is not likely to be based on economic or financial calculation alone. Convenience and reliability of supply will probably be of major concern. The projected scenarios envision residential/commercial use of coal to double from 12 million tons in 1975 to 25 million in the year 2000, a growth that is virtually immaterial to the national energy picture, but perhaps environmentally significant in some regions.

## Distribution of Coal Combustion and Production

The national levels of consumption shown in table 3 will not be uniform across the country. Most coal will be burned within a few hundred miles of the coalfields, as it is now. The regional distribution of use is shown in figure 2 and listed in table 4. Production by State is shown in table 5. Even more than for the national levels, these projections are not estimates but indications. The actual distribution will depend on relative changes in regional cost of production and transportation, success in controlling emissions of high-sulfur coal, customers' perceptions of the reliability of delivered supply, and other unpredictable factors. Policy decisions affecting these factors can produce interregional shifts of several hundred million tons annually by 2000. For instance, the trend towards Western coal is expected here to moderate after 1985 because of the full scrubbing regardless of sulfur content required by the Clean Air Act Amendments of 1977 (discussed in chapter IV, pages 167-175).

The manpower required to produce this coal (assuming no change in mine productivity) is shown in table 6.

## Conclusions

The coal projections discussed here appear to be achievable under the present legal/regulatory and economic climate. If actual use

Figure 2.—Coal Combustion Distribution

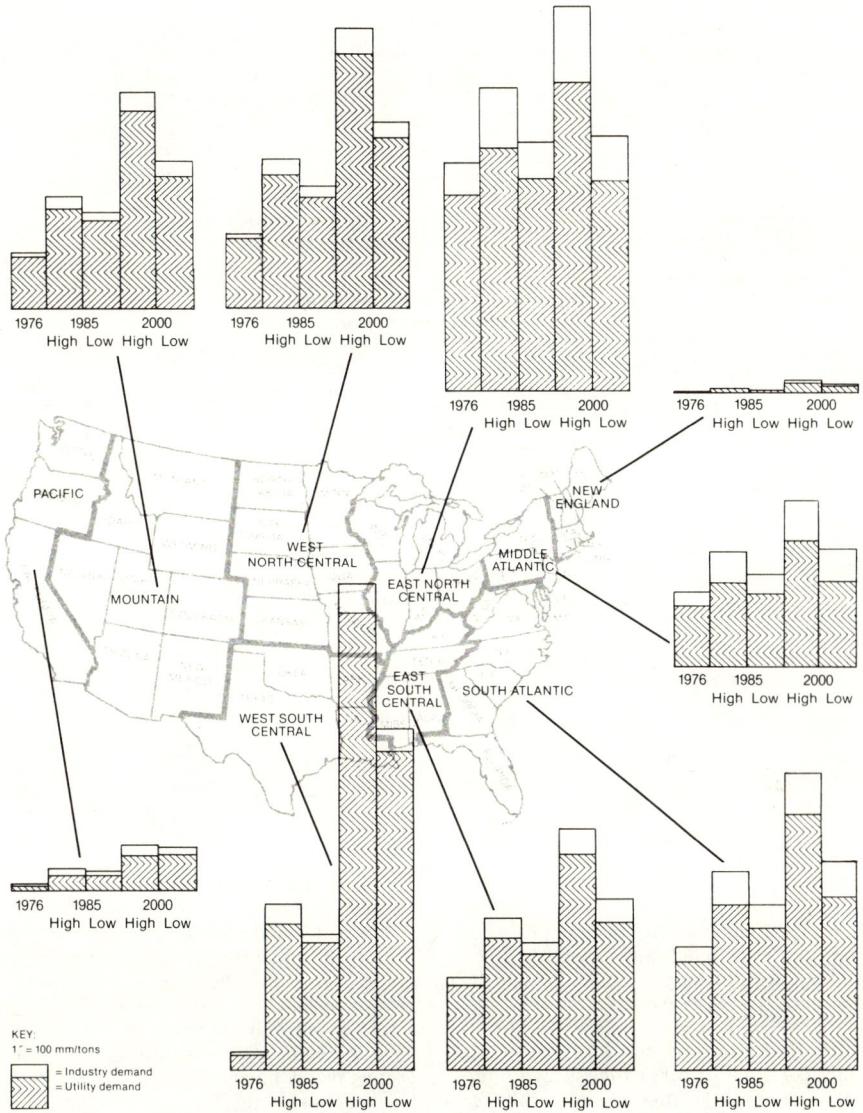

KEY:
1″ = 100 mm/tons

☐ = Industry demand
▨ = Utility demand

NOTE: Bureau of Census geographic regions as reported in EEI Statistical Yearbook

SOURCE: Office of Technology Assessment

## Table 4.—Powerplant and Industry Coal Combustion
### (million/tons)

| Region* | 1976 Utility | 1976 Industry | 1985 High utility | 1985 High industry | 1985 Low utility | 1985 Low industry | 2000 High utility | 2000 High industry | 2000 Low utility | 2000 Low industry |
|---|---|---|---|---|---|---|---|---|---|---|
| New England | 0.8 | 0 | 3 | 0 | 2 | 0 | 7 | 2 | 5 | 1 |
| Middle Atlantic | 45 | 10.85 | 64 | 23 | 56 | 15 | 96 | 30 | 65 | 25 |
| South Atlantic | 83.3 | 10.85 | 124 | 25 | 108 | 16 | 193 | 24 | 130 | 27 |
| East South Central | 64.7 | 6.33 | 101 | 15 | 88 | 10 | 165 | 20 | 112 | 17 |
| West South Central | 12.7 | 1.81 | 112 | 15 | 98 | 6 | 344 | 21 | 234 | 18 |
| Pacific | 4 | .4 | 11 | 5 | 10 | 3 | 27 | 7 | 27 | 5 |
| Mountain | 39 | 3 | 76 | 10 | 66 | 6 | 151 | 14 | 102 | 10 |
| West North Central | 52 | 3 | 99 | 12 | 86 | 6 | 193 | 18 | 131 | 12 |
| East North Central | 145 | 24.4 | 185 | 45 | 161 | 28 | 234 | 57 | 159 | 35 |
| Total | 446.5 | 60.64 | 775 | 150 | 675 | 90 | 1410 | 200 | 965 | 150 |

*New England - Maine, New Hampshire, Vermont, Massachusetts, Connecticut, Rhode Island
Middle Atlantic - New York, Pennsylvania, New Jersey
South Atlantic - Maryland, Delaware, West Virginia, Virginia, North Carolina, South Carolina, Georgia, Florida
East South Central - Kentucky, Tennessee, Mississippi, Alabama
West South Central - Arkansas, Louisiana, Oklahoma, Texas
Mountain - New Mexico, Arizona, Utah, Colorado, Nevada, Idaho, Montana, Wyoming
Pacific - California, Oregon, Washington
West North Central - North Dakota, South Dakota, Minnesota, Iowa, Nebraska, Kansas, Missouri
East North Central - Wisconsin, Illinois, Indiana, Michigan, Ohio

## Table 5.—Coal Production Projections
### (million of tons per year)

| | Surface 1977 production[a] | Surface Planned capacity[b] | Surface 1985 Low | Surface 1985 High | Surface 2000 Low | Surface 2000 High | Underground 1977 production[a] | Underground Planned capacity[b] | Underground 1985 Low | Underground 1985 High | Underground 2000 Low | Underground 2000 High | Total 1977 production[a] | Total 1985 Low | Total 1985 High | Total 2000 Low | Total 2000 High |
|---|---|---|---|---|---|---|---|---|---|---|---|---|---|---|---|---|---|
| **Appalachia** | | | | | | | | | | | | | | | | | |
| Alabama | 15 | 3 | 12 | 15 | 10 | 15 | 7 | 14 | 15 | 18 | 27 | 40 | 22 | 27 | 33 | 37 | 55 |
| Georgia | — | 1 | — | 1 | — | 1 | — | — | — | — | — | — | 0 | 1 | 0 | 1 | |
| Kentucky east | 51 | 6 | 36 | 40 | 40 | 49 | 41 | 34 | 50 | 58 | 85 | 115 | 92 | 86 | 98 | 125 | 164 |
| Maryland | 3 | 2 | 3 | 4 | 2 | 3 | — | 2 | 1 | 2 | 2 | 3 | 3 | 4 | 6 | 4 | 6 |
| Ohio | 32 | 3 | 20 | 25 | 30 | 40 | 14 | 12 | 18 | 20 | 32 | 50 | 46 | 38 | 45 | 62 | 90 |
| Pennsylvania | 45 | 3 | 30 | 33 | 15 | 20 | 38 | 20 | 40 | 47 | 80 | 100 | 83 | 70 | 80 | 95 | 120 |
| Tennessee | 6 | — | 4 | 5 | 3 | 4 | 5 | 1 | 3 | 4 | 4 | 5 | 11 | 7 | 9 | 7 | 9 |
| Virginia | 12 | — | 7 | 9 | 5 | 8 | 26 | 6 | 21 | 24 | 25 | 40 | 38 | 28 | 33 | 30 | 48 |
| West Virginia | 21 | 5 | 18 | 23 | 25 | 35 | 74 | 43 | 77 | 87 | 125 | 187 | 95 | 95 | 110 | 150 | 222 |
| Total Appalachia | 185 | 23 | 130 | 155 | 130 | 175 | 205 | 132 | 225 | 260 | 380 | 505 | 390 | 355 | 415 | 510 | 680 |
| **Midwest** | | | | | | | | | | | | | | | | | |
| Missouri | 7 | — | 3 | 4 | 5 | 10 | — | — | — | — | — | — | 7 | 3 | 4 | 5 | 10 |
| Illinois | 24 | 10 | 23 | 30 | 39 | 55 | 30 | 30 | 39 | 53 | 75 | 110 | 54 | 62 | 83 | 114 | 165 |
| Indiana | 27 | 11 | 25 | 32 | 21 | 32 | 1 | — | — | — | 10 | 25 | 28 | 25 | 32 | 31 | 57 |
| Kentucky west | 28 | 4 | 21 | 24 | 28 | 35 | 23 | 9 | 20 | 25 | 33 | 42 | 51 | 41 | 49 | 61 | 77 |
| Oklahoma | 5 | 1 | 3 | 5 | 2 | 3 | — | 2 | 1 | 2 | 2 | 3 | 5 | 4 | 7 | 4 | 6 |
| Total Midwest | 91 | 26 | 75 | 95 | 95 | 135 | 54 | 41 | 60 | 80 | 120 | 180 | 145 | 135 | 175 | 215 | 315 |
| **West** | | | | | | | | | | | | | | | | | |
| Arizona | 11 | 3 | 11 | 13 | 8 | 11 | — | — | — | — | — | — | 11 | 11 | 13 | 8 | 11 |
| Colorado | 8 | 31 | 21 | 30 | 45 | 64 | 4 | 30 | 18 | 20 | 25 | 30 | 12 | 39 | 45 | 70 | 94 |
| Montana | 29 | 63 | 55 | 65 | 130 | 200 | — | — | — | 5 | 5 | 10 | 29 | 55 | 60 | 135 | 210 |
| New Mexico | 11 | 59 | 38 | 50 | 75 | 120 | — | 1 | — | 1 | 2 | 5 | 11 | 38 | 43 | 77 | 125 |
| North Dakota | 12 | 53 | 37 | 47 | 80 | 135 | — | 4 | 2 | 3 | 3 | 4 | 12 | 39 | 45 | 83 | 139 |
| Texas | 17 | 58 | 43 | 54 | 85 | 140 | — | — | — | — | — | — | 17 | 43 | 48 | 85 | 140 |
| Utah | — | 16 | 10 | 10 | 8 | 10 | 9 | 39 | 29 | 33 | 39 | 49 | 9 | 38 | 43 | 47 | 59 |
| Washington | 5 | 1 | 5 | 6 | 9 | 15 | — | — | — | — | — | — | 5 | 5 | 6 | 10 | 17 |
| Wyoming | 45 | 268 | 196 | 220 | 260 | 310 | — | 3 | 1 | 3 | 5 | 10 | 45 | 192 | 207 | 265 | 340 |
| Total West | 141 | 553 | 415 | 495 | 700 | 1,005 | 13 | 77 | 50 | 60 | 80 | 110 | 154 | 460 | 510 | 780 | 1,115 |
| National total | 417 | 602 | 620 | 745 | 925 | 1,315 | 272 | 250 | 335 | 400 | 580 | 795 | 689 | 955 | 1,145 | 1,505 | 2,110 |

[a] 1978 Keystone Coal Industry Manual, p. 666 (estimated).
[b] Ibid., pp. 674-685; data adjusted to eliminate present production capacity.

does not rise to these levels, it is more likely to be from lack of demand rather than restricted supply, but several factors could induce the latter situation, as discussed in chapter IV. Demand for coal could remain below these levels either because energy demand has been successfully curtailed or other fuels prove unexpectedly bountiful. Neither situation calls for remedial action, though attention could still be directed to reducing the negative impacts of coal (see chapters V and VI).

If it is deemed necessary to achieve projections higher than these because of disappointing oil and gas discoveries or inadequate growth by nuclear and other energy sources, attention may have to be directed to loosening environmental and other restrictions or accelerating development of technologies that better accommodate them. Coal's growth rate may also need acceleration beyond that dictated by the immediate market conditions, possibly because of an unexpected sharp drop in oil or gas availability. Then greater efforts of coercion or incentives would be required, possibly coupled with a streamlining of the process of getting mines into production.

These scenarios are used as guidelines in the following chapter. Detailed analyses of the implications, such as emission quantification, have not been made, as the results would not vary greatly from published analyses.

## Table 6.—Coal Mine Employment Forecasts

| State | Surface 1985 Low | Surface 1985 High | Surface 2000 Low | Surface 2000 High | Underground 1985 Low | Underground 1985 High | Underground 2000 Low | Underground 2000 High | Total 1985 Low | Total 1985 High | Total 2000 Low | Total 2000 High |
|---|---|---|---|---|---|---|---|---|---|---|---|---|
| **Appalachia** | | | | | | | | | | | | |
| Alabama......... | 3,290 | 4,080 | 2,720 | 4,080 | 9,430 | 11,320 | 16,970 | 25,150 | 12,700 | 15,400 | 19,690 | 29,230 |
| Georgia ......... | — | 240 | — | — | — | — | — | — | — | 240 | — | — |
| Kentucky ........ | 6,725 | 7,470 | 7,470 | 9,160 | 20,010 | 23,210 | 34,010 | 46,010 | 26,740 | 30,680 | 41,480 | 55,170 |
| Maryland ........ | 575 | 770 | 380 | 575 | 475 | 950 | 950 | 1,420 | 1,050 | 1,720 | 1,330 | 1,995 |
| Ohio ............ | 3,210 | 4,010 | 4,810 | 6,420 | 9,990 | 11,100 | 17,760 | 27,750 | 13,200 | 15,110 | 22,570 | 34,170 |
| Pennsylvania...... | 6,070 | 6,680 | 3,040 | 4,050 | 22,090 | 25,960 | 44,180 | 55,230 | 28,160 | 32,640 | 47,220 | 59,280 |
| Tennessee........ | 940 | 1,180 | 710 | 940 | 1,380 | 1,840 | 1,840 | 2,300 | 2,320 | 3,020 | 2,550 | 3,240 |
| Virginia.......... | 1,480 | 1,910 | 1,060 | 1,700 | 11,090 | 12,670 | 13,200 | 21,110 | 12,575 | 14,580 | 14,260 | 22,810 |
| West Virginia...... | 4,410 | 5,630 | 6,120 | 8,570 | 42,480 | 47,990 | 68,950 | 103,160 | 46,890 | 53,620 | 75,070 | 111,730 |
| Total Appalachia. | 26,680 | 31,970 | 26,310 | 35,495 | 116,945 | 135,040 | 197,860 | 282,130 | 143,630 | 167,010 | 224,170 | 317,625 |
| **Midwest** | | | | | | | | | | | | |
| Missouri......... | 590 | 790 | 990 | 1,970 | — | — | — | — | 590 | 790 | 990 | 1,970 |
| Illinois .......... | 3,960 | 5,170 | 6,720 | 9,470 | 12,440 | 16,910 | 23,920 | 35,090 | 16,400 | 22,080 | 30,640 | 44,560 |
| Indiana.......... | 3,510 | 4,490 | 2,950 | 4,490 | — | — | 2,820 | 7,060 | 3,510 | 4,490 | 5,770 | 11,550 |
| Kentucky (west).... | 2,960 | 3,380 | 3,950 | 4,930 | 6,280 | 7,850 | 10,360 | 13,180 | 9,240 | 11,230 | 14,310 | 18,110 |
| Oklahoma ........ | 845 | 1,410 | 560 | 850 | 1,010 | 2,020 | 2,020 | 3,030 | 1,855 | 3,430 | 2,580 | 3,880 |
| Total Midwest ... | 11,870 | 15,240 | 15,170 | 21,710 | 19,730 | 26,780 | 39,120 | 58,360 | 31,595 | 42,020 | 54,290 | 80,070 |
| **West** | | | | | | | | | | | | |
| Arizona.......... | 660 | 780 | 480 | 660 | — | — | — | — | 660 | 780 | 480 | 660 |
| Colorado ......... | 1,920 | 2,740 | 4,110 | 5,850 | 7,840 | 8,720 | 10,900 | 13,070 | 9,760 | 11,460 | 15,010 | 18,920 |
| Montana.......... | 1,800 | 2,130 | 4,260 | 6,550 | — | — | 2,270 | 4,540 | 1,800 | 2,130 | 6,530 | 11,090 |
| New Mexico....... | 3,090 | 4,070 | 6,100 | 9,760 | — | 490 | 970 | 2,430 | 3,090 | 4,560 | 7,070 | 12,190 |
| North Dakota...... | 1,775 | 2,250 | 3,840 | 6,480 | 910 | 1,360 | 1,360 | 1,820 | 2,685 | 3,610 | 5,200 | 8,300 |
| Texas ........... | 2,340 | 2,940 | 4,630 | 7,630 | — | — | — | — | 2,340 | 2,940 | 4,630 | 7,630 |
| Utah ............ | 1,670 | 1,670 | 1,330 | 1,670 | 9,520 | 10,830 | 12,800 | 16,080 | 11,190 | 12,500 | 14,130 | 17,750 |
| Washington....... | 760 | 910 | 1,360 | 2,270 | — | — | 860 | 1,720 | 760 | 910 | 2,220 | 3,990 |
| Wyoming ......... | 1,210 | 1,350 | 1,660 | 1,910 | 420 | 1,260 | 2,110 | 4,210 | 1,625 | 2,610 | 3,710 | 6,120 |
| Total West ...... | 15,200 | 18,840 | 27,710 | 42,780 | 18,690 | 22,660 | 31,270 | 43,870 | 33,900 | 41,500 | 58,980 | 86,650 |

Chapter III

# COAL TECHNOLOGY AND TECHNIQUES

# Chapter III.—COAL TECHNOLOGY AND TECHNIQUES

Chapter III

# COAL TECHNOLOGY AND TECHNIQUES

This chapter describes the various conditions and technologies that exist in the coal system from resources in the ground through use and disposal of wastes. The section below provides an overview and summary. Subsequent sections expand on the various elements of the system.

## PATTERNS OF USE

Coal is burned to produce heat, which in turn is used to generate steam for process heat or the production of electricity. Alternatively the heat may be used directly in industrial process systems or space heating. Coal also can be used to effect chemical reactions as in the reduction of iron ore or the production of lime, or indirectly as a source for synthetic gaseous or liquid fuels, but such uses are not considered in this report.

Although the direct use of coal in the utility, industrial, commercial, and residential sectors is commercially feasible using current technology, other factors may limit its use. The total energy system—from resources in the ground, through extraction, transportation, and end use—must be examined to identify constraints on and impacts of its use. Figure 3 illustrates the variety of conditions and technological options that must be considered in the direct use of coal.

Coal is an abundant, widely distributed resource in the United States. It is variable in composition, moisture, sulfur, ash, and trace element content. Because of geological and geographic differences, mining and transportation methods vary. In Appalachia, most coal is mined underground with a variety of mechanized equipment. Most Midwestern and Western coal is surface-mined, using gigantic shovels or draglines to remove the overburden and expose the seam. Railroads are the dominant transportation means. In the East and Midwest, barges are important supplements. Alternatively, the coal may be converted to electricity at the minesite and the energy transported to market by wire.

The major factors influencing the design of combustion facilities are the size of the unit, the type of coal to be burned, and the environmental standards it must meet. Thus it is difficult and sometimes impossible to switch to a different coal. Also the technology of the utility sector is often inappropriate for the industrial or commercial installation.

Because of the many interrelated factors involving coal, geographic conditions, and technologies, control equipment and methods often must be "fine-tuned" to individual circumstances. Reclamation techniques must be adjusted to topography, climate, and future desired land use and social values. Control of sulfur oxide ($SO_x$) emissions from the burning of coal can be achieved by removing sulfur before, during, or after combustion. The most effective and economical design for a particular case depends on such matters as the mode of operation of the combustion unit, the sulfur concentration in the coal, the availability of quality absorbents, etc. Similarly, the efficiency of particulate control technology can be affected by changes in the sulfur content of the coal or the physical and chemical composition of the ash.

New technologies for extracting, transporting, "cleaning," upgrading, and burning coal are continuously emerging, as are environmental controls for every phase.

To illustrate the combinations that exist in the overall coal-energy system, three hypothetical examples are depicted for the utility sector as end-user. Each user is an electric powerplant of 800 MW operating 70 percent of the time.

**Figure 3.—Coal System Components**

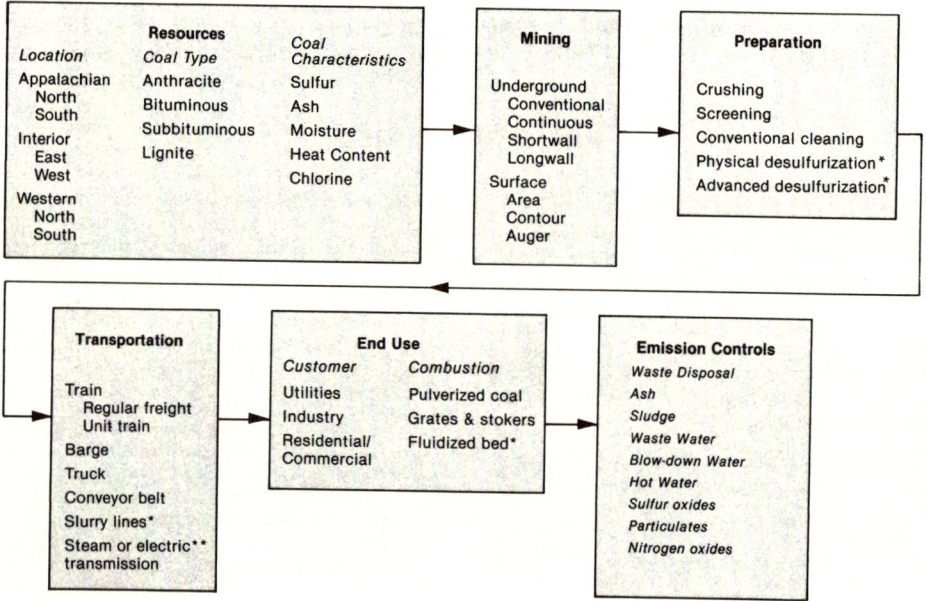

| Resources | | | Mining | Preparation |
|---|---|---|---|---|
| Location | Coal Type | Coal Characteristics | | |
| Appalachian | Anthracite | Sulfur | Underground | Crushing |
| North | Bituminous | Ash | Conventional | Screening |
| South | Subbituminous | Moisture | Continuous | Conventional cleaning |
| Interior | Lignite | Heat Content | Shortwall | Physical desulfurization* |
| East | | Chlorine | Longwall | Advanced desulfurization* |
| West | | | Surface | |
| Western | | | Area | |
| North | | | Contour | |
| South | | | Auger | |

| Transportation | End Use | | Emission Controls |
|---|---|---|---|
| | Customer | Combustion | Waste Disposal |
| Train | Utilities | Pulverized coal | Ash |
|   Regular freight | Industry | Grates & stokers | Sludge |
|   Unit train | Residential/ | Fluidized bed* | Waste Water |
| Barge | Commercial | | Blow-down Water |
| Truck | | | Hot Water |
| Conveyor belt | | | Sulfur oxides |
| Slurry lines* | | | Particulates |
| Steam or electric** | | | Nitrogen oxides |
| transmission | | | |

*This system undergoing research and development in the U.S.
**Can be considered an alternative to coal transportation in getting the energy to the final point of use.

SOURCE: Office of Technology Assessment.

## Eastern Deep-Mined Coal Burned at an East Coast Plant

The plant requires about 1.85 million tons of coal per year with a heat content of 13,500 Btu/lb. Any one of the five largest West Virginia underground mines could currently be the single source. Such a supply requires mining at the rate of about 275 acres per year if the seam is 8 feet thick. After the coal is brought to the surface it is crushed, screened, washed, and dried at a preparation plant near the minesite. About 700 persons are employed for extraction and preparation. The price of the coal leaving the preparation plant is $30/ton.

Next, the coal is loaded onto railroad hopper cars in a unit train. The train consists of 100 cars and makes three 250-mile round trips each week between the preparation site and the powerplant. Transportation adds about $5 to $8 per ton to the price. At the powerplant, a 60- to 90-day supply is normally stored to assure a continuous feed to the boilers during any short-term supply interruption. Three-fourths of a pound of coal is burned to produce a kilowatthour of electricity. Ten percent of the coal is ash. In the combustion step about 20 to 40 percent of the ash settles and is removed from the bottom of the furnace. The lighter residual ash is entrained in the flue gases and carried out as "fly ash," more than 99 percent of which is collected in particulate control equipment. In some plants, a portion of the ash is used for cement and the remainder disposed as solid waste. The plant is 20 miles from the main electric load center. A transmission line of 66,000 volts is used to deliver the power to the distribution network, where the voltage is appropriately reduced for ultimate use in the industrial, commercial, and residential sectors.

## Eastern Strip-Mined Coal Burned at a Mine-Mouth Plant

This coal has a lower heat content (11,300 Btu/lb) than that in the previous example, so 2.2 million tons are needed annually. It is produced at a mine by simultaneously exploiting three superimposed seams. The top seam is 9 feet thick and lies under 100 feet of overburden. The second and third seams are 4 and 6 feet thick and lie below intervening layers of 10 feet of shale. Eighty-five acres of coal are removed yearly. A total of 250 miners are employed in two operating shifts. The coal is crushed, screened, and transported 15 miles by truck to the powerplant. Although the powerplant is similar to that in the previous example,

it is larger to accommodate the greater volume of coal with high moisture and ash content. The ash is returned to the mine. The higher capital cost of the powerplant is offset by the lower operating cost of the fuel — $20/ton.

## Western Strip-Mined Coal Burned at a Midwestern Plant

This coal has only 8,500 Btu/lb, so 3 million tons/yr are required. A 30-foot seam is exploited by 90 miners working in two shifts. The overburden is 125 feet. Mining and reclamation proceed at the rate of 50 acres/yr. The extracted coal is crushed and screened at the mine and delivered to the powerplant daily by a 100-car unit train. As the 1,000-mile trip requires 50 hours each way, excluding loading and unloading time, six trains are continuously in service and spare equipment is provided for maintenance periods.

These examples, although hypothetical, are similar to actual cases. Table 7 lists quantitative values in some actual cases. The diversity of suppliers, customers, types of coal, means of delivery, and contracts revealed in the table illustrates the variability in a coal-based energy system.

### Table 7.—Examples of Coal/Lignite Deliveries at U.S. Powerplants During 1977

| Plant and unit no. | State | Cap. MW | Strip[a] | Deep[a] | Btu/[b] lb | Wt. sulf. % | Wt. ash % | Wt. moist. % | Coal train miles | Conveyor feet | Barge miles | Contract yrs. | Deliv. price[c] $/ton | Origin, State, and mines |
|---|---|---|---|---|---|---|---|---|---|---|---|---|---|---|
| Walsh, #1 . . . . . . . . . . | Tex. | 528 | 1.75 | — | 8,250 | 0.30 | 5.0 | 32.0 | 1,451 | — | — | 25 | 16.69 | Wyoming, Amax, Gillette |
| Monticello, #1-2 . . . . . | Tex. | 1,150 | 2.92 | — | 6,300 | 0.70 | 14.3 | 35.0 | 11 | — | — | 25 | 4.30[d] | Texas, Winfield |
| Chesterfield, #5-6 . . . | Va. | 1,000 | 1.25 | .22 | 12,500 av. | 1.20 | 10.0 | 12.0 | 450 | — | — | 25 | 37.00 | E. Kentucky, St. Paul |
| Kincaid, #1-2 . . . . . . . | Ill. | 1,212 | — | 2.80 | 10,500 | 3.70 | 9.0 | 16.8 | — | 1,130 | — | 34 | 19.50 | Illinois, Peabody #10 |
| Will County, #1-4 . . . . | Ill. | 1,093 | 2.50 | — | 9,600 | 0.46 | 4.5 | 23.2 | 1,248 | — | 175 | 20 | 23.70 | Mont. & Wyo. Decker & Big Horn |
| Powerton, #5-6. . . . . . | Ill. | 1,700 | 1.70 | 1.70 | 10,550 | 3.68 | 11.8 | 13.7 | 410 | — | — | 4 | 13.00 | Illinois, Monterey (Deep) Captain (Strip) |
| Chalk Point, #1-2 . . . . | Md. | 660 | .41 | .13 | 12,995 | 1.71 | 12.2 | 6.3 | 330 | — | — | 25 | 28.00 | Centr. W. Penn., some Md. |
| Bruce Mansfield, #1 . | Pa. | 825 | — | 2.92 | 11,500 | 3.10 | 15.0 | 10.0 | — | — | 80 | 25 | 23.00 | Ohio, Quarto, Wheeling |

[a] In millions of tons.
[b] As received by utility.
[c] Estimated average price during 1977.
[d] The $4.30 per ton quoted does not include transportation over the utility's 11-mile self-operated rail system.

# COAL RESOURCES

Coal deposits already identified in the United States contain enough energy to supply the Nation's entire present energy demand for

more than 500 years. It is estimated that other deposits, as yet unidentified, are equivalent in magnitude. By comparison, the known reserves of domestic crude oil would last less than 20 years at the same rate of depletion. The amount of coal that can be extracted is less than the quantity residing underground. Some deposits are too small or inaccessible to warrant the financial investment required to bring a mine into operation. Others underlie surfaces that cannot be disturbed. Finally, all the coal in any particular deposit cannot be extracted by even the best mining technology and equipment because of safety or economic considerations.

These concepts are illustrated in figure 4 and defined as follows:

- Coal Resources are the total coal deposits regardless of whether they can be mined or recovered. Some but not all of the resources are known. There is no assurance that the unknown quantities actually exist. This estimate is based on the geological similarity of potential fields to known coal-bearing areas.
- Coal Reserves are deposits assumed to be commercially minable by virtue of their seam thickness and accessibility. The magnitude of reserves can be estimated with considerably more confidence than that of resources.
- Recoverable Reserves are the estimated tonnage of coal that can be produced with existing mining technology. Reserves that cannot be mined for legal, political, social, or other reasons are not included.

The recoverable reserves represent less than 50 years of the Nation's present total energy consumption, and this figure is subject to downward revision if more stringent regulations are enacted. Upward revision will result if the estimated resources can be identified, or if new techniques increase the recoverable fraction. Coal may never be required to supply energy at such a high consumption rate (75 Quads of energy imply about 3.75 billion tons/ yr), although it did supply almost 80 percent of a much smaller demand in 1910. It can, however, substitute for a large share of the decline in the production rate of oil and gas. Unlike oil and gas, coal can be a major source of energy through the 21st century.

The estimates of recoverable reserves in figure 4 are for deposits no deeper than 1,000 feet and at least 28 inches thick if they are to be mined underground, or with depth-to-seam thickness ratios of 10 or less for surface mines. These limits are set for reasons of mining economics. Large quantities of coal exist in otherwise minable deposits that do not meet these qualifications, but little incentive now exists to mine them because others are less expensive to mine. Until the energy expended for recovery exceeds the recoverable energy, no intrinsic reason exists why these less attractive deposits cannot eventually be exploited after more favorable seams are exhausted. Average depths of coal mines in Europe exceed 1,000 feet. One

**Figure 4.—U.S. Coal Resources and Reserves**

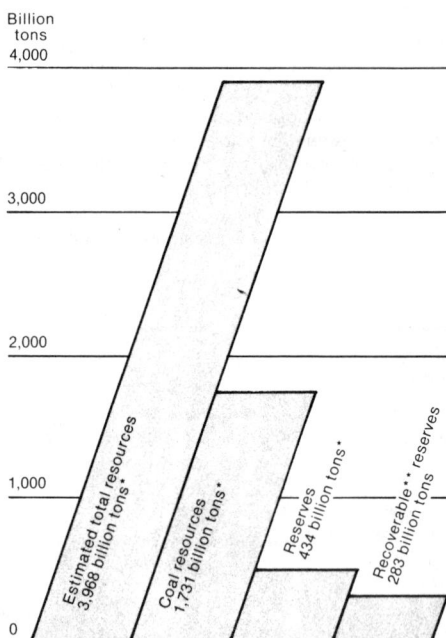

SOURCE: *Paul Averitt, Coal Resources of the United States, Jan. 1, 1974, Geological Survey Bulletin 1412.

**Based on 57 percent recoverability in underground mines and 80 percent in surface mines.

mine in the U.S.S.R. is reported to operate deeper than 2,500 feet. Even in the United States, one Colorado coal mine is at a 3,000-foot depth and an Alabama coal mine at 1,900 feet. Several Appalachian seams of less than 28 inches are mined underground. Exploitation of these reserves presently uncounted as recoverable could substantially extend the useful lifetime of coal.

## Types of Coal

Coal is divided into five classes: anthracite, bituminous, subbituminous, lignite, and peat. Peat is the earliest stage in the formation of coal. Heat and pressure force the moisture and hydrocarbon from the peat until progressively higher ranks of coal are formed. Thus, anthracite contains the lowest percentage of moisture and the highest percentage of carbon. The chemical and physical properties of the other types lie between anthracite and peat. The composition, heat content, and locale of typical ranks of coal are shown in table 8. High-volatile bituminous coals have agglomerating characteristics (tendency to fuse when heated), as do higher ranks of coal generally. The lowest rank coals, subbituminous and lignite, are nonagglomerating. In general, moisture and volatile content decreases and fixed carbon increases with increased rank.

Boilers and furnaces can be designed to burn almost any coal, but may not be readily adaptable to other coals even of the same types. The characteristics that affect equipment design are the heating value, ash melting characteristics, hardness of both the coal and its mineral matter (for grinding), and mineral composition, which affects corrosion and fouling tendencies. Coal beneficiation can improve undesirable characteristics and reduce the differences among coal types.

The sulfur content of coal became increasingly important when limits were imposed by the Clean Air Act of 1970 on emissions from combustion. New or modified powerplants starting operation between 1974 and 1978 are limited to emissions of 1.2 lbs sulfur dioxide ($SO_2$)/million Btu of input. (More stringent

## Table 8.—Coal Ranks and Typical Characteristics

| | Coal rank | | | | Coal analysis, bed moisture basis | | | | | |
|---|---|---|---|---|---|---|---|---|---|---|
| Class | Group | | State | County | M | VM | FC | A | S | Btu |
| I. Anthracitic .................. | 1. Meta-anthracite | | Pa. | Schuylkill | 4.5 | 1.7 | 84.1 | 9.7 | 0.77 | 12,745 |
| | 2. Anthracite | | Pa. | Lackawanna | 2.5 | 6.2 | 79.4 | 11.9 | 0.60 | 12,925 |
| | 3. Semianthracite | | Va. | Montgomery | 2.0 | 10.6 | 67.2 | 20.2 | 0.62 | 11,925 |
| II. Bituminous ................. | 1. Low-volatile bituminous coal | | W.Va. | McDowell | 1.0 | 16.6 | 77.3 | 5.1 | 0.74 | 14,715 |
| | | | Pa. | Cambria | 1.3 | 17.5 | 70.9 | 10.3 | 1.68 | 13,800 |
| | 2. Medium-volatile bituminous coal | | Pa. | Somerset | 1.5 | 20.8 | 67.5 | 10.2 | 1.68 | 13,720 |
| | | | Pa. | Indiana | 1.5 | 23.4 | 64.9 | 10.2 | 2.20 | 13,800 |
| | 3. High-volatile A bituminous coal | | Pa. | Westmoreland | 1.5 | 30.7 | 56.6 | 11.2 | 1.82 | 13,325 |
| | | | Ky. | Pike | 2.5 | 36.7 | 57.5 | 3.3 | 0.70 | 14,480 |
| | | | Ohio | Belmont | 3.6 | 40.0 | 47.3 | 9.1 | 4.00 | 12,850 |
| | 4. High-volatile B bituminous coal | | Ill. | Williamson | 5.8 | 36.2 | 46.3 | 11.7 | 2.70 | 11,910 |
| | | | Utah | Emery | 5.2 | 38.2 | 50.2 | 6.4 | 0.90 | 12,600 |
| | 5. High-volatile C bituminous coal | | Ill. | Vermillion | 12.2 | 38.8 | 40.0 | 9.0 | 3.20 | 11,340 |
| III. Subbituminous ........... | 1. Subbituminous A coal | | Mont. | Musselshell | 14.1 | 32.2 | 46.7 | 7.0 | 0.43 | 11,140 |
| | 2. Subbituminous B coal | | Wyo. | Sheridan | 25.0 | 30.5 | 40.8 | 3.7 | 0.30 | 9,345 |
| | 3. Subbituminous C coal | | Wyo. | Campbell | 31.0 | 31.4 | 32.8 | 4.8 | 0.55 | 8,320 |
| IV. Lignitic ....................... | 1. Lignite A | | N.D. | Mercer | 37.0 | 26.6 | 32.2 | 4.2 | 0.40 | 7,255 |
| | 2. Lignite B | | | | | | | | | |

Data on coal (bed moisture basis)
M = equilibrium moisture, %; VM = volatile matter;
FC = fixed carbon, %; A = ash, %; S = sulfur, %;
Btu = Btu/lb, high heating value.
Calculations by Parr formulas.
  Adapted from Babcock and Wilcox, *Steam, Its Generation and Use,* p. 5-11.

regulations forthcoming from the Clean Air Amendments of 1977 will eliminate the use of low-sulfur coal as the sole means of compliance for new powerplants.) Maximum sulfur in coal to meet this emission standard without other controls is approximately 0.6 percent for coal with a heating value of 10,000 Btu/lb. For each 1,000 Btu/lb higher or lower, the allowable sulfur content is changed about 0.06 percent (i.e., the sulfur limit for coal with 12,000 Btu/lb would be about 0.72 percent). Some powerplants will be limited to more stringent limits, as described in chapter IV. Other users, such as industrial facilities, may be subject to different standards. The availability of coal by region as a function of $SO_x$ emissions is shown in figure 5.

A large number of other substances have been identified in coal. Many elements were drawn up from the soil into the vegetation that formed coal. Later streams and ocean poured in over the rotting beds of vegetation carrying in mud, sand, and other minerals. Thus, coal can contain varying degrees of many elements. Typical elements in the mineral matter of coal are shown in table 9. The quantities listed are averages based on many samples. The number of samples is indicated at the heading of each column. As with sulfur content, elemental composition varies widely, not only among samples from different beds but also among those from the same mine. All of these elements are found elsewhere in the environment, but are known to be damaging to the environment or to human health at certain concentrations and mode of entry into organisms. The potential for the damage, if any, cannot be reliably predicted because the mineral matter of most coals has not yet been fully analyzed, and the pathways of these trace elements through the biosphere after combustion are generally not known. The possible effect of trace elements in the environment is discussed in chapter V.

## Coal Formations

Coal generally occurs in major structural basins as shown in figure 6. The actual coalbeds are within these basins. Most beds are broad and thin with most of the coal within 3,000 feet of the surface. A few in the Rocky Mountain region were canted very steeply with the upheaval of the mountains during their formation. Some reach depths of 30,000 feet.

Coalbeds exhibit the structural vagaries of geological changes that have occurred since their inception. An example of what can be expected in the structure of a bed is shown in figure 7. Note that the coal is not uniform in thickness in the roof or floor condition—nor even continuous. These variations present a changing pattern of mine problems and even limit the amount of reserves that can be safely and economically recovered. Other beds, however, are startlingly level and geometrically uniform for miles in every direction. In the Appalachian region, an area of ancient eroded mountains, outcrops can often be found circumscribing adjacent mountains. Some beds are split into several layers with intervening rock layers. Beds quite unrelated geologically can be stacked atop each other.

The most desirable commercial beds are those that are thick, uniform, and very extensive. The Pittsburgh bed is minable over an area of 6,000 square miles in Maryland, West Virginia, Pennsylvania, and Ohio with a thickness of up to 22 feet. The edges thin out over a much broader area. This bed presents a striking continuity over this vast region and has yielded more than 20 percent of the U.S. coal mined to date.[1] Other Eastern beds also cover several thousand square miles with coal 2 to 10 feet thick. Midwestern beds can be even thicker and broader. Herrin (No. 6) in Illinois, Indiana, and western Kentucky covers 15,000 square miles with a thickness of up to 14 feet. The largest bed in the United States is the Wyodak in the Powder River Basin of Wyoming and Montana. The average thickness is 50 to 100 feet and 150 feet has been observed.

## Coal Distribution[2]

Coal deposits are dispersed over much of the country. The major coal provinces are

---

[1]*Geological Survey Bulletin,* 1412.
[2]This section is drawn largely from "Coal Mining," by the National Research Council, National Academy of Science, 1978.

**Figure 5.—Regional Recoverable Reserves That Burn With Emissions Equal or Less Than Indicated**

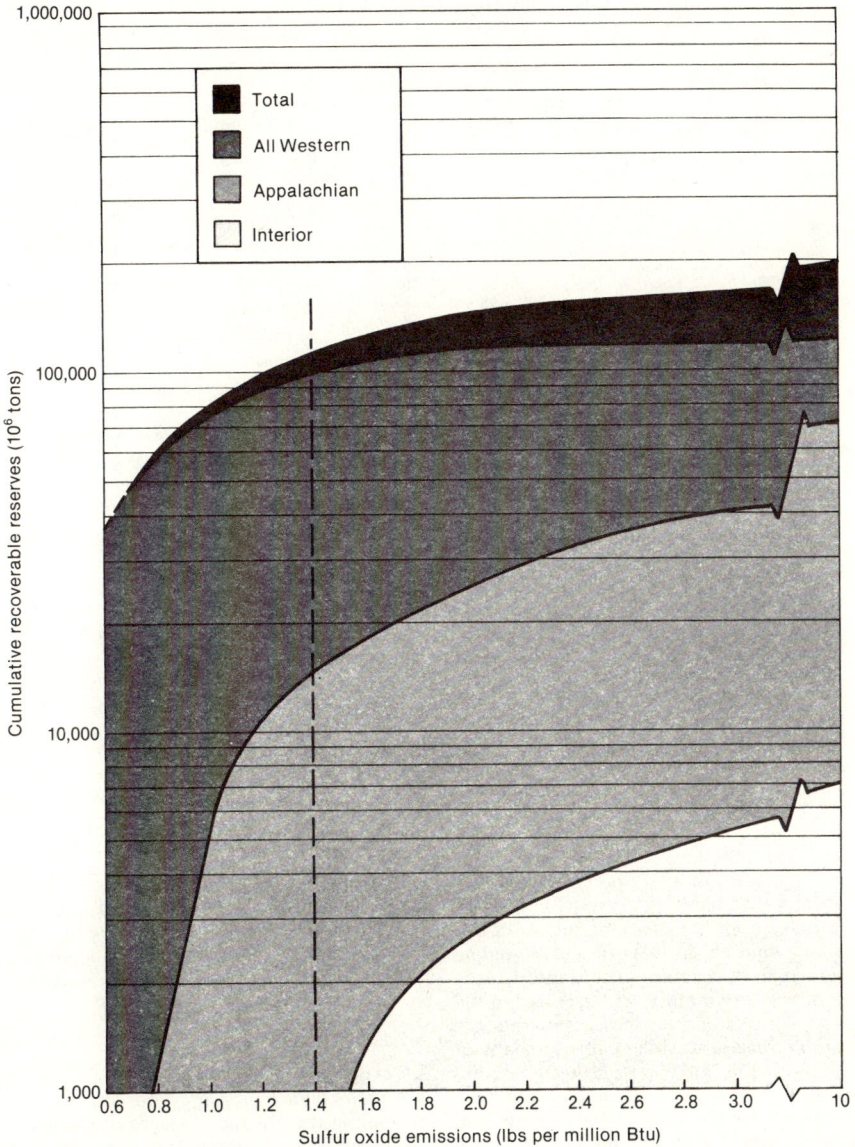

**Table 9.—Geometric Mean (Expected Values) of 36 Trace-Elements in 601 Coal Samples From Four Major Coal-Producing Regions, based on sequential ppm (parts per million) values recorded from Eastern, Appalachian coals**

| Element | East (331) + | Interior (194) + | West (93) + | Texas (34) = | Average (601) samples |
|---|---|---|---|---|---|
| Mn | 200 | 72 | 34 | 51 | 89.2 |
| Ba | 70 | 30 | 300 | 150 | 137.7 |
| Sr | 70 | 30 | 100 | 150 | 112.5 |
| F | 60 | 58 | 37 | 91 | 61.5 |
| Zr | 30 | 10 | 15 | 50 | 26.2 |
| B | 20 | 50 | 70 | 100 | 60 |
| V | 20 | 20 | 7 | 30 | 19.2 |
| Li | 18.8 | 7 | 4.3 | 14 | 11 |
| Cu | 16 | 16.3 | 7.4 | 20 | 14.9 |
| Cr | 15 | 10 | 3 | 15 | 10.7 |
| Ni | 15 | 18 | 2 | 15 | 12.5 |
| Zn | 12.8 | 58 | 12.8 | 28 | 27.9 |
| As | 11 | 12 | 2 | 5 | 7.5 |
| Pb | 10.9 | 19 | 4.3 | 2.8 | 9.2 |
| Ga | 7 | 2 | 3 | 7 | 4.7 |
| Y | 7 | 7 | 3 | 15 | 8 |
| Co | 5 | 7 | 1.5 | 5 | 4.6 |
| Se | 3.5 | 2.8 | .5 | 5.8 | 3.1 |
| Nb | 3 | .7 | 3 | 2 | 2.2 |
| Sc | 3 | 3 | 1.5 | 5 | 3.1 |
| Th | 2.8 | 1.6 | 2.4 | 3 | 2.4 |
| Be | 2 | 1.5 | .3 | 2 | 1.4 |
| Mo | 2 | 2 | 1.5 | .7 | 1.5 |
| Al% | 1.3 | .77 | .59 | 1.6 | 1.1 |
| Si% | 1.2 | 1.4 | 1.1 | 4.2 | 2 |
| Fe% | 1 | 2.3 | .45 | 1.6 | 1.3 |
| U | 1 | 1.4 | .7 | 2.4 | 1.4 |
| Sb | .8 | .8 | .4 | .7 | .7 |
| Yb | .7 | .7 | .3 | 1.5 | .8 |
| Cd | .3 | .12 | .2 | .2 | .2 |
| Hg | .14 | .1 | .06 | .13 | .1 |
| K% | .13 | .11 | .028 | .15 | .1 |
| Ca% | .093 | .5 | .92 | .6 | .5 |
| Ti% | .074 | .04 | .037 | .11 | .1 |
| Mg% | .052 | .063 | .245 | .17 | .1 |
| Na% | .025 | .026 | .1 | .009 | |

NOTE: As, F, Hg, Sb, Se, Th, and U values used to calculate the statistics were determined directly on whole-coal. All other values used were derived from determinations made on coal wash.
SOURCE: U.S. Geological Survey open-file report 76-468, 1976, pp. 56, 223, 312, and 341.

shown in figure 6. About 13 percent of the entire country has coal beneath the surface. West Virginia and Illinois are about two-thirds underlain, while North Dakota and Wyoming are more than 40 percent. The deposits vary widely in size and quality, as described in the previous sections. It is useful to separate the recoverable reserves by their likely means of extraction, as the impacts of strip mining are quite different from those of underground mining. Table 10 lists the recoverable reserves and 1976 production by mining method for each State in units of tons and heat content. The

reserves in each State are large enough that production could be increased substantially even though several States (Virginia, Tennessee, Alabama, and Kansas) have depleted more than 40 percent of their original reserves.

## Eastern Regions

**Appalachian Region:** Coal production in the Appalachian region may be characterized as being in the advanced stages of maturity (i.e., having annually supplied 60 percent or more of total national production since coal mining

**Figure 6.—Coalfields of the Conterminous United States**

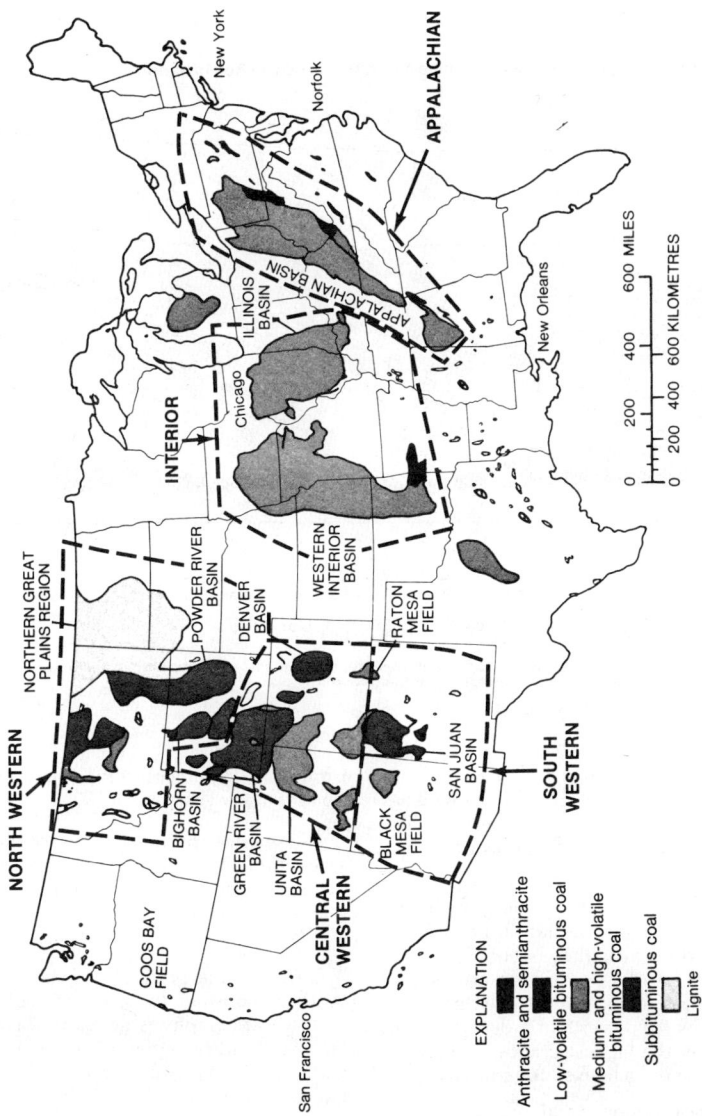

SOURCE: P. Averitt, *Coal Resources of the United States, Jan. 1, 1974,* U.S. Geological Survey
Bull. 1412, at 5 (1975).

**Figure 7.—Incline No. 1, Herrin (No. 6) Bed in Illinois**

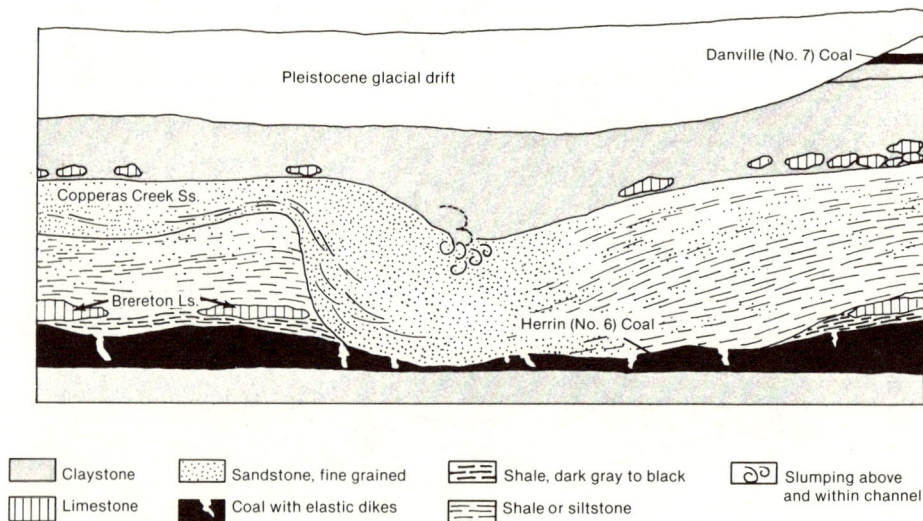

SOURCE: W. H. Smith, et al., 1970, Depositional Environments in Parts of the Carbondale Formation—Western and Northern Illinois—Francis Creek Shale and Associated Strata and Mazon Creek Brota. *Illinois State Geological Survey Guidebook*, Series 8, p. 5.

began in the United States). The coal is obtained from a large number of mines of widely varying sizes and from a large number of coalbeds, many of which occur in areas of steep terrain.

The remaining recoverable underground reserves are large, but substantial portions may remain unmined because they occur in uneconomical seams. Surface mine reserves are probably insufficient to support even present production rates for more than a decade. The coal is generally the highest grades of bituminous, but it often has a high sulfur content.

**Eastern Interior Region:** Coal production in the eastern interior region involves a small number of individual mines with relatively high annual production rates. Indiana, western Kentucky, and Illinois have annually supplied

from 20 to 25 percent of total national production for many years.

The greatest portion of the total remaining recoverable reserves for underground mining occurs in beds underlying already mined-out seams. Very large quantities remain, especially in Illinois.

Surface mine production has remained essentially constant for a number of years. As surface topography is generally favorable for large-scale surface mining, it is likely that this method will be further pursued (although at reduced annual rates from individual mines) in thinner beds as reserves in the principal beds become exhausted.

Prospects are favorable for substantial expansion of production from underground mines. Surface mines should be able to maintain their present rate for many years. The coal tends to have a slightly lower heat content than Appalachian coal, and the sulfur content can be quite high.

**Western Interior Region:** Underground production has entirely ceased in Arkansas, Kansas, and Oklahoma and has been very limited for many years in Iowa. While surface production has remained approximately constant in recent years, the amount produced is not sig-

nificant. The coalbeds are persistent but thin throughout the region, except in Iowa and the southern portion of Oklahoma, where some beds are of moderate thickness. Were it not for local markets, it seems doubtful that much coal would be mined in this region. Some coal occurring under difficult mining conditions in the southern portions of Oklahoma and Arkansas is of metallurgical and foundry grade and is being mined and shipped out of the region.

Production in this region is unlikely to increase significantly in the near future.

**Table 10.—Recoverable Reserves as of January 1, 1976**

| State | Underground heat content quadrillion Btu | Surface heat content quadrillion Btu | Total | Underground minable million tons | Surface minable million tons | Total | 1976 production Underground million tons | Surface million tons | Total |
|---|---|---|---|---|---|---|---|---|---|
| Ohio | 180 | 107 | 287 | 7,500 | 4,900 | 12,400 | 16.2 | 29.3 | 45.5 |
| Pennsylvania | 418 | 28 | 446 | 16,700 | 1,200 | 17,900 | 43.8 | 39.9 | 83.7 |
| Kentucky E | 136 | 84 | 220 | 5,200 | 3,600 | 8,800 | 41.5 | 48.0 | 89.5 |
| Virginia | 53 | 17 | 70 | 2,000 | 700 | 2,700 | 24.0 | 12.8 | 36.8 |
| W. Virginia | 516 | 100 | 616 | 19,100 | 4,100 | 23,200 | 88.4 | 20.5 | 108.9 |
| Maryland | 14 | 3 | 17 | 500 | 100 | 600 | 0.2 | 2.5 | 2.7 |
| Alabama | 27 | 26 | 53 | 1,000 | 1,100 | 2,100 | 7.4 | 14.2 | 21.6 |
| Tennessee | 9 | 6 | 15 | 400 | 300 | 700 | 4.1 | 4.7 | 8.8 |
| Total Appalachia | 1,353 | 372 | 1,725 | 52,400 | 16,000 | 68,400 | 225.6 | 171.9 | 397.5 |
| Illinois | 682 | 257 | 939 | 30,300 | 12,700 | 43,000 | 31.0 | 27.0 | 58.0 |
| Indiana | 117 | 29 | 146 | 5,100 | 1,400 | 6,500 | 0.4 | 23.7 | 24.1 |
| Kentucky W | 118 | 69 | 187 | 4,800 | 3,200 | 8,000 | 22.5 | 28.3 | 50.8 |
| Total E. Interior | 917 | 355 | 1,272 | 40,200 | 17,300 | 57,500 | 53.9 | 79.0 | 132.9 |
| Arkansas | 4 | 3 | 7 | 100 | 100 | 200 | — | 0.6 | 0.6 |
| Iowa | 23 | 8 | 31 | 1,000 | 400 | 1,400 | 0.3 | 0.5 | 0.5 |
| Kansas | — | 19 | 19 | — | 800 | 800 | — | — | — |
| Missouri | 18 | 57 | 75 | 800 | 2,900 | 3,700 | — | 5.4 | 5.4 |
| Oklahoma | 18 | 8 | 26 | 700 | 300 | 1,000 | — | 3.3 | 3.3 |
| Total W. Interior | 62 | 95 | 157 | 2,600 | 4,500 | 7,100 | 0.3 | 9.5 | 9.8 |
| Montana | 898 | 794 | 1,692 | 40,400 | 39,700 | 80,100 | — | 26.1 | 26.1 |
| N. Dakota | — | 118 | 118 | — | 8,100 | 8,100 | — | 11.1 | 11.1 |
| Wyoming | 421 | 404 | 825 | 18,000 | 19,000 | 37,000 | 0.6 | 30.3 | 30.9 |
| S. Dakota | — | 4 | 4 | — | 300 | 300 | — | — | — |
| Colorado | 159 | 61 | 220 | 7,100 | 3,000 | 10,100 | 3.4 | 6.1 | 9.5 |
| Utah | 91 | 5 | 96 | 3,600 | 200 | 3,800 | 7.9 | — | 7.9 |
| Arizona | — | 5 | 5 | — | 300 | 300 | — | 10.2 | 10.2 |
| N. Mexico | 29 | 42 | 71 | 1,200 | 2,000 | 3,200 | 0.9 | 8.9 | 9.8 |
| Texas | — | 52 | 52 | — | 2,500 | 2,500 | — | 14.2 | 14.2 |
| Washington | — | — | — | 600 | 400 | 1,000 | — | 3.9 | 3.9 |
| Alaska | — | — | — | 3,100 | 600 | 3,700 | — | 0.7 | 0.7 |
| Total West | 1,598 | 1,485 | 3,083 | 74,000 | 76,100 | 150,100 | 12.8 | 106.9 | 119.7 |
| GRAND TOTAL | 3,931 | 2,307 | 6,238 | 169,000 | 113,900 | 282,900 | 292.6 | 367.3 | 659.9 |

SOURCE: Based on Bureau of Mines data.

## Western Regions

**Surface-Minable Regions:** The Western States contain the most recoverable surface-minable reserves (75 billion tons) and are virtually untouched (except for a few large active operations). The beds generally are thicker than beds being surface-mined elsewhere in the country (excluding Pennsylvania anthracite). In Washington and Arizona, surface-minable reserves are essentially those remaining in the surface mines now active in each State; in South Dakota and Utah, surface-minable reserves are relatively small, and in Colorado, surface-minable reserves are largely confined to the two northwestern counties.

The sparsity of surface mining in these States has been due to their remoteness from large markets. In anticipation of future need, intensive prospecting and acquisition by lease or purchase was initiated about 10 years ago, and many large individual mines already are operating or have been planned. Contracts for large annual shipments from some of these unit areas to existing or contemplated new powerplants have been negotiated, while other units have been set aside for prospective gasification or liquefaction plants. Plans for new or connecting transportation facilities have been completed or are being developed.

**Underground Reserves:** Excluding Washington, New Mexico, and Alaska (where significant recovery of underground reserves is con-sidered dubious), and North Dakota and Texas (where the lignite beds are considered unsuitable for underground mining), the Western States of Colorado, Utah, Montana, and Wyoming contain more than 69 billion tons of recoverable underground reserves. Although important portions of the coalfields of Utah, Colorado, and Wyoming have undergone substantial depletion, many relatively unmined areas remain. Some of these may be comparatively difficult to mine because of the degree of dip or the excessive thickness of overlying cover, but in many areas mining conditions are favorable.

Significant increases in underground production in Utah and Colorado can be expected in the next few years. Similar interest in the underground reserves in Wyoming and Montana may also develop, although underground recovery of as much as 50 percent of the thicker beds may be difficult with present technology, which is generally limited to bed thicknesses not exceeding 15 feet. If or when additional production becomes necessary, there is no reason why underground production could not be conducted simultaneously or subsequent to surface mining.

Most Western coal is subbituminous or lignite. It has a substantially lower heat content and a higher moisture content than Eastern coal. The sulfur fraction is typically low.

# MINING AND PREPARATION

A wide variety of mining techniques exist to exploit the different types of coal seams described in chapter II. The two general methods, surface (i.e., strip) and underground, are used under quite different geological conditions, and a number of mining techniques are utilized. The ultimate choice of method and specific technique depends on a number of factors such as geological conditions, safety, productivity rate, skills available in the labor force, economics, etc. This section presents a brief description of the technologies, the factors that suggest a particular methodology, and the resulting impacts.

## Surface Mining

Surface mining involves exposure of the seam by removal of the overlying soil and rocks (overburden). The four basic methods are area, open-pit, contour, and auger mining. Until about 1965 surface mining of coal was not considered feasible unless the overburden-to-seam thickness ratio was 10:1 or less. Thus, to justify removing 50 feet of overburden, the coal seam would have to be 5 or more feet thick. Since 1965, this ratio has been increasing and, depending on the nature and structure of the overburden, coal within 150 feet of the surface may now be economically recoverable, even when the overburden-to-seam thickness ratio is as much as 30:1.[3]

Area mining, the major strip-mining method used on Midwestern and some Western coal lands, involves the development of large open pits in a series of long narrow strips (usually about 100 feet wide by a mile or more in length). It is the preferred method in flat terrain where the coal seam is parallel to the surface, as it is for many Western coals. Before mining begins, access roads and maintenance and personnel facilities must be constructed. This phase may require 5 years and about 200 workers. If all other components (preparation plant, transportation, and customer facility) are in place, the process can be compressed to as little as 1 year with a much greater demand for manpower. The actual mining starts with the cutting of a trench across one end of the strip, using bulldozers or a dragline. Top soil is reserved for reclamation, and the remaining overburden is placed in a spoil bank. Blasting is often needed to fracture the overburden and coal. A loading shovel lifts the coal into large trucks, which take it to the crusher. The scraper or dragline is moved to the next position and opens a new trench, transferring the overburden to the mined-out trench. This cycle is repeated to the limit of the mine boundaries. Reclamation is accomplished by smoothing the spoil, covering with top soil, and revegetating. Figure 8 illustrates the procedure.

Open-pit mining is somewhat similar except that it involves the preparation of a larger area, perhaps 1,000 by 2,000 feet wide. The overburden is moved around in the pit by truck and shovel to uncover the coal seams. This technique is used primarily for the very thick Western seams.

Contour strip mining is most commonly

[3]*Energy Alternatives* (University of Oklahoma, May 1975), pp. 1-21.

practiced where deposits outcrop from rolling hills or mountains, particularly in Appalachia. This method consists of removing the overburden above the bed by starting at the outcrop and proceeding along the contour of the bed in the hillside. After the deposit is exposed and removed by this first cut, additional cuts are made into the hill until the ratio of overburden to product brings the operation to a halt. A variant on this technique is mountaintop removal, when the seam lies sufficiently near the top.

Contour mining creates a shelf, or "bench," on the hillside. The highwall borders it on the inside, the downslope on the other side. Before the enactment of the Surface Mining Control and Reclamation Act of 1977, or prior State requirements, the overburden, or spoil material, sometimes caused severe environmental problems. Unless controlled or stabilized, this spoil material can cause severe erosion, landslides, and silting of streams. Contour mining is illustrated in figure 9.

The mining equipment used in contour mining is generally smaller than that used for area mining and consists of bulldozers, shovels, front-end loaders and trucks, all of a scale to fit on a relatively narrow ledge.

Auger mining is employed in hilly terrain where the slope above the coal is too steep or high to allow normal contour mining. The usual procedure is to contour mine as far into the hill as practical, then auger mine further. Huge drills, with cutting heads up to 7 feet, are driven horizontally up to 200 feet into the coal seam.

The types of equipment used in surface mining are shown in figures 8 and 9. These items of equipment (except auger) are used to achieve similar basic steps in each surface mining method. The steps and equipment used are:

1. Topsoil removal using bulldozers and loaders. Trucks convey it to a stockpile.
2. Overburden removal, using blasting to loosen it if it is rock or shale, and scooping it out with bulldozers, stripping shovels, bucketwheel excavators, or draglines. Walking draglines are used in some contour mines, but their main use is in area mines. They have buckets up to 200

Figure 8.—Strip Mining Method

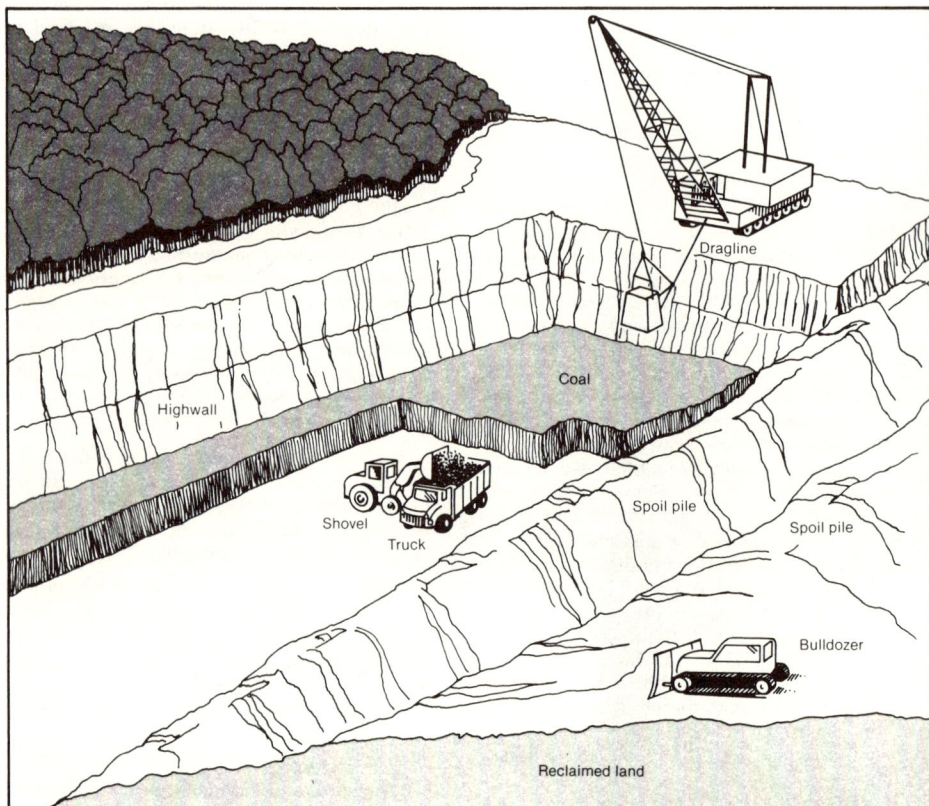

cubic yards and boom lengths of 180 to 375 feet. Stripping shovels can be used in any type of surface mine. Bucket wheel excavators can be used with stripping shovels when the overburden is soft. From 1 to 40 tons of overburden must be moved for every ton of coal mined.

3. Coal fracturing is usually done by blasting, though some types can be scooped up directly.

4. Coal loading and hauling, using mine shovels or front-end loaders and trucks.

The trucks are usually used only to take the coal to a nearby processing facility. After crushing, and sometimes cleaning, the coal is stored at the tipple awaiting transport to market.

5. Backfilling the overburden is generally done simultaneously with removing it from a fresh area. It is graded and compacted with bulldozers.

6. Topsoil is returned and spread.

7. The area is revegetated.

Photo credit: OTA

**Western surface mining. Dragline (background) exposes coal seam while front-end loaders (foreground) dump it into trucks for haulage to rail spur or powerplant**

Photo credit: Colorado School of Mines

**Dragline bucket—immense size of bucket
is self-revealing**

**Stripper shovel, eastern Ohio**

Photo credit: OTA

**Figure 9.—Typical Contour Stripping Plus Auger Method**

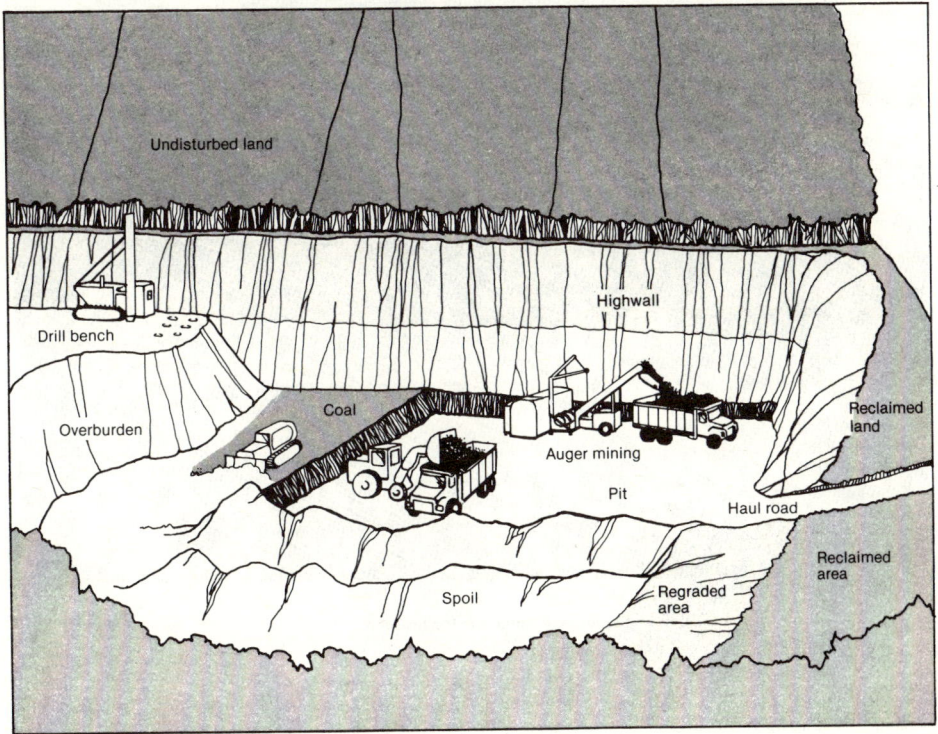

## Underground Mining

Underground mining is considerably more complex than surface mining. Rather than simply scraping away the overburden and trucking away the coal, underground miners must work with the thick overburden above them, connected to the outside world by shafts and passageways sometimes thousands of feet long. Problems nonexistent or unimportant on the surface loom large underground: roof support, ventilation, lighting, drainage, methane liberation, equipment access, and coal conveyance, among others, must be dealt with.

As with surface mines, the first step is the construction of access roads and necessary aboveground facilties. The next step is to construct a portal—the passageway to the seam. The selection of location and type of portal (generally shaft or slope) depends on the particular site. The mining plan must consider costs of access and construction, personnel transit, and coal haulage among other factors. From the portal, parallel entries are driven into the coal to provide corridors for haulage, ventilation, power, etc. Cross-corridors then reach to the sides of the mine, leaving pillars in a checkerboard pattern to support the roof. The

**Contour strip mining in Martin County, Ky.**   *Photo credit: OTA*

deeper the mine, the bigger the pillars must be relative to the mined-out areas. In some cases, less than 50 percent can be removed. In such cases it may be desirable to mine the pillars as the equipment retreats back towards the main corridors. The roof must then be supported by other means or allowed to collapse.

Equipment used in underground mining ranges from relatively simple up to highly automated and productive machinery. The oldest method, still in occasional use in very small mines, is primarily hand labor. The coal is undercut at the bottom of the face, and blasting holes are drilled into the face above. Explosives shear the coal loose by forcing it down into the cut. It is then hand loaded into shuttle cars. This method has been almost entirely replaced by conventional mining: mech-

anized undercutting, drilling, and loading. Undercutting is accomplished by huge chain saws protruding from the bottom of self-propelled vehicles. These cut a slot about 6 inches high, perhaps 10 feet deep, into the coal and then across the face, perhaps 20 feet. This machine moves to the next face while the drilling machine takes its place. The coal drill is a self-propelled vehicle with a long auger attached to a movable boom. It drills holes above and as deep as the cut. One hole is required for a face area 3 to 4 feet high and 4 to 5 feet wide. The presence of rock in the coal may mandate more holes. Blasting (called shooting) is done with chemical explosives or compressed gas. The latter is considered safer, but it is slower because only one hole can be shot at a time. The coal is loaded by a machine that slides it onto a conveyor belt and dumps it into a shut-

**Shoveling loose coal at Pittston's Buffalo Mining Co., operation on Buffalo Creek, W. Va., 1975**

tle car. These cars take the coal either to a change point, where it is transferred to a conveyor belt or mine car, or directly out of the mine. Roof bolting also is an integral part of the operation in order to maintain structural integrity of the roof. A bolting machine drills holes in the roof and inserts anchor bolts, which firmly attach the roof to stronger overlying layers of rock through expansion shells or resin. Bolts are generally required on a 4- by 4-foot array. This is one of the most dangerous activities in the mine. All these operations produce a large amount of dust and liberate methane trapped in the coal. Exposed areas are rock-dusted to prevent coal dust explosions. In all seams, frequent methane testing is required at the face. Continuous monitoring is required under some conditions. Some machines must be hooked to a water supply for a spray to con-

trol dust. In addition, time must be spent moving equipment, changing bits, hooking up electric power, and performing maintenance. A typical conventional mining sequence is shown in figure 10.

Continuous miners were developed to combine the operations of conventional mining, thereby further increasing productivity. Many different types of continuous miners are available, but all operate on the same basic principle. A rotating head digs into the coal and dislodges it while arms scoop it up onto a conveyor belt for loading. Thus, the multiple cyclic transfers of machines from face to face in conventional mining are greatly reduced. Most continuous miners still must withdraw to allow roof bolting and ventilating duct extension. The continuous mining method is illus-

## Figure 10.—Underground Mining

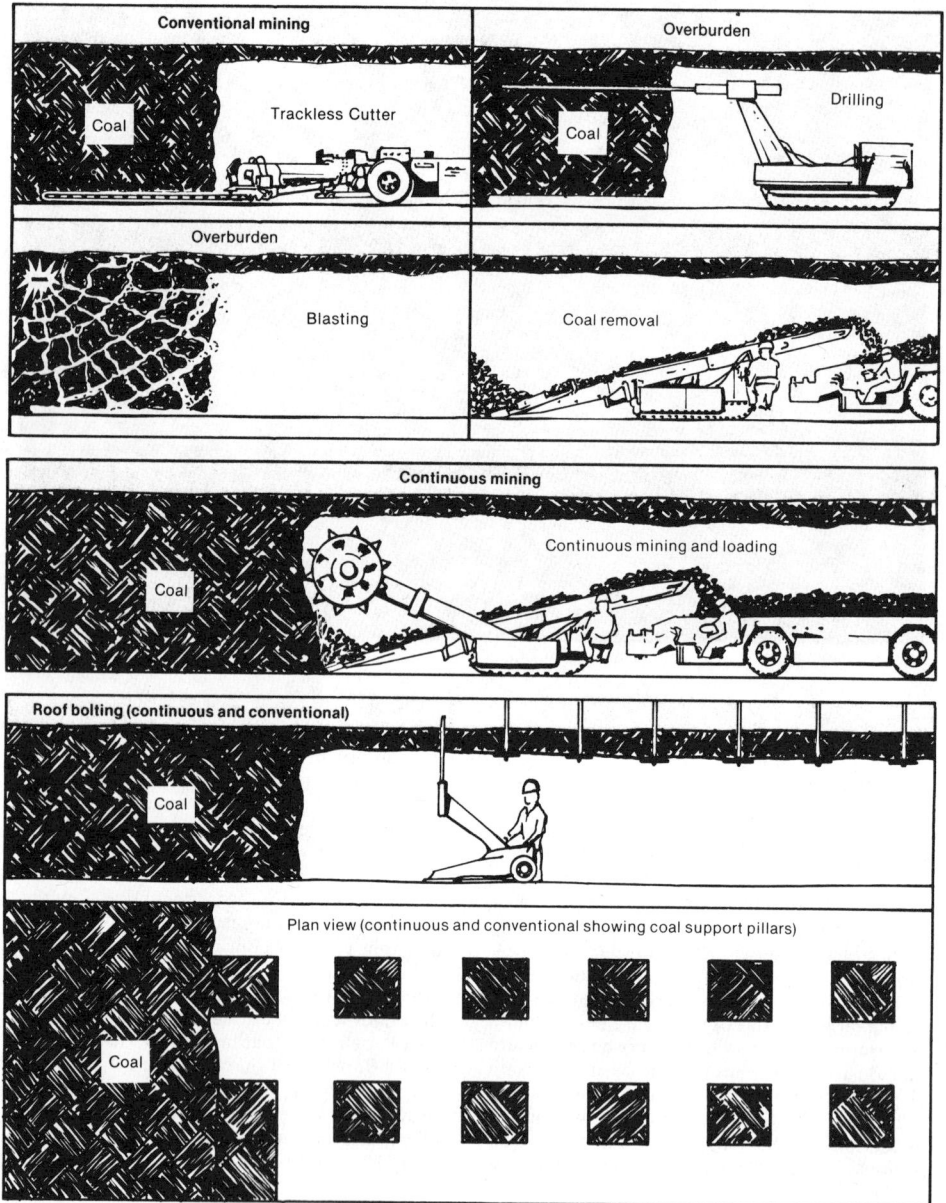

**Conventional mining**

Coal

Trackless Cutter

Overburden

Coal

Drilling

Overburden

Blasting

Coal removal

**Continuous mining**

Coal

Continuous mining and loading

**Roof bolting (continuous and conventional)**

Coal

Plan view (continuous and conventional showing coal support pillars)

Coal

Photo credit: OTA

**End of shift in 30″ coal at Cedar Coal Co., Logan County, W. Va.**

trated in the lower portion of figure 10. The continuous miner's useful activity is limited to 20 to 30 percent of the time. Haulage of coal away from the face is a major constraint on operations.

Longwall mining is a significantly different approach than both conventional and continuous mining work in the room and pillar system described above. Longwall mining has been used in Europe for many years and is now gaining popularity in the United States. Corridors 300 to 600 feet apart are driven into the coal and interconnected. The longwall of the interconnection is then mined in slices. The roof is held up by steel jacks while the cutter makes a pass across the face. The roof jacks are advanced with the shearer to make a new pass. The roof collapses in the mined-out area behind the jacks. Practically all the coal can be

extracted by this process. Figure 11 shows the major elements. A variant of the longwall technique is known as the shortwall. It is conceptually similar, but the shorter side of the rectangle to be mined is attacked. This rarely exceeds 200 feet.

Ventilation is a major problem in modern mines because the highly productive equipment produces high levels of dust and methane. Dust is reduced by water sprays at the working face. (Further discussion about the impact of dust can be found in chapter VI.) Methane must be diluted with air to less-than-flammable limits to avoid the possibility of underground explosions. In a few mines methane is drained from the coal seam before mining. The operation recovers methane (natural gas) for use as fuel and reduces the danger of mine explosions. The legal ownership of the gas recovered before mining is not clear, as mineral rights to the coal do not necessarily convey rights to the gas or the gas may emanate from undeterminable regions beyond the ownership boundary. Generally the methane content of the coal increases at greater depths because the methane has less opportunity to diffuse to the surface over geologic time. Thus, as the shallower seams are depleted and the deeper mines are developed, the safety problem will worsen and the recovery opportunities increase. It has been estimated that 200 cubic feet of methane are released for every ton of coal mined. Hence the potential contribution to our gas supply is not trivial if the legal questions can be resolved.[4]

## Coal Preparation

Coal preparation modifies the mined coal to help meet the customers' needs in terms of size, moisture, mineral concentration, and heat content. A preparation plant of some sort is an integral part of most large mines and may often serve a number of mines. In some cases, particularly in the West, the coal is only crushed. About half the raw coal in the Nation is cleaned to reduce impurities such as mineral

---

[4]"Degasification of Coal Beds—A Commercial Source of Pipeline Gas," *AGA Monthly*, January 1974.

Longwall mining at a Consolidation Coal mine

matter and rocks. Before the emphasis on environmental concerns these inert materials were reduced to improve heat content of the coal, reduce deposits in the furnace, reduce the load on particulate removal equipment, and improve overall operating performance of the furnace. Because the mineral matter contains sulfur and trace elements, coal washing reduces these as well. Recently coal preparation has focused on greater removal of these particular contaminants. Advanced cleaning for desulfurization is discussed in the section on *Future Technologies* of this chapter.

Crushing (or comminution) is performed on 90 percent of all bituminous coal to provide a product of uniform size and increase the exposure of impurities for removal. In 1975, 374 mil-

Figure 11.—Longwall Mining System

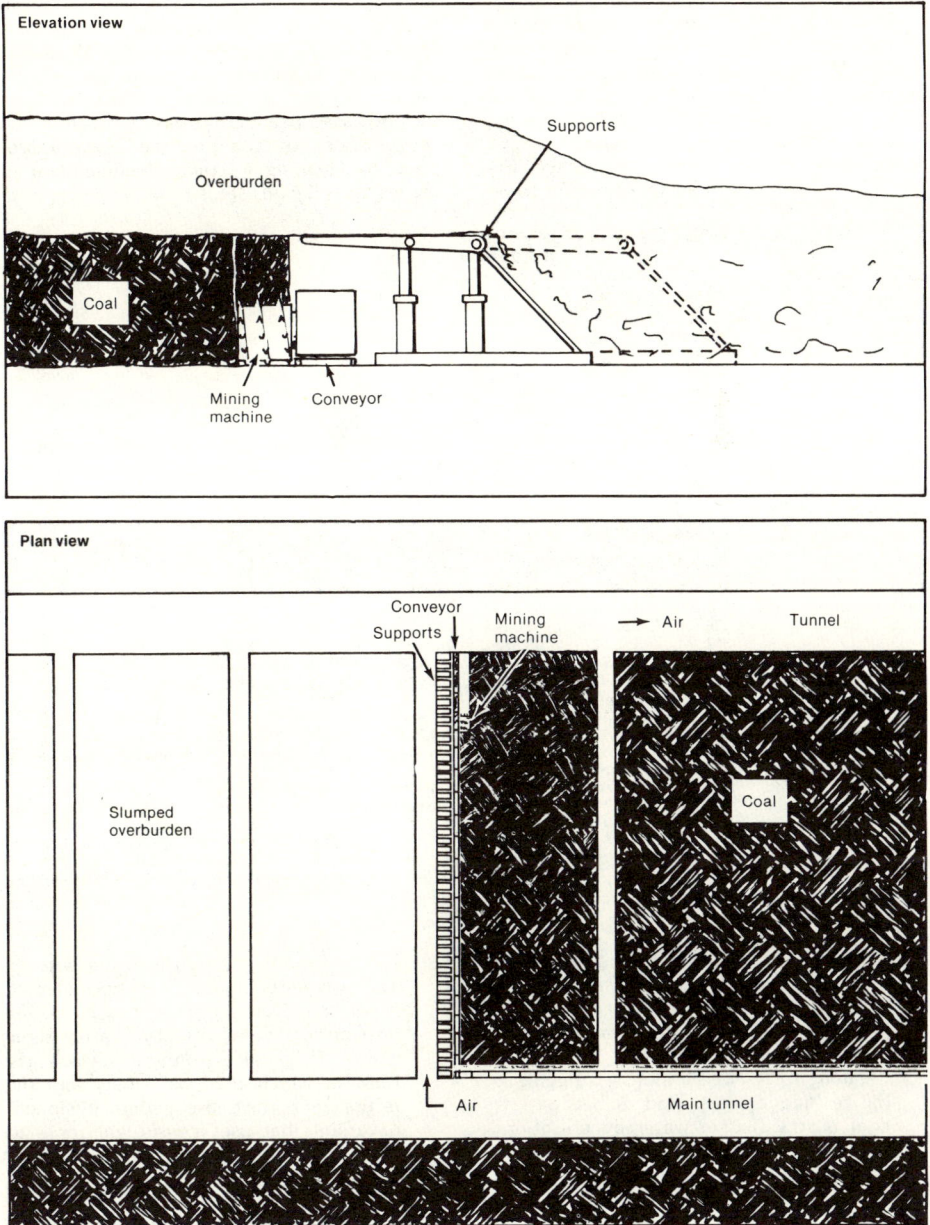

lion tons, or 41.2 percent of total production of raw coal was cleaned at a plant. This resulted in a net production of 267 million tons of processed coal and 107 million tons of refuse. The refuse fraction has been increasing, as shown in table 11. Two prominent phases of increase are notable. The first, which lasted until about 1962, was mostly due to increased mining automation that reduced selectivity of mining coal between mineral partings at the face. The second, which is still in evidence, seems to be due to the mining of lower quality coal. This rise also coincided with a decrease in the total amount of coal being cleaned. Utilities began finding cleaning less worthwhile, in part because of the growth in mine-mouth plants, which reap no transportation benefits from the reduced weight of clean coal. Also, much of the increased production has come from the West, where most coal produced is not appropriate for cleaning. A typical cleaning plant includes the following steps:

**Table 11.—Mechanically Cleaned Bituminous Coal and Lignite[a]**
**(thousand short tons)**

| Year | Total raw coal moved to cleaning plants | Cleaned by wet methods | Cleaned by pneumatic methods | Total cleaned | Total production | Percent of total production mechanically cleaned | Refuse resulting in cleaning process | Percent refuse of raw coal |
|------|------|------|------|------|------|------|------|------|
| 1940 | 115,692 | 87,290 | 14,980 | 102,270 | 460,771 | 22.2 | 13,422 | 11.6 |
| 1945 | 172,899 | 130,470 | 17,416 | 147,886 | 577,617 | 25.6 | 25,013 | 14.5 |
| 1950 | 238,391 | 183,170 | 15,529 | 198,699 | 516,311 | 38.5 | 39,692 | 16.6 |
| 1955 | 335,458 | 252,420 | 20,295 | 272,715 | 464,633 | 58.7 | 62,743 | 18.7 |
| 1956 | 359,378 | 268,054 | 24,311 | 292,365 | 500,874 | 58.4 | 67,013 | 18.6 |
| 1957 | 376,546 | 279,259 | 24,768 | 304,027 | 492,704 | 61.7 | 72,519 | 19.2 |
| 1958 | 320,898 | 240,153 | 18,882 | 259,035 | 410,446 | 63.1 | 61,863 | 19.3 |
| 1959 | 337,138 | 251,538 | 18,249 | 269,787 | 412,028 | 65.5 | 67,351 | 20.0 |
| 1960 | 337,686 | 255,030 | 18,139 | 273,169 | 415,512 | 65.7 | 65,517 | 19.4 |
| 1961 | 328,200 | 247,020 | 17,691 | 264,711 | 402,977 | 65.7 | 63,489 | 19.4 |
| 1962 | 339,408 | 252,929 | 18,704 | 271,633 | 422,149 | 64.3 | 67,775 | 20.0 |
| 1963 | 362,141 | 269,527 | 19,935 | 289,462 | 458,928 | 63.1 | 72,679 | 20.0 |
| 1964 | 388,134 | 288,803 | 21,400 | 310,203 | 486,998 | 63.7 | 77,931 | 20.1 |
| 1965 | 419,046 | 306,872 | 25,384 | 332,256 | 512,088 | 64.9 | 86,790 | 20.1 |
| 1966 | 435,040 | 316,421 | 24,205 | 340,626 | 533,881 | 63.8 | 94,414 | 21.7 |
| 1967 | 448,024 | 328,135 | 21,268 | 349,402 | 552,626 | 63.2 | 98,624 | 22.0 |
| 1968 | 438,030 | 324,123 | 16,804 | 304,923 | 545,245 | 62.5 | 97,107 | 22.2 |
| 1969 | 435,356 | 315,596 | 19,163 | 334,761 | 560,505 | 59.7 | 100,593 | 23.1 |
| 1970 | 426,606 | 305,594 | 17,855 | 323,452 | 602,936 | 53.6 | 103,159 | 24.2 |
| 1971 | 361,168 | 256,892 | 14,506 | 271,401 | 552,192 | 49.1 | 89,766 | 24.9 |
| 1972 | 398,678 | 281,119 | 11,710 | 292,829 | 595,387 | 49.2 | 105,850 | 26.5 |
| 1973 | 397,646 | 278,413 | 10,505 | 288,918 | 591,737 | 48.8 | 108,728 | 27.3 |
| 1974 | 363,334 | 257,592 | 7,557 | 265,150 | 603,406 | 43.9 | 98,184 | 27.0 |
| 1975 | 374,094 | 260,289 | 6,704 | 266,993 | 648,438 | 41.2 | 107,101 | 28.6 |

[a] U.S. Bureau of Mines Yearbook, various years.

1. Comminution (crushing) to liberate impurities and provide a product of appropriate size.
2. Sizing (to ensure uniformity of size) with screens or fluid classification.
3. Cleaning to remove impurities. Jigging is the technique most used. A bed of raw coal is stratified in water by pulsations that move the light particles (coal) to the top and the heavy refuse to the bottom. The two products can then be collected separately. Dense-medium processes, the next most popular, effect a sharper separation. The coal is immersed in a heavy fluid in which the coal floats and the refuse sinks. The dense-medium cyclone is a variant that uses centrifugal force to assist the separation. Other techniques

are concentrating tables, froth flotation and pneumatic methods. These are described in appendix II of volume II.

4. Dewatering and drying reduce shipping weight and problems of handling and combustion. Dewatering, the mechanical separation of water, is done with screens, filters, and centrifuges of various types. Drying uses heat to remove water. It is relatively expensive and can cause air pollution problems, which explains why the practice has been declining.

5. Water clarification is the process of removing enough suspended coal from the water that the water can be recycled. This is usually done in settling tanks assisted by flocculents.

The coal preparation plant may itself become a significant source of pollution even though it serves to reduce pollution where the coal is burned. Refuse piles contain sufficient coal to smoulder, causing noxious emissions, but this can be prevented by proper management. Pneumatic cleaning methods and thermal dryers are sources of particulate matter, and both are being phased out or equipped with baghouses. The wash water has the potential of releasing large quantities of impurities similar to acid mine drainage. As mentioned

above, the modern practice is to recycle the water. Completely closed loops are coming into practice. Leaching from the refuse piles is prevented by compaction and layering.

## Capital Costs

Before a mine can be opened, a considerable amount of capital must be committed, and the operator must be sure of obtaining the appropriate manpower, material, and equipment. New mines are quite expensive to open. Costs vary significantly from site to site. A 1976 estimate gave the following results in dollars per annual ton capacity.[5]

|  | Appalachia | Illinois Basin | Western |
|---|---|---|---|
| Surface | 47 | 50 | 18 |
| Underground | 66 | 48 | 48 |

These figures are sharply higher than earlier estimates, largely because of hyperinflation in the heavy equipment market. Thus a new 2-million ton/yr surface mine in Illinois would cost $100 million. As described in chapter II, this mine would just meet the demand of one large electric utility unit. In 1976 it would have been the 46th largest mine in the country.

# TRANSPORTATION AND TRANSMISSION

Once mined and processed, coal is transported to another site for combustion. Trucks and belt conveyors serve nearby "mine-mouth" facilities. Highways, railroads, waterways, and slurry pipelines are the transportation mode for longer distances. Energy from mine-mouth powerplants can be transmitted by high-voltage transmission cable. The cost of transportation is a major determinant of the availability of specific mines or combustion facility sites. Further, the revenues earned from transporting coal will be very important to the transportation industry. This section describes methods of long-distance transportation of the coal and transmission of electric power.

The following description of rail, water, and truck transport has been excerpted from a major section on coal in a comprehensive discussion of current energy transportation systems and movements issued recently by the Congressional Research Service.[6] This report and its accompanying maps are commended to the

[5]George W. Land, "Capital Requirements for New Mine Development," Third Conference on Mine Productivity, Pennsylvania State University, April 1976.

[6]National Energy Transportation, vol. I and accompanying maps, prepared for Senate Committees on Energy and Natural Resources and on Commerce, Science, and Transportation (Washington, D.C.: Congressional Research Service, 1977), maps prepared by U.S. Geological Survey, Committee Print, Publication 95-15.

reader. Additional material on unit trains and coal slurry pipelines is taken from a previous OTA report.[7]

## Railroads

About two-thirds of the coal produced in the United States is transported in railroad cars. Coal is moved by rail almost entirely in open-topped hopper cars or gondola cars, which can be unloaded by turning the car completely over, by opening ports, or more rarely, by unloading from the top. The cars are generally loaded from the top at mine sidings or coal-loading installations central to a number of mines. Coal from the mines is lifted into silos or onto storage piles, travels by gravity or conveyer, and then moves through a chute into the car.

The average hopper car carries about 75 tons of coal. Older cars average 55 tons, and the newer ones move 100 tons at capacity. The complete cycle of loading, hauling to the unloading point, unloading, and returning for another cargo averages 13 days for each car. Although this is a shorter turnaround time than is experienced for any other type of bulk movement, inactivity is still a major factor in the economics of coal movement by railroad.[8] The size of hopper cars that can be accommodated depends on roadbed conditions and the weight the track can tolerate. Western coal unit trains tend to use 100-ton hopper cars; Eastern shipments often require the older 55-ton hoppers.

Unloading facilities for hopper cars, like loading facilities, are large-scale, permanent installations. They may take the form of trestles running over storage piles, into which coal is dumped from the bottom of moving cars, or they may be rotary dumpers, which tip the entire car over, spilling its contents. Cars with swivel couplings need not be uncoupled to unload by rotating. Some loading and unloading facilities can handle thousands of tons per hour. As a coal trade journal reported: "A typical requirement today would be to load three trains per day, each consisting of 100 cars of 100 tons net loading each; provide one and one-half trainloads of storage on hand at the start of each loading and provide a loading rate of 4,000 tons/hr."[9]

Track conditions dictate the types of equipment and the loads that can be hauled. No general compilation of conditions is available because they vary substantially with maintenance, weather conditions, and other factors, but it is generally accepted that younger track, track in the West, and track operated by the more solvent railroads tend to be in better condition; older track, track in the Northeast, and track owned by railroads that are in financial trouble tend to be in worse shape.

A particular type of rail service for major shippers of coal and other bulk commodities is provided by unit trains. This type of train, designed to take advantage of economies of scale, generally carries a single commodity in dedicated service between two points in sufficient volume to achieve cost savings. The cars are designed for automated loading and unloading, and the train is operated by procedures that minimize switching and time-consuming delays in freight yards.

A typical coal unit train consists of six 3,000-horsepower locomotives and 100 hopper cars with carrying capacities of 100 tons each. Roughly two such trains per week are therefore required to deliver 1 million tons of coal per year depending on the results involved. Speeds vary considerably depending on track conditions, but 20 to 50 miles per hour is a common range.

## Waterways

The "hardware" of the shipment of coal by barge or other water carrier includes the tugs,

[7]A Technology Assessment of Coal Slurry Pipelines, (Washington, D.C.: Office of Technology Assessment, 1978).

[8]Railroad Freight Cars Requirement for Transporting Energy, 1974-85, prepared for the Federal Energy Administration under contract by Peat, Marwick, Mitchell, and Co. Table of turnaround times at p. II-11.

[9]"From Mines to Market by Rail . . . The Indispensable Transport Mode," Coal Age, v. 7, pp. 102-112 at p. 103. McGraw-Hill Publications. Provides a good technical description of unit train loading and unloading.

Coal tipple on Cabin Creek, W. Va.

or self-propelled vessels, which push the "tow," composed of as many as 36 barges, usually of 1,500-ton capacity,[10] from a loading dock to an unloading dock. A great variety of sizes, shapes, drafts, and power capability characterizes the towboat fleet.

Modern steel barges are of numerous designs to handle differing commodities. Their bows and sterns are usually formed to fit into a neat tow that presents an unbroken, low-friction surface to the water. A jumbo-sized (195 by 35 feet) open barge cost $70,000 in 1968 and probably costs more than $125,000 today; the towboats cost many times more.[11]

The loading facilities used for barges resemble some rail loading facilities. They generally entail a conveyor belt carrying coal from a stockpile to a movable chute, which can dump the coal into waiting barges along a pier. Unloading facilities are much different, however: barges must be unloaded without the aid of

---

[10]*National Transportation Trends and Choices* (Washington, D.C.: U.S. Department of Transportation), p. 256, 258.

[11]*Transportation of Mineral Commodities on the Inland Waterways of the South Central States,* Bureau of Mines information circular 8431 (Washington, D.C.: U.S. Department of the Interior, 1968), p. 13.

**Unit coal trains being loaded at Martin Lake Plant, East Texas, La.**

gravity, usually by a crane equipped with a clam shell bucket. The unloaded coal is generally moved by belt conveyors to storage yards. One modern unloader consists of two revolving scoops, which can clean out a barge in three passes.[12] The cost of such facilities is in the millions of dollars.

On the Great Lakes, oceans, and coastal tidewaters, single self-propelled bulk cargo carriers are used. Three Great Lakes vessels being built to carry Western coal to a Detroit utility plant will be 1,000 feet long and will each carry 62,000 tons of coal. They will be self-unloading, using long conveyor belts traversing the length of the ships.[13] The average lake carrier has about 20,000-ton capacity.

The inland waterways themselves have been constructed and maintained by the Federal Government, with the exception of the New York State Barge Canal. Since the first appropriation to the U.S. Army Corps of Engineers for this purpose in 1824, billions of dollars have been spent to make possible the large-scale commercial traffic now using these internal waterways. In general, the improvement of a waterway to make it navigable involves deepening and widening the channel by dredg-

---

[12]"Using Waterways to Ship Coal . . . No Cheaper Way When Destination is Right," *Coal Age,* vol. 79, No. 7, July 1974. Description of unloading plant for J. M. Stuart Generating Station on Ohio River, pp. 122-123.

[13]Ibid., pp. 126-127.

ing and constructing dams and locks. Erosion control along sections of the riverbank and operation of the dams and locks are also part of the Federal role. There were 255 navigation locks and dams being operated by the Corps of Engineers as of June 1, 1976. The size of the locks available on the river systems is the primary limiting factor on river traffic. They are sized to accommodate an expected volume of traffic that generally tapers off toward the source of the river system. Thus in the region of the headwaters fewer tows can be managed; this makes it less economical to operate the carriers and less cost-beneficial to maintain the waterway or justify improvements.

## Highways

The fastest growing means of coal transportation is by truck. As opposed to the average haul of 300 miles by railroad, and 480 miles by barge, the average highway shipment of coal is only 50 to 75 miles.

The public highway trucks carry a total of 15 to 25 tons each. The standard diesel tractor is used to pull one and sometimes two trailers, depending on weight limitations. Much coal movement to nearby powerplants takes place on private roads using vehicles too large for public highways; some of these vehicles can carry up to 150 tons of coal.[14] The trucks may be loaded from a fixed chute, but are more often loaded by shovels or front-end loaders directly at the side of the coal seam in a surface mine. Most trucks are dumpers. Truck shipments are sometimes used in connection with deep-mine operations, but most often are connected with strip-mine operations.

## Slurry Pipelines

Transport by this mode involves three major stages:

1. grinding the coal and mixing it with a liquid (generally water) to form the slurry;
2. transmission through the pipeline; and
3. dewatering the coal for use as a boiler

[14]Ibid., p. 81.

Photo credit: OTA
**Coal-truck traffic, Kentucky**

fuel, for storage or for transloading to another transportation mode.

Problems involve condemnation rights to certain property, inter-basin transfer of water, and the discharge or use of the separated (and contaminated) water.

Coal is assembled from a mine or group of mines at a single point where mixing, cleaning, or other beneficiation may take place, and where the slurry is prepared. Preparation begins with impact crushing, followed by the addition of water and further grinding to a maximum particle size of one-eighth of an inch. More water is then added to form a mixture that is about 50 percent dry coal by weight, and the resulting slurry is stored in a tank equipped with mechanical agitators to prevent settling. Water is not necessarily the only slurry medium, and oil—as well as methanol derived from the coal itself — has been proposed.

The slurry from the agitated-storage tanks is introduced into a buried steel pipe and propelled by pumps located at intervals of 50 to 150 miles, depending upon terrain, pipe size, and other design considerations. The slurry travels at about 6 ft/sec. Ideally, the flow is maintained at a rate that minimizes power requirements while maintaining the coal in suspension. Once started, the flow must continue uninterrupted or the coal will gradually settle and possibly plug the pipe. Considerable technical controversy exists over the likelihood of plugging and the ability to restart the pipeline after given periods of downtime. To prevent this type of settling, the operating pipeline at Black Mesa, Ariz., has ponds into which the pipeline can be emptied in the event of a break or other interruption.

At the downstream end of the pipeline, the slurry is again introduced into agitated tank storage, from which it is fed into a dewatering facility. Dewatering is accomplished by natural settling, vacuum filtration, or by centrifuge, and the finely ground coal still suspended in the water can be separated by chemical flocculation. Additional thermal drying is generally required before use. The reclaimed water can be used in an electric-generating station's cooling system to condense steam — or it could theoretically be recycled in a return pipeline.

**Schematic of Slurry Pipeline System**

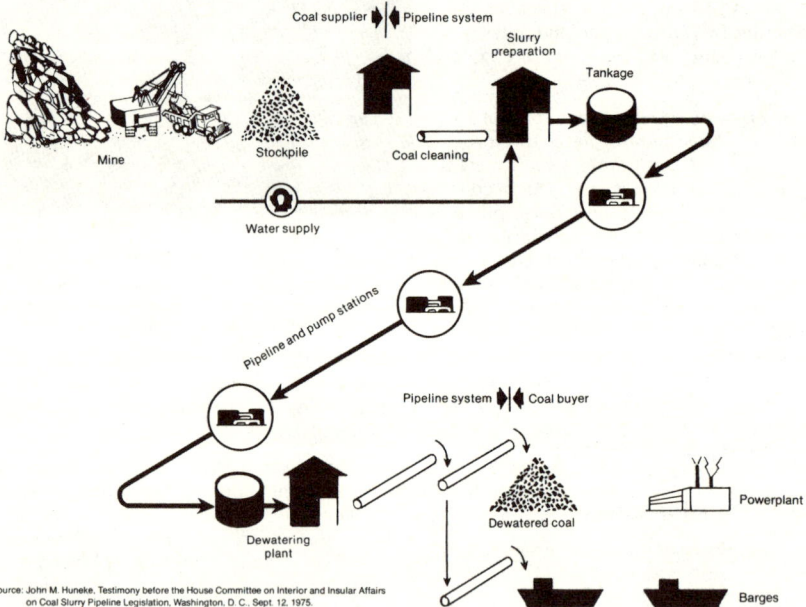

Source: John M. Huneke, Testimony before the House Committee on Interior and Insular Affairs on Coal Slurry Pipeline Legislation, Washington, D. C., Sept. 12, 1975.

A potential economic advantage of this technology is that the volume of coal that can be transported increases approximately as the square of the pipeline diameter (which means that a 2-foot diameter line can carry four times the amount of coal as a 1-foot diameter line), while construction, power, and other operating costs increase more slowly. If throughput volumes are high enough to take advantage of this economy of scale, and if the pipeline is long enough to recover the cost of gathering, preparing, dewatering, and delivering the coal at the termini, the pipeline can compete with unit trains.

## Electric Transmission

The principal means of transmitting large blocks of electric power is by high-voltage alternating current (AC) transmission. Electrical transmission will become more important as the number of mine-mouth plants increases. Electricity is usually generated at voltages of 12,000 to 22,000 volts (12 to 22 kV). However, it is not economical to transmit electric energy at these generated voltages because the high current levels lead to excessive resistive losses in the transmission system. The attainment of higher and higher transmission voltages continues to be a major goal. An AC transmission system requires, therefore, transformers to step up the voltage from the generator and to step it down where it feeds into the distribution system at the load. In addition, there are towers, conductors, and devices to protect the system in the event of short circuits and power surges. The conductors are designed to minimize effects of high electric fields that occur because of the high voltages. These effects, which include corona formation and radio and television interference, increase energy loss and may create disturbances for the population near the powerline.

As of 1976, there were about 360,000 miles of high-voltage (69 kV or higher) transmission lines in the United States. These are the candidate voltages for long-distance, mine-mouth plant transmission lines. Currently the highest operating transmission voltage in the United States is 765 kV. The upper voltage has in-

creased by a factor of more than 2.5 since 1950. Technical and environmental problems complicate development of transmission systems beyond 765 kV.

High-voltage direct current (DC) systems for long-distance power transmission are becoming increasingly competitive relative to high-voltage AC. Although only one is operating in the United States (on the west coast), other nations have more experience. The principal disadvantage of DC transmission is the difficulty in varying the voltage levels as is done with AC systems and transformers. Consequently, it is necessary to convert the voltage from AC to DC and back again at each end of the DC line. This conversion process adds considerably to the cost of a DC system. Until now, this has greatly limited its use, but new developments promise to facilitate the process. The great advantage of DC is that the line is capable of carrying considerably more power, about 1.5 to 2 times more, than an AC system of the same voltage. In addition to the increased energy density, DC lines produce less radio interference and the corona loss is considerably smaller than equivalent capacity AC lines.

Research is currently underway to increase the energy and economic efficiency of high-voltage electric transmission and to develop underground systems, gas-insulated lines, and low-resistance (superconducting lines). Systems from 1,100 to 2,300 kV are being tested. The principal technical problems are the development of adequate insulators and the ability to protect against surge voltages. Other uncertainties are the possible environmental effects of very high electric fields.

Underground transmission exists in this country to only a very limited extent. The principal roadblock is cost, owing to the need for cable insulation and the difficulty of installing and cooling cables underground. There is continuing interest in high-voltage underground transmission, however, because of the technical, environmental, and social problems of expanding overhead networks. The utility industry is conducting R&D to lower costs and improve performance, but the introduction of underground high-voltage transmission as a cost-competitive technology is many years away.

**Southwestern Electric Power Company, Shreveport, La.; Welsh Power Plant, East Texas; units #1 and #2**

Gas-insulated and low-resistance lines similarly are undergoing long-term development programs.

## Transportation Patterns and Cost

Coal is seldom mined where the energy is needed, so transportation is an important element of coal use. Table 12 indicates, by State of origin and destination, the tonnages transported in 1976 to electric utilities, which consume about three-quarters of all domestic steam coal. It also shows that the average ton moved 368 miles in the Nation. Figure 12 shows that railroads carry most of this coal, and this share is increasing.

Two coal slurry pipelines have been constructed in this country; one is in operation, carrying 4.8 million tons per year 273 miles from northeastern Arizona to southern Nevada. (The other closed after the competing railroad introduced unit train service and lowered its rates.)

Unit trains and slurry pipelines are clearly appropriate only for very large users of coal. Smaller users generally rely on less regular deliveries by train or truck, sometimes through an intermediary retailer. Similarly, a small mine operator could not fill a unit train, but must load a few cars at a time and depend on

**Table 12.—Origin and Destination of Coal Deliveries to Electric Utilities in 1976**
**(in thousand tons)**

| Destination | Origin | | | | |
| --- | --- | --- | --- | --- | --- |
| | Northern Appalachia[a] | Southern Appalachia[b] | Midwest[c] | Southwest[d] | West[e] |
| Alabama | — | 18,094.1 | — | — | — |
| Arizona | — | — | — | 6,528.6 | 234.9 |
| Colorado | — | — | — | — | 7,088.9 |
| Delaware | 765 | — | — | — | — |
| Florida | — | 4,324.6 | 1,553.4 | — | — |
| Georgia | — | 13,438.7 | 1,163.9 | — | — |
| Illinois | 10 | 1,778.7 | 21,784.4 | — | 11,217.9 |
| Indiana | 132.7 | 5,086.8 | 21,989.1 | — | 3,270.7 |
| Iowa | — | 83.8 | 3,070.5 | — | 3,553.2 |
| Kansas | — | — | 2,259.9 | — | 1,172.5 |
| Kentucky | 44 | 22,271.4 | 3,036.6 | — | — |
| Maryland | 770 | 29 | — | — | — |
| Michigan | 12,019.3 | 7,144.4 | 434.3 | — | 1,710 |
| Minnesota | 8.6 | 148.1 | 1,436.4 | — | 8,911.5 |
| Mississippi | — | 1,056.5 | 520.3 | — | — |
| Missouri | — | 534 | 19,043 | — | 979.1 |
| Montana | — | — | — | — | 2,316 |
| Nebraska | — | — | 12.5 | — | 1,839.6 |
| Nevada | — | — | — | — | 820.5 |
| New Hampshire | 753.2 | — | — | 4,174 | — |
| New Jersey | 1,894.8 | 637.9 | — | — | — |
| New Mexico | — | — | — | 8,019.7 | — |
| New York | 5,253.7 | 484.4 | — | — | — |
| North Carolina | 2,908.6 | 16,678.6 | .1 | — | — |
| North Dakota | — | — | — | — | 6,884.9 |
| Ohio | 38,436.9 | 8,510.2 | 16.6 | — | 9,758.6 |
| Oklahoma | — | — | — | — | 598.8 |
| Pennsylvania | 37,225.5 | 186.4 | — | — | — |
| South Carolina | 3.7 | 5,491 | — | — | — |
| South Dakota | — | — | 27.0 | — | 2,459.7 |
| Tennessee | 481 | 19,549 | 1,002.3 | — | — |
| Texas | — | — | — | 18,867 | 903.9 |
| Utah | — | — | — | — | 1,869 |
| Virginia | 665.6 | 4,643.3 | — | — | — |
| Washington | — | — | — | — | 3,600 |
| West Virginia | 23,307 | 4,813.3 | — | — | — |
| Wisconsin | 695.3 | 1,984.4 | 5,079.3 | — | 3,389.9 |
| Wyoming | — | — | — | — | 8,459.4 |

[a]Maryland, Ohio, Pennsylvania, and West Virginia.
[b]Alabama, Georgia, Kentucky, Tennessee, and Virginia.
[c]Illinois, Indiana, Missouri, and Oklahoma.
[d]Arizona, New Mexico, and Texas.
[e]Colorado, Montana, North Dakota, Utah, Washington, and Wyoming.

the railroads to fit them into their schedules. Transportation is thus a major factor in the high price charged by retailers, as discussed in chapter II.

Of the average cost of coal delivered by rail, transportation represents about 20 percent.[15] Rough approximate costs of barge, rail, and truck transport are 0.4-, 1-, and 5-cents/ton-mile

respectively,[16] and slurry pipelines can cost less than rail under some specific conditions. These figures do not include an element of public subsidy for barges and trucks in the form of Government highway and waterway expenditures.

Fuel is required for transportation, amounting on the average to 540 to 680 Btu/net-ton-mile for barge, 536 to 791 Btu for rail, and

[15]U.S. Coal Development—Promises, Uncertainties (Washington, D.C.: U.S. General Accounting Office, 1977), p. 5.5

[16]Ibid.

**Figure 12.—Coal Distribution by Transportation Mode and End Use, January-September 1977**
**(Million short tons)**

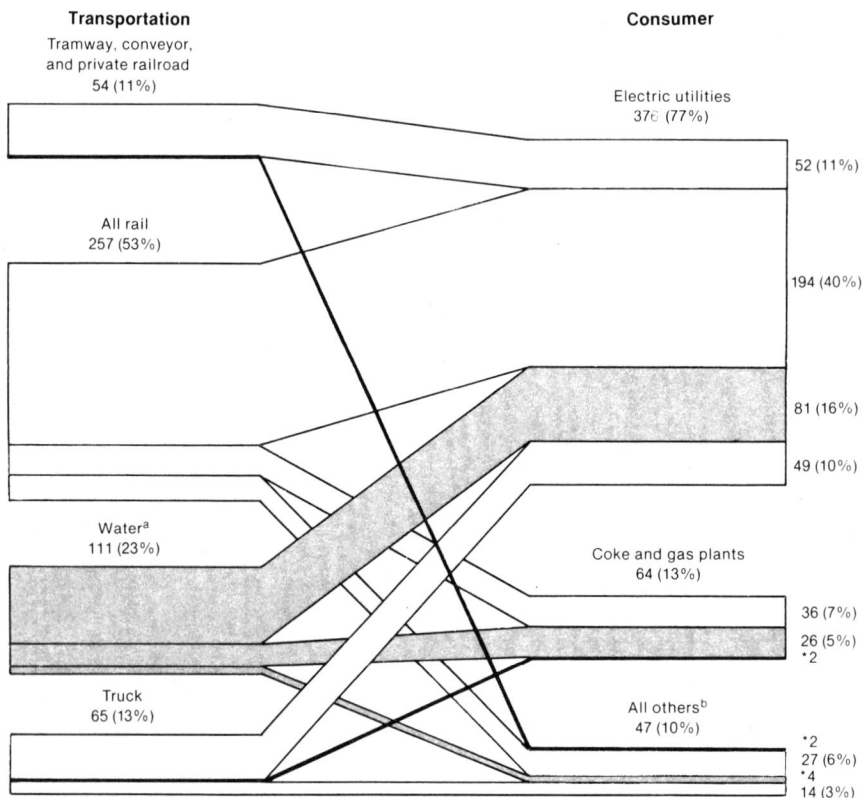

**Transportation**

Tramway, conveyor,
and private railroad
54 (11%)

All rail
257 (53%)

Water[a]
111 (23%)

Truck
65 (13%)

**Consumer**

Electric utilities
370 (77%)

52 (11%)

194 (40%)

81 (16%)

49 (10%)

Coke and gas plants
64 (13%)

36 (7%)

26 (5%)
*2

All others[b]
47 (10%)

*2
27 (6%)
*4
14 (3%)

*Less than 1 percent.
[a]Includes shipments partly transported by other modes; excludes oversea exports.
[b]Includes retail.

SOURCE: Department of Energy, *Bituminous Coal and Lignite Distribution, January-September 1977*, Feb. 6, 1978

2,518 to 2,800 Btu for highway shipment.[17] Unit trains are somewhat more efficient than rail generally, and slurry pipelines, although more energy intensive than unit trains, consume electricity derived from a variety of fuels, including coal, while the other three modes cur-

[17]Congressional Research Service, 1977, op cit., p. 85

rently require diesel fuel from petroleum. A ton of 8,500 Btu/lb coal hauled 1,000 miles at 600 Btu/net-ton-mile requires the equivalent of 3½ percent of its energy value, usually in diesel fuel. Hence, importation of low-sulfur coal can use more energy than flue-gas desulfurization applied to local high-sulfur coals. (See the section on *Air-Pollution Control Tech-*

*nology* of this chapter for a further discussion on flue-gas desulfurization.)

The regional pattern of production and distribution is also sensitive to transportation costs. One study[18] simulated the purchasing behavior of the electric utility industry under two hypothetical scenarios. In one, the future cost (in constant dollars) of mining coal was assumed to increase by 25 percent while transportation cost declined in the same proportion. The other case embodied the reverse assumption, that transportation increased by 25 percent while mining declined by the same relative amount. The second case resulted in a predicted production level of Appalachian coal in the year 2000 that was twice that predicted by the first case. Western coal exhibited the reverse behavior, amounting to twice the production in the first case as in the second. The pattern was also sensitive to other considerations, like the extent to which power-plants constructed after 1983 with mandatory flue-gas desulfurization would be used only for peakloads. Traditionally, electric companies use their lowest cost plants for baseload and successively add higher cost units for peak demand. In this way they provide customers with lowest cost energy. If new plants with scrubbers become more costly than old plants using low-sulfur Western coal, the pattern of coal origin and distribution may be expected to adjust accordingly.

One of the factors affecting the economies of powerplant siting is the cost of delivering the power to the load center. If a number of suitable sites are available between the mine and the load center, the utility may choose remote siting, with power transmission replacing coal transportation. Typical costs of transmission as a function of voltage are:

| | Construction cost per mile | Capacity | Cost per kWh (500 miles) |
|---|---|---|---|
| 345 kV . . . . | $213,000 | 750 MW | 0.43 cent |
| 500 kV . . . . | 267,000 | 1,500 MW | 0.29 cent |
| 765 kV . . . . | 400,000 | 3,000 MW | 0.23 cent |

DC lines are less expensive than AC for the same capacity because the lower voltage allows smaller rights-of-way, towers, and conductors. As previously stated, however, DC terminals are more expensive because of the conversion equipment. Hence DC is considered only for long lines (at least several hundred miles). New technology, such as very high-voltage AC or DC or new types of DC terminals, may change the equation. Underground transmission will be several times as expensive as overhead. DC may enjoy a relative advantage if underground transmission is required.

As for comparison with other modes, a recent OTA study on coal slurry pipelines estimated costs of transporting coal by both rail and pipeline for four different routes in the United States. The cost estimates ranged from $7 to $8/ton for distances of 500 miles. Assuming an energy content of 20 million Btu/ton and a heat rate of 9,000 Btu/kWh for the powerplant at the end of the pipe-rail line, energy delivery cost is about 3.0 to 3.5 mills/kWh. Costs vary considerably but in general, electric transmission is competitive with rail or pipeline for distances of up to 500 to 600 miles. Beyond that point, rail and slurry costs increase at a slower rate than electric transmission costs. Federal policies can have a substantial effect on relative pricing advantages through differing tax advantages and subsidies.

## COMBUSTION TECHNOLOGY AND METHODOLOGY

Three major factors influence the way coal is burned. The first is the size of the facility to be constructed, with utilities usually building the biggest, industry substantially smaller, and the residential/commercial sector the smallest.

[18]Teknekron, Inc., *Projections of Utility Coal Movement Patterns: 1980-2000*, OTA, 1977.

The latter units are responsible for such a small fraction of coal burned that they are not described here. A full description is included in appendix III of volume II. The second factor is the set of environmental standards the facility is required to meet. The rapidly changing regulatory climate is radically changing coal use patterns and forcing the development of new technologies. Finally, the characteristics of the coal to be burned control some design parameters as discussed earlier in the section on *Coal Resources.*

## Utility Boiler Furnaces

Utility boiler furnaces generally have a design firing rate greater than 250 million Btu/hr, equivalent to about 10 tons of bituminous coal per hour. The largest utility units burn as much as 500 tons of coal each hour at full load. Industrial boilers usually range downward from the smaller utility units, from 20 tons to about 3 tons of coal per hour. Large coal-fired utility powerplants are much the same in principle today as they were in the 1920's when pulverized coal was first used to generate steam. Today's boiler furnaces are much larger, produce steam at much higher temperatures and pressures than half a century ago, and are more efficient in converting the energy in coal into steam. But the concept of grinding the coal to very small size and blowing it with air into large furnace cavities, where the cloud of coal dust burns much like a fuel gas, is still the way most coal is burned today.

### Pulverized Coal-Fired Boiler Furnace

The principles of this most widely used steam generator are relatively simple. Raw crushed coal is fed continuously into pulverizers, where the coal is dried and ground so that at least 70 percent will pass a 200-mesh sieve to form a combustible cloud of coal. Air, heated by the spent products of combustion before they enter the stack, is blown through the pulverizers, drying the coal and conveying it to the burners in the boiler furnace. At the burners, additional heated air is mixed with the airborne coal stream to provide a slight excess of oxygen to ensure complete combustion. The coal burns in a long, luminous flame in the huge furnace cavity at temperatures in the flame of at least 2,700° F. Heat energy from this flame is transferred largely by radiation to the relatively cool furnace walls. The heat is transferred to water, which boils to generate steam. The steam is separated from the unevaporated water in a steam drum at the top of the boiler. The water is returned by "downcomers" to the bottom of the boiler, where it is distributed by "headers" to the lower end of the furnace wall tubes to pass once more through the heated zone of the furnace.

Steam, separated from the unevaporated water, flows through superheater tubes of low-alloy steel located at the furnace outlet. These banks of tubes may be heated partly by radiation from the pulverized-coal flame, but usually mostly by convection from the hot furnace gases, now cooled to about 2,000° F. This raises (superheats) the temperature of the steam to about 1,000° F as it leaves the boiler for the steam turbine. After expanding in part in the turbine and converting some of its heat energy into shaft horsepower, the partially cooled steam may be returned to the boiler where it passes through another bank of boiler tubes that reheat the steam. The steam at various stages of the turbine is bled in part to auxiliary equipment to heat water or to power apparatus and finally to the condenser, which reduces the final pressure at the exit of the turbine to below atmospheric pressure. The condensed water is pumped back through a series of heaters and an economizer—a bank of boiler tubes located downstream of the superheater and reheater. Here some of the heat in the flue gas that otherwise might be wasted raises the temperature of the boiler feed water, which passes once more into the boilerdrum.

Flue gas leaving the economizer still contains appreciable heat energy, which is recovered in part by passing it through an air heater, generally a massive array of mild-steel sheets arranged in bundles or "baskets," which rotate successively between ducts carrying the flue gas to the stack and air to the furnace. Tubular heat exchanges have been widely used as air heaters, but rotating plate-type regenerative heaters described above are common. In either type, metal heated by the flue gas passes the heat to the incoming air, which usually is raised to temperatures between 300° and 600° F. Because the overall efficiency of a steam-generating unit as a whole will increase about 2.5 percent by raising this air temperature by

100° F, it is obvious that the highest practical temperature is sought.[19]

Flue gas leaving the air heater contains particles of fly ash. Electrostatic precipitators, under ideal conditions, can capture 99.9 percent of this fly ash. The flue gases consist of carbon dioxide, nitrogen, oxygen, and lesser amounts of nitrogen oxides, sulfur oxides, carbon monoxide, and hydrocarbons. Flue-gas desulfurization systems, mostly based on scrubbing the flue gas with chemically active solutions or with slurries of lime or limestone, are being heavily promoted for controlling $SO_x$ emissions.

Work now underway on alternative energy sources and fuel systems will not be sufficiently advanced in the next 10 years to have a major impact on electric power generation methods using fossil fuels or their derivatives; the use of pulverized coal-fired boiler furnaces may be expected to dominate the fuel-burning public-utility field during that time.

### Characteristics of Pulverized Coal Furnaces

Furnaces for burning pulverized coal can be very large. Configuration varies among the four major boiler manufacturers, but basically a furnace is a huge, hollow, rectangular box made up of vertical water-filled mild-steel tubes welded into panels to constitute the furnace walls. For one typical boiler furnace providing steam to an 800-MW turbine generator, the height of this cavity is 160 feet, the width 45 feet, and the depth 90 feet. A vertical water-wall panel divides the furnace into two halves, each with a cross-section of 45 by 45 feet. Pulverized coal is blown tangentially into each half of this split furnace at the corners by four arrays of burners, aimed near the center of the furnace to produce a rotating vortex flame pattern with a vertical axis. In some furnaces, the burners can be tilted upward or downward to adjust the location of the flame and to control steam temperature. At full load, essentially the entire bottom half of the furnace is filled with flame, with the upper portion providing time to complete the combustion reactions. Rough-

ly half the total heat released when the coal is burned is picked up by furnace wall tubes.

Other furnace configurations call for all the burners to be arranged on one wall, firing horizontally into the furnace, or on opposite walls to provide opposed firing. Many such variations are used to control the admission of air and coal to the furnace, to develop good turbulence for complete combustion, to minimize problems with slagging and fouling of furnace walls and heat-receiving surfaces by the ash in the coal, to utilize each square foot of furnace walls as effectively as possible as heat-receiving surfaces, and to permit control of superheated temperature within the narrow range demanded by the steam turbine.

The largest pulverized-coal-fired boiler, which was contracted in 1976, is rated at 1,281 MW. It will provide nearly 10 million lbs steam/hr at a pressure of 3,845 lbs/in² and at a temperature of 1,010° F with reheat to 1,000° F. The surface area of the furnace walls is 146,000 ft², with 192,000 ft² in the superheater, 255,000 ft² in the reheater, 494,000 ft² in the economizer, and over a million ft² of surface in the air preheater.[20]

### Supercritical Boilers

A development of considerable importance in recent years has been the design of utility boilers to operate above the critical pressure of 3,208 lbs/in². Above this pressure, water behaves as a single-phase fluid, and there is no distinction between liquid and steam, as there is at lower pressures. In supercritical boilers, water from the economizer enters the boiler and is heated as it passes first through the wall tubes and then through the superheater, without any change in phase, as by "boiling" for subcritical boilers. There is no steam drum in a supercritical boiler. Supercritical boilers are also known as "once through" boilers, because feed water is pumped into one end of the boiler and "steam" exits from the other. The outlet steam temperatures depends only on the rate at which feed water is supplied to the boiler and the amount of heat picked up in the

[19]*Steam. It's Generation and Use* (Babcock & Wilcox, 1972), p. 13-4.

[20]"1976 Annual Plant Design Report," *Power*, November 1976, p. S-4.

furnace. Control of water quality is critical in supercritical boilers, for without a steam drum boiler, water that has accumulated impurities entering with the feed water cannot be "blown down" occasionally.

Supercritical units generally are large. In 1972, contracts were let for six such boilers, the smallest rated at 520 MW. In 1973, there were four supercritical boilers sold; in 1974, seven; in 1975, six; in 1976, nine, and in 1977, three.[21] There is an apparent trend downward, with 30 percent of the new boilers listed being supercritical in 1971, 17 percent in 1976, and 9 percent in 1977. Such trends generally develop as the utilities find by experience that reliability, operating problems, maintenance, startup, or economic factors influence their choice of equipment. There does not appear to be any one reason for this falling off of interest in supercritical boilers, although material limitations are a major factor.

## Slagging and Fouling

Probably the single greatest problem in the operation of coal-fired boiler furnaces is the accumulation of coal ash on boiler surfaces. All coal contains noncombustible inorganic mineral matter. Some of it was an integral part of the original vegetation that accumulated to form coal seams; however, most of the mineral matter associated with coal accumulated as sediment from mineral-laden water that percolated through the coal deposits over eons of time. The chemical composition of the mineral matter varies widely and there are no "typical" concentrations. About 95 percent of all the noncombustible material in coal is made up of kaolinite ($Al_2O_3 • 2SiO_2 • 2 H_2O$), pyrites ($FeS_2$), and calcite ($CaCO_3$).[22] Many variations occur, particularly in the clay minerals; probably a hundred different minerals have been identified in coal.

This mineral matter causes three main problems when burned in larger boiler furnaces: 1) buildup of ash on furnace wall tubes; 2) accumulation of small, sticky, molten particles

[21]"Annual Plant Design Surveys," Power, 1972-77.
[22]William T. Reid, "External Corrosion and Deposits: Boilers and Gas Turbines" (Elsevier, N.Y., 1971), p. 52.

of ash on superheater and reheater tube banks; and 3) corrosion.

Some of the mineral forms in coal melt in the pulverized-coal flame, where the temperature may reach 3,000° F. These tiny molten droplets pick up other mineral forms by collision in the highly turbulent zone just beyond the flame, forming larger drops of molten "slag." Some of these reach the wall tubes of the furnace and, by some mechanism still not fully understood, adhere to that metal surface, even though the metal is cooler than about 800° F. Gradual buildup of these slag droplets eventually forms a solid layer of slag. Because coal-ash slag does not conduct heat readily, this layer of slag decreases the amount of heat reaching the wall tube, and hence lowers the quantity of steam produced at this point. Periodically, as the load changes, and there is a differential expansion between the wall tubes and the slag layer, huge sheets of slag may peel off the walls. But to control slag more effectively, "wall blowers" are provided that use high-velocity jets of air or of steam, and occasionally streams of water, to dislodge the slag and leave clean furnace walls. Occasionally, slag accumulations may get so large that manual cleaning is required, or large pieces of slag that have fallen to the bottom of the furnace cannot be removed without breaking them by hand. A furnace outage is then necessary, a costly step that is avoided as far as possible.

Fouling occurs when very small particles of ash are carried to the bundles of tubing making up the superheaters and reheaters. These fly ash particles collect on the tube surface, not only insulating the metal so that not enough heat is transferred to raise the steam temperature to design levels, but also accumulating so much that it plugs the normal gas passages. "Soot blowers" are provided to remove such ash deposits periodically, but some coal ash may form dense adherent layers very difficult to remove except by manual cleaning.

The third problem is corrosion of the wall tubes beneath a layer of slag or deposits. This has been a serious problem in the past and remains a potential threat to uninterrupted boiler operation.

## Cyclone Furnaces

An innovation in boiler design in the late 1930's was the "cyclone" furnace, a water-cooled, slightly tilted horizontal cylinder with combustion air blown in tangentially to provide a high level of turbulence to burn the centrally admitted crushed coal. Because of the high combustion intensity, temperature exceeds 3,100° F and the ash in the coal forms a liquid slag covering the inside of the cylinder and dripping from the lower end. Larger particles of the crushed coal are caught by this slag layer and burn as these pieces of coal are contacted by the rapidly rotating tangential air stream.

Advantages of the cyclone furnace are the capture of a major part of the ash in the coal as molten slag, the use of crushed rather than more costly pulverized coal, and the smaller furnace required for a given steam output. With cyclone furnaces, 70 to 80 percent of the ash is converted to slag, thereby greatly decreasing the fly ash in the secondary furnace, compared with 20 percent ash recovery in a dry-bottom, pulverized-coal-fired furnace. Crushed coal is simpler to handle and is appreciably cheaper than pulverized coal. And because of the very high combustion intensity in the cyclone, the associated secondary furnace serving as a heat-recovery system for several cyclone furnaces can be significantly smaller than a conventional furnace of the same output.

Shortcomings of the cyclone revolve mainly around ash characteristics of the coal; the viscosity of the molten slag must allow flow on a horizontal surface. Because of the very high temperatures in the combustor, $NO_x$ formation is appreciable and difficult to control. Hence, in recent years, cyclones have not been popular. But the growing demand for low-sulfur Western coal, often containing coal ash that forms such fluid slag at moderate furnace temperature as to cause serious problems with slagging in dry-bottom furnaces, could bring cyclone and other slag-tap furnaces back into the market.

## Industrial Boiler Furnaces

Industrial boilers are generally rated by horsepower or steam output, ranging from about 10,000 lbs steam/hr to as much as 1 million lbs steam/hr. The largest industrial boiler contracted in 1977 will generate 840,000 lbs steam/hr, burning oil and gas. The largest coal-burning industrial boiler ordered in 1977 is rated at 550,000 lbs steam/hr, with bark as a supplementary fuel.[23]

Based on earlier data[24] for the various sizes of industrial boilers fired with coal, traveling-grate, and underfeed stokers are most popular in smaller units, accounting for about 41 percent of the steam generated in boilers with output less than 20,000 lbs steam/hr, and about 29 percent of the boilers rated under 100,000 lbs/hr. Spreader stokers dominate the midsize industrial field, accounting for roughly half of the steam generated in coal-fired boilers rated at 200,000 lbs/hr and smaller. Pulverized coal boilers dominate the largest industrial field, providing 78 percent of the steam generated in units larger than 200,000 lbs/hr.

Natural gas and fuel oil have been the preferred fuel for industrial boilers in smaller sizes, in part because the cost and space requirements of such plants is much lower than for coal firing. With the present unsettled fuel supply situation, more companies may be willing to invest the greater costs for solid-fuel boilers to take advantage of a more secure source of energy.

Many industrial boilers are shop-assembled; fabricated under conditions favoring low-cost production methods. The largest size is dictated by shipping limitations. Such shop-assembled boilers are completely assembled and enclosed in a steel casing, eliminating even the need for applying insulation or housing in the

[23]"1977 Annual Plant Design Survey," *Power*, November 1977, p. 5-12.
[24]"Evaluation of National Boiler Inventory," Environmental Protection Agency Technical Series, EPA-600/2-75-067, October 1975.

field. Although generally fired with fuel oil or natural gas, similar packaged units can be designed with grates and stokers to be fired with coal. A coal-fired packaged boiler was designed and built in the 1950's, but only a few units were installed because the growing availability then of low-cost natural gas and the lesser attention by a fireman needed with gas firing proved to be more economical. With a change in fuel availability and costs, packaged boilers fired with coal may see greater use.

## Pulverized Coal-Fired Industrial Boiler Furnaces

Industrial boilers fired with pulverized coal are basically similar to utility units except that the industrial boilers are smaller, steam output is less, and steam pressure seldom exceeds 1,500 lbs/in². Because of costs for auxiliaries such as pulverizers, the lowest practical steam output for pulverized coal-firing is about 200,000 lbs/hr. The same problems with slagging and fouling occur in large industrial boilers as in utility steam generators, and the plant is more sophisticated and difficult to operate than other industrial furnaces.

## Spreader Stokers

These stokers are very popular for steam demand below that supplied by pulverized coal. A spreader stoker can burn a variety of coals, it responds rapidly to load changes, and it is more economical in larger sizes than other kinds of stokers.

The concept of the spreader stoker dates back to 1822, as a way of feeding coal to a grate without opening the furnace door.[25] The same principle is used today. A mechanical distributing device, either an assembly of rotating paddles or a series of air jets arranged at the end of the furnace, throws crushed coal into the combustion zone of the furnace. The smaller particles of coal burn in suspension while the heavier ones fall onto a moving grate, where they burn with air supplied upward through the grate. The grate moves slowly so that the ash from the coal is discharged at

the end. Adjustment of the distribution assures that the coal falling on the grate is properly positioned to burn out completely by the time the coal reaches the ash-discharge position. Some spreader stokers use stationary sectioned grates that are dumped section by section into the ashpit as ash accumulates.

Roughly half the coal burns in suspension, so there may be problems with emission of fly ash and unburned carbon with spreader stokers. To minimize loss of boiler efficiency from unburned carbon, fly ash collected in hoppers is reinjected into the furnace. Erosion of boiler tubes may be troublesome.

## Traveling-Grate Stokers

These consist of a horizontal moving grid of iron links connected to form an endless belt driven by sprockets to move a fixed bed of coal through a furnace. Coal is fed at one end to a depth of 4 to 6 inches, ignites as it enters the hot furnace by radiation from a furnace arch and by recirculation of burning particles of coal from farther in the furnace, burns as it moves from the inlet, and eventually is dumped as ash into the ashpit at the end of the grate. The speed of the grate is adjusted to ensure complete combustion, and air is admitted through the grate in zoned sections to control combustion.

Traveling-grate stokers can burn a wide variety of fuels, ranging from coal to garbage. They are not well suited to coals that cake strongly.

## Underfeed Stokers

Whereas in traveling-grate stokers the coal and the combustion air move in opposite directions (countercurrent flow), in underfeed stokers the coal and the air move together in the same direction into the combustion zone. Coal is forced upward through a conical retort topped with nozzles through which the air is blown. This establishes the zone where ignition occurs, the burning coal then moving onto stationary perforated grates, where combustion is completed.[26] Temperatures are high and the coal ash fuses into lumps. Unlike traveling-

---

[25]*Combustion Engineering* (New York: Combustion Engineering, Inc., 1966), p. 18-2.

[26]Ibid., p. 18-8.

grate stokers, strongly caking coals are handled readily in underfeed stokers.

Single-retort underfeed stokers are used in small boilers, generally under about 30,000 lbs steam/hr. For greater output, multiple-retort stokers have been built at ratings up to 500,000 lbs steam/hr. These operate on the same principle of concurrent movement of coal and air but with many engineering changes to accommodate the large fuel beds.

### Vibrating-Grate Stokers

These stokers depend upon oscillating grate bars, or flexing grates successively, to move coal into a sloping fuel bed. As with traveling-grate stokers, air moves upward through the grates, so this is a special case of overfeed burning. The same principle is used in grates for burning municipal refuse. Vibrating-grate stokers appear to be more popular in Europe than in the United States, possibly because our strongly caking bituminous coals are more troublesome than the lower rank coals overseas.

# AIR POLLUTION CONTROL TECHNOLOGY

There are currently six "criteria" air pollutants—pollutants for which National Ambient Air Quality Standards (NAAQS) have been promulgated: sulfur oxides ($SO_x$), nitrogen oxides ($NO_x$), particulates, carbon monoxide (CO), hydrocarbons, and photochemical oxidants. Emissions of the first three can be reduced in the direct combustion of coal by three methods. One is to remove contaminants before combustion, the second is capture them during combustion, and the final is to remove them from the flue gases after combustion. The effectiveness and cost of removing any one is in part dependent on the method and sequence of removal of the others. Because of the wide variation in the composition and makeup of the contaminants in coal and because boilers are designed especially for the fuel being fired, no universal method can be prescribed that would be best for all new or existing plants.

The level of emissions of CO and hydrocarbons in direct coal combustion is a result of the completeness of combustion, regardless of the coal quality. Oxidants are not directly produced by combustion but are the reaction projects of hydrocarbons and $NO_x$ in the atmosphere.

Lowering the emissions of the first three criteria pollutants by precombustion cleanup and by new combustion technologies is described later in this chapter. The next section focuses on control technology for their removal from flue gas.

## Particulate Control

Stationary combustion sources burning coal produce bottom ash, cinders, and slag, which are removed from the furnace itself, and fly ash, which is entrained in the flue gas.

Four types of systems are capable of removing particulates from flue gas: mechanical or cyclone collectors, electrostatic precipitators, wet scrubbers, and fabric filter baghouses.

### Cyclone Collectors

Mechanical or cyclone collectors work on the principle of gravitational, centrifugal, or inertial forces to separate the particles from the gas. One type consists of an enlarged chamber, which slows the gases and allows the particles to settle. Another form is cylindrical with a tangential entrance that causes the gases to swirl rapidly. The particles, heavier than the gases, migrate by centrifugal force to the walls of the vessel and drop to the bottom. The clean gases exit at the top.

These units are simple, relatively maintenance free, reliable, low in pressure loss, relatively low in both capital and operating cost, and independent of operating temperature. They are particularly effective in removing large particles and can handle high dust loadings. However, their performance is sensitive to variable dust loading and gas velocity.

The efficiency of well-designed units ranges from 80 to 90 percent, depending upon particle size. Because of their low cost but inadequate effectiveness to reduce small particles, they are sometimes used as first-stage cleaners in conjunction with other devices that are more efficient and expensive. Their use may lower the size and cost of the follow-on unit.

## Electrostatic Precipitators

Electrostatic precipitators are widely used in large stationary combustion sources to remove particulates from flue gas. Gas containing particulates is passed horizontally between a number of high-voltage discharge electrodes and electrically grounded collecting plates. Dust particles are charged by the ions from the discharge electrode and migrate to the grounded collection plate, where they adhere and agglomerate. As they do so, some become heavy enough to fall into the collection hoppers. Other dust particles are moved downward by mechanical rapping of the collection plates. The particles are then drawn from the hoppers and sent to a central particulate collection area. After passing the multiple system of electrodes, the clean gas exits the unit.

Electrostatic precipitators are capable of collection efficiencies greater than 99.5 percent, and in some cases remove more than 99.9 percent of the dust from the flue gases. Pressure drop through electrostatic precipitators is very low, resulting in low operating cost. Capital costs for large utility boilers burning medium- to high-sulfur coal are on the order of $25/kW.

Electrostatic precipitators are sensitive to changes in the properties of the particles and the flue gas. The changes can significantly affect their collection efficiencies; therefore, their design must compensate for normal variations in fuel composition and flue-gas temperature.

Electrostatic precipitators designed for high-sulfur fuels are adversely affected in their particulate collection efficiency by the use of low-sulfur fuels. For example, the Capitol Power Plant in Washington, D.C., was originally designed to burn 3 percent sulfur coal. Because of recent air pollution regulations the plant now burns 1 percent sulfur coal. This reduction in sulfur resulted in a change in the electrical properties of the particulates produced, which dropped electrostatic precipitator performance from 99 percent to about 76 percent.

High-sulfur coal has been the standard fuel for the power industry. To comply with $SO_x$ emission regulations the industry is using greater quantities of low-sulfur coal. The ash from low-sulfur coal is usually of high resistivity; with coal of a sulfur content of less than 1 percent, the naturally formed sulfur trioxide ($SO_3$) is seldom sufficient to reduce resistivity of the fly ash to a level (about $5 \times 10^{10}$ ohm-cm) that permits the electrostatic precipitator to function normally.

There are several methods of overcoming these difficulties. Existing facilities can add to the size of their precipitators (provided they have the space, which is not always the case) to compensate for the increased resistivity, or chemically change the resistivity of the particles by "flue gas conditioning," described below. New facilities have these and other options available. Since resistivity varies inversely with temperature, one option is to collect the particles in a hotter environment. Precipitators are normally placed on the cool, exhaust side of the air preheater (the heat exchanger that transfers heat from the flue gas to the air being drawn into the boiler to allow combustion), where average temperatures are 250° to 300° F. "Hot side" precipitators collect particulate matter on the intake side of the preheater, where the gases are at 500 to 700° F and the resistivity of the particles is in the same range as that of higher sulfur coals at "cold side" temperatures. For the Navajo Power Station in Arizona, use of hot-side precipitators with low-sulfur coal caused turnkey capital costs for meeting the particulate New Source Performance Standards (NSPS) to rise from $25/kW — the cost of a "cold side" precipitator used with medium- to high-sulfur coal — to about $45/kW; the cost of simply using a larger precipitator would have been about $50/kW.[27] Other options would have been to use control technologies such as wet scrubbers or baghouses that depend on the physical rather than the electrical properties of the fly ash.

Flue-gas conditioning may be an attractive option for existing plants because it minimizes the additional machinery that must be added to a facility. Flue-gas conditioning processes inject artificially produced $SO_3$ or some other substance into the flue gas ahead of the precipitator. The "conditioner" becomes attached to the particles in the flue gas, reduces the particles' resistivity and thereby restores collection efficiency without increasing the size of existing precipitators or the need for installing oversized new ones. One manufacturer claims that the incremental cost of using flue-gas conditioning to meet particulate emission limitations when using low-sulfur coal is approximately one-tenth that of alternatives.[28] Inexpensive (about 3 cents/ton of coal burned) $SO_3$ injection systems have been installed in at least 60 powerplants and are reported to be working reliably.[29]

Designers of new powerplants generally have not used flue-gas conditioning as a solution to resistivity problems; they have used hot-side precipitators or larger cold-side precipitators. Conditioners are apparently considered as "last resort" solutions if the precipitators fail to meet emission requirements. The Environmental Protection Agency (EPA) has tested the effects of various conditioning agents and has found some evidence that conditioning can add some pollutants to the flue gas while improving the particulate removal efficiency; for example, in some instances a significant percentage of $SO_3$ injected into the gas was emitted. However, the effect does not occur in every system and perhaps may be eliminated by better design.

### Wet Scrubbers

Wet scrubbers utilize water or other liquids to "rain out" the particles in the flue gas. Many variations of the wet scrubber exist, each with specific advantages. Although successful installations on large powerplants exist, the economics are not favorable except for low-sulfur coals, and most utilities appear to prefer the more traditional precipitators. Turnkey costs are in the range of $50/kW.[30] Operating energy costs become quite high if fine particulates removal is required.

### Fabric Filter Baghouses

Baghouses contain fabric bags suspended vertically. The particulate-laden gas passes through the fabric and the dust is filtered out. Removal efficiency is usually in excess of 99.9 percent. Removal efficiency of fine particulates can be greater than with other devices, although the small size needed to achieve high efficiency of these particulates increases pressure drop and operating costs. Although baghouses have a long history of use outside the utility industry, their general use on boilers has been slowed by problems of fabric clogging and chemical damage, poor high-temperature operation, and high maintenance costs, in addition to their expensive high-pressure drop. Because bag replacement represents a major maintenance cost, the recent development of high-temperature bags made of teflon-coated fiberglass has made baghouses more attractive to utilities. More than 50 baghouse systems have now been installed or ordered at coal-fired powerplants. The next few years of experience with several large new facilities are likely to be critical to their future.

## Sulfur Oxide Control

### Flue-Gas Desulfurization—Utilities

The removal of $SO_x$ from stack gases is termed flue-gas desulfurization (FGD) and uses devices commonly referred to as scrubbers. The function of the scrubber is to bring the $SO_x$ laden flue gases into contact with a liquid that

---

[27]Personal communication with Les Sparks, U.S. Environmental Protection Agency, Research Triangle Park, N.C.

[28]*Flue Gas Conditioning for Electrostatic Precipitators* (Santa Anna, Calif.: WAHLCO, Inc.).

[29]William E. Archer, "Fly Ash Conditioning Update," *Power Engineering*, vol. 81, No. 6, June 1977, pp. 76-78.

[30]Personal communication with Les Sparks, U.S. Environmental Protection Agency.

will selectively react with the $SO_x$. FGD processes are generally characterized by the chemical absorbent that is utilized, such as lime, limestone, magnesium oxide, double alkali, sodium carbonate, alkali flyash, and ammonia. The processes are further characterized as throwaway or regenerative. In the throwaway processes, the absorbent and the $SO_x$ react to form a product of little or no market value; it is disposed of as a sludge or solid. By contrast, the regenerative processes recover the absorbent in a separate unit for reuse in the scrubber and generally produce a product with market value (such as elemental sulfur or sulfuric acid). However, if the product has little or no market value, it must be discarded as in the throwaway processes.

FGD is used mainly by the utility industry. Although there are a number of non-utility FGD installations, their percentage of the total power generating capacity is not significant. In the U.S. utility industry, the technology is predominantly used on coal-fired boilers, whereas its use on oil-fired units is prevalent in Japan.

Lime-limestone scrubbing, a throwaway process, is presently the dominant technology for flue gas desulfurization in the United States. A basic flowsheet for lime-limestone scrubbing is shown in figure 13. The absorbent, lime or limestone, is introduced into the reaction tank along with scrubber effluent slurry and clear liquor recycle from the dewatering section. The mixture, after holding in the tank for a few minutes to allow reactions to take place, is pumped to the top of the scrubber and sprayed down into the gas. A bleed slurry stream is tapped off from the scrubber effluent and dewatered to give the product sludge.

Some 11,500 MW of coal-fired utility boiler capacity have been fitted with FGD, as compared to roughly 200,000 MW that possibly could be so equipped. Another 17,700 MW are under construction and 27,200 MW are planned, giving a total of 56,400 MW scheduled to be in operation by 1985, by which time there will be about 300,000 MW of coal-fired capacity that could use FGD (assuming existing units are retrofitted). Thus about 15 percent of the coal-fired capacity now operating or under construction will have FGD.

The 1977 Clean Air Act Amendments passed by Congress have considerably changed the situation for new plants; a percentage reduction in $SO_2$ emission will be required that effectively rules out use of low-sulfur coal as a sole means of compliance for new boilers. In view of this, it seems likely that all new boilers for which construction is started after early 1978 will require FGD. The new capacity to be installed in the 1980-2000 period is expected to be on the order of 200,000 MW (although there is considerable uncertainty about this projection).

## Input Requirements

FGD systems require considerable amounts of raw materials, energy, water, and manpower. If the effect of the 1977 amendments is to greatly expand FGD use, then these requirements could have an impact on other sectors of the economy.

1. *Raw Materials.* Although projections are difficult, there is some indication that limestone consumption (either as is or calcined to produce lime) could approach 100 million tons/yr by 2000. The limestone industry should have no trouble in supplying the required amount of material, assuming (as available evidence indicates) that the lower grades of limestone are acceptable.

2. *Water.* FGD systems require from 0.5 to 1.5 tons of water/ton of coal burned. By the year 2000, this could amount to as much as 1 billion tons, or slightly more than 700,000 acre-feet/yr. In the west, FGD water use must be considered a significant addition to total water demand.

3. *Energy.* FGD systems require energy both as electrical power to drive equipment and as thermal energy to (when necessary) reheat the flue gases that have been cooled in the scrubber and have lost some of their buoyancy. The former ranges from 1.0 to 2.5 percent of the boiler capacity (depending on the FGD process). Reheat energy requirement is also variable, depending on the degree of reheat attempted. A recent EPA study indicates that total FGD energy requirements would

Figure 13.—Lime-Limestone Scrubbing System

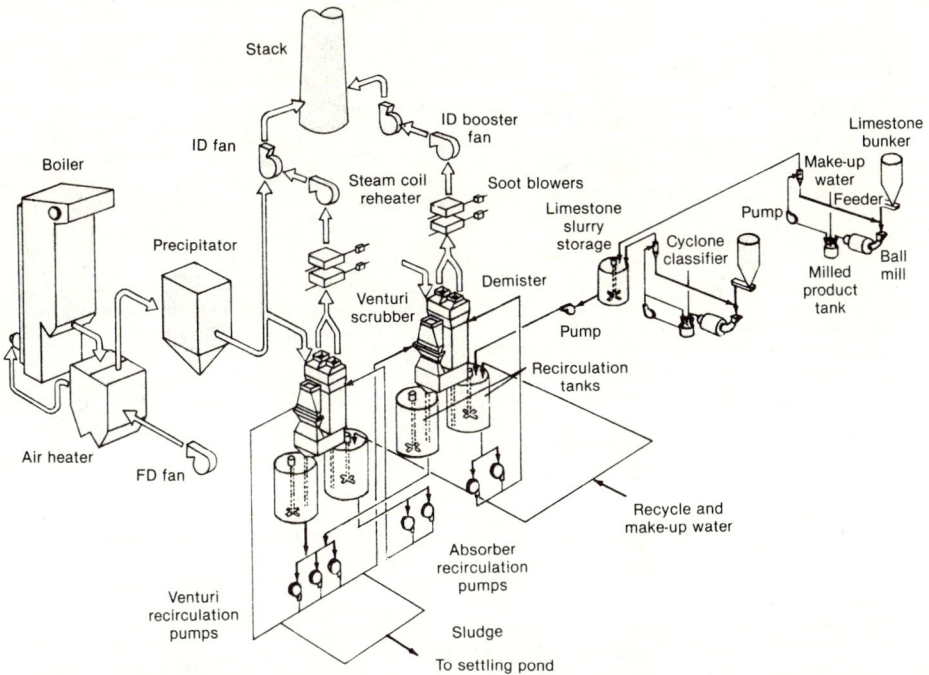

SOURCE: OTA, based on EPA data.

increase U.S. power consumption over a noncontrol situation by 0.7 percent in 1987 and 1.0 percent in 1997.

4. *Manpower.* The operating manpower required in the year 2000 may be on the order of 10,000 workers, which does not seem significant. The main problem is inability of vendors and manufacturers to keep pace with scrubber demand. The consensus is that designers and fabricators probably will not produce any bottlenecks but that without special measures there may not be enough skilled workers (welders, electricians, boilermakers, and others) to keep construction projects on schedule. This applies particularly to industrialized areas that already have a shortage of skilled labor.

Cost

Although other impacts may be minor or uncertain, it is clear that the capital and operating costs of FGD systems will have a substantial impact. The systems are expensive to build and operate, and overall cost will be a significant percentage of power production costs.

Many site-specific factors affect FGD cost, and "average" costs must not be assumed to apply to a particular plant. For new systems, the capital cost will "typically" be in the $80 to $120/kW range in 1975 dollars, with an annualized operating cost (capital charges plus operating and maintenance) of 4 to 7.5 mills/kWh. In extreme cases the cost may be 10 mills/kWh or greater. Assuming an average of $100/kW for capital cost, the 200,000 MW projected to come online between 1985 and 2000 would require a capital outlay of $20 billion (based on "current dollars"). If a cost of 5.5 mills/kWh is assumed, boiler system operating costs would be increased by about $6 billion/yr. These estimates, which are rough at best, indicate that a powerplant's installation of FGD may add 14 percent to investment cost and 18 percent to annualized operating cost if the alternative is no attempt at control of $SO_2$ emissions. Site-specific conditions could vary these costs over a wide range. Back-fitting cost would be even higher.

## Alternative FGD Processes

Although conventional lime-limestone scrubbing is now preferred, better FGD processes might be developed and could have some impact on inputs and costs. The alternative nearest to commercial adoption appears to be double-alkali scrubbing, which gains a major advantage by reducing the scaling experienced with limestone.

One other recovery process, sodium scrubbing-thermal stripping (Wellman-Lord), has been tested on a large scale. Current installations use natural gas as a reducing agent, but the industry is testing coal as an alternative agent. Disposal of sodium sulfate will be a problem. Several other recovery methods have promise but have not been developed far enough to have much significance at the present time.

## Flue-Gas Desulfurization—Industrial

Flue-gas desulfurization was first applied to industrial boilers in the United States at the General Motors plant in St. Louis, Mo. Two systems were installed on coal-fired boilers in 1972. At the close of 1977, 35 systems had been or were being constructed at 15 industrial sites. Twenty-four systems were operational, with the remaining scheduled to begin operation within a year. Of the 24, 18 were retrofitted to existing boilers and the remaining 6 were installed on new boilers.

Thirteen of the operational systems use sodium hydroxide or sodium carbonate to absorb $SO_2$. One uses lime-limestone. The other 10

operational units are double-alkali systems. Eleven additional double-alkali units are scheduled to start within the year.

The sodium-based units produce a liquid waste containing dissolved sodium compounds, which are either sent to a pond for evaporation, or treated and disposed into a municipal sewer. The double-alkali and the lime-limestone systems produce a dewatered calcium sulfite/sulfate material claimed to be suitable for landfill; however, leaching may still be a problem.

Operating availability of the industrial systems is reported to be greater than for FGD systems on utility boilers. However, the problems that occur are similar to those associated with prototype FGD systems on utility boilers. They include corrosion, erosion, poorly sized equipment due to lack of operating experience, and poor pH and gas flow control.

Capital costs range from $25 to $115/kW. Capital costs are reported as lower than for utility installations because capital costs for pretreatment and disposal are not generally included in industrial systems. Although none has been reported to date, operating costs should be higher.

## Nitrogen Oxide Control

Nitrogen oxides ($NO_x$) are formed during the combustion of fuels. The nitrogen originates from the nitrogen content of the fuel as well as the nitrogen in the combustion air. Factors in the formation of $NO_x$ are flame temperature, amount of excess air in the flame, and the length of time the combustion gases are maintained at the elevated temperature and subsequently quenched. An increase in flame temperature and excess air favors an increase in the formation of $NO_x$. A rapid cooling of $NO_x$ by relatively cool boiler tubes tends to stabilize the $NO_x$ that was formed. Typical $NO_x$ concentration in the flue gas is 500 to 1,000 parts per million (ppm) for coal, 200 to 500 ppm for oil, and 120 to 200 ppm for natural gas.

Currently, the common practice for lowering $NO_x$ emissions is to modify the design and/or operating conditions of combustion equipment.

Some new furnaces have been designed for two-stage combustion. In the initial combustion stage, combustion is partially completed with less air than is needed for complete burning. In the second stage downstream the remaining fuel is burned completely, with secondary air injected through strategically located ports. The technique has been successfully tried on a few coal-fired boilers as well as applied to oil- and gas-fired units.

The location and spacing of burners influence $NO_x$ formation. Tighter spacing has a tendency toward greater $NO_x$ formation, probably due to higher temperatures. Burners located in the corners of furnaces produce tangential flames with a somewhat lower flame temperature. Tangential corner-fired boilers may produce about half the $NO_x$ that is generated in either a front wall-fired or an opposing-fired furnace.

Modification of operating conditions has also proven effective in reducing $NO_x$ formation. The common techniques consist of lowering excess combustion air, recirculating the flue gas, and injecting steam or water into the firebox. Reducing the excess air reduces the quantity of atmospheric nitrogen available for $NO_x$ formation. Flue-gas recirculation and steam or water injection reduce flame temperature, which is an important factor in decreasing $NO_x$ production. Combinations of these strategies have succeeded in lowering $NO_x$ emissions from large utility boilers by 40 to 50 percent.

A promising technology for reducing $NO_x$ emissions is a low-$NO_x$ burner currently under development by EPA. Single-burner tests using low-sulfur coal have yielded an 85-percent reduction in $NO_x$ emissions. There appears to be a good chance that similar reductions can be achieved with a multiple-burner system (all large boilers have multiple burners). However, full demonstration probably cannot be achieved before the mid-1980's.

This technology is of particular interest because, if proved successful, it will offer an efficient control of $NO_x$ for low cost on new plants (the new burners may not be more costly than current designs), and allow a moderate cost retrofit to existing plants if this is considered desirable.

Desulfurization of fuels may provide a concomitant, although small, reduction in the nitrogen content of the fuel. The lower nitrogen content of the fuel can also reduce $NO_x$ formation during combustion.

Research and development (R&D) on the control of $NO_x$ emission by stack-gas cleaning is currently being conducted. Unfortunately, none of the processes is commercially available for coal-fired units.

A number of dry processes are being developed. In one, ammonia is injected at 1,400° to 1,800° F, which causes the conversion of 40 to 60 percent of the $NO_x$ to nitrogen and water. In other processes various solids are utilized to effect NO removal. The more prominent solids consist of copper oxide sorbent, base metal catalysts, and activated carbon. In Japan the capital costs of the dry processes range from $60 to $100/kW and have a removal efficiency between 40 and 70 percent. To date these processes have been developed for oil-fired units, and problems caused by catalyst poisoning could be expected in their application to coal-fired plants.

A number of wet processes for the simultaneous removal of $SO_x$ and $NO_x$ are being developed. One process utilizes magnesium oxide as an absorbent. Other processes rely on oxidants such as chlorine dioxide, ozone, and potassium permanganate to oxidize NO to $NO_2$ for easier absorption. Capital costs are said to be in the range of $60 to $90/kW in Japan.

# FUTURE TECHNOLOGIES

With coal the dominant fossil fuel for electrical generation, a great many R&D efforts are being aimed at future ways of burning solid fuels more effectively, and of substituting coal for fuel oil and natural gas in applications where these clean fluid fuels have been widely used. Improvements in the "quality" of mined coal are being made, new combustion schemes are being evaluated, direct energy-conversion systems based on coal are being investigated, and serious attention is being given to combined cycles based on coal as the energy source. These new technologies have much to offer, but some of the problems will elude the best efforts of fuel technologists for many years. Nevertheless, if coal is to become a more attractive fuel under stringent environmental regulations, new technologies are required, especially for the smaller industrial users.

## Coal Preparation

Upgrading mined coal to improve coal's usefulness or to minimize environmental problems is an old art. At one time, coal beneficiation began at the face in the coal mine, where the miner had some control over what he loaded into his mine car. But today's high-speed mining machines cannot distinguish between coal and the rock partings within coal seams. All goes to the surface. Handpicking rock from coal moving on a belt or a shaking table constituted coal preparation at one time. Today, coal cleaning methods based on gravity separation of the lighter coal particles from the heavier rocks and pyrites is becoming more important as the cost of coal increases in the marketplace and environmental restrictions are imposed on emissions from coal-burning installations. Ordinary coal cleaning to remove ash and rocks, etc., wherein sulfur removal is incidental, is described in the section on *Mining and Preparation.* Advanced washing, which can in some cases remove a substantial amount of sulfur, is considered here. More sophisticated chemical beneficiation methods also are being developed. Further information

on coal cleaning and desulfurization is provided in appendix IV of volume II.

### Mechanical Coal Cleaning

Mechanical cleaning processes are based on differences in specific gravity or surface characteristics of the materials being separated. They can be designed to remove a large fraction of the pyritic sulfur, generally the major part of the sulfur in high-sulfur coals. Pyritic sulfur occurs as discrete particles. It is much heavier than coal, with a specific gravity of 5.0, compared to coal's 1.4. Hence when raw coal is immersed in a dense medium, the coal floats and the pyrites sink. This process is now used to remove shale and rocks, etc. (specific gravities from 2 to 5) but pyrite is more dispersed and finer crushing of the coal than is now generally practiced is required to free it for removal.

Two facilities for fine coal sulfur reduction are about to begin commercial operation. The Homer City Coal Preparation Plant in Pennsylvania uses dense-medium cyclones to produce a primary coal stream of less than 0.6 percent sulfur and an intermediate sulfur coal stream to be used in areas outside of critical air basins. The other plant, near Barnesboro, Pa., uses a two-stage froth flotation process to reduce pyritic sulfur by up to 90 percent. Both plants are described in more detail in appendix IV of volume II. Neither of these processes, nor some other mechanical and chemical cleaning processes intended for pyritic sulfur, affect the organic sulfur. Hence, they will not be usable without other control measures by utilities in new plants. They may be of considerable interest in smaller or existing facilities that have less stringent sulfur removal requirements. The applicability of these processes will be determined by their cost and the environmental regulations. To date, no reliable cost figures are available.

### Advanced Cleaning Processes

Several physical and/or chemical treatments

have been proposed for improved pyritic sulfur removal. These are:

1. High-gradient magnetic separation (HGMS)—separation of pyrite by exploiting its magnetic properties.
2. Magnex process—a "pretreatment" process allowing better magnetic separation.
3. Meyers process—a chemical leaching of pyrite from the coal.
4. Otisca process—washing with a heavy liquid rather than a water suspension.
5. Chemical comminution—a "pretreatment" process that chemically breaks down the coal to smaller sizes.

These processes will be quite expensive. Their future role is unclear because they will not remove much more sulfur than the advanced mechanical cleaning methods. Appendix IV of volume II evaluates them in greater detail.

Processes that attack organic sulfur in addition to pyritic sulfur include:

1. Ledgemont oxygen leaching process—dissolution of pyrites and some organic sulfur using a process simulating the production of acid mine water.
2. Bureau of Mines/DOE oxidative desulfurization process—a higher temperature, air instead of oxygen variation of the ledgemont process.
3. Battelle hydrothermal process—leaching of pyrites and organic sulfur under high pressure.
4. KVB process—gaseous reaction of the sulfur with nitric oxide.

These processes are claimed to remove substantially all of the pyritic and 25 to 70 percent of the organic sulfur. All are still in the laboratory stage, so cost data are only conjectural. At present, costs seem to be many times those associated with physical coal cleaning.

## Solvent-Refined Coal

Solvent-refined coal (SRC) involves dissolving crushed coal in a suitable solvent at moderately high temperature and pressure, treating the solution with hydrogen to remove an ap-

preciable part of the sulfur, filtering the hot solution to remove the insoluble coal-ash minerals, and then driving off the solvent and recovering the demineralized, low-sulfur product. Two pilot plants have been running since 1973, a 6-ton-per-day unit at Wilsonville, Ala., and a 50-ton-per-day plant at Tacoma, Wash. Together they have investigated the response of a wide variety of coals to SRC processing, and have produced sufficient product for burning tests. Commercial-scale burning tests were made at Georgia Power. General conclusions drawn from these combustion tests indicate that SRC is a premium fuel in regard to heat value, ash, and sulfur content. Commercial-size SRC plants have been proposed.

Sustained operation for periods as long as 75 days has been achieved at Wilsonville, with some 3,700 hours of operating time in its first year.[31] In an early stage, Kentucky and an Illinois coal were treated with 90 to 95 percent recovery. The raw coals contained 8.9 to 11.1 percent ash and 3.1 percent sulfur. The SRC product averaged 0.16 percent ash and 0.96 percent sulfur. The designs for a demonstration of a commercial-size module have been proposed by the private sector in conjunction with DOE.

## Low-Btu Gas and Combined Cycles

Coal can be gasified to produce a low-Btu gas. Since the gas cannot economically be stored or shipped more than a few miles before combustion, it is effectively a form of direct combustion of coal. The gas is cleaned before burning so that no emission controls are required at the combustion facility. Low-Btu gas can be burned directly in a boiler to produce steam for industrial use or for the production of electricity in a conventional steam turbine. Alternatively the gas generator can be integrated with a combined cycle plant.

The first option of direct firing of existing boilers with clean, low-, or intermediate-Btu gas is higher in overall capital and operating

[31]"Status Report of Wilsonville Solvent Refined Coal Pilot Plant," *EPRI Interim Report 1234,* May 1975, 40 pp. plus app.

cost than conventional coal firing with stack gas cleanup, as indicated in recent studies by the Tennessee Valley Authority (TVA) for the Electric Power Research Institute (EPRI). In addition, the option will probably be uneconomical for many industrial plants. Nevertheless, gas producers may be practical for the production of chemicals or other industrial uses.

The second option is more attractive for power generation. With this system crushed coal, water, and compressed air are fed to a gasifier. The hot gases that are produced pass through a purification system to remove chemical contaminants and particulate matter. The clean gases power a gas turbine engine that produces electricity and compressed air for the gasifier. The turbine exhaust gases travel through a waste heat boiler driving a steam electric generator before being exhausted to the atmosphere.

Although the state of the art of turbine design will allow efficient combined cycle operation, gas temperature to the turbine must be as high as possible to obtain further efficiency gains. Turbine research at up to 3,000° F gas temperature is being conducted, but the temperature of the commercially available metal blades is limited by metallurgical problems to 1,000° F. Blade temperature is kept low by elaborate schemes to circulate cooling water or air through the rotating blades. Research is underway on blades that will withstand higher temperatures and on improved cooling means. Prospects for improved efficiency are good.

The low-Btu gasifier integrated with a combined-cycle turbine system has been identified as promising from the standpoint of cost of electricity, overall efficiency, and limitation of emissions. New combined cycle plants fueled by low-Btu gas may be 10 percent more efficient in converting fossil fuel to electricity than are the most efficient conventional powerplants.

## Fluidized-Bed Combustion

Fluidized-bed combustion (FBC) is an important technological alternative for industrial applications and perhaps coal-based power gen-

eration. Its basic principle involves the feeding of crushed coal for combustion into a bed of inert ash mixed with limestone or dolomite. The bed is fluidized (held in suspension) by injection of air through the bottom of the bed at a controlled rate great enough to cause the bed to be agitated much like a boiling fluid. The coal burns within the bed, and the $SO_x$ formed during combustion react with the lime-

Photo credit: Department of Energy

This small-scale model of the coal-fired, fluidized bed, gas turbine system is used to study the behavior of the fuel mixture of coal and limestone in the production of electrical power

stone or dolomite to form a dry calcium sulfate. To maintain high combustion efficiency, high heat transfer rates, and efficient capture of $SO_x$, the bed temperatures are maintained between 1,500° and 1,800° F. Production of $NO_x$ is minimized because of the low excess air environment of the combustion zone and the low generating temperature of the fluidized bed compared to conventional combustion, where the temperature may be in excess of 3,000° F. In addition, in some combustion processes burning certain coals, the ash melts during combustion to form molten slag and clinkers, which foul the boiler tubes and the furnace walls. At the temperature of the fluidized bed more of the ash remains solid, tending toward unimpeded, uniform operation. Banks of boiler tubes are in contact with the fluidized bed. The rate of combustion and production of heat in the bed is rapidly transferred to and through the tubes to produce steam while the bed is maintained at a constant temperature. As bed temperatures must remain in a narrow range during operation, low load is difficult. Consequently, the first commercial units are expected to consist of a series of relatively small cells, only some of which would operate during periods of low load.

The advantages visualized for FBC include:

- the flexibility to burn a wide range of rank and quality of coals,
- a higher heat transfer rate than in conventional boilers, which reduces the requirements for boiler tube surface and furnace size and also lowers capital costs,
- an increased energy conversion efficiency through the ability to operate without the power requirements needed for a scrubber,
- reduced emissions of $SO_x$ and $NO_x$,
- a solid waste more readily amenable and acceptable to disposal than that from a wet-scrubber applied to conventional boilers, and
- the potential for operation at an elevated pressure sufficient to use with a combined gas-turbine/steam-turbine cycle for generating electricity at higher efficiency.

FBC may make particular sense for small commercial and industrial facilities if emission standards for smaller sources are relatively stringent.

The fluidized bed can be adapted to a variety of modes to produce heat and power. Two variations are prominent—atmospheric and pressurized operation. The atmospheric FBC can be used for generating electricity or for process or space heating. The pressurized FBC is slated for use with a combined cycle system of gas and steam turbines.

A major concern with atmospheric FBC relates to adequate removal of fine particulate matter prior to releasing the flue gas to the atmosphere. The agitation of the ash and limestone in the bed results in their breakdown into fine particles that readily entrain in the high-velocity flue gases. Attention is focused on the effectiveness of cyclones, electrostatic precipitation, and fabric filters for particulate removal. The ash entrained in the flue gases may be high in unburned carbon. The disposal of such material would result in lower combustion efficiency and perhaps disposal problems. To prevent this loss of potential fuel, atmospheric FBC units may employ a carbon burnup cell. By passing the entrained particles through this cell, which operates at a higher temperature and with greater excess air than the primary FBC cells, the residual carbon is consumed. Considerable concern regarding corrosion and erosion of the submerged tubes also exists.

R&D work is being conducted with atmospheric FBC in the United States and abroad. In the United States a 30-MW atmospheric FBC pilot plant began operations in 1976 at a utility site in West Virginia. Atmospheric FBC is also about to be demonstrated for the production of process heat and steam in a number of industrial configurations.

In the pressurized variation of FBC, the combustion occurs in conditions similar to those in the atmospheric version except that the furnace is maintained at 4 to 16 atmospheres of pressure. The elevated pressure compresses the flue gases, resulting in a dramatic reduction of furnace size from that of a comparably rated atmospheric FBC. Although the development of pressurized FBC is less advanced than that of atmospheric, it has a number of potential advantages: the efficiency of combustion and the capture of $SO_x$ are higher; less absorbent (limestone or dolomite) is required for the same sulfur capture; the furnace is smaller for the same coal throughput, and $NO_x$ emissions are expected to be lower. Its primary advantage, however, is in the potential for improved plant efficiency by means of combined-cycle operation. With no major advances in gas turbine technology, an increase in efficiency of 5 percent is possible for power generation.

With the integration of the pressurized FBC with combined-cycle operation, the problem of particulate removal becomes critical, as the hot pressurized flue gases are expanded through a gas turbine to recover a portion of their energy. The solids suspended in the flue gases are particularly corrosive and erosive of gas turbine blades. More efficient particulate removal systems capable of operating at high temperatures need to be demonstrated to overcome this obstacle to higher efficiency by the pressurized FBC system.

Development of combined-cycle FBC systems is underway. A 13-MW combined-cycle pilot plant is being built to begin operation in 1980 at Woodridge, N.J. The International Energy Agency (IEA) is constructing a test facility at Grimethorpe with joint participation of the United Kingdom, the Federal Republic of Germany, and the United States. Initial operations are expected in early 1979. The program is expected to span 7 years including construction, and commercial utility application is not anticipated until about 1995.

Although the fluidized-bed principle has many attractive advantages, it is still an immature technology that must demonstrate a competitive position with the pulverized coal-fired boiler. The fluidized bed may require additional process steps and have new problems not common with today's conventional utility boilers.

## Magnetohydrodynamics (MHD)

The interest in MHD stems mainly from high expected thermal efficiency for an entire system including a conventional steam cycle. In MHD generators, a stream of very hot gas (roughly 5,000° F), flows through a magnetic field at high velocity. Because the gas at high temperatures is an electrical conductor, an electrical current is produced through electrodes mounted in the sides of the gas duct. A natural gas-fired MHD generator has been tested in the U.S.S.R., and coal-fired systems are still under development.

Coal-fired MHD poses special problems because of the ash in coal. At the high tempera-tures in the combustor, most of the constituents in ash are vaporized. As the gas cools in passing through the system, the ash condenses. The electrodes and the walls of the MHD channel must be cooled, thus coal-ash slag can accumulate on those surfaces, even at the near-sonic velocity of the gas stream. This slag layer affects the way electrical current flows through the gas stream from one electrode to the other, hence the characteristics of the coal ash are important.[32]

It is unlikely that MHD will be widely used in this century, for the remaining technological problems are formidable. Progress is being made, however, and MHD could become an important adjunct to new coal-fired power stations.

## Fuel Cells

In principle, a fuel cell consists of two electrodes immersed in a conducting electrolyte. One electrode, the anode, is flooded with hydrogen, which reacts with ions in the electrolyte, typically a solution of potassium hydroxide, to release electrons. These electrons flow through an external circuit—the electrical load—to the cathode, flooded with oxygen, where the electrons react with the electrolyte to form the ions that can react with hydrogen. Hence the fuel cell is simply a battery, consuming chemical reactants and producing electricity. Unlike ordinary primary batteries, however, fuel cells continue to produce electrical energy as long as hydrogen and oxygen are supplied to their electrodes.

The voltage of a single fuel cell is low, typically 0.8 V under load, but the power output can be high, as much as 100 W/ft$^2$ of electrode area. The thermal efficiency of hydrogen and oxygen fuel cells is excellent, well over 90 percent at room temperature and at low load. With some fuels, the thermal efficiency is greater than 100 percent, showing that a fuel cell can produce additional electrical energy from the heat absorbed from its surroundings. Efficiency of power production remains high

[32]"In-Channel Observations on Coal Slag," EPRI Contract 468-1 (Stanford University, 1976).

for widely varying loads, which is a very desirable characteristic.

Although hydrogen is ideal for fuel cells, carbon-based fuels, such as distillate and natural gas, as well as producer gas from coal, or other fuel gases containing carbon and hydrogen, are more readily attainable. In such systems, an alkaline electrolyte, as in hydrogen fuel cells, is not suitable, and a $CO_2$-rejecting electrolyte is needed. Phosphoric acid has been used as an electrolyte in fuel cells at about 400° F, but a mixture of molten carbonate salts, at temperatures up to about 1,400° F, allows the direct electrochemical oxidation even of natural gas.

Fuel cells can serve in a reverse fashion as an electrolyzer to produce hydrogen from off-peak power, to store the hydrogen until a peak demand develops, and then to convert that hydrogen back into electricity. A shortcoming is that fuel cells produce DC, not AC, requiring an inverter to match 60-cycle powerlines. That technology exists today.

Another advantage of fuel cells is that, above some moderate size, unit costs remain about the same as the size of the installation changes. This means that fuel cell "substations" serving a small group of consumers, say 50 to 100 residences, or a typical industrial plant, could substitute for central-station powerplants and their extensive electrical distribution networks. The fuel cell substations would get their energy as hydrogen, pipelined from a coal gasification plant located in a remote area. The fuel-cell electrical-generating plant would emit no pollutants, probably could be operated with no direct supervision, and might also serve its nearby customers with hot water from its waste-heat recovery system. A potential problem is the necessity to supply extremely clean fuel. The technology does not now exist to economically clean low-Btu gas from coal to sufficient purity. Hydrogen would be easier to use. The problems now are mainly of an engineering nature, utilizing available data on fuel cells to design, construct, and operate a fuel-cell installation able to hold its own economically and reliably with other energy-conversion systems based on coal.

Chapter IV
# FACTORS AFFECTING
# COAL PRODUCTION AND USE

# Chapter IV.—FACTORS AFFECTING COAL PRODUCTION AND USE

Chapter IV
# FACTORS AFFECTING
# COAL PRODUCTION AND USE

Chapter II shows a range of projections for coal use. Actual growth will depend on the decisions of users and producers. The major factors affecting these decisions are the cost, convenience, and availability of coal relative to competing fuels.

Unlike oil and gas, coal will not be subject to absolute resource constraints or the resulting scarcity-induced price increases over the next few decades. But the availability of coal depends on much more than its presence in the ground. The legal right to mine a specific reserve must be established through ownership or leasing arrangements. A mining company must decide whether it can sell its product profitably for the life of the mine. A number of studies and extensive planning must be carried out, and a host of permits secured to comply with various regulatory processes. Arrangements must be made for adequate capital, equipment, and labor. Once mining starts, recent laws and regulations (e.g., the Surface Mine Control and Reclamation Act (SMCRA) and the Coal Mine Health and Safety Act (CMHSA)) affect methods and costs. Labor-management conflicts can limit coal availability and color customers' perceptions of future reliability. Transportation of coal is a problem in the East and will demand attention in the West.

On the demand side, the Clean Air Act makes coal combustion more complicated and costly and presents new problems of waste disposal. Smaller users may have difficulty in physically accommodating the necessary equipment or obtaining regular coal supplies. Converting current oil- and gas-fired equipment to coal will be especially difficult. Public opposition to particular sites for surface mines or coal-burning facilities may cause substantial delays or force expensive plant modifications. Other regulatory requirements do not impose constraints on local combustion as severe as those of the Clean Air Act, but their cumulative impact will also make combustion more complicated and costly.

The factors listed above increase either the cost or the difficulty of producing and using coal. (The environmental, health, and social benefits from these restrictions are discussed in chapters V and VI.) But powerful incentives are at work, on the other hand, for users to turn to coal. Given favorable market conditions, many of the potential constraints on coal may never materialize. Others, such as Government regulation, are not directly subject to market pressures and may slow the growth in coal demand, especially in very high-growth scenarios.

This chapter analyzes the factors that affect the components of the coal system outlined in chapter III. Analysis provides a framework for policy discussion of the problems of increasing coal output and consumption and for determining the effect on coal development of measures designed to ameliorate its negative impacts.

## COAL AVAILABILITY

Coal is plentiful enough on a national basis to meet even rapidly growing demand. As described in chapter II, a massive increase of Western coal production is underway because of its low production cost and low-sulfur characteristic. Unlike Eastern reserves, however, which are almost all privately owned and more accessible, about 65 percent of all Western reserves are owned by the Federal Government and can be mined only under Federal lease. Much of the remaining Western coal is owned by States or Indian tribes. As more than half of all domestic coal reserves are found in the West, Federal leasing policy is an important factor in determining long-run coal availability. Federal leasing law is analyzed in chapter

VII. This section discusses the effect of current leasing policy on Western coal production.

The Interior Department's Bureau of Land Management (BLM)—the agency responsible for administering Federal leases—reports that about 5 percent[1] of federally owned coal resources has been competitively leased. This now represents 17.3 billion tons of known coal reserves.[2] About 9 billion tons is subject to existing applications for preference-right (noncompetitive) leases, obtained when a prospector demonstrates that commercial quantities of coal have been found in an area previously not known to have any.

Since 1971 a Federal leasing moratorium has been imposed and the amount of leased reserves has remained relatively constant (only 30,460 additional acres leased). Short-term (3-year) leasing criteria were developed to allow current coal producers to obtain mining rights on adjacent Federal lands, but a successful suit by the National Resources Defense Council (NRDC) resulted in the criteria for short-term leases being limited to those required to maintain an existing operation or to meet existing contracts. The moratorium was imposed because most leaseholders were not mining their leases. This gave rise to the charge that Federal coal reserves were being held primarily for speculative purposes. In the mid-1970's, coal production from leased reserves stepped up considerably as rising coal prices and increasing demand made Western coal economical and attractive. Western coal production has more than tripled since 1971. About 166 million tons were mined there in 1977. One recent study reports that total output from the 67 active Federal coal leases in 1977 was 52.4 million tons, a 241-percent increase since 1973.[3] Those leases represented about 14 percent of the total of all Federal leases. Federal coal lands under lease contributed about 31 percent of Western coal production in 1977. Coal production on Indian lands doubled between 1973 and 1977 to almost 23 million tons.

Because coal leases are often not contiguous, some operators have found it difficult to assemble the 20 to 30 years' worth of reserves needed for a mine large enough to be economical. The moratorium and the NRDC suit have created uncertainty among operators, who may defer opening a new mine until they can lease Federal coal adjacent to their current holdings. Some operating mines must also know soon whether they will be allowed to move into adjacent areas or should plan to close down when present leases are exhausted. Regardless of what the Carter administration chooses to do about leasing, prolonged uncertainty is a constraint on rapid Western coal expansion. A continuation of the moratorium past 1980 will affect coal development if demand approaches current forecasts. In any case, if leasing were to be resumed, regional environmental impact statements would have to be prepared before mining could begin, a process that may take more than a year. Additional delay in the form of court challenges to renewed leasing can also be expected. Despite these constraints, sufficient Western coal has been leased to meet anticipated increases in demand through the 1980's. However, currently leased reserves would not be adequate to support expected coal production levels in the 1990's. The long leadtime required to put a mine into operation requires a resumption of leasing in the early 1980's to meet these levels.

There is little chance that a significant number of new leases will be offered before the 1980's, according to the best available information.[4] The terms of any future leasing are unresolved. If leasing is reinstituted, the new criteria may limit the amount of coal available. If the administration chooses to initiate a leasing policy soon, this coal would probably be commercially available after 1985.

---

[1]An Analysis of Existing Federal Coal Leases (Washington, D.C.: U.S. Department of the Interior, Bureau of Land Management, 1976), p. 6.

[2]Reserve Data System Report No. 2A (U.S. Department of the Interior, Geological Survey, Dec. 31, 1976).

[3]James Cannon, Mine Control: Western Coal Leasing and Development (New York, N.Y.: Council on Economic Priorities, 1978), p. 7.

[4]"Completion of Interior's Coal Leasing Program Targeted for Mid-1980," U.S. Department of the Interior News Release, Oct. 25, 1977.

# INDUSTRY PROFILE

Coal is mined by companies ranging from major corporations to one-family operations. The future of the coal industry depends in large part on how these companies react to changing conditions. This section describes these companies and how they market their coal. The competitive situation in the industry is examined. Finally, the mechanisms for setting prices for coal are analyzed.

## Ownership and Markets

The coal industry has evolved from a large number of small- and medium-sized independent companies to a small number of very large companies (most of whom are subsidiaries) and a large number of small independents. Most major, noncaptive producers are now owned by energy companies or conglomerates. Of the top 15 coal producers in 1977, as shown in table 13, only two were independent. Five were captives of steel companies or utilities, three were subsidiaries of conglomerates, and five were owned by integrated energy com-

panies. Sixteen years ago, all major coal companies were independent except for those few owned by industrials.

The terms "commercial" and "captive" arose in an earlier time when companies could be differentiated by their markets and whether their product competed freely. "Commercial" is a term that no longer means quite what it did in the 1930's and 1940's. Then, a commercial operator sold to a variety of customers and often in several markets. As the utility market ascended after 1950, big operators negotiated long-term contracts with major utilities, a trend that is still increasing. In 1976, 86 percent of all coal shipped to utilities was under long-term contract. Sometimes these contracts span 20 years or more and, in effect, turn "commercial" coal into dedicated or "captive" coal, as the coal no longer competes in an open market. Such coal is in market competition only when the buyer is evaluating bids.

Most modern coal-supply agreements contain price-escalation provisions that allow up-

### Table 13.—Top 15 Coal Producers and Parent Companies, 1977

| Coal company | 1977 Tonnage | Status[a] | Controlling company |
|---|---|---|---|
| 1. Peabody Group | 65,425,088 | c | Peabody Holding Co. (Newmont Mining Co.; Williams Cos.; Bechtel Corp.; Fluor Corp.; Equitable Life Assurance Society; and Boeing) |
| 2. Consolidation | 47,994,000 | e | Continental Oil Co. |
| 3. AMAX | 28,127,161 | c | AMAX, Inc. (SOCAL owns 20.6% of AMAX stock) |
| 4. Island Creek Group | 16,749,859 | e | Occidental Petroleum |
| 5. Pittston | 14,309,049 | i | Pittston Corp. |
| 6. U.S. Steel | 13,959,000 | s | U.S. Steel |
| 7. Arch Mineral | 12,600,000 | e | Ashland Oil (48.9% stock ownership) and Hunt Oil (48.9%) |
| 8. NERCO Group | 11,988,906 | u | Pacific Power and Light Co. |
| 9. Bethlehem Mine | 10,609,970 | s | Bethlehem Steel |
| 10. Peter Kiewit | 10,298,630 | c | Peter Kiewit Corp. |
| 11. American Electric Power | 10,223,000 | u | American Electric Power |
| 12. Western Energy | 9,773,700 | u | Montana Power Co. |
| 13. Old Ben | 9,720,447 | e | SOHIO |
| 14. North American | 8,905,203 | i | North American Coal Corp. |
| 15. Pittsburg & Midway | 8,202,640 | e | Gulf Oil Corp. |
| Total | 278,886,654 | | |

[a] Status = c-conglomerate; e-energy company; i-independent; s-steel company; u-utility.

SOURCE: *1978 Keystone Coal Industry Manual.*

ward price adjustments to offset the seller's cost of inflation, overhead, new labor agreements, and taxes. Some contracts have "market reopener" provisions, which allow either party to reopen the contract's negotiated price when the market price for substantially similar coal sold from relevant market areas rises or falls. Typically, a coal supply agreement will contain a force majeure clause, which excuses either party from meeting its obligations when unforeseen or uncontrollable events—such as labor disputes, equipment breakdowns, faults in the coal seam, or new laws—frustrate performance. Most long-term contracts run full term with various price adjustments along the way.

Coal is also sold in a "spot" market. Unlike term contracts, spot sales are totally the creature of short-term, supply and demand forces. Most spot-market suppliers are smaller companies operating small mines, though some big mines sell excess production this way. Many spot sellers enter the market when prices rise. Between 1973 and 1978, the number of mines increased 33 percent from 4,650 to almost 6,200 as the average price per ton rose 140 percent from $8.53 to $20.50. Spot prices rose faster and went higher than contract coal because of the perceived fuel shortage created by the 1973 OPEC embargo and other factors. Because many long-term contracts have reopener clauses pegged to spot prices, industrywide coal prices can be pushed up by very short-term or unique price pressures in the spot market. Although reopener clauses are supposed to work both ways, recent experience suggests that contract prices flow up more readily than down.

Captive producers historically were organized as wholly owned subsidiaries of steelmakers, auto manufacturers, or utilities to assure the parent company of a steady supply of a certain kind of coal. Usually, most captive-produced coal was (and is) sold to the parent company. In 1973, utilities mined about 8.9 percent of their total burn; in 1977, 14.5 percent. Many of the giant new strip mines in the West are utility captives. The Federal Power Commission (now absorbed into the Department of Energy (DOE)) estimated "captive" (utility) coal production will triple from

1975 to 1985, reaching 145.1 million tons per year, about 18.8 percent of the projected 770 million tons of coal to be consumed by the electric utilities in 1985."[5] DOE estimated that utilities controlled 11.6 billion tons of recoverable coal reserves as of December 31, 1975.

A utility reaps many advantages from mining its own coal. Supply is made more dependable. Protection is gained against noncost-related price increases. Tax shelters are available. Leverage can be exercised in negotiations with independent coal suppliers. Prices may be adjusted to achieve the "potential for greater return on equity than afforded by regulated utility operations."[6]

The economic implications of this "vertical integration" in the coal industry are disputed and have not been studied adequately. Utilities argue that vertical integration allows them to effect supply reliability and cost control—both to the benefit of the consumer. In return, critics say that utilities sometimes hold back their captive production in order to justify rate increases. A second charge is that utilities have little incentive to keep down production costs in their captive mines as long as they can be passed through to electricity consumers. Consumer advocates say some utilities pay more for their own coal than do utilities without captive production. Utility profits are increased through this inflationary process, it is said. Further analysis is beyond the scope of this report.

Horizontal integration, the ownership of coal companies by companies that produce other forms of energy, has received more attention than vertical integration. In 1963 Gulf Oil took over Pittsburg & Midway Coal Co., beginning what became substantial energy-company (oil and gas producers) investment in coal-producing companies and reserves. The Congressional Research Service reported that 77 percent of all coal producers mining more

---

[5]Electric Utilities Captive Coal Operations, (U.S. Department of Energy, Federal Power Commission, June 1977), p. 17.
[6]Ibid.

than 3 million tons annually were controlled by noncoal companies.[7]

Heated debate over the significance of energy-company ownership of coal producers and reserves was sparked by the unprecedented increase in coal prices, which accelerated after the 1973 OPEC oil embargo. Has horizontal integration of energy production affected coal supply, price, profits, investment, competitiveness, and markets? Did oil-owned coal suppliers, for example, raise their coal prices after the embargo to keep high-priced oil competitive with coal in the Atlantic coast market? Does oil and gas ownership of coal reserves mean anything for the future? Conclusive answers to such questions cannot be offered because the data are unavailable. Some of the needed information may be considered proprietary by coal subsidiaries and parent corporations. It is often difficult to isolate the variable of oil/gas ownership as being the only cause of coal production and price patterns, as many other factors are also at work. However, the general scope of horizontal ownership can be sketched.

Coal production by oil and gas company subsidiaries totaled 166.6 million tons in 1976 (table 14), or 25 percent of national production. Of this, about 125 million tons was steam coal. As much as 35 percent of noncaptive steam coal is currently mined by oil- and gas-owned coal producers.

Energy-company production is expected to increase its share of the total in the years ahead. All horizontally integrated producers plan to increase production. In addition, a number of other major energy companies, such as Sun Oil Co., Kerr-McGee, ARCO, Shell Oil Co., Natural Gas Co., and Mobil Oil,[8] are in the process of opening large surface mines in the West.

Horizontally integrated energy companies account for about 40 percent of all planned new capacity (table 15). Some of this capacity will not be realized, but under the projection of chapter II, about 260 million to 340 million tons of new coal will be mined by horizontally integrated energy companies in 1986, almost all steam coal. To that sum should be added the 125 million tons of coal already being mined by such companies. Therefore, integrated energy companies will mine about 385 million to 465 million tons of steam coal in 1986. This will represent about 48 percent of the total domestic consumption of coal used for energy purposes (table 16).

The top seven companies (in table 15) have 69 percent (233 million tons of a total of 336 million tons) of this planned capacity. These companies are: AMAX (70.5 million),[9] ARCO (32.4 million), Kerr-McGee (30 million), EXXON (27.6 million), Consolidation Coal (25.6 million), Pittsburg & Midway (25 million), and Shell (21.6 million). It is noteworthy that three of these seven—ARCO, Kerr-McGee, and Shell —produced no coal in 1977, and EXXON produced no more than 3 million tons.

The consequences of energy-company ownership of coal are a matter of dispute. The National Coal Association argues that production has increased following acquisition; capital investment in coal subsidiaries has risen, and oil/gas technology and management expertise have benefited the subsidiaries. Critics take opposite positions and fear that an energy oligopoly may emerge capable of manipulating all energy supply and prices for all fuels.

By 1969, four major coal producers—Consolidation Coal, Island Creek, Old Ben, and Pittsburg & Midway—had been acquired by major oil companies. Three of the four mined less coal in 1976 than in 1969. (Together, Consolidation, Island Creek, and Old Ben mined 103.2 million tons in 1969 compared with 83.2 million tons in 1976.)[10] P & M's output rose from 7.6 million to 7.9 million tons. Other oil-owned coal companies increased output after

[7]*National Energy Transportation, Vol. 1 — Current Systems and Movements*, prepared by the Congressional Research Service for the Committee on Energy and Natural Resources, May 1977, pp. 93-94.

[8]*The 1978 Keystone Coal Industry Manual* (New York: McGraw-Hill, 1978).

[9]See note in table 14.

[10]William T. Slick, additional material, testimony before the Subcommittee on Antitrust and Monopoly of the Committee on the Judiciary, U.S. Senate *The Energy Industry Completion and Development Act of 1977, S. 1927* pt.1, 95th Cong., 1st sess., August 2, 4, p. 241.

### Table 14.—Oil and Gas Ownership of Coal Producers
### (In millions of tons produced, 1976)

| Coal company | Parent company | 1976 production |
|---|---|---|
| **Group 1** | | |
| Consolidation Coal | Continental Oil | 55.9 |
| AMAX Coal | Standard Oil of California[a] | 23.1 |
| Arch Mineral | Ashland Oil-Hunt Oil | 18.0 |
| Island Creek | Occidental Petroleum | 17.6 |
| Old Ben | SOHIO | 9.7 |
| Eastern Associated Coal | Eastern Gas & Fuel | 8.0 |
| Pittsburg & Midway | Gulf Oil | 7.9 |
| Zeigler Coal | Houston Natural Gas | 5.2 |
| Falcon Coal | Diamond Shamrock | 5.2 |
| MAPCO | MAPCO | 3.9 |
| Valley Camp | Quaker State | 3.6 |
| Monterey Coal | EXXON | 2.8 |
| Youghiogheny & Ohio | Panhandle Eastern | 2.1 |
| Southern Utah Fuel | Coastal States | 1.0 |
| Hawley Fuel | Belco | .9 |
| Braztah | McCulloch Oil | .5 |
| Tesoro Coal | Tesoro | .4 |
| Tosco Mining | TOSCO | .4 |
| Belva | International Mining and Petroleum | .3 |
| Husky Industries | Husky Oil | .1 |
| Total | | 166.6 |

| | | |
|---|---|---|
| Total 1976 production | 678.7 | |
| Energy share | 25% | |

| | | |
|---|---|---|
| 1976 Noncaptive Production | 569.7 | |
| Energy share | 29% | |

[a]AMAX Coal is a wholly owned subsidary of AMAX Inc., 20.6 percent of whose stock is owned by Standard Oil of California. Standard has attempted to extend its control over AMAX Inc., but has encountered resistance. Spokesman for AMAX coal points out that it is not a subsidary of Standard Oil. Twenty-percent ownership, however, does convey the possibility of considerable influence though not outright control. OTA acknowledges that there is a good deal of uncertainty and sensitivity regarding the ownership of AMAX Coal, and include it in this table with the above caveat.

SOURCE: Keystone Coal Industry Manual, 1978; William T. Slick (senior vice president of EXXON) additional material submitted to the Subcommittee on Antitrust and Monopoly of the Committee on the Judiciary, U.S. Senate, The Energy Industry Competition and Development Act of 1977, S. 1927, Part 1, 95th Cong., 1st. sess., August 2, 4, p. 251; and National Coal Association, Implications of Investments in the Coal Industry by Firms from Other Energy Industries, September 1977.

### Table 15.—New Steam-Coal[a]Capacity of Horizontally Integrated
### Energy Companies[b] by 1986 (In millions of tons)

| State | Tonnage | State | Tonnage |
|---|---|---|---|
| **Alabama** | | **Illinois** | |
| ARCO | 2.4 | AMAX | 11.4 |
| | | Arch Minerals | 2.0 |
| **Colorado** | | Consol | 2.4 |
| Consol | 1.0 | Monterey (EXXON) | 3.6 |
| Empire Energy (Houston Natural Gas) | 2.4 | Old Ben | 4.0 |
| Zapata/Getty Oil | 2.0 | Shell | 1.8 |
| | | Zeigler | 5.7 |
| Total | 5.4 | Total | 30.9 |

**Table 15.—New Steam-Coal[a] Capacity of Horizontally Integrated
Energy Companies[b] by 1986 (In millions of tons)
(Continued)**

| State | Tonnage | | State | Tonnage |
|---|---|---|---|---|
| **Indiana** | | | **Texas** | |
| AMAX | 9.1 | | Shell | 6.0 |
| Old Ben | 2.6 | | **Utah** | |
| | | | Braztah | 3.5 |
| Total | 11.7 | | Coastal States | 5.5 |
| | | | Consol | 4.6 |
| **Kentucky** | | | Valley Camp | 4.0 |
| Island Creek | 1.2 | | | |
| Martiki (MAPCO) | 3.0 | | Total | 17.6 |
| Pontiki (MAPCO) | 2.0 | | **West Virginia** | |
| Pittsburg & Midway | 1.0 | | Island Creek | 5.6 |
| | | | Valley Camp | 2 |
| Total | 7.2 | | | |
| | | | Total | 7.6 |
| **Maryland** | | | | |
| Mettiki (MAPCO) | 1.8 | | **Wyoming** | |
| | | | AMAX | 45.0 |
| **Montana** | | | Arch | 7.2 |
| AMAX | 5.0 | | ARCO | 30.0 |
| Consol | 5.0 | | Carter (EXXON) | 24.0 |
| Shell | 10.0 | | Consol | 5.0 |
| | | | El Paso | 5.0 |
| Total | 20.0 | | Kerr-McGee | 30.0 |
| | | | Mobil | 8.0 |
| **New Mexico** | | | Pittsburg & Midway | 19.0 |
| Arch | 6.0 | | Sunoco | 14.0 |
| Pittsburg & Mdwy. | 5.0 | | Shell | 4.0 |
| | | | | |
| Total | 11.0 | | Total | 191.2 |
| | | | | |
| **North Dakota** | | | GRAND TOTAL | 335.75 |
| Consol | 17.3 | | | |
| Husky | .6 | | | |
| | | | | |
| Total | 18.0 | | | |
| **Ohio** | | | | |
| Y & O (Panhandle Eastern) | 4.6 | | | |
| **Pennsylvania** | | | | |
| Consol | 0.3 | | | |

SOURCE: *1978 Keystone Coal Industry Manual.*

[a]Not included is the capacity dedicated for gasification and that planned by Utah International, Falcon, and St. Joe Minerals. This represents 47.5 million tons.

[b]See footnote (a) in table 14 for AMAX.

**Table 16.—Estimated Energy-Company Share of Noncaptive Utility Consumption, 1985**

| 1985 | Total production[a] | Total domestic energy consumption[b] | Total energy-company share of consumption[c] | Utilities consumption | Noncaptive utility consumption[a] | Total energy-company share of noncaptive utility consumption |
|---|---|---|---|---|---|---|
| Case A | 955 | 790 | 48% | 675 | 548 | 69%[d] |
| Case B | 1,050 | 875 | 48% | 725 | 592 | 71% |
| Case C | 1,145 | 955 | 48% | 775 | 630[e] | 73% |

[a] Millions of tons. OTA estimates.

[b] Domestic energy consumption includes: utilities, industry, residential/commercial, and addition to stocks. It excludes nonenergy consumption (metallurgical coal) and exported production.

[c] Horizontally integrated producers are estimated to mine about 410 million tons by 1986. These percentages are calculated by dividing 410 million tons by the three case estimates of total domestic energy consumption.

[d] OTA estimates that up to 10 percent of the 410 million tons of energy-company production may slip from the utility market. On this assumption, OTA has calculated energy-company shares of noncaptive utility consumption using 369 million tons (410 million less 41 million tons—10 percent) as 1985 energy-company output.

[e] Based on an estimate of 770 million tons of utility consumption, Federal Power Commission/Department of Energy estimated 145.1 million tons would be captive through 1985. Case A and Case B estimates were scaled down proportionately by OTA.

acquisition; Arch Minerals is a case in point. The significance of the ownership variable is unclear given this mixed record of gain and loss. Independents, steel-owned captives, and conglomerate-owned producers also showed output declines. AMAX, Arch Minerals, and the utility-owned captives—Pacific Power and Light, American Electric Power, and Western Energy—increased tonnage steadily. No clear lesson can be drawn from these data because other factors—strikes, whether a company mined underground or surface, quality of supervision, equipment maintenance, markets, declining productivity, seam characteristics and the like—obviously affected company output. It would be unusual if the oil-owned companies deliberately restricted their own production in order to force up coal prices so as to maintain oil's competitiveness. The explanation for the 1970's coal price spiral may ultimately be found in another aspect of coal-company behavior.

It is equally difficult to reach conclusions about whether oil companies are investing oil-generated capital into coal subsidiaries, or whether they are putting coal earnings into other corporate activities. Company-by-company data are often either not available or not comparable, owing to differences in accounting methods. Industry argues that "oil and natural gas companies owning coal firms have invested large amounts of money in coal production."[11] This statement is difficult to confirm. Industry spokesmen do not spell out what portion of capital investment was generated by the coal subsidiary and what, if any, was generated and diverted from the oil parent. It is also impossible to estimate what the level of capital investment in coal would have been had these acquired producers remained independent. In some cases, oil companies may have taken coal-generated capital away from their coal operations and shifted it to their oil enterprises. Occidental Petroleum drew 60 percent of Island Creek Coal's income in 1975 into parent-company activities, and SOHIO used Old Ben's coal profits to develop Prudhoe Bay

and construct the Trans-Alaska Pipeline, according to one analysis.[12]

Recent-entry oil companies—those that have taken over small producers or acquired large Western reserves—have invested their own capital in their coal subsidiary. The reason is obvious: insufficient coal-generated capital was available to finance development. However, consistent or persuasive evidence that oil capital is expanding established coal production beyond what might normally be expected cannot be found. While the capital expenditures of oil-owned coal companies increased in the 1970's, their share of coal industry investment fell. In 1971, 20 oil and gas owners invested $203 million, or 44 percent of total coal industry investment of $457 million.[13] In 1976, these companies invested $568 million, or 32 percent of the total $1,773 billion invested. This pattern suggests that energy-owned coal affiliates were not investing capital as rapidly as the industry as a whole. Yet without access to corporate decisions, it is impossible to determine how much the ownership variable had to do with the lower rate of investment growth.

In sum, two conclusions can be drawn with confidence about the effect of energy-company ownership of coal. First, the company-specific data that would answer many questions are not available. This information is essential to analyzing the effects of horizontal integration. Second, from the available evidence a strong case cannot be made for coal's being advantaged by oil/gas ownership. Production patterns are mixed and causal relationships are hard to decipher. Proof is not available that coal supply or prices have or have not been manipulated to advantage oil or gas. Nevertheless, if increasing coal supply makes coal mining less profitable, incentives may be created to regulate fuel supplies or prices. Energy companies may not embrace these incentives.

---

[11]Implications of Investment in the Coal Industry by Firms From Other Energy Industries (National Coal Association, September 1977), p. 19.

[12]Dr. Thomas Woodruff, prepared statement before the Senate Subcommittee on Antitrust and Monopoly, The Energy Industry Competition and Development Act of 1977, pp. 348-349.

[13]Implications of Investments in the Coal Industry by Firms from Other Energy Industries, p. 28.

Patterns of reserve ownership may shape the future structure of the coal industry, especially if Federal leasing does not resume soon. Reserve ownership affects major market factors: supply, price, and the ability of potential coal producers to enter the field. Six of the top ten reserve holders are wholly or partially owned by energy companies: Continental Oil, EXXON, El Paso Natural Gas, Standard Oil of California, Occidental Petroleum, and Mobil. Energy-company representatives argue that coal reserves are widely dispersed, that the share of total reserves controlled by oil companies is not subtantial,[14] that large reserve holdings are necessary for long-range development plans, and, finally, that supply, price, and competition have not been adversely affected by energy-company ownership.[15] As the Federal Government still controls most reserves, industry spokesmen say any potential anticompetitive situations can be controlled by Federal policy.

The research has not been done that would confirm or deny the actual market implications of coal-reserve control by horizontally integrated energy companies. The major charge against energy companies has been that by speculating and waiting higher coal prices, they did not develop their Western reserves in the late 1960's and early 1970's. Had Western coal been readily available in these years, prices might not have increased. It is worth noting that the major purchasers and leasers of coal reserves in the last decade have been energy companies. Conglomerates do not have significant holdings. As energy companies add to their reserves, it may be increasingly difficult for new entrants to acquire enough long-term reserves to justify the high start-up costs of mining. Smaller companies may not be able to compete with bigger companies because of their capital constraints.

It is worth noting that the major purchasers and leasers of coal reserves in the last decade have been energy companies. Other conglomerates do not have significant holdings.

## Competition

The competitiveness of coal producers is determined by the number of suppliers, degree of production concentration, and ownership structure of reserves and production.

The 6,161 mines operating in 1976 were owned by more than 3,000 individual firms, about 70 percent of which produced less than 50,000 tons. About 600 of these are said to be completely independent producers or producer groups.[16]

The top 15 coal producers mined 279 million tons in 1977, or 41 percent of all domestic production (688 million tons). This list is evolving as utility captives and western mining companies expand more rapidly than eastern underground companies. No one company dominates the industry, but regionally, Consolidation is the key company among eastern-based producers, Peabody in the Midwest, and AMAX in the West.

Coal is not a concentrated industry at the national level. Its top four producers accounted for about 23 percent of total national output in 1977, well below the 35- to 50- percent share that is usually considered to represent a "moderately concentrated oligopolistic core that can produce price manipulation and excess profits."[17] The national concentration ratio understates meaningful concentration because coal does not have a national market. When coal production is divided into four regional markets, the concentration ratio for the top-four producers in 1977 were: Appalachia, 22.3 percent; Midwest, 65.1 percent; northern Plains, 37.7 percent; and Southwest, 64.1 percent.[18] West of the Appalachian fields, moderate to high concentration exists. Even regional markets can be deceptively large. For example, a single Appalachian market does not really exist— northern Appalachian coal is not sold in southern Appalachian markets; east

[14]*The Energy Competition Act*, p. 109 ff.
[15]Ibid., (testimony of William T. Slick) pp. 117-118.

[16]William T. Slick, Jr., (senior vice president, EXXON, U.S.A.), prepared testimony, Ibid, p. 179.
[17]Joe S. Bain, *Industrial Organization*, second edition (New York: John Wiley and Sons, 1968), p. 142.
[18]*Competition in the Coal Industry* (U.S. Department of Justice, May 1978), p. 63. These market percentages do not distinguish between steam and met coals. That distinction is important in the Appalachian market.

Kentucky coal does not sell in the Northeast. Further, high-grade metallurgical coals do not generally compete against steam coals. Captive coal, moreover, does not generally compete against noncaptive coal.[19] Similarly, 1977's 54 million tons of export coal—most of which was metallurgical—does not compete domestically, except when foreign demand slumps. A true assessment of concentration in the coal industry must disaggregate production data and look at concentration in much more precisely defined markets.

DOE estimated that the top four noncaptive suppliers will provide about 47 percent of new noncaptive, electric utility supply now under contract in 1985.[20] The top eight noncaptive producers will supply about 71 percent of this tonnage. Production from the new key coal States, Wyoming and Montana, will be more concentrated: 70 percent for the top four and 92 percent for the top eight.[21]

The long-term contracts discussed above also bear on competition. In effect, these contracts reduce both the real amount of coal available in the marketplace and the number (and needs) of customers. The Federal Trade Commission noted:

In manufacturing industries, production concentration is also an indicator of the present supply alternatives open to potential buyers. If a firm produces 10,000 units of output per year, it can be assumed that up to 10,000 units are available to any qualified buyer. This is not the case in the coal industry. Due to the prevalence of long-term contracts, the annual production of a coal company may not be a valid measure of the quantity of coal available to potential buyers from that producer.[22]

These contracts account for about 86-percent of coal's utility sales. With some exceptions, once a contract of this sort is concluded

another supplier does not compete for that utility's needs. The role of long-term contracts will continue to be substantial. DOE estimates that "slightly more than two-thirds, 243 million tons, of the 1985 coal requirements for . . . new generating units (emphasis added) have already been committed to contracts, in the majority to long-run contracts."[23] The weighted average contract length is 19.5 years for coal mined east of the Mississippi and 26 years for Western coal.

Competition is also regulated by how long-term, coal-supply agreements are negotiated. Although the Justice Department study finds that coal supply is competitive, its description of the "competition" for long-term agreements suggests a modification of normal market forces:

Bid prices frequently are used by utilities just as a means of screening for the "best match" suppliers. The actual schedule of delivery prices and other complex contract terms then are arrived at through direct negotiation and bargaining with the leading candidate(s). Sometimes the utility will pay little attention to or not even solicit bids; instead, it will go directly into negotiations with a company which it believes to be in the best position to provide a reliable source of supply.[24]

The proportion of coal that is contracted for under what may amount to sole-source conditions is not known. Lawyers involved in drafting coal-supply agreements indicate that de facto sole-source procurement is common, particularly for new powerplants and those burning Western coal. Accurate analysis of the competitive state of the U.S. coal industry requires examination of the effect of long-term contracts and negotiating procedures.

The final question concerns the possible anticompetitive effects of interlocking directorates and concentrated stockownership.[25] Most

[19]In 1977, captive production amounted to about 17 percent of all coal output, or more than 120 million tons. Of that, utilities mined almost 69 million tons; steel companies 45 million tons; and industrial users, 6 million tons. 1978 Keystone Coal Industry Manual, pp. 662-663.
[20]Alexander Gakner (Chief, Branch of Fuel and Environmental Analysis, Federal Power Commission (now Department of Energy), testimony, The Energy Industry Competition Act, p. 53.
[21]Ibid.
[22]The Structure of the Nation's Coal Industry, 1964-1974 (U.S. Federal Trade Commission, 1977), p. 66.

[23]Gakner, testimony, The Energy Competition Act, p. 48
[24]Competition in the Coal Industry, p. 82.
[25]Interlocking Directorates Among the Major U.S. Corporations, U.S. Senate, a staff study prepared by the Subcommittee on Reports, Accounting, and Management of the Committee on Governmental Affairs, January 1978, and Voting Rights in Major Corporations, U.S. Senate, a staff study prepared by the Subcommittee on Report, Accounting and Management of the Committee on Governmental Affairs, January 1978.

coal producers, like much of American business, use "outside" boards of directors to guide corporate policy. Directors are often chosen because their expertise or primary affiliation will help a company do business profitably. Many mining companies and manufacturers include representatives of major financial institutions on their boards. Major stockholders—which may include families, financial institutions, and other corporations—are also represented. *Direct interlocks*—where one individual serves as a director of at least two corporations—are common between a coal producer and a capital supplier. One coal producer may be indirectly interlocked with others when a director of each sits on the board of a third corporation. *Indirect interlocks* among coal producers, coal consumers (utilities and industrials, especially), and capital suppliers are common. It is also common to find representatives of major financial institutions sitting on the boards of competing coal producers.

Distribution of voting rights differs from corporation to corporation. In those cases where a single family does not dominate a company, bank trust departments, insurance companies, and mutual funds often own the biggest blocks of stock.[26] Where stockownership is dispersed, holdings below 5 percent can constitute corporate control in some intances.[27] If a single company—a bank, for example—owns substantial voting rights in several coal producers, some analysts argue that the potential for anticompetitive behavior is present.

Consolidation Coal—the Nation's second largest coal producer—is fairly typical of the ownership patterns among energy-owned coal producers. Consol is a subsidiary of Continental Oil. Continental's board was tied to 12 coalreserve holders or coal producers, 9 coal consumers, and 20 capital suppliers through at least one indirect director interlock as of 1976.[28] Continental shared a director with Bankers Trust of New York, Continental Illinois, and Equitable Life Assurance. (Equitable is a major stockholder in Peabody Coal,

the Nation's biggest coal producer and fourth largest reserve holder.) Seven of the 20 capital suppliers were among the company's top 20 stockholders.[29] Continental's biggest stockholder, Newmont Mining Company (3.29 percent) is also the principal stockholder in Peabody Coal (27.5 percent). Capital suppliers with whom Continental shares a director are major stockholders in other leading coal producers and reserve holders.[30]

It is fair to ask a simple question at this point: What do these interlocks mean? The fairest answer is equally simple: We are not certain. The research needed to confirm or deny the significance of this ownership network has not been done. Although the potential for antitrust abuse exists where corporations with common interests are interlocked, no coal industry case study has been done to determine whether this potential for abuse has been used. Similarly, the implications of coal's stockholding distribution have not been studied.

## Price, Profitability, and Productivity

Meeting U.S. coal-production goals depends on price behavior in the utility market. Coal prices must be sufficiently high to provide investment incentives. Increased production then becomes a question of whether long-run price behavior promises a rate of return sufficient to induce the diverse owners of coal companies to invest in coal production.

Price forecasts for 1985 and beyond vary according to consuming region, Btu-content, and sulfur content. The forecasts range from a national average of roughly $22 per ton for high-sulfur coal to $34 per ton (1975 dollars) for low-sulfur coal, with the highest prices predicted for consumers along the Atlantic coast and in the East-Central region.[31] These cost and price projections assume a 15-percent discounted cash flow rate-of-return and imply that prices will rise sufficiently to provide the investment

[26]*Voting Rights*, p. 1 ff.
[27]Ibid., p. 2.
[28]See vol. II, app. VI for complete listing of interlocks.

[29]*Voting Rights*, U.S. Senate, pp. 88-89.
[30]See appendix for complete discussion.
[31]Foster Associates, *Fuel and Energy Price Forecasts*, vol. II (Palo Alto, Calif., Electric Power Research Institute, 1977), pp. 124-125.

resources necessary to meet production goals. However, if the realized rate of return does not meet investor expectations, capital is likely to flow into noncoal investments. Energy companies and conglomerates can choose between alternative investments whereas a nondiversified coal company cannot. It is possible that leading coal producers might attempt to encourage price increases and maximize their rates of return by regulating supply. Deregulation of—and higher prices for—oil and gas should lead to higher coal prices, although the increases may not be identically proportional. Higher prices should encourage coal investment and enable operators to spend more on health and safety, environmental protection, and community improvement.

The interaction among price, production costs, profit, and capital investment is central to expanding coal supply. The relationship among these factors changed dramatically in the 1970's. All rose substantially over pre-1970 levels, but production did not rise proportionately and productivity fell. The price per ton quadrupled between 1968 and 1975 while labor costs doubled in that period.[32] The unit labor cost share dropped from 58.5 percent in 1950 to 20.3 percent of per ton value in 1974. It probably amounts to 25 to 30 percent of value today due to increases in wages and benefits and the leveling off of the coal price rise. Rising prices helped boost industry profits. Coal's return on net worth exceeded 11 percent only once in the 1950-73 period, but it approached 30 percent in 1974 following the OPEC embargo.[33] From 1971 through 1974, net coal income of 24 leading coal producers rose 298 percent from $128.4 million to $639.5 million, most of which came in 1973-74.[34] Coal's unprece-

dented level of profitability in the mid-1970's was due less to rising production than to the price increases that followed the OPEC embargo and to a temporary surge in worldwide demand for met coal.

A precise picture of the financial status of the industry is difficult to develop because much of the necessary data has not been available to the public. Parent companies have not been required to separate financial data for their coal subsidiaries in their reports to the Securities and Exchange Commission. A check of the financial data for the Pittston Corp. (primarily a West Virginia-based metallurgical coal producer) and North American Coal (an eastern producer of met and steam coals) adds case-study data to the gross statistics noted above (see table 17). Despite declining productivity and slipping production, both companies recorded increased net income after 1973. Net income per employee increased in both cases, suggesting that income productivity and labor productivity do not necessarily coincide. Between 1970 and 1977, Pittston's net income per employee rose 252 percent and North American's gained 73 percent (current dollars). Consolidation Coal, the second biggest coal producer and a subsidiary of Continental Oil, showed coal revenues being five times greater in 1977 than in 1970, although revenues for 1975 through 1977 were relatively unchanged.[35]

Return on investment and profits slacked in 1978 because of the coal strike and demand softness for certain coals. For most companies, this should prove to be a temporary phenomenon.

[32]Joe Baker, Coal Mine Labor Productivity: The Problem, Policy Implications, and Literature Review (Oak Ridge, Tenn., Oak Ridge Associated Universities, 1978), pp. 7-8.
[33]Charles Rivers Associates., Coal Price Information (Palo Alto, Calif., Electric Power Research Institute, 1977) p.4-40. Based on data collected by Citibank of New York.
[34]T. T. Tomimatsu and Robert Johnson, The State of the U.S. Coal Industry, Information Circular 8707 (Washington, D.C.: U.S. Department of the Interior, Bureau of Mines, 1976), pp. 17-19.

[35]Continental Oil Company, Annual Report, 1977, p. 36.

**Table 17.—North American Coal and Pittston Corporations Profitability and Productivity**

| | 1970 | 1971 | 1972 | 1973 | 1974 | 1975 | 1976 |
|---|---|---|---|---|---|---|---|
| **North American Coal** | | | | | | | |
| Net operating profit (NOP) | Not available | | | | | | |
| Net Income (NI)[e] | $1,894 | $1,248 | $2,629 | $4,452 | $4,923 | $7,827 | $7,286 |
| Employees | 2,949 | 3,700 | 4,357 | 4,661 | 4,970 | 6,501 | 6,569 |
| Production[b] | 9,633 | 8,432 | 11,480 | 10,783 | 9,900 | 9,831 | 10,866 |
| Productivity[c] | 14.3 | 10.9 | 11.7 | 10.3 | 9.7 | 6.5 | 7.3 |
| NOP/Employee | Not available | | | | | | |
| NI/Employee[d] | $642 | $337 | $603 | $955 | $991 | $1,204 | $1,109 |
| **Pittston** | | | | | | | |
| Net operating profit (NOP)[e] | $62,860 | $54,977 | $33,228 | $32,368 | $193,362 | $333,185 | $222,510 |
| Net income (NI) | $39,442 | $43,437 | $28,585 | $15,341 | $107,446 | $200,146 | $146,372 |
| Employees[f] | 16,347 | 17,028 | 17,390 | 16,980 | 17,100 | 17,800 | 17,520 |
| Production[g] | 20.5 | 20.1 | 20.6 | 18.8 | 17.4 | 18.6 | 17.1 |
| Productivity[h] | 3.5 | 3.3 | 3.3 | 3.1 | 2.8 | 2.4 | 2.7 |
| NOP/employee | $3,845 | $3,229 | $1,911 | $1,906 | $11,308 | $18,718 | $12,700 |
| NI/employee | $2,413 | $2,551 | $1,644 | $903 | $6,283 | $11,244 | $8,485 |

[a] In thousands of dollars.
[b] In thousands of tons. Figure does not include coal sold but not mined by North American.
[c] Productivity calculated by dividing tonnage by employees, and then dividing the resulting product by the average number of days worked in a given year. Expressed as tons per worker per shift.
[d] Calculated by dividing net income by employees.
[e] NOP and NI in thousands of dollars.
[f] About half of Pittston's employees are found in its Coal Division, although they account for more than 90% of the corporation's income.
[g] Production in millions of tons.
[h] Productivity in tons per worker per shift.

SOURCE: Securities and Exchange Commission and National Coal Association.

# LABOR PROFILE

Some generalizations are unavoidable in describing a work force, but the differences among coal workers are often as important as their similarities. Coal workers include miners (both surface and underground), coal-mine construction workers (at least 7,000), preparation plant and tipple workers, and those who work in mine related repair shops.

Coal workers represent a wide range of cultures, ages, education, political perspectives, income levels, and attitudes. Of the 237,000 coal workers in 1977, about 2,000 were women.[36] (As late as 1970, no women were known to be working in the mines.) Perhaps 10,000 coal workers are black. Another 2,000

[36]More than 100 coal companies are now negotiating to reach a settlement with a women's rights group—The Coal Employment Project—that has sued them, alleging employment discrimination.

or 3,000 are Indian. Mexican-American miners are found in the West, and miners of Mexican descent work underground in the mines of southern West Virginia. In northern Appalachia, the grandfathers of many miners emigrated from Italy, Poland, Russia, Czechoslovakia, and Hungary in the early years of this century. Other miners are descended from earlier miner immigrants coming from Scotland and Wales.

Coal workers are found scattered through two-thirds of the continental United States. Most work east of the Mississippi River and most of those labor in the long diagonal of the Appalachian coalfields. Others work in Utah, Colorado, Wyoming, Montana, and New Mexico. Their jobs vary. Some operate multimillion-dollar earth excavators; others load coal by shovel. Some sweat all day; others never do. Many are highly skilled; others aren't. Their work environments range from air-conditioned, bucket-seated dragline cabs to 20-inch coal seams where the machine operator lies nearly on his back most of the day. Most coal labor—about 160,000 persons— belong to the United Mine Workers of America (UMWA), which sets many of the wage and benefit standards for the entire industry. Several thousand belong to the Progressive Mine Workers (PMW), Southern Labor Union (SLU), Operating Engineers, and other unions. More than 40,000 probably belong to no union at all, either from choice or lack of opportunity.

One of the most important distinctions between miners is whether they work above or below ground. About 60 percent of all coal production was mined in surface operations in 1977, but 68 percent of all miners worked in underground mines. The perspective of a UMWA presidents has been that of the dee miner. This experience has shaped coal's co lective bargaining from the beginning.

Important differences are found betwee the two groups. Accident frequency, for exam ple, is significantly lower for surface miner although some kinds of surface mines are le safe than some deep mines. Surface miners ar generally older, work more days annually strike less, and are paid more than their unde ground counterparts. The average age of mos UMWA deep miners was 35 in 1976 compare with 41 for most UMWA surface miners.[37] I 1975 and 1976, an industrywide sampl showed one strike per year at surface mine

[37]Information supplied by the Bituminous Co Operators' Association.

Last shift comes out at Eccles #5 before the 1977 contract strike

Photo credit: Douglas Yarrow

ompared with an average of more than three
trikes at underground mines.[38] Surface miners
work 10 to 20 more days each year than under-
round miners. Finally, most UMWA under-
round miners averaged $14,170 in annual in-
ome in 1975 compared with $19,456 for most
UMWA surface miners. In 1976, UMWA under-
round miners earned an average of $15,203
ompared with UMWA surface miners at
20,643[39] (figure 14).

Survey research of 400 randomly selected
West Virginia miners in November 1977 found
hat 46 percent said their before-tax income
was $10,000 to $15,000, and 38 percent re-
orted $15,000 to $20,000.[40]

Although UMWA had roughly 100,000 underground miners in 1976, representing about
3 percent of all underground miners, it had
only about 11,500 surface miners, representing
1 percent of all surface miners.[41] The UMWA
hare of underground tonnage is probably
omewhat higher as its members tend to work
n the bigger, more productive underground
mines. The smallest underground mines are
generally not organized by UMWA. However,
he UMWA share of surface production is
probably less than 21 percent because most of
ts surface miners work in comparatively less
productive eastern surface mines.

A second important set of distinctions
among coal workers is the changing age distribution of the work force. As noted, the
average age of UMWA underground miners
was 35 in 1976. The median was 33. (Median
age for UMWA surface miners was about 37.)
In contrast, the median age for UMWA workers in 1964 was 48. The age distribution is
changing rapidly. In 1974, 53 percent of most
UMWA deep miners were under 35 years of
age and 22 percent were 50 or older (figure 15).
By 1976, 50 percent were under 35 and only 15
percent were older than 50. What had been an

aging, underground work force, heavily
weighted toward men in their 40's and 50's,
had moved toward a heavy imbalance between young and old by the mid-1970's. The
age distribution of UMWA surface miners—
and probably surface miners generally—is
more evenly distributed. In 1974, 34 percent of
most UMWA surface miners were under 35,
and 20 percent were 50 or older.

In the 1950's and 1960's, the mine work force
aged gradually as tight demand and mechanization limited new recruits to a trickle. This
resulted in a more experienced work force.
Productivity was high. Training was almost
nonexistent because there were few new
miners to train. This changed quickly in the
1970's. Almost 50,000 experienced miners retired or left because of black lung disability.
New contract provisions and new safety requirements, plus expanding demand, brought
another 100,000 new workers into the industry
between 1969 and 1978. The combination of
retirement and work force expansion meant
that the work force was becoming younger and
less experienced but better educated. In 1974,
61 percent of UMWA deep miners had less
than 6 years of mine service; in 1976, 63 percent had less than 6 years.[42] Forty-nine percent
of UMWA surface miners in 1974 had less than
6 years service, and 47 percent had less than 6
years in 1976.[43]

A third distinction among coal workers is
regional. Of the 190,000 miners recorded in
1975, 94 percent worked east of the Mississippi
River. Fully 97 percent of the 136,000 underground miners recorded that year worked in
the East along with 87 percent of the 55,000
surface miners. However, the West produces
more coal than its share of miners indicates because of the higher productivity of western surface mining systems. Fewer than half of the
34,000 to 43,000 western miners expected by
1986 will work underground. A majority of the
new miners in the East will work underground.
If current trends continue, western coal miners
will be substantially nonunion and mostly surface. The eastern work force will be mostly
underground and predominately UMWA. The

---

[38]Jean M. Brett and Stephen B. Goldberg, Wildcat
Strikes in the Bituminous Coal Mining Industry: A Preliminary Report (Northwestern University, 1978), p. 23.

[39]Data supplied by BCOA.

[40]Don R. Richardson Associates, A Survey of West
Virginia Coal Miners, 1977, p. 26.

[41]OTA made these estimates based on data supplied by
BCOA and MSHA.

[42]Data supplied by BCOA.

[43]Ibid.

**Figure 14.—Distribution of Employees by Annual Earnings at Deep and Surface Mines**

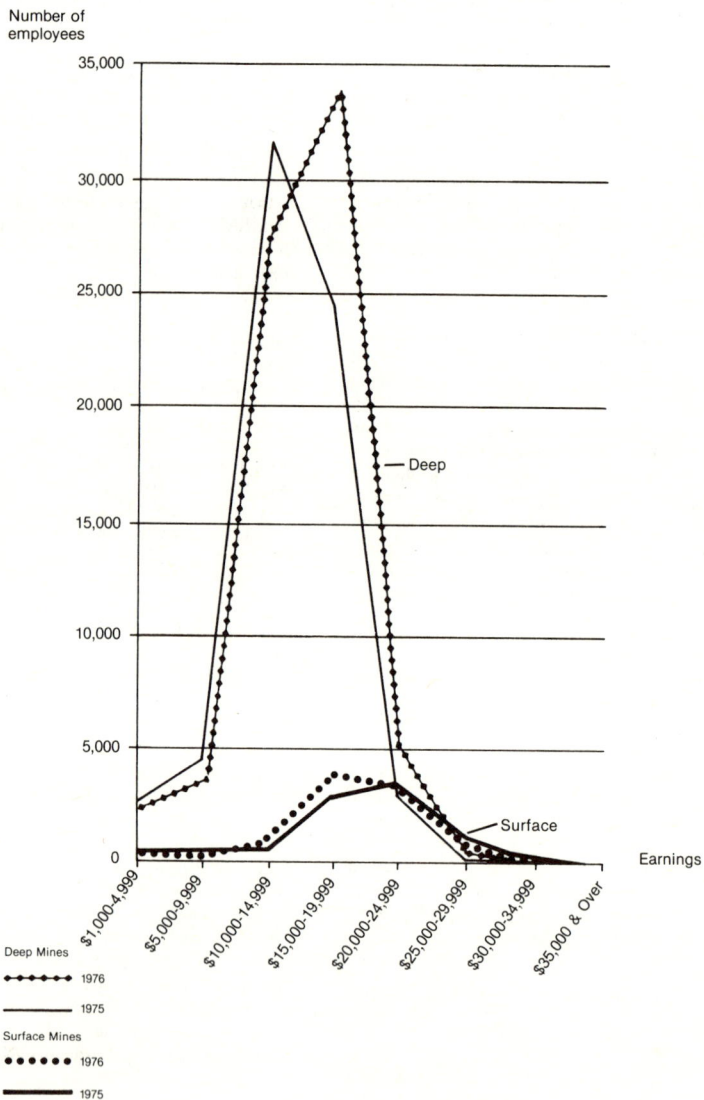

Number of
employees

35,000

30,000

25,000

20,000

15,000 — Deep

10,000

5,000

Surface

0                                                                 Earnings

$1,000-4,999
$5,000-9,999
$10,000-14,999
$15,000-19,999
$20,000-24,999
$25,000-29,999
$30,000-34,999
$35,000 & Over

Deep Mines
●●●●● 1976
——— 1975

Surface Mines
●●●●● 1976
——— 1975

SOURCE: Income data supplied by the Bituminous Coal Operators' Association, 1978.

**Figure 15.—Distribution of UMWA Employees by Age—All Mines<sup>a</sup> 1974-76**

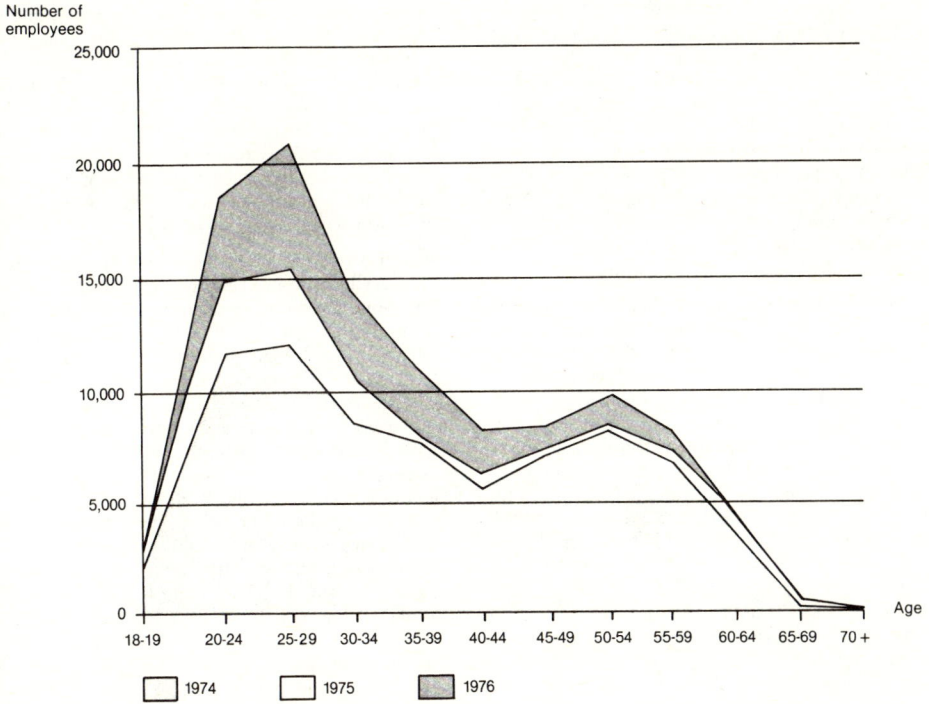

SOURCE: Age data supplied by the Bituminous Coal Operators' Association, 1978.

few cases of moving eastern miners and mine management to the West have not been very successful.

The fourth and perhaps most significant distinction is unionization. Those who have not spent time in the UMWA coalfields have a hard time understanding the intensity of the miners' feeling for UMWA. Whether they are disgusted or delighted with its performance, they feel it is theirs, built by them to serve their interests. As harsh as its internal critics are, most believe that unionization "is the only thing standing between them and the coal company."

Non-UMWA miners are located principally in southern Appalachia (southwest Virginia, east Kentucky, and east Tennessee) and west of the Mississippi. Few demographic or attitudinal data exist about them. Often they make as much money as—or more than—UMWA miners. This relatively recent phenomenon was made possible by the higher profitability of mining after the OPEC embargo. Most non-UMWA mines are relatively new. They do not carry pension obligations to older miners and retirees that UMWA-organized companies carry. Few of their workers have even retired, and many of the smaller, non-organized companies have small pension obligations, if any.

In big, non-UMWA mines health insurance plans and other benefits are comparable to UMWA standards. In a few western surface mines, the benefits are more comprehensive. What many nonunion miners lack contractually are job protections, a grievance procedure, and some safety rights (including the right to elect a safety committee and the right of the individual to withdraw from danger). Non-UMWA miners also lack a national voice. The UMWA, whatever its failings, speaks for the Nation's coal miners in Washington, D.C., Non-UMWA miners are not represented, although they benefit from legislation or regulatory initiatives UMWA is able to push through.

The extent of unionization varies in the West. The majority of mines in the Carbon and Emery Counties area of Utah is organized by UMWA. On the other hand, only the captive Wyodak Mine associated with the Wyodak Power Plant in Campbell County, Wyo., is organized (International Brotherhood of Electrical Workers (IBEW)). North Dakota mines are organized by several unions. Consolidation Coal and North American Coal have UMWA contracts. Mines of the Knife River Coal Mining Company are organized by the PMW. The Operating Engineers, the predominant union in Colorado, does not represent mines in North Dakota. The captive Baukol-Noonan Mine is organized by the IBEW. The UMWA is a major force in parts of the West. It is the dominant union in Utah underground mines. It represents mines on the Navajo Reservation. It has lost some of its power to other unions. Only 2,000 to 3,000 surface miners are UMWA-organized in the West. Most Western States have right-to-work laws that limit unionization.

These demographic and social indicators reflect major structural changes that occurred in the coal industry over the last decade. In these years, surface production moved ahead of underground production. The locus of national coal output moved westward. Oil companies took over major coal producers, and most of their effort centered on developing western surface mines. The proportion of coal workers who are UMWA members dropped from between 75 to 80 percent in the late 1960's to around 65 to 70 percent today.[44] UMWA tonnage, however, has fallen to about 50 percent of the total production owing to the increase in nonunion, surface-mined coal.

Still, the heart of the coal industry is in the East. And it is there that one must go to explain labor-management instability and the attitudes that cause it. The 1960's allowed management to become complacent about its labor needs. Mechanization and layoffs meant a large pool of experienced miners. Wildcat strikes were infrequent because of the labor surplus and organizational policy. Coal producers had few incentives to provide better working conditions. Their ranks were not hospitable to policy innovators. An illusion set in that all was well. It deprived management and labor of a vital spirit that both need for continued health. Industry found that its lower management consisted of men too old, too timid, or simply too set in their ways to become good leaders. But as long as coal demand stagnated and prices were more or less constant, the industry's stability was not upset.

[44]OTA estimates that 160,000 coal workers are represented by the UMWA of a total coal-worker laborforce of 237,000 (which includes an estimated 7,000 coal construction workers).

# COLLECTIVE BARGAINING: A STORMY HISTORY

Labor relations in the U.S. coal industry have been characterized by suspicion, acrimony, violence, and, occasionally, cooperation. Throughout this century the unwillingness of miners to accept the conditions of their work has led to chronic unauthorized work stoppages, turmoil, and prolonged contract strikes. The number, duration, and cost of wildcat strikes have increased since 1970.[45] The absence of equitable and stable labor-management relationships has hampered coal production and supply reliability. Coal miners' income, benefits, and health care systems, have been cut back. On the other hand, many miners say wildcat strikes are often the only way to force an employer to deal quickly—and presumably favorably—with one of the miners' concerns. The roots of today's conflict lie deep in the past. But it is simplistic to believe that current practices do not contribute. Many of the sources of discontent are deeply embedded in coal economics and management policies. These may not change easily. Finally, even though U.S. energy policy counts heavily on coal, the advisability of Federal intervention in the industry's labor-management relations is not clear.

Collective bargaining and labor relations generally have been shaped by the economics and structure of the industry. How miners were treated by operators was often determined by how operators treated each other and how coal was treated in the marketplace. Free enterprise has never produced stable labor relations in coal. Competition, rather than strengthening industry stability, tended to destroy it. Chronic labor unrest has been re-

lated to the cost-cutting pressures of competition amid demand stagnation. Periodically, coal strikes erupted and dragged on for months or even years. Strikes of national consequence occurred in 1902, the early 1920's, the late-1940's, and the mid-1970's. Often they occurred when demand was strong. When demand surged and high prices encouraged marginal operators to begin mining, the market was quickly oversupplied and a slump soon followed. Miners felt it opportune to press their demands during these booms, and when demand slacked, they struck to maintain their gains. Understanding the boom-bust cycle as they did, operators resisted labor's demands during both the short booms and the long busts.

Apart from market factors, a second source of labor-management strife originates from the attitudes and conditions resulting from the nature of the mine workplace and the work process. The work environment of underground coal mining is unique. Danger is inherent, although controllable. Work has been made easier by mechanization, but the workplace has not been made more pleasant. Modern coal miners end their shifts wet, dirty, often chilled, with mine dust embedded in every pore. Often they work on their knees; standing or kneeling in cold, oily water for hours at a time; listening to the "working" of the mountain above them. A roof that looks safe may doublecross them in a second. Wariness and caution are essential.

Unlike assembly-line work, mining requires workers—as individuals and as part of an interdependent team—to control much of their own work. Miners must constantly adapt their work to ever-changing environmental conditions. The mining process requires a good deal of individual judgment and peer-coordination. The pace of work is determined to a great extent by the miner's minute-by-minute evaluation of physical conditions and machinery. Because so much of mining is a matter of judgment, miners and section supervisors often disagree. Disputes stem from the pace of work; who is to do what; what needs—or doesn't

---

[45]Between 1974 and 1976, more than 400 additional strikes and 300,000 more days-idle were recorded in bituminous coal mines than between 1970-73. In 1976, 1,132 bituminous coal strikes were recorded during the term of the contract, which accounted for one-fourth of all industrial disputes recorded that year. About 20 percent of all U.S. workers participating in work stoppages in 1976 were bituminous coal miners, most of whom, presumably, were UMWA members. See Linda H LeGrade, *Collective Bargaining in the Bituminous Coal Industry*, Report 514, U.S. Department of Labor, Bureau of Labor Statistics, November 1977.

need—to be done; what precautions must be taken; and what kind of work miners can legitimately be asked to do. Miners and foremen develop routines of interaction that are both adversarial and cooperative. Both take pride in "running" a lot of coal during their shifts. But this shared goal is often subverted when management asserts authority in ways that miners perceive to be arbitrary or as violating their job rights. UMWA miners have developed elaborate work rules and codifications of their rights to protect themselves from perceived management transgressions. If one of the commonly understood rules is breached, miners close ranks and resist.

Workplace-generated attitudes lend themselves to constant confrontations. First, miners are proud of doing useful and dangerous work. Pride is linked to self-confidence; both are combined with a certain truculence against being told how to do their work. Second, the danger of the workplace readies miners for turbulent conflict outside of it. Third, the work process trains miners to work collectively. This interdependence carries over to conflicts with management. A wrong done to one miner is interpreted as a threat to all. Shared dangers and interdependent work produce group cohesion when the group is faced with a common threat. It creates a "them and us" attitude. It enables miners to stick together through prolonged strikes.

The social environment of the coalfields is a third factor contributing to the volatility of labor-management relations. The basic form of social organization in coal's early years was the company town. In these communities, mine operators owned or controlled everything—housing, medical care, schools, the law, churches, and commerce. Until the 1930's and 1940's, miners were often forced as a condition of their employment to accept wages in scrip instead of U.S. currency and to buy only at the company store. The miner's perception of work victimization was compounded by this same perception in a company-controlled community. The work routine regulated coal-camp existence. Mine operators established the quality of community life. Social equality among mining families reflected the equality

of the workplace. When work conflict arose, it quickly enveloped the camp's entire social system. Mining was not only a job, it was a way of life for workers and their families. Although the coal-camp system was dismantled a generation ago (when mechanization cut employment and spendable income to the bone, thus making the closed system unprofitable), many of the facilities are used today, and the psychology of the system remains.

Finally, industrial conflict plagues coal because of the past. Coal camps were experiential hothouses; each perceived wrong, each dispute became part of the community's history. The bitterness could never dissipate. Children of miners absorbed—and continue to absorb—the attitudes of their parents. There is no quick remedy for this historical sensitivity; it must be accepted.

The history that follows is necessary to understand present day labor-management relationships and how they will or will not be affected by Federal policy.

## The Early History

Unlike other basic industrial activities, coal production did not grow steadily over the first 70 years of this century. The industry's capacity to produce did not change significantly after 1918. As recently as 1974, the operators produced only slightly more than they had in 1918 and less than in 1947. Demand stagnation forced the industry to be acutely cost-conscious. This often translated into protracted opposition to unionization and continued pressure to reduce labor costs. Both policies occasioned many strikes and much bitterness in the 1920's and 1930's. A high level of market competition among hundreds of suppliers intensified the cost-cutting pressures within the industry. Because demand fluctuated but did not grow, operators were less concerned about mining more coal and more concerned about continued lowering of their labor costs to maintain competitiveness. The implications for harmonious labor relations are apparent.

Coal's industrial relations changed in the 1930's and 1940's. Until that period, the majority of operators had never accepted the principle of unionism. Federal legislation had not protected collective bargaining. But Depression-spurred legislation—Norris-LaGuardia, the National Industrial Recovery Act (NIRA), and the Wagner Act—encouraged collective bargaining and outlawed certain antiunion practices. The New Deal also rescued the coal industry through its price-fixing, production-stabilizing codes. UMWA became the bargaining agent for most coal miners in 1933 following a whirlwind organizing campaign. Although UMWA regularly negotiated contracts with associations of northern and southern operators, collective bargaining was characterized by strikes, threats, denunciations, Federal seizure of the mines, and ritualized hostility. Despite the gains won by UMWA president John L. Lewis, the industry's fortunes generally improved in this period while wages, as a percentage of sales, declined. In the late-1940's, coal demand plummeted as railroads switched to diesel locomotives and residential customers shifted to oil or gas. Market pressure forced Lewis and the operators to modify their adversarial relationship.

To stabilize labor-management relations, each side had to be able to speak with a single voice. Although many hundreds of operators mined coal in the 1940's, political leadership of the industry was assumed by Consolidation Coal Co. under George Love. Consolidation was the product of a 1945 merger between the Hanna-Humphrey-controlled Consolidation Coal Co. of Cleveland, Ohio and the Mellon-owned Pittsburgh Coal Co. The merger made Consolidation the leading coal producer in the late-1940's. From that base, Love reached out to the major steel companies to establish a single industrywide organization to negotiate with Lewis. This took shape in 1950 as the Bituminous Coal Operators' Association (BCOA).

John L. Lewis had established unquestioned control over UMWA by this time. His political opponents had been expelled or neutralized in the 1920's. The rank and file had lost its right to ratify contracts. The union's internal organization was under Lewis' personal direction. Lewis and Love recognized that the demand-

limited situation in 1949-50 threatened both sides with a devastating circle of oversupply, wage cuts, and layoffs—the very pattern that had brought them close to ruin 20 years before. The self-interest of each led to considering ways of saving the other.

## Cooperation: 1950-72

Observers of coal's collective bargaining have different interpretations of events after 1950. Neither UMWA nor BCOA has sponsored or endorsed official histories of this era. However, several doctoral dissertations, journalistic accounts, and one major research effort[46] were published in the last few years. All report an undeniable shift from hostility to partnership between the union and the big operators following the 1950 contract. They trace the new dynamics of collective bargaining to economic pressures resulting from soft demand. The relationship of John L. Lewis to the coal operators reversed itself in the 1950's, and these observers criticize Lewis for some of his judgments and policies. Although this interpretation of the past annoys some in industry and labor, it is, in fact, the only body of interpretation available and has not been refuted. In this light, it should be noted that representatives of the operators and UMWA disagree with elements of the narrative that follows.

Labor relations were turned upside down after 1950. Between 1950 and 1972, Lewis and his successor, W. A. (Tony) Boyle, fashioned a strike-free partnership with the major coal operators. All earlier coal stabilization mechanisms—Federal intervention, self-regulation, and union regulation of supply—had failed. But a fourth—labor-management alliance—did not.

BCOA, which represented about half of the industry's output and all of its major companies, had two goals in the 1950 bargaining. On one hand, BCOA's Love sought to stabilize coal production and make it predictable by eliminating strikes and overproduction. On the

---

[46]Melvyn Dubofsky and Warren Van Tyne, *John L. Lewis: A Biography* (New York: Quadrangle/New York Times, 1978).

other, he hoped that BCOA members could sew up the blossoming utility market, the key to coal's future. Both objectives depended on the ability of BCOA companies to increase productivity and cut labor costs to maintain their competitiveness with oil and gas and eliminate low-priced coal suppliers.

Love and Lewis saw mechanization as the way to increase productivity and reduce labor costs. In underground mining, mechanization meant eliminating hand loading. Machines could produce three times as much coal with 10 workers as 86 workers in a hand loading section. Mechanization also meant more surface mining, which was more efficient than even underground continuous miners. Surface-mined production increased from 24 percent of total output in 1950 to about 60 percent in 1977. Industrywide productivity almost tripled between 1950 and 1969. However, coal-mine employment fell 70 percent between 1950 and 1969, from 415,482 to 124,532.

The crucial factor in mechanizing coal was John L. Lewis. If he had chosen to delay it, the industry would have suffered. But Lewis championed mechanization, and had done so at least since the 1925 publication of his only book, where he wrote:

The policy of the United Mine Workers of America will inevitably bring about the utmost employment of machinery of which coal mining is physically capable.

Fair wages and American standards of living are inextricably bound up with the progressive substitutions of mechanical for human power. It is no accident that fair wages and machinery will walk hand in hand.[47]

Lewis reiterated his views on mechanization in the early 1950's:

We decided that question [the UMWA's position on mechanization] long years ago . . . in return for encouraging modernization, the utilization of machinery and power in the mines and modern techniques, the union . . . insists on a clear participation in the ad-

vantages of the machine and the improved techniques.[48]

Mechanization was so linked in Lewis' thinking to higher wages and benefits that he simply shrugged off unemployment and other social costs as the price of industrial rationalization. With the UMWA supporting mechanization, big BCOA operators were able to cut their labor costs, secure long-term contracts, and maintain their profitability. Much of the new machinery was financed by using the contracts themselves as collateral. UMWA also loaned money directly to a number of companies to finance capital investment.

UMWA itself began reevaluating Lewis' policy on mechanization in 1973:

A question which has persisted throughout UMWA history is how closely related are the welfare of the miner and of the coal industry. The coal operators have always argued that what's good for Consolidation Coal Co. is good for every mine worker, and that miners should therefore refrain from asking for too big a share of the profits. During most of his career, John L. Lewis knew better. When he argued for creation of the Welfare Fund, when he argued for better safety laws, when he argued for price controls during wartime wage controls, he repeated his confidence that the coal industry had the wealth to meet the human needs of those who supported it.

But in the 1950's he went against that long-time judgment. Faced with mechanization of the mines, Lewis, in his own words, 'decided it's better to have a half a million men working in the industry at good wages, high standards of living, than it is to have a million men working in the industry in poverty and degradation.'

It was a firm decision, and mechanization with its drastic reduction of the work force was carried out. Unfortunately, no provision was made for the hundreds of thousands of men who were put out of work. There were no benefits, no retraining programs, no new industry brought in. A great many miners were forced to take their families to northern cities

[47] John Lewis, The Miners' Fight for American Standards, (Indianapolis, Ind.: Bell Publishing Co., 1925) pp. 108-109.

[48] Justin McCarthy, A Brief History of the United Mine Workers of America (Washington, D.C.: United Mine Workers of America, 1952), pp. 7-8.

like Detroit and Chicago, where most did not fit in and did not want to.[49]

Lewis exacted a price from BCOA for supporting mechanization. Hourly wages were increased—albeit modestly—for working miners in each round of bargaining after 1950. The major concession Lewis received was the establishment of the UMWA Welfare and Retirement Fund, which provided pensions and near-comprehensive, first-dollar medical coverage for UMWA members and their families. The Fund was started in 1946, when Lewis persuaded Interior Secretary Julius Krug (who was administering the Nation's mines after they had been seized) of its merit. A modest tonnage royalty was levied. But the operators fought the plan and its funding formula for the next 5 years. Love committed the big operators to support the Fund in 1950. A tonnage royalty was fixed, and it did not rise between 1952 and 1971. Although the Fund was supposed to be distinct from UMWA, Love gave Lewis de facto control of the three-person board of trustees when he accepted Josephine Roche, a long-time Lewis confidante, to serve as the neutral trustee. The Fund did much good work in coalfield health care in the 1950's and 1960's. Clinics were organized. A chain of hospitals was built. Preventive services were encouraged. Controls over cost and quality of health services were established. But the financial resources of the Fund were always dependent on the level of production of the BCOA companies that paid the royalties. When demand fell in the late 1950's and early 1960's, the Fund—living on an unchanged tonnage royalty—was caught short. It had to sell its hospitals and withdraw medical benefits from thousands of miners who had been "mechanized" out of their jobs. When the Fund's records began to show startling increases in respiratory disease among its beneficiaries in the 1960's, Roche and Boyle refused to demand dust controls or an industrywide dust standard. Lewis and Boyle might have negotiated a higher tonnage payment to cover Fund

needs, but that risked increasing the pressure on BCOA members who were being battered by low prices and demand stagnation.

UMWA finally raised such criticisms of the Fund in 1973:

> The history of the [bituminous] Funds since their early days is a mixed one.... Many persons have been paid, but many men were cut out by arbitrary and unfair rules while the hard coal [anthracite] pension dropped to $30 per month. Medical services were provided for miners and their families, but were taken away from disabled miners and widows. For almost 20 years the royalty stayed at 40 cents, while up to $90 million of the soft coal Fund's money was kept in non-interest-bearing checking accounts at the union-owned National Bank of Washington.[50]

The scope of the Fund's work grew increasingly constricted over the years. Under the 1978 contract, medical insurance for working miners was switched to private carriers, leaving the Fund to administer health benefits and pensions only for retired miners.

As labor costs were lowered in the 1950's and 1960's, the industry also externalized production costs. Social costs and externalized costs were rarely assessed in these years. Reclamation standards for surface mines, for example, were not enacted until the late-1960's. Dust controls were not required in underground mines until 1970, when coal workers' pneumonconiosis (CWP) was recognized as a disabling occupational disease by Federal legislation. Federal safety standards were minimal. Industry was not expected to bear the public costs of the unemployment produced by mechanization. Systematic air pollution controls had yet to be enacted; coal suppliers sold relatively "dirty" coal to utilities. Finally, coalfield communities generally imposed small tax burdens on local mine operators and coal-reserve owners, fearing that even the slightest additional economic pressure would make local employers uncompetitive. When demand picked up in the late 1960's and early 1970's, the industry was coincidentally expected to begin paying many of these social costs through compliance with environmental

---

[49]*The Year of the Rank and File: Officers' Report to the United Mine Workers of America 46th Constitutional Convention,* United Mine Workers of America, 1973, p. 68.

[50]*The Year of the Rank and File,* p. 67.

regulations and health and safety standards. The deficit of community services that had accumulated during the depressed 1950's and 1960's handicapped fast growth in the 1970's, when coal towns struggled to meet the needs of hundreds of new miners.

The Love-Lewis contract of 1950 sought not only to stabilize labor but to impose order on coal suppliers. Surplus miners increased costs of production while surplus operators drove down prices. Both were seen as problems to be solved. Lewis shared the BCOA's attitude toward the marginal independent suppliers, of whom he said in 1950:

> The smaller coal operators are just a drag on the industry. The constant tendency in this country is going to be for the concentration of production into fewer and fewer units . . . more obsolete units will fall by the board and go out of production.[51]

Lewis used UMWA resources to further the competition-limiting ends of his partnership with BCOA. His organizing drives in the 1950's focused on the small companies exclusively. It appears that the real purpose of the union's campaign of dynamite and sabotage was less to organize these companies than to eliminate them. While battling the non-BCOA operators, Lewis signed "sweetheart" contracts with a number of BCOA members. These secret agreements allowed the favored operator to pay less than union-scale wages or suspend royalty payments to the UMWA's Welfare and Retirement Fund. Coal companies often had difficulty finding money to finance mechanization. Lewis solved this problem for certain BCOA companies by lending them $17 million from the UMWA-owned National Bank of Washington and the Fund.

The small operators fought back against the UMWA-BCOA squeeze. The 1950 contract established a single industrywide wage scale (thus advantaging the most efficient companies), which was an economic handicap to small suppliers. Other contractual provisions devised over the years had the same intent and effect. Small operators brought a number of

antitrust suits against UMWA and BCOA in the 1960's. Two succeeded in winning conspiracy verdicts. In *Tennessee Consolidated Coal Company*, the Supreme Court said the "union and large coal operators, through their National Wage Agreement and its Protective Wage Clause, conspired in violation of the Sherman Antitrust Act to drive small operators out of business."[52] Two months later, the Court affirmed a $7.2 million triple damages judgment awarded to South-East Coal Company against UMWA and Consol. The Court agreed with South-East Coal that the two had engaged in a "conspiracy . . . designed to force South-East and other small coal producers in eastern Kentucky out of the bituminous coal business."[53] South-East's brief charged that the BCOA "was formed specifically to eliminate smaller operators."[54]

The partnership did little to benefit the union's rank and file. Mechanization threw several hundred thousand miners out of work and cut many off from medical and pension benefits. The annual income of those miners who continued working in the 1950's and 1960's did not keep pace with workers in comparable industries such as steel and motor vehicles. Increasing productivity did not lower the frequency of mine fatalities among underground and surface miners. Injury frequency did not improve between 1950 and 1970. Underground mechanization greatly increased noise and dust levels. Unregulated dust conditions produced black lung disease in thousands of miners by the end of the 1960's. Finally, the partnership seems to have required the political disenfranchisement of UMWA's rank and file. The terms and consequences of the partnership probably could not have borne the scrutiny of democratic unionism. The UMWA under Arnold Miller reviewed this touchy subject in this manner:

> Under W. A."Tony" Boyle, who followed John L. Lewis and Thomas Kennedy, the

[51]Lewis quoted in Thomas Bethell, *Conspiracy in Coal*, (Huntington, W. Va.: Appalachian Press, 1971), p. 17.

[52]Tennessee Consolidated Coal Company, 72 L.R.R.M. 2312, 1970.
[53]*South-East Coal Company* v. *Consolidation Coal Company*, 75 L.R.R.M. 2336, 1970.
[54]"UMW and Coal Company Sued," *Coal Age*, June 1966, p. 52.

UMWA leadership grew more and more out-of-touch. Boyle maintained the clamp on dissent and the close collaboration with industry of the late Lewis years, but in the three contracts he negotiated he simply could not deliver as Lewis had done.[55]

But the ordinary coal miner had no way of knowing about union loans to operators, sweetheart contracts, and other sleights-of-hand. He could see that Lewis ran the union autocratically, but so great was his trust in John L. that demands for rank-and-file contract ratification or local election of district officers never elicited much support.

The 1950-70 period is often recalled as an era of "labor peace" and "labor stability." True, there were no contract strikes against the BCOA between 1950 and 1971 (and comparatively few wildcat strikes), but the peace appears to have benefited BCOA and UMWA at the expense of small operators and many coal miners. For non-BCOA companies, market uncertainties and cost pressures—which eliminated 4,779 mines between 1950 and 1973—and UMWA organizing campaigns, can hardly be remembered as a golden age. For many miners, "labor stability" meant unemployment, dust, disease, and fear.

But the UMWA-BCOA partnership did accomplish what it set out to do. Mechanization and "labor peace" boosted productivity and helped BCOA companies to ride out the hard times. Competition was stabilized by eliminating marginal suppliers and through long-term contracts. It was also regulated by a deliberate, coordinated merger movement that began in 1954 among the major companies. When the dust settled, each of the biggest companies had combined with another major producer. With both labor and operators stabilized, big producers were able to increase profits despite a 20-year price freeze and stagnant demand. In 1955, for example, Consolidation Coal and Eastern Associated reported net profits of $12 million and $2 million, respectively. Profits rose to $20 million and $4 million, respectively, in 1960, and to $33 million and $8 million in 1965. The alliance did

one other thing. It created forces within the workplace and labor force that led eventually to the dismantling of the alliance itself. The instability that has characterized the coal industry in the 1970's is part of the process of ending the Lewis-Love partnership.

## The Rebellion: 1969-77

The rebellion of UMWA miners over the terms of the alliance was front-page news in 1969. Beginning with health and safety, the revolt soon expanded to union reform, Fund policies and collective bargaining. Two events—the West Virginia black lung strike (1969) and the 78-victim Farmington mine disaster—propelled coal health and safety problems directly into the public consciousness. In the process, the structure of labor-management relations began to be unveiled.

Coal workers' pneumoconiosis is a progressive, incurable and, in its last stages, fatal disease caused by prolonged exposure to coalmine dust. (See chapter VI.) Although medical authorities in England had recognized CWP as occupationally related in 1942, most American doctors refused to agree. The UMWA did not demand or fund extensive research into the disease in the 1950's, even though the new continuous miners were increasing dust. The Fund did, however, try—unsuccessfully—to persuade the American medical establishment that CWP was a distinctive disease of the trade. Expensive dust-control programs and company-paid compensation for respiratory disability would have undercut efforts to lower production costs. By the late 1960's, respiratory disease among working and retired miners was widespread. Slowly, miners began to link their disability to the dust they "ate" on the job. For years, coalfield doctors had told them that coal dust was not harmful, and some even said it prevented tuberculosis. West Virginia was especially ripe for a black lung protest as 80 percent of its production came from underground mines and almost one-third of the Nation's miners worked there.

Compensation legislation was passed there in February 1969 after a month-long wildcat

---

[55]*The Year of the Rank and File,* p. 70.

strike that idled 42,000 miners, some of whom marched on the West Virginia Capitol carrying coffins. The final version did not incorporate many of the innovative provisions of the original bill, which the Black Lung Association (BLA), an ad hoc group of miners and their allies, had supported.

The West Virginia protest pointed up UMWA's apparent lack of concern for occupational health. The BLA became one base in the political movement by rank-and-file miners over health and safety conditions. To the BLA were added hundreds of disabled and retired workers and their widows who objected to restrictive Fund policies that denied them health and pension benefits.

Mine safety was thrust into the national arena when a Consolidation Coal Co. mine at Farmington, W. Va., blew up in November 1968, killing 78. UMWA president Tony Boyle appeared at the mine and said: "As long as we mine coal, there is always this inherent danger . . . but Cononsolidation Coal was one of the best companies to work with as far as cooperation and safety are concerned." John Roberts, Consol's public relations director, agreed: "This is something we have to live with."

Others did not share their fatalism. The disaster prompted Joseph Yablonski, a UMWA official, to challenge Boyle for the presidency. Strong mine-safety legislation was introduced in Congress that addressed the safety and health problems dramatized by Farmington and the black lung uprising. The UMWA's complacency toward the disaster had discredited Boyle and mobilized the reformers.

The Yablonski campaign merged the rebellion over health and safety with growing dissatisfaction over the absence of union democracy. Yablonski urged improved health and safety, democratization, a merit system, mandatory age-65 retirement for officers, a better grievance procedure, a higher Fund royalty, and an end to nepotism. The demand for union democracy was a reaction to Lewis and Boyle's authoritarianism, which had become a prerequisite for maintaining the alliance. Yablonski lost the election, 80,577 to 46,073, on December 9, 1969. Three weeks

later, he, his wife, and daughter were assassinated by gunmen hired by Boyle and paid from union funds. Because of "flagrant and gross" violations of Federal law, the 1969 election was overturned by a Federal judge and a rerun scheduled for 1972.

During the 1969 Campaign, Federal health and safety legislation was being hammered out in Congress. A reasonably strong bill emerged from the year-long debate. It toughened safety standards, gave the Interior Department broad regulatory and enforcement powers, set dust standards, and provided compensation for black lung victims. The UMWA opposed the strongest legislative measures, as did the operators.

UMWA safety activists and union reformers were bolstered by rank-and-file disenchantment over the contracts Boyle signed in 1968 and 1971. Boyle touted the 1971 agreement as the $50-a-day package. But the $50 came only in the last year of the 3-year agreement and applied only to a small number of miners. Boyle had finally broken the 20-year freeze on the $0.40-tonnage royalty (raised to $0.80), but most of the new revenue went to cover the cost of the pension increase he had contrived during the 1969 campaign. Other benefits changed little.

As the dissidents escalated their campaign against Boyle, some came to understand the prerequisites of labor peace that Lewis had worked out with Love. Some began to see Boyle's "corruption" more as an exaggerated consequence of the union-industry partnership than as a character flaw. This perspective is reflected in the writings of UMWA officials and staff following Boyle's ouster.

The reformers organized themselves into the Miners for Democracy (MFD) in 1972 and nominated Arnold Miller for president. Miller, a disabled miner from Cabin Creek, W. Va., had worked out with the BLA and other reform elements since 1969. He leaned heavily on the advice and skills of a dozen or so young, liberal nonminers to organize the MFD campaign. The MFD slate defeated the Boyle-led incumbents handily in December 1972. The reform movement effected many changes in Miller's first 5-year term of office. Union programs in safety, political action, internal communications, lobbying, and research were begun or strengthened. New leadership was placed in the Fund. District elections were democratized. The rank and file obtained the right to approve or disapprove negotiated contracts. Organizing—the lifeblood of any union—was stepped up. Major wage and benefit gains were secured in the 1974 contract. Yet the 1973-77 period was also one of tremendous rank-and-file unrest. Strikes and absenteeism increased. As the old partnership between UMWA and BCOA dissolved, "labor peace" was replaced by "labor instability." Miners began demanding the right to strike (which they had in the late-1940's and which Lewis conceded in 1950). Some of the discontent was focused on Miller's handling of UMWA. It came from UMWA liberals who thought he was not moving fast enough in many areas and from conservatives or Boyle stalwarts who disliked the changes he was making. Strikes and intra-UMWA turmoil escalated after the 1974 contact.

### 1974-78: An Overview of Labor-Management Relations

The factors discussed above explain the predisposition of miners and mine management to lock in what Lewis once called "a deadly embrace." But the political and workplace struggles that came to the Appalachian coalfields in the mid-1970's were qualitatively and quantitatively different from the Lewis-led contract strikes of the 1930's and 1940's. The growing rebelliousness of UMWA miners culminated in the 3½-month-long contract strike in the winter of 1977-78. The roots of this confrontation went back to the MFD victory in December 1972. Freed from what the MFD reformers called "the tragedy and corruption of the Boyle years,"[56] miners found the change in union leadership stirred their long suppressed yearnings for more say not only over UMWA affairs but also over their working conditions and communities. Democratizing the UMWA also politicized hundreds of miners and led them to believe that they could make other social changes.

But neither UMWA nor the workplace lived up to these new expectations. Many miners

---

[56]*The Year of the Rank and File,* p. 71.

were frustrated by what they saw as failures of the UMWA leaderhip to fulfill its early promises despite the gains of the 1974 contract. Operators and miners clashed over control of the workplace when companies failed to adjust to the new demands of their employees. The conflict over mine health and safety shifted from the hearing rooms of Washington to the individual mine face as UMWA and local mine safety committees pushed for greater protection.

Management's struggle with its employees has been fought over many issues during these years. Absenteeism and disputes over the grievance procedure, compulsory overtime, and job rights repeatedly disrupted coal production. Miners saw management as trying to take away job protections that they had won through the 1974 contract, Federal law, and local custom. Management saw UMWA job rights as an impediment to steady output. Operators believe Federal health and safety regulations and UMWA work rules enabled miners to encroach on traditional management prerogatives. This, coupled with what they perceive to be increasing Federal regulatory "harassment," impedes profitable mining, the industry says.

The expectations raised by union democracy were applied by miners to their communities as well. High wages were hard to reconcile with the frequently dismal quality of community life in the coalfields. Why, miners asked, did the roads have to be so bad? Why wasn't there decent housing? Why no water and sewer hookups? Why the poor schools? Why were the politicians unable or unwilling to change this situation? The sense of deprivation miners felt in their communities was carried into the pits, where it emerged as hair-triggered combativeness over working conditions.

Wildcat strike activity was relatively constant between 1969 and 1974, accounting for more than 550,000 worker-days-idle annually. But, unauthorized work stoppages and lost time jumped dramatically in 1975-77. Part of the increase is attributable simply to the fact that the work force increased 84 percent between 1969 and 1977. A second explanation is that miners were less afraid to assert themselves now that demand seemed to be expanding, labor was in demand, and wages were rising. A third explanation looks to the broad range of unmet rising expectations miners developed in the course of union reform, safety legislation, and political and social changes in the coalfields. Finally, when other channels of communication with mine management and politicians seem to miners to be unproductive, strikes are used as a way for one, often-unrepresented group to get its message across.

A recent study[57] of wildcat strikes in Appalachian coal mines found—among other things—the following:

1. Some companies had no strikes in 1976; others had as many as 17 strikes per mine. Even within the same company, some mines have many more strikes than others.
2. In a sample of four underground mines, it was found that miners at low-strike mines consistently reported better relations with their section foreman than did miners at high-strike operations.
3. Miners at high-strike mines believed it was necessary for them to strike to get management to talk with them. Many miners believed striking would help resolve disputes in their favor. Miners at high-strike mines reported that management was generally uncooperative with the union and unwilling to settle a dispute when the union had a good case.
4. Ninety percent of the miners believed that one reason for wildcat strikes was the excessive delay in the grievance procedure. Arbitrators were distrusted.
5. Local union officials do not lead strikes. They are led by rank-and-file miners. District presidents do not appear to have the political strength to take any measures against wildcat strikes.
6. The findings suggest that wildcat strikes are related to management practices in dealing with employee matters.

---

[57]Jeanne M. Brett and Stephen B. Goldberg, p. 23 ff.

7. Many of the problems of the grievance procedure, which may contribute to strikes, can be reduced if management and the union resolve more of their disputes at the minesite, rather than through recourse to arbitration.

Wildcat strike activity dropped in 1978. But it is premature to conclude that coalfield peace will reign. The effects of the 4-month contract strike that ended in March left UMWA miners in debt and anxious to work. Thousands have been working on-again, off-again because of poor market conditions and a 10-week strike by railroad workers on the N&W line. Absenteeism declines when the mines are working short weeks, and rises when 6-day weeks and lots of overtime are the rule. Further, miners can't strike if they are not working. A truer indication of the mood of the work force should develop in 1979 and beyond. It would be an error to believe that steady growth in demand for coal will necessarily lead ro good labor-management relations. Growth may, in fact, lead to more militant demands and less concern for cooperating with management.

## 1974 UMWA Contract

The 1974 BCOA-UMWA negotiations took place in new historical circumstances. The reform UMWA rejected partnership with BCOA. Demand was growing. Profits had risen for 4 straight years as a result of price increases and the 1973 OPEC embargo. Net coal income rose from $128.4 million to $639.5 million—a 398-percent increase—for 24 major companies, the Bureau of Mines reported. Many major coal companies had been absorbed by energy companies or conglomerates, whose greater resources raised UMWA negotiating goals. The union came to the 1974 bargaining table with two basic purposes: 1) to make up ground lost over the previous 20 years, and 2) to work out new terms for future labor-management relations.

BCOA conceded an unprecedented wage and benefit package in 1974. Wages were increased more than ever before. A cost-of-living escalator and paid sick leave were granted for the first time. Miners were given the right to withdraw individually from conditions any one of them judged to be an "imminent danger." Paid vacation days were increased significantly, but a two-track pension plan was introduced. Current retirees received a $75 to $100 monthly raise to $225 to $250, depending on whether they received Federal black lung compensation. But the pensions of miners who retired after 1975 were calculated according to a sliding-scale formula based on age at retirement and years of employment. Their average pension was $425 a month. UMWA negotiators believed this was the first step in phasing in a benefits-according-to-service principle they thought fairer than the flat payment. Newly retired miners liked their higher benefit levels, but much bitterness arose among the 80,000 pensioners locked into the lower, flat rate.

BCOA agreed to continue providing first-dollar medical coverage for working and retired miners and their dependents. Health care and pensions had been administered by a single UMWA Welfare and Retirement Fund until 1974, when it was divided into four separately funded but jointly administered trusts. The two "1950" plans provided health care and pensions to miners who had retired before 1976 and were financed by a tonnage-based royalty. The 1974 plans were funded on a tonnage and hours-worked basis. They provided benefits to working miners and new retirees. Inasmuch as benefits were tied into production-related financing formulas, no benefits were guaranteed. In the event that future production did not meet the income estimates used by the 1974 contract negotiators, the Funds would be unable to meet their obligations.

This happened in the spring of 1977. Although the 1974 contract had increased Funds' income 62 percent in 2 years, revenues were still not adequate to continue services to the more than 800,000 beneficiaries.[58] Overly opti-

mistic production projections, inflation, lack of adequate cost controls, the inability of the UMWA to organize new mines, wildcat strikes, and unexpectedly high growth in the beneficiary population—all forced the Funds to retrench. Either benefits to miners or payments to health providers had to be cut. The Funds chose to cut benefits by instituting deductibles, which ended the tradition of first-dollar coverage established in 1950. Miners had to pay up to $500 per family for medical services. The Funds also ended negotiated retainers with about 50 coalfield clinics in favor of fee-for-service reimbursement. This had the effect of forcing the clinics to cut services. Miners throughout the East stopped work for 10 weeks to protest these cutbacks. BCOA refused to refinance the Funds because it felt that wildcat strikes had caused the cash crisis.[59]

BCOA also came to 1974 collective bargaining with a new strategy. The old union-management partnership that had glued the big operators together since 1950 was discarded. The Lewis-Love alliance might be dead, but the operators hoped major wage and benefit concessions would buy labor stability.

The BCOA strategy did not work. The UMWA leadership was not strengthened by the BCOA's concessions in the 1974 contract. In fact, the plan backfired. If anything, the improvements secured by the union raised the expectations of the membership more than ever. The contract was ratified by 44,754 to 34,741. The two-tier pension plan created animosity and bitterness. The new grievance procedure did not settle workplace disputes expeditiously. Thousands of cases were appealed to arbitrators. Many cases were not decided for more than a year. Unauthorized work stoppages continued to increase in 1975 and 1976 despite contract gains. The number of days lost in wildcat strikes in 1975 and 1976 more than doubled the 1972-73 experience. Many miners felt the 1974 contract should have given them the right to strike over unsettled

---

[58]"Report of the Health and Retirement Committee," *Proceedings of the 47th Consecutive Constitutional Convention of the United Mine Workers of America,* Sept. 23-Oct. 2, 1976, (United Mine Workers of America) pp. 348 ff.

[59]Although wildcat strikes contributed to the cash shortage, a story in the *Wall Street Journal* (Dec. 3, 1977) correctly pointed out that production lost to such strikes amounted to only 3 percent of projected income, and was not the principal cause of the Funds' financial troubles.

grievances, a right that Federal court decisions had eliminated in the early 1970's. Rather than solidify Miller's standing with the UMWA, the 1974 contract divided him from part of his constituency. Miller's efforts to discipline wildcat strikers further alienated some of his membership. Some observers see the 1977-78 coal strike as an extension of the rank-and-file dissatisfaction that had been building within the UMWA and against the operators since 1974. UMWA spokesmen, on the other hand, interpret the turbulence of the 1974-78 period more as an acting out of newly discovered strength unleashed by the reform victory than as a reflection on Miller's leadership.

## 1977-78 Negotiations

BCOA was most concerned in the 1977 negotiations over reasserting management authority over labor at the minesite and curbing wildcat strikes, absenteeism, and certain work rules. Related to this was the matter of productivity, which had fallen steadily since 1969. BCOA negotiators abandoned their 1974 hope that Miller could be transformed into an instrument of labor pacification akin to John L. Lewis. From that reluctant conclusion, the operators were

> . . . grasping for something in 1977-78 negotiations. We realized the UMWA international had lost control of their people. We were grasping for ways to put stabilization into a contract.[60]

BCOA hoped to effect "labor stability" —a phrase the industry coined—by winning these specific demands in 1977:

1. the unarbitrable right of the employer to discharge an employee for alleged striking;
2. cash penalties against employees for unexcused absences and striking;
3. the employer's unarbitrable right to terminate a new employee during his first 30 days of employment;
4. the employer's right to set up production incentives;

5. the employer's right to schedule production or processing work on Sundays;
6. the employer's right to change shift starting times at his discretion;
7. reduction to 45 days of the 90-day protection period for new employees (who are not allowed to operate face machines or work beyond sight and sound of an experienced employee), and
8. increased restrictions on the mine safety committee.[61]

BCOA also insisted on restructuring UMWA's health and pension system. BCOA demanded that the 1974 health trust (for working miners) be replaced by private medical insurance arranged through individual operators. UMWA was to lose its influence over coalfield medical care by this change. BCOA also insisted that coinsurance be made permanent. Coinsurance would save BCOA members $70 million to $75 million annually. Fund expenditures would go down as well. BCOA would not consider equalizing pension levels or significantly increasing benefits. The union did not demand pension equalization even though the 1976 UMWA Convention resolved that the union's "Bargaining Council . . . give high priority in the next national contract negotiations to equalizing pensions."[62]

Finally, the operators did not like the idea that miners felt UMWA was providing their pensions and medical services. Employee loyalty was not created by carrying a "UMWA health card" or receiving a "UMWA pension" each month. Operators saw no advantage to them in the Funds' subsidizing several dozen clinics that everyone casually referred to as "the miners' clinics." By switching to a strict fee-for-service system based on operator-arranged insurance, BCOA hoped to save money and build employee identification.

It is not easy to discern Miller's 1977 negotiating goals and strategy. Certainly, UMWA negotiators believed they were in a

---

[60]Allen Pack, "President of Cannelton Discusses Coal Issues," *Charleston Gazette*, Apr. 12, 1978, p. 1C.

[61]"Memorandum of Summary of Major Improvements and Changes Contained in the Settlement Between the United Mine Workers of American and the Bituminous Coal Operator's Association," Feb. 6, 1978.

[62] *Proceedings of the UMWA's 47th Constitutional Convention,* pp. 357-358.

weaker position than in 1974. Miller entered bargaining after narrowly winning re-election in June 1977 with 40 percent of those voting in a three-way race. UMWA lacked a research department and was not well prepared for bargaining. Perhaps the union's biggest disadvantage was coal's market situation. Utilities had doubled their normal stockpiles. Demand for Appalachian metallurgical coal was weak. Prices for both steam and met coals were generally static and, in some cases, falling. BCOA operators and customers could absorb a long strike, particularly as the long-term effects were appraised as outweighing the short-term losses.

According to newspaper accounts at the time, Miller had two principal goals: winning the right to strike at the local level and restoring first-dollar-coverage health benefits. The UMWA won neither, ultimately. Wildcat strikes had bedeviled Miller for 3 years, so skeptics and others surmised that he planned to trade the right-to-strike for softening of the BCOA's "labor stability" package. The right-to-strike was not included in any of the three negotiated agreements. Miller sought the restoration of health benefits to precutback levels and their full guarantee over the course of the contract. Apparently, high priority was not given to opposing deductibles, private insurance plans, and clinic cutbacks.

By contrast, the bargaining priorities outlined by the more than 2,000 delegates to the UMWA convention in 1976 were clear and aggressive. Most of all, they did not want to lose any of the ground gained in 1974. That meant no "labor stability" package, no deductibles, and no reduction in benefits. The right to strike had mixed support among the membership. The convention strongly endorsed it for use in "local issues threatening the safety, health, working conditions, job security, and other fundamental contract rights." Convention delegates also endorsed:

1. rank-and-file participation in the negotiating process;
2. across-the-board wage increases;
3. abolition of compulsory overtime;
4. more personal and sick leave days;

5. a 1-cent-per-hour increase for every 0.2 increase in the Consumer Price Index;
6. no maximum (cap) to the Cost of Living Adjustment;
7. seniority based on length of service alone (rather than on the company-determined ability to perform the work and seniority); and
8. modifications of the grievance procedure.[63]

Health and safety was another priority. The delegates called for:

1. a full-time, company-paid health and safety committeeman at each mine;
2. a minimum of one UMWA miner trained as an emergency medical technician on every operating section on every shift at each mine;
3. automatic, continuous dust sampling;
4. a UMWA miner employed as a full-time dust weighman "with the authority to enforce dust standards;
5. strengthened language concerning the shutdown powers of the mine safety committee;
6. no arbitration of safety committee decisions;
7. tougher language protecting the right of the individual miner to withdraw from hazardous conditions;
8. a 90-day training period for new miners, and a 30-day time limit on arbitrators' decisions on safety and health disputes.

None of these was included in any of the tentative agreements. Obviously, the union's negotiators were not expected to win this entire shopping list of convention-endorsed objectives. But the scope and quality of the changes envisaged by the rank-and-file delegates suggested they expected a feisty, offensive posture from their negotiators in the 1977 talks. Perhaps the UMWA team lived up to this sentiment in private discussions with the BCOA, but miners found the contracts presented to them in February and early March 1978, did not measure up.

---

[63]See "Collective Bargaining Committee Report," *Proceedings of the 47th Consecutive Convention.*

UMWA miners were off the job for 109 days in the winter of 1977-78. Two months after their contract expired on December 6, 1977, the union's 39-member bargaining council rejected the first tentative agreement Miller and BCOA had worked out. A month later, miners rejected a second draft, 2-1. As this occurred, newspapers were filled with administration predictions of impending power shortages and strike-caused unemployment of up to 3½ million workers. No power shortages occurred, and only one State—Indiana—seriously enforced mandatory power cutbacks. The Bureau of Labor Statistics in the Labor Department was reporting to the White House only 25,500 strike-related layoffs at the height of the strike.[64] Still, the administration sought and received a temporary back-to-work order under Taft-Hartley procedures invoked on March 6. The prospect of increased Federal intervention in their affairs brought the two sides together, and they hammered together a third agreement. By that time, the Federal judge who had issued the temporary Taft-Hartley order refused to extend it on the grounds that the administration had never proved the existence of a national emergency. On March 24, 56 percent of those miners voting accepted the contract. The third contract softened some of the BCOA's demands embodied in the earlier drafts. But it fell short of rank-and-file objectives in many areas. Why did miners accept it?

First, many were beginning to suffer economic hardship. The union had no national strike fund, and the lengthy wildcat strikes of 1977 had cut into miners' savings just months before expiration of the 1974 contract. By mid-March there was an inverse relationship between a miner's militance and the number of mouths he or she had to feed. The contract represented income.

Second, there was widespread distress about health and pension benefits. As the strike wore on, fears increased that the entire pension pro-

gram might collapse. (Benefits had been suspended at the outset of the strike.)

Third, miners were increasingly vocal in expressing their lack of confidence in their own negotiators. Summarized, their argument seemed to be: "We can stop the operators from taking away what we won before, but we can't make our negotiators get what we want." This belief led to a sense of fatalism about what could be accomplished by continued striking.

Fourth, the ratification process itself weighed in favor of getting a settlement sooner rather than later. Miners understood that ratification involved at least 10 days. Rejection of the third contract would mean more delay before negotiations resumed, and then further delay before new terms were agreed on.

All of these factors affected the ratification vote. It is important to understand that in casting his ballot a miner was not necessarily taking the final step in a rational process of assessing the objective advantages and disadvantages of the contract he had been asked to consider.

What explains the protracted inability of the negotiators to agree to a contract that the miners would accept? One answer lies in the ability of each side to endure a long shutdown and the expectation that by doing so the final offer could be a net improvement over any earlier terms. A second is BCOA's demand for unprecedented changes in the status quo. The magnitude of these changes, together with the barely disguised threat to resort to company-by-company bargaining, was perceived by miners as an all-out attack on their way of life, its culture, and its protections. Third, the perceived cleavage between Miller and his membership encouraged BCOA to demand drastic contract revisions. Had the union side been united, the operators might have compromised sooner. Fourth, had BCOA not insisted on "labor stability" penalties, it is arguable that miners might have accepted BCOA's other demands more quickly as part of a package. However, miners felt the BCOA's stability demands were a reassertion of the operator's wish to do as they pleased with their employees. The stalemate came to focus on the

---

[64]The administration's spurious estimates of unemployment and power shortages were examined by the General Accounting Office in a report to Cong. John Dingell, entitled, "Improved Energy Contingency Planning is Needed to Manage Future Energy Shortages More Effectively," Oct. 10, 1978.

Photo credit: Douglas Yarrow

UMWA coal miners protesting changes in medical benefits in Washington, D.C., during a month-long wildcat strike that shutdown most Appalachian production, 1977

shared perception that "rights" were at stake: the employee's right to job protection, safety, and security; and the operator's right to manage his business according to his best judgment. When each side perceives that its rights are challenged, wars of attrition are common. Fifth, it appears that because of poor market conditions, metallurgical coal producers and some other eastern operators were not terribly hurt by a long strike. Had there been no strike, hundreds of mines would have shut down or laid off workers because of excessive utility stockpiles, static prices, and soft markets.

BCOA was not pleased with the final version, although it had won major changes in the 1974 contract. Miners saw the final product as less punitive than earlier drafts but not as good

as the 1974 agreement in many respects, particularly with respect to health benefits.

## The 1978 Contract Terms

### Discipline

The harsh language of the first two drafts concerning management's right to fine and discipline wildcat strikers was deleted. In its place, however, is a memorandum of understanding that continues decisions made by the Arbitration Review Board (ARB). The ARB decision decided in October 1977, gives employers the right to discharge or selectively discipline employees who advocate, promote, or participate in a wildcat strike. ARB 108 is less punitive than the proposed "labor stability" provisions as employees can appeal grievances through arbitration.

No significant work stoppages have occurred since the 1978 contract went into effect. Some mines report less absenteeism. Are these short-term phenomena, or are they trends that are likely to continue and if so, to what extent are they traceable to the provisions of the new contract?

The second question is the easier. The recent decrease in work stoppages and absenteeism probably cannot be credited solely to the new contract because, with one exception, no new terms bear directly on this issue. The sole exception occurs in the article governing settlement of disputes—the grievance procedure. A small but significant change was written into the first step. A section foremen now has the authority to settle a complaint at the minesite within 24 hours. His decision no longer sets any precedents in the handling of other grievances. Formerly, mine management delayed conceding a point in one matter for fear that it would be binding for the duration of the contract. Although individual miners seem to be getting grievances settled faster than before, redundant disputes over the same points may be occurring. A spot check of different districts in June 1978 produced no definitive information about dispute frequency but did confirm that foremen had more latitude on grievance handling. The officers of the union's largest and most strike-prone district (District 17 in southern West Virginia) reported then a very sharp drop in the number of grievances being referred to union representatives. The layoffs and short weeks that idled thousands of miners in Appalachia in the past year probably discourage grievances being filed and wildcat strikes from happening. If demand perks up in these areas, grievances may rise proportionately.

Whether the reduction in wildcat strikes is a short-term phenomenon is a much more difficult question. Miners are still recovering from a strike that lasted much longer than expected. They have been concerned with retiring large debts and replacing whatever savings they had built up. Like other Americans, they have also been troubled by continuing inflation. Perhaps

the best explanation, again, is simply that miners can't strike if they don't work. Almost 20,000 Appalachian miners were laid off in the summer of 1978 because of soft metallurgical-coal markets, productivity-boosting plans, and the strike against the N&W railroad. About half that number were working irregularly in the winter of 1978-79. On the other side, many of BCOA's larger members appear to have made a special effort to resolve grievances as promptly as possible. Some claim that because of the new clause, more grievances are being resolved in the miner's favor. In any case, experience with the 1978 contract is insufficient to permit drawing many meaningful conclusions about whether it will lead to a sustained reduction in the number and frequency of wildcat strikes and absenteeism.

In the long run, wildcat strikes and absenteeism are unlikely to be brought under lasting control until the union's internal situation stabilizes and the industry adopts a more generally enlightened attitude toward labor relations. It is too soon to know whether either will occur.

## Productivity Incentive Plans

The UMWA had traditionally opposed plans to increase production by offering cash bonuses for tonnage over a stated quota. The union argued against bonus plans on two grounds. First, if bonus plans succeeded in significantly increasing a worker's real spendable income, the union would find negotiation increasingly difficult for substantial across-the-board wage gains. The union feared that wages would be shifted from an hourly basis to piece rates, which is considered regressive. Second, UMWA argued that bonus systems encourage risky short-cuts, lack of proper equipment maintenance, and speed up—all leading to more accidents and disease. BCOA made incentive systems a central demand in its negotiating. The post-1969 decline in productivity might be turned around, the operators thought, if cash bonus plans were adopted. UMWA agreed to this change.

The new contract language provides for a majority vote (among those miners voting) before an operator can adopt a bonus plan.

Once accepted, it cannot be rescinded by the employees. Each plan would provide monetary incentives for production or productivity increases. (The contract language does not specify what the cash bonus is pegged to). One of the conditions of the plan is that it "does not lessen safety standards as established by applicable law and regulations." Some operators and some miners may not interpret safety "standards" as being the same as safety performance (that is, fatality and injury rates and numbers). It is possible that accident frequency could rise, but so long as management standards or number of injuries did not change, the bonus system would not be voided. Nothing is included about health standards or experience. The cash bonus system will be popular with many miners, but its adverse impact on workplace health and safety may be significant.

Operators have adopted two different approaches to increasing productivity in recent months. Consolidation Coal has won acceptance at four mines of plans that tie the level of the cash bonus to increased tonnage and number of injuries. It is too early to tell if this approach will raise output without compromising safety. Some union officials are worried that both miners and operators will fail to report injuries in order to preserve the cash bonuses. A second approach taken by another operator seeks to raise productivity by eliminating nearly 2,100 jobs, representing about one-sixth of its workforce in West Virginia. A company spokesman said that reduction was not prompted by poor market conditions; rather "it's strictly to improve productivity." This company hopes to maintain the same level of output after the layoffs, thus increasing productivity.[65]

Aggregate productivity statistics over the next 2 to 3 years will tell relatively little about the effects, if any, of the new incentive clause. Meaningful evaluation will require close monitoring of individual mines where incentive plans are put into effect to determine relationships between these plans and changes in productivity, accident frequency, and dust levels.

Health Care

The most dramatic change in the 1974 contract involved the refinancing and control of UMWA's health plan. Under the 1978 agreement, all working miners and recent retirees shifted to company-specific, private insurance plans. A Funds medical plan will exist for pre-1976 retirees only. Coinsurance—up to $200 per year per family—was instituted. All services, including those provided by the clinics, will be reimbursed on a fee-for-service basis. Health benefits are supposedly guaranteed, but uncertainty remains in some areas about what these guarantees mean in practice. The imposition of deductibles for working miners and retirees ends first-dollar coverage, which had been the rule since 1950. Because they apply only to physician care and medication, some health analysts believe miners will be discouraged from buying preventive physician care. The experience of other medical systems—the Stanford University Group Health plan and the Saskatchewan plan—suggests that the introduction of coinsurance reduces demand for physician services.[66] The entire health-benefits package is now seen by many miners as having been seriously compromised. They were widely dissatisfied with the administration of the Funds, and the current situation is commonly perceived as substantially worse. Strikes over the insurance reimbursement issue have been narrowly averted several times over the last year. The dismantling of the Funds is also an emotional issue that is unlikely to dissipate during the term of the 1978 agreement. For management and union leadership alike, it will remain a difficult problem. Strikes over coalfield health care are likely.

Occupational Health and Safety

None of UMWA's occupational health and safety demands survived the bargaining. The 1974 language continues in force. At each UMWA mine, a health and safety committee is elected (generally three miners). Committee members are entitled to special training. They have the right to close sections of the mine

---

[65]*Coal Industry News,* Sept. 18, 1978.

[66]Anne A. Scitovsky and Nelda McCall, "Coinsurance and Demand for Physician Services: Four Years Later," *Social Security Bulletin,* May 1977.

when they find conditions of "imminent danger." If their employer challenges their judgment, Federal or State inspectors may be called in to settle the question, or the issue may be taken through arbitration. The committee's action can be reversed only after the fact. This provision, stronger than most labor agreements, dates back to 1946, when John L. Lewis won it when the mines were under Federal control. If an arbitrator upholds an employer's accusation that a committee member acted "arbitrarily and capriciously," the miner may be removed from the committee but cannot be fired or otherwise disciplined for his official actions.

These provisions contrast starkly with those generally in effect at non-UMWA mines, where the safety committee, if one exists at all, is often reluctant to enforce standards for fear of jeopardizing its members' jobs.

## Workplace Relations

The basic issues that created friction under previous agreements were not resolved and can be expected to cause difficulties in the future.

For example, management retains the right to operate mines 6 days a week (a BCOA proposal to permit operation 7 days a week failed). The compulsory sixth shift is a primary source of discontent among miners at operations where the 6-day week is scheduled. Compulsory shift rotation is another source of widespread discontent. Wildcat strikes have erupted over it. Some miners believe management routinely awards preferred jobs on the basis of favoritism, and that the contract language does not protect a miner's right to be awarded a job on the basis of seniority and ability to perform the work. Some of these conflicts might be resolved if each side were willing to meet and work out a compromise. It is still too early to tell whether this will happen under the 1978 agreement.

## Education and Training

Education and training have been foreign concepts in the coal industry through much of its history. Both are inextricably tied to improving workplace relations. The significance of the changes that have taken place in the industry's work force over the past decade are difficult to exaggerate. The industry laid off 70 percent of its work force between 1950 and 1970, then turned around and began hiring again in record numbers when demand picked up. Today, most miners are in their 20's and 30's. The supervisory work force is also overwhelmingly young. In some mines, informal alliances between young foremen and young miners have reportedly occurred in response to perceived intransigence by upper management. Low-level supervisory skills are no longer sufficient. Most industry spokesmen acknowledge the shortage of experienced mine foremen who are able to maintain good rapport with hourly employees. Some companies are trying to upgrade supervisory skills; others hope to get by with what they have. The likely safety and production payoffs of more capable foremen cannot be understated.

The 1978 contract provides for employer-developed orientation programs of not less than 4 days for inexperienced miners and not less than 1 day for experienced workers. The contract specifies that each program should emphasize health and safety. The trainee period—during which a new miner cannot operate mobile equipment or work beyond sight and sound of another miner—was cut from 90 to 45 days.

Education and training for miners can take many forms. A sample list of priority objectives might include:

• Machine-specific training beyond basic requirements of State and Federal law.
• Training in dispute settlement directed at both supervisors and members of the UMWA Mine Committee.
• Training for the Health and Safety Committee to familiarize them with occupational health and safety hazards and Government regulations.
• Interchangeable skills for work crews within the mine. The benefits from such training are obvious. Safe working habits are encouraged. The impact of absenteeism is blunted when a miner can step into

the job of an absent worker without exposing himself and other crew members to hazards arising from unfamiliarity with a machine or work procedure. Recent Federal regulations on miner training are a step in this direction, but critics argue that they are inadequate in many respects.

## Wages

Most miners approved the wage package. Unadjusted standard wage rates rose from 1977's $55.68 per day for the top-paid miner to $74.32 per day in the first year. The lowest paid underground miner went from $50.38 to $64.78 per day in the first year. The gap between lowest- and highest-paid miners is $8.64 per day, as it was in the previous contract.

The 1974 agreement initiated a capped cost-of-living escalator pegged to increases in the Consumer Price Index. In the final adjustment period (November 1977), the maximum inflation adjustment was $0.98 per hour. The new contract includes no inflation adjustment mechanism in the first year, and a $0.30 maximum per hour raise in the second and third years. Although the 3-year, hourly wage increase amounts to about 30 percent, the increase in real income is likely to be closer to 3 percent after inflation, deductibles, and higher tax payments are taken into account.

## Union-Industry Relations

In the last round of negotiations, some coal operators were encouraged by the Carter administration to break from the BCOA bargaining structure and conclude contracts with UMWA miners on a company or geographical basis. Proponents of this innovation argue that decentralization would reduce the impact of contract strikes because no common contractual expiration date would exist. Some also believe that the more profitable companies could buy "labor stability" through comparatively "fat" packages, leaving less-profitable companies to settle as best they can. Many newer companies—as well as some miners—feel the pension obligations of older BCOA companies limit the size of wage gains that can be negotiated. The other side of the debate argues that regional or company-specific bargaining will create more problems than it promises to solve. They point out that this kind of bargaining historically has destablized this industry. The signing of a number of agreements (each of which embodies a different set of wages and benefits) would create strike conditions because of the differences. It is improbable that miners at one mine will work very long for less than miners at an adjacent mine. It is also likely that miners who strike when their contract expires will "picket out" other mines in order to increase economic pressure on their employer. This chain reaction could go on indefinitely as staggered expiration dates are reached. It is debatable whether decentralization would produce a net reduction in time lost to strikes. Had the 3½-month strike actually imperiled the country, solid ground would have been established for a radical restructuring of the UMWA-BCOA structure. But as a national emergency never existed—and is even less likely to in the future—it does not appear to be imperative to change the status quo.

## Community Development

Community development problems are beyond the scope of the collective bargaining process, but they must be addressed more effectively over the term of the 1978 agreement or the basic unrest that has characterized coalfield life will not abate. The most significant difference in strike activity in the 1970's from earlier periods was that the former sometimes was triggered by nonworkplace social and economic conditions, such as a gasoline rationing plan in West Virginia, controversial school textbooks, unsafe coal-waste impoundments, Federal court decisions, pending Federal black lung legislation, and cutbacks in health benefits. Before the 1970's, strikes were rarely caused by matters other than those arising over working conditions or contract terms. Recent field studies suggest that inadequate, unstable living conditions contribute to absenteeism, wildcat strikes, and lower productivity. The President's Commission on Coal, which began its investigation in September 1978, is exploring the relationship between living conditions and work relations, as well as possible strategies for improving the quality of coalfield life.

## Non-UMWA Labor

Roughly 30 percent of 1977's 230,000 coal workers did not belong to UMWA. Some belong to other unions, such as the International Operating Engineers or PMU. The rest belong to company-approved associations or to no labor organization at all. In the East, most non-UMWA labor is found in small, mostly non-union mines in eastern Kentucky, Virginia, and Tennessee. PMU is based in Illinois. In the West, miners generally are either unorganized or non-UMWA. With a few exceptions, UMWA mines there are underground operations.

Still, UMWA exerts a good deal of indirect influence over non-UMWA operations. In the 1977-78 strike, much non-UMWA production in the East was shut down, too. Wages and benefits in non-UMWA mines are often pegged to UMWA rates. Non-UMWA operators will often offer higher wages than the UMWA scale in order to keep UMWA from organizing its miners. Three UMWA contract provisions appear most objectionable to non-UMWA operators: job security protections, safety and

health rights, and benefits for retired miners. If UMWA succeeds in upgrading its western surface-mine agreements to prevailing non-UMWA standards, it will probably have better organizing success among western surface miners as nonwage issues take on increasing importance.

## Conclusion

The 1977-78 shutdown was one episode in a long-running coalfield drama; it was not the final act. Labor peace is not unreasonable for a coal operator to expect, but it is unlikely to be achieved by the approach adopted by BCOA in the recent talks. Too much bitter history and too many contemporary conflicts exist to enable threats and penalties to promote stability, let alone peace. Any long-term approach requires recognition that many of the problems have become embedded in the American system of coal mining—its production process, work relations, work environment, communities, and culture. Changing the consequences of this system—absenteeism, strikes, distrust, arbitrariness—means changing the components of the system that produce conflict.

# REGULATORY RESTRICTIONS ON MINING

Federally mandated permits and operating methods have become significant factors for the mining industry. The principal requirements arise under the Surface Mining Control and Reclamation Act (SMCRA), the Mine Safety and Health Act, and the Clean Water Act. These are discussed in detail in chapter VII. This section addresses the ways these regulations affect mine operators as well as their potential to affect the supply of coal. The major areas of concern are the increased leadtime required to open a new mine, increased capital and operating expenses, the designation of certain areas as unsuitable for mining, and a sense of "harassment" within the industry.

A longer leadtime for opening new mines could result from the need for additional planning, design work, and permits to comply with

Federal regulations. Permits are required under SMCRA, the Clean Water Act, and the Resource Conservation and Recovery Act (RCRA). SMCRA mandates State permit systems in accordance with Federal guidelines that include comprehensive performance standards for surface mining operations and for the surface effects of underground mining. These standards are intended to prevent adverse environmental impacts such as ground and surface water contamination, degradation of land quality, and subsidence. Mine operators must demonstrate, as a prerequisite to obtaining a mining permit, that the land can be restored to a postmining land use equal to or greater than the premining use. The Clean Water Act and RCRA also mandate State permit systems in accordance with Federal guidelines. Under the Clean Water Act,

Federal effluent limitations designed to achieve national water quality goals apply to all active mining areas (surface and deep) including secondary recovery facilities and preparation plants. Solid waste disposal standards under RCRA are designed to control open dumping; substantial constraints could be imposed on the disposal of mine wastes if they were declared to be hazardous. Additional design restrictions may be imposed by the requirements of the Mine Safety and Health Act. In addition, the planning stage for leases involving federally owned coal or for mines on Federal lands may include the preparation of an environmental impact statement under the National Environmental Policy Act (NEPA) or the Federal leasing program.

Before the implementation of these environmental and health and safety regulations, a mine could be opened in little more time than that required to conduct geologic surveys, assemble the tract, order equipment, and prepare the site. This process typically might have required 5 years. However, permitting and review and other mandated planning have added significantly to this leadtime. Coal operators now estimate that the opening of some new mines could require 8 to 16 years from initial exploration to full production, depending on the characteristics of the mine site and the resources of the mining company (see figure 16). However, if many of the required activities proceed simultaneously, and if full compliance with applicable legislation is achieved at the outset, the leadtime should not be increased substantially.

However, if greatly increased leadtimes become the norm, they could constrain coal supplies for the high-growth scenario by the late 1980's, when a rapidly expanding coal industry would need new mines not yet in the planning stage. Supplies should be more than adequate for all scenarios until then. In addition, substantial delays in the opening of new mines could alter the structure of the industry. Because utilities will not contract for coal not needed for many years, coal mine planning would have to begin without an identified market. Larger mining companies can maintain fully permitted nonoperating mines for poten-

tial customers, but smaller companies could lose their historic ease of access into the market because the planning and permitting costs would be incurred so far in advance of a return. The flexibility of coal supplies also could be reduced greatly because only those mines that are well into the planning stage could be considered viable suppliers.

Increased capital and operating expenses for surface mines would result primarily from reclamation and other environmental protection requirements. For underground mines both environmental and occupational health and safety requirements increase production costs. Before passage of SMCRA, surface mine reclamation costs averaged $3,000 to $5,000 per acre. Depending on seam thickness, this cost translates to a range of about $0.20 to $1.00 per ton. Industry estimates of the increased costs attributable to SMCRA vary widely, depending on the characteristics of the site.

Any cost increases probably will not limit the industry's ability to supply coal, but they will make coal more expensive and could force small, capital-short companies out of business. In addition, coal operators could have difficulty obtaining reclamation bonds for areas that are difficult to reclaim because of the stringent, detailed SMCRA requirements for these sites. Federal programs are available to help small companies meet these increased costs, but it is unclear whether they will be adequate.

A third concern raised by regulatory restrictions is the designation of areas as unsuitable for mining. Except for valid existing rights, SMCRA prohibits surface mining on Federal lands valuable for recreation or other purposes (such as national parks, wildlife refuges, wilderness areas, wild and scenic rivers) and on much of the national forest lands. In addition, SMCRA requires the States to institute planning processes for designating areas unsuitable for all or certain types of surface mining. These include areas where reclamation would not be technologically or economically feasible; where mining would be incompatible with existing land use plans; where it would adversely affect important historic, cultural, scientific, and aesthetic values; where it would result in

Figure 16.—Permitting Network for Surface Mines

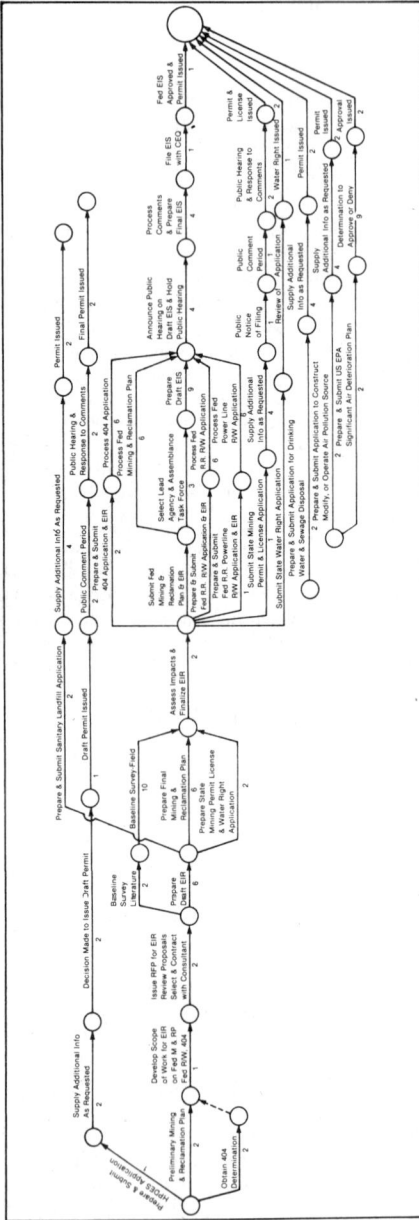

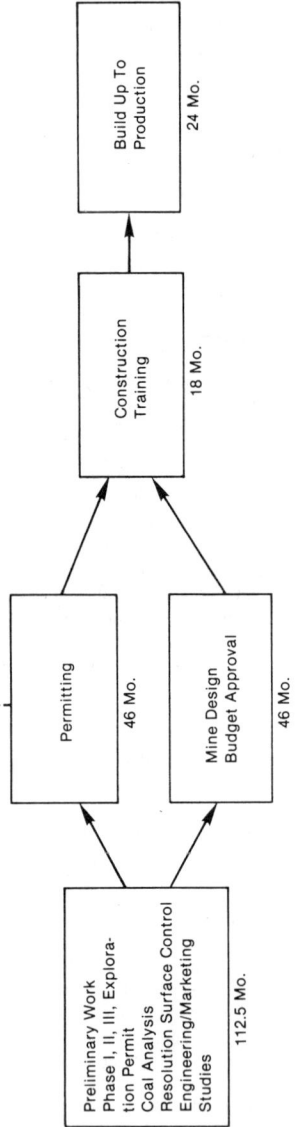

substantial loss or reduction in long-range pro- ductivity of water supplies or food or fiber products; or where it would endanger life or property in areas subject to flooding or un- stable geology. Until these State planning processes have been established, it is not clear to what extent they could limit coal supplies.

The final factor related to Government regu- lations is the sense of "harassment" en- gendered within the industry by Federal reg- ulation of what historically had been the in- dustry's prerogatives. This hostility can result from the concerns discussed above (delays in mine openings, increased costs) or from inter- ference in mining practices by Federal inspec- tors, and it can occur among mine operators as well as between operators and miners or the public. Mine operators may perceive frequent inspections as harassment, especially if the operators feel that they are making a rea- sonable effort to comply with the regulations or that the regulations are counterproductive. For example, mine health and safety inspectors may be seen by management as allied with labor or by labor as cohorts of management, resulting in increased tension between these

groups. Similarly, SMCRA inspectors may be perceived by mine operators as allied with the "environmentalists" and vice versa. On the other hand, all three groups—mine operators, miners, and environmentalists—may feel that the inspectors are incompetent and do not pro- tect any interests adequately.

An additional source of hostility within the industry is the automatic citation provision of the Mine Safety and Health Act, which re- quires an inspector to cite every violation re- gardless of its severity or its potential for im- mediate correction. Automatic citations and civil penalties are intended to deter noncom- pliance. However, as with frequent inspec- tions, they may be perceived as harassment by operators who feel they are making a good- faith effort to comply with regulations. On the other hand, miners claim, that Federal inspec- tors do not always issue the automatic cita- tions because their experience tells them that the specific circumstances do not warrant it. Probably most inspectors use judgment in issu- ing citations; some are too lenient, others are too literal.

# PRODUCTIVITY

Productivity is a measure of the efficiency of an industrial process. It expresses a relation- ship between a unit of output and the effort that goes into it. In coal mining, productivity is expressed as tonnage mined per worker per shift.

Production is not the same as productivity. Production refers to total output or tonnage; it says nothing about the efficiency of a coal mine or industry. The two concepts are related. Coal production can rise through increasing productivity; that is, by more efficiently using labor, capital, management, technology, and raw resources. But output will also increase without any improvement in productivity when more units of production are employed. Increasingly productive companies usually re- flect an ever more efficient use of economic resources. When a company is able to raise its

productivity, it should be able to lower its pro- duction costs, better compensate its labor, sell more cheaply, and show a better return on in- vestment. Although it is often imperative for an individual company to raise its productivity each year, that imperative may be much less strong for the industry as a whole. Where sup- ply (and capacity to produce) exceeds demand as it does in the coal industry, higher produc- tivity leading to more output may hurt coal suppliers who already have more coal than they can sell.

Mining productivity—output per worker shift—expresses efficiency in terms of labor productivity. Coal (labor) productivity is cal- culated by dividing total tonnage mined by the number of employees. This formula can be misleading. Actually, productivity reflects the efficiency of how many factors work together to produce a ton of coal. When coal mining was a labor-intensive industry and labor costs accounted for as much as 70 percent of total production costs, this way of measuring efficiency made sense. Then, each miner was paid on a piecework basis—so many cents per ton mined—and mining systems required many workers, little capital, and simple tools. Today, coal mining—both surface and underground systems—is capital intensive, and productivity rises or falls according to how many variables interact, labor being only one. A more useful indicator of mining productivity today might be derived by measuring capital efficiency (in constant dollars), mining system efficiency, individual machine efficiency, or total-factor efficiency.

The decline in coal productivity since 1969 has received much attention. That decline has been dramatized because it reversed a steeply rising productivity curve that had characterized the industry since 1950. In that year, productivity was calculated at 6.77 tons per worker shift compared with 19.9 tons in 1969. Productivity almost tripled over that 20-year period. But annual *production* never matched the 1950-51 level until the late-1960's. Rising productivity did not result in increased production. (In fact, more bituminous and anthracite coal was mined in 1920—655.5 million tons—at a productivity level of 4 tons per worker shift than was mined in 1975—654.6 million tons—at 13½ tons per worker shift.) Productivity improvement in the 1950's and 1960's did, however, allow the biggest coal companies to survive a protracted demand slump. The spectacular rise of coal productivity in these years was the result of mechanization of underground mining, increased surface mining, the absence of environmental controls on surface mining, and inadequate mine safety standards. Productivity improved statistically because 70 percent fewer miners were working in 1969 as were in 1950 while production increased slightly. As the numerator (tonnage) remained relatively constant in these two decades, the denominator (workers) dropped steadily, resulting in ever higher productivity.

In the 1970's, production went up, but employment rose at an even faster rate. Conse-

quently, productivity declined. Underground productivity fell from 15.6 tons per worker shift in 1969 to 8.7 tons in 1977. Surface mining productivity dropped from 35.7 tons to about 26 tons. Production rose from about 560 million tons in 1969 to about 688 million tons in 1977, a 23-percent gain. But employment increased 84 percent from 125,000 in 1969 to almost 230,000 in 1977.

Industrywide productivity data must be used with caution because of numerous inconsistencies and uncertainties in the numerator (tonnage) and denominator (workers) of the productivity formula.

Tonnage, for example, is reported differently by different companies—and even within the same company. Productivity is affected by whether operators report tonnage "as sold" or "as mined." The latter represents raw coal as it comes directly from the mine face. The former represents both raw coal (sold as is) and cleaned coal. Productivity falls in proportion to the amount of cleaned coal reported because wastage (which has been increasing—see chapter II) is subtracted from "as mined" output. Because tonnage data include different definitions of output, productivity statistics may not be measuring efficiency in the same way.

The data also do not distinguish between steam and metallurgical coals. Met coal and steam coal prices were about the same in the 1960's, but met prices have been roughly double steam prices in the 1970's. High prices enabled inefficient met coal mines to operate profitably. The effect of high met prices has been to lower industrywide productivity.

Another data weakness involves the concept of tonnage itself. Coals differ in quality, and tonnage simply measures weight. If, for example, the numerator in the productivity formula were Btu instead of tonnage, underground and eastern productivity would rise relative to surface mine and western productivity.

The raw data may also be misleading because coal companies count their workers differently. Some count only hourly employees; others include office workers. Obviously, the larger the denominator (workers), the lower the productivity rate. Another weakness involves the definition of a working shift. Actual shift times range from 7¼ hours to as much as 12 hours. Generally, surface miners work more overtime than underground miners. The typical surface shift may be longer than the average underground shift. In this fashion, long-shift mines would show higher productivity than short-shift operations, other things being equal. This would tend to show surface-mine productivity higher and underground productivity lower on an aggregate basis.

Perhaps the single most serious flaw in the data is that they often grossly combine the performance of vastly different mining systems. Coal mines range from huge investments producing 10 to 15 million tons annually to tiny "dogholes" punched into a hillside, producing 1,000 tons or less. Shovel size runs from 130 yd³ down to a No. 4 hand shovel. Coal is moved by 650 hp jumbo trucks as well as one pony-power railcars. The industry is made up of very different parts, and in analyzing productivity the differences may be more relevant than the similarities.

The most obvious distinction is between surface and underground mining. Surface mines are roughly three times as productive as deep mines. The productivity advantage of surface mining is inherent in its extraction technology and geologic conditions. As more and more surface-mined production comes from the tremendously efficient open-pit mines in the West, surface mining productivity should rise and to a lesser extent, industrywide productivity as well. Although OTA estimates that the 60/40 ratio of strip to deep production is not likely to change over the next 20 years, industrywide productivity would rise considerably if this ratio does increase (as it has since 1950).

Even within similar mining systems, variations in productivity occur according to seam thickness and accessibility, mining techniques, and equipment size, among other factors. Underground productivity rates range from 3¼ tons from a thin-seam in southern West Virginia to 30 tons or more in a thick-seam. Similarly, surface productivity ranges from about 20 tons in east Kentucky to 100 tons or more in Wyoming.

A second crucial distinction needs to be drawn between new mines and old mines. The productivity of new underground mines averaged about 18 tons per shift compared with 11 tons for existing deep mines in 1974.[67] Similarly, new strip mines averaged about 70 tons per worker shift—and in some Western operations approached 150 tons—compared with about 36 tons for existing surface mines.[68] Industrywide productivity will increase as new mines replace old mines.

To be analytically useful, productivity data must compare like things—mining systems, machines, geologic conditions, etc. Mine-by-mine data that include all the variables involved in productivity are needed. The absence of such data impedes sound policymaking.

The productivity decline since 1969 must be seen in light of these data problems; the unique factors causing the productivity increase between 1950 and 1970, and the equally unique factors causing the uneven gains in tonnage and employment since 1970.

The rise in productivity in the 1950-69 period was the result of a combination of conditions that are unlikely to be repeated. Significant advances in underground and surface mining technology maintained output levels while sharply reducing labor needs. (Indeed, the labor cost per ton of coal dropped from 58.5 percent of value in 1950 to 38.3 percent in 1969.[69] Further, the environmental and safety costs of mining were not well regulated by State or Federal law. Productivity rose but at a social cost. The combination of shrinking demand for labor and the cooperative arrangements between UMWA and BCOA kept strike activity to a minimum. Finally, poor market conditions in the 1950's and 1960's weeded out several thousand small, inefficient mines, thus boosting industrywide productivity. Unemployment, black lung, environmental damage, and accidents were the hidden costs of rising productivity.

[67]*Coal: Task Force Report for Project Independence* (U.S. Department of the Interior, November 1974), p. 29.
[68]Ibid.
[69]Joe Baker, "Coal Mine Labor Productivity," p. 8.

What happened after 1969 to reduce productivity? The explanation for the decline cannot be pinned to a single cause. A number of demographic, economic, and regulatory developments contributed.

First, innovations in mining technology topped out both in reducing labor and in expanding output. Continuous-miner technology has not changed significantly since 1969 and, by that year, almost 96 percent of underground output was mechanically loaded. The only productivity enhancing change in underground mining has been the introduction of longwall systems,[70] which mined about 4 percent of all underground coal in 1977. Although surface mining expanded its share of national production, its technology has not changed significantly in scale or concept.

Second, as coal prices rose to unprecedented levels, hundreds of small, inefficient mines opened, adding many employees but relatively little tonnage to the industry's aggregate.[71] Inflated coal prices enabled these mines to operate profitably despite their relative inefficiency. Conditions and policies that encourage marginal mining will lower aggregate productivity. As long as coal prices remained constant in the 1950's and 1960's, productivity rose steadily. But when prices rose sharply in the 1970's, productivity fell.

Third, clean air legislation and relatively steady metallurgical demand in the 1970's meant coal had to be cleaned, sometimes cleaner than before 1969. More than 100 million tons of waste (the equivalent of 15 percent of 1977's total output) was cleaned from run-of-the-mine output. Although this material

is taken from the ground, much of it is not counted in productivity calculations. For a number of reasons, deep-mined coal is more often and more thoroughly cleaned than surface-mined coal, which further lowers deep-mine productivity. Productivity at individual mines may be substantially affected by this factor. For example, one underground mine OTA visited was routinely discarding 56 percent of material taken from the mine in order to fulfill coal-quality specifications in its supply contract. Productivity at this mine was about 3½ tons per worker shift. If its production had not had to be cleaned, productivity would have doubled. Throughout the 1970's, the percentage of refuse-to-raw coal has inched upward. This reflects either dirtier raw coal or stricter customer specifications, and possibly both. Assuming that the Environmental Protection Agency (EPA) finally promulgates a 0.2-lb/85-percent sulfur removal limitation, the amount of coal cleaned is likely to increase as the utility market expands. Also, as the more marginal seams are mined, the amount of waste extracted in mining may increase. Both factors retard productivity gains.

Fourth, the labor force has changed dramatically since 1969. Both miners and supervisors are young, often inexperienced and inefficient. Almost half of the experienced 1969 work force retired in the 1970's and 65 percent of today's work force has been hired in the 1970's. The industry has had trouble transferring practical wisdom of its older miners to this new generation. Absenteeism has also reduced productivity. Few hard data exist on the actual extent of absenteeism or its root causes. One way of offsetting absenteeism is to train miners to have interchangeable skills so that each can do every job on the crew. But few coal companies do this. Nor do many companies employ "utility" workers as the automakers do, whose job it is to substitute for absentee workers. The industry may be able to improve productivity by adopting such changes if permitted by union work rules. Finally, an estimated 5,000 miners were hired after 1974, when the new UMWA contract required helpers on underground face equipment. Helpers safen the operation of mobile equipment by moving high-voltage trailing

[70]Longwalls can mine more coal with fewer miners, but they are expensive and workable only in certain geologic conditions. About two-thirds of the 77 units operating in 1977 were in metallurgical mines, where high product price justified high capital investment.

[71]The biggest single employment category in 1975 was miners in underground and surface mines producing fewer than 25,000 tons annually. 46,695 miners in these mines (about 24 percent of all miners) produced only 14.1 million tons (about 2 percent of all output). Calculated from, Injury Experience in Coal Mining, 1975, Informational Report, 1077 (Department of Labor, Mining Safety and Health Administration, 1978), p. 77.

cables out of the way, watching roof and rib conditions, helping the machine operator maneuver, and being around in case of trouble. Helpers also receive training in machine operation from experienced miners during their apprenticeship, which allows them — after 120 training days — to operate equipment when the regular operator is not present. The addition of 5,000 helper jobs lowered productivity statistics. But the industry may recoup these statistical losses in the future by having a pool of trained machine operators ready to step in at a moment's notice. Although the productivity decline has its qualitative side (such as inexperience or anti-company attitude), the simple quantitative aspect — more miners — is its real cause.

Fifth, disabling injuries lower productivity to the extent they disrupt production teams and require extra personnel. Disabling injuries have risen since 1969. In that year, 8,358 underground injuries and 967 surface injuries were recorded. In 1977, underground disabling injuries had risen 40 percent to 11,724 and surface injuries had gone up 132 percent to 2,246 (see Workplace Safety, table 35). The amount of lost time from disabling injuries is substantial. Average severity of an injury refers to the average number of calendar days lost by an injured worker. In 1977, underground miners experienced 146 permanent partial injuries costing an average of 297 lost days and 11,575 temporary total injuries with an average cost of 73 work days.[72] Surface miners experienced 66 permanent partial injuries with an average loss of 529 days each and 2,211 temporary total injuries with an average loss of 58 days each. Together, underground and surface injuries resulted in the loss of 1,051,489 calendar days in 1977, which meant 751,079 lost production days. That represents 3,391 lost worker years.[73] In one form or another, an equivalent amount of labor had to be hired to substitute for these injured workers to maintain production. This represented extra personnel and extra dollar costs to mine operators, as well as reduced productivity. To put 3,391 lost worker years in a comparative perspective, it represents more

than 30 percent of time lost from wildcat strikes in 1977.

Sixth, the 1970's saw extensive legislative and administrative regulation of the coal industry. States began to require surface operators to reclaim land and control landslides and water runoff. This forced the industry to hire more workers and buy new equipment. This added effort was "socially productive" from a national and public perspective but lowered the economic productivity of the individual operator and the industry. Neither the Federal Office of Surface Mining nor industry associations could supply data on the percentage of the Nation's 64,000 surface miners that were engaged in reclamation work.

A major regulatory initiative that affected underground mining was the 1969 Coal Mine Health and Safety Act. Federal health and safety standards forced operators to devote more personnel and equipment time to ventilation, rockdusting, methane and dust monitoring, roof control, maintenance, and retrofitting machinery with safety features. Safer operating procedures require equipment to be moved in and out of the working place more frequently than before. Had these practices not been adopted, it is reasonable to suppose that coal's safety record would have been substantially worse. Some new personnel were needed to handle the employee-training and administrative aspects of the Act. Other governmental regulations required new safety and environmental personnel. Office workers — professional and clerical — were needed to ob-

---

[72]Data supplied by MSHA's Data Analysis Center, January 1979. Excludes permanent total injuries.

[73]Calculated in the following manner: Multiply number of injuries by the average number of days lost to get total number of lost days. To find the number of lost work days, subtract two-sevenths of the total number of lost days (representing weekend time) from the total number of lost days. Divide this sum by the average number of working days each year — 220 (underground), 230 (surface) — to get the number of work-year equivalents lost to disabling injuries. For underground mining, this formula finds 888,337 lost calendar days reduces to 634,539 lost work days, or 2,884 lost worker years. For surface mining, there were 163,152 lost calendar days or 116,539 lost work days, the equivalent of 507 lost worker years. Adding underground and surface lost worker years gives a total of 3,391 lost worker years from mine injuries in 1977. Data for workers in mine construction, preparation plants, etc. are not included.

tain permits, develop plans and administer them at the minesite. In 1969, for example, operators reported 2,640 office workers of all kinds at minesites; that figure rose to 7,037 in 1977, an increase of 167 percent. Presumably, all of these management employees do necessary work. But they lower a mine's productivity when they are included as workers in the productivity equation. It has not been possible to determine how many officeworkers are routinely counted in productivity calculations, but anecdotal evidence suggests the number is substantial.

Other factors may contribute to declining productivity. Operators may have trouble getting capital. Delays in replacing mining equipment may reduce output. Underground haulage technology has not kept pace with cutting technology. These systems often break down. In older mines haulage and ground-control problems tend to be more troublesome. Transportation problems from mine to market slow down production when logjams develop. Primitive coalfield social conditions and public service deficiencies have been cited in several recent studies as contributing to high absenteeism and hostile labor-management relations, which lower productivity.

## What of the Future?

The decline in coal mine productivity has probably bottomed out. Gradual improvement over the next decade should occur if many of today's trends continue. Some of these trends are mentioned here. Western surface mining will account for an increasing share of total production. Its productivity ranges from 50 to 150 tons per worker shift. Many of the personnel needed to handle the requirements of Federal environmental and safety regulations are already hired. Major legislative initiatives in both areas requiring many more salaried or hourly employees are not likely. The big hiring surge of young workers has already taken place. They are likely to spend the next 20 years or so in mining as highly skilled, experienced employees. Marginal, inefficient mines will probably close if coal prices stabilize and big companies expand. New big mines with high productivity rates will open;

old, smaller mines will close. Finally, it is likely that many operators will try to improve their training programs and labor relations, which should help raise productivity.

However, increasing productivity may be a mixed blessing. Strategies to enhance productivity directly affect the health and safety of coal workers and the quality of community life in the coalfields. Production can be boosted through one of four productivity-enhancing strategies, each of which carries a different set of workplace and social consequences.

These strategies are:

1. Use fewer workers to mine more coal.
2. Speed the pace of work so that each miner produces more tonnage.
3. Invest capital in more efficient equipment to enable each worker to produce more.
4. Make the work process more efficient by redesigning jobs, improving morale and motiviation or improving coordination in the production cycle.

The first two approaches can have substantial health, safety, and social costs, depending on how the work force is expected to maintain production levels. If labor-saving equipment is purchased, health and safety may be improved because fewer workers will be exposed to dangerous conditions. However, if the remaining employees are simply expected to work faster or spend less time taking safety precautions, accident frequency or health conditions may worsen. One major West Virginia operator recently announced the layoff of 2,100 workers to improve productivity. This operator who hopes to maintain output with fewer workers consistently shows extremely high disabling-injury rates, and the safety and health consequences of this change may be substantial. Inefficient sections of these mines may close down entirely, or "nonproductive" jobs related to mine housekeeping, rock dusting, and ventilation work may be eliminated.

Speedup strategies (#2) may involve heavy safety and health costs. Recently, a number of mines have adopted cash bonus systems whereby a miner can receive extra income if a tonnage quota is exceeded. One union official in the United Kingdom, where a similar plan was begun in 1977, reported a 50-percent increase in fatalities at mines that switched to the incentive plans. The American experience with these plans is too recent to assess. Some U.S. plans tie the cash bonus to both safety (defined as disabling injuries) and productivity. Clearly, this is a commendable approach. But the plans may encourage miners and management to undercount injuries in order to preserve the bonus payoff. Even plans that tie production to safety ignore possible health costs. If miners work faster at producing coal, they may take less time to maintain workplace conditions that protect their health. As there is no way to measure a worker's immediate respone to increased dust or noise, bonus plans may be encouraging workers to buy short-term extra dollars in return for their long-term health.

Investing capital for greater productivity (# 3) can take two forms. First, a mine operator may upgrade his mining system by replacing existing machines with newer, more efficient models. The costs and benefits of this investment—in productivity, health and safety—are generally well-known. A second form of productivity investment is the purchasing of a new product that has recently been commercialized. The productivity advantages of new technology are usually more easily estimated than any possible health and safety costs. Congress stated clearly that one of the purposes of the 1969 Act was " . . . to prevent newly created hazards resulting from new technology in coal mining." The industry is aware that new technology sometimes may increase productivity at the expense of safety. Joseph Brennan, president of BCOA, noted that "any new mining technology developed to aid in boosting productivity should also incorporate safety as one of its key features."[74] It is likely that the safest productivity strategy would be one that removed the most workers from the workplace, but the unemployment implications of this approach would be immense.

Although rising *productivity* brings economic benefits to management and labor, greater *production* is the primary goal of a na-

tional energy policy. Increasing coal production is related to—but is not necessarily the same as—raising productivity. Each objective entails a different set of economic and social implications. Price, for example, affects productivity and production differently. High coal prices lower industrywide productivity but raise production; low prices do the reverse. Federal policy has yet to specify the desired blend of production and productivity. Private policy, if the 1978 UMWA-BCOA contract is any indication, appears to have shifted toward productivity enhancement as a way of increasing production. Before decisions are made that commit Federal policy to either direction, major research is needed into the relationships among productivity, safety and market conditions on a mine-by-mine basis, together with a re-evaluation of productivity measurements themselves. Most Federal research personnel do not feel that a radical technological breakthrough—something like in-situ combustion for power generation or a surface-controlled, automated underground mining system—are in the offing. If it were, Federal and local officials would face the social dislocation of thousands of unemployed miners and possibly the economic devastation of hundreds of coal mining communities akin to that of the 1950's and 1960's. The state of coal mining technology is such that production can double or triple without any major technological innovations. Specific aspects of the mining cycle can benefit from technological innovation, but viewed as a whole, no compelling or urgent need exists to commercialize new technologies to meet anticipated production goals.

Restructuring the production process and better training and education (#4) may offer the most promise for productivity improvement over the next decade. Little research and experimentation has been done on better ways of organizing work and managing production crews. Similarly, little attention has been given to determining the most effective safety and job training and education programs. Both efficiency and safety are likely to increase in relation to the relevance of their training and the involvement miners feel they have in its practical application.

---

[74]"Productivity and the BCOA," *Coal Age*, July 1975, p. 97.

# WATER AVAILABILITY

Water is an essential resource for almost all phases of energy development. The availability of sufficient water to meet energy needs is more than a function of the presence of large quantities of water; it depends on the political acceptability of using the physically available supply, the economic competition from other users, the legal system controlling water rights, environmental factors, and other influences.

Irrigated agriculture is the major consumptive water use in the United States. Most agricultural water is consumed west of the Mississippi. Improved irrigation efficiency can reduce consumptive use of water, with a portion of this savings made available to the energy sector in the West. The Soil Conservation Service (SCS) estimates that in a dry year 195 million acre-feet of water is diverted for irrigation, of which 103 million acre-feet of depletion is charged to irrigated agriculture. Of this, 78 million acre-feet (76 percent) is consumptively used by the growing crop and the remaining 24 million acre-feet is lost to "incidental" causes.

SCS estimates that, by the year 2000, 8-million acre-feet of the water that is presently "lost" each year can be salvaged by improving irrigation efficiency. (A bonus of improved efficiency would be a 48-million acre-feet reduction in withdrawals and a similar reduction in return flows, yielding a substantial improvement in water quality.) Depending on its location, part of this water could become available to energy developers. However, agriculture pays such a low price for water ($15 to $40/-acre-foot) that little incentive exists for its efficient use.

Besides improving irrigation efficiency, a number of other agriculture water conservation alternatives exist, including:

- cropswitching (from water-intensive crops, such as alfalfa, to less intensive crops such as wheat, oats, and barley);
- improvement of cultivation procedures;
- removing marginal lands from production; and

- removal of water-using natural vegetation.[75]

None of these alternatives is applied universally; some have potentially undesirable side effects. Crop switching depends on available markets and proper farming conditions. Removing marginal lands from production can lead to unemployment and other social problems. Natural vegetation ("phreatophyte") removal can harm wildlife habitat and erode affected lands.

A major policy problem is that incentives to conserve agricultural water have been so low that minimal research and analysis have been done to illuminate the tradeoffs among alternative water use policies. This lack of knowledge, coupled with extreme resistance to any change in water policy, has hampered intelligent reform of water management practices for agriculture.

Meanwhile, without management reform to make water more readily available, the energy industry can afford to pay far more for water than a farmer can. In the Western States, some owners are selling their agricultural land and its associated water rights for coal mining and energy development. If this process of converting agricultural land and water resources to the energy sector continues and accelerates, it may become an important political issue in the West. It seems improbable that State governments will allow wholesale replacement of agriculture by energy. It is clear, however, that the heavy subsidies paid to marginal farmers by selling them water at a small fraction of actual costs is a very expensive policy.

One possible compromise would be to establish a partnership between farmers and the energy sector with the cooperation of the State governments. The energy sector could finance water conservation measures and use the water that is saved. Although cost estimates are unavailable, the cost seems likely to be considerably less than other alternatives being

---

[75]S. Plotkin, et al., "Water and Energy in the Western Coal Lands," Water Resources Bulletin, February 1979.

considered, such as the diversion of water from the Oahe reservoir in South Dakota to Gillette, Wyo., at an estimated cost of $700 to $900/per acre-foot, or the changing of technology in electric power from wet-cooling to dry-cooling systems at an estimated minimum equivalent cost of $900 per acre foot.[76]

Inter-basin transfers are often hampered by legal and institutional complaints that have emotional overtones. Congress has the power either to prohibit or require an interstate inter-basin water transfer. As part of the Colorado River Basin Project Act (Public Law 90-537, also known as the Central Arizona Project Act), Congress in 1968 declared a 10-year moratorium on studies on the importation of water into the Colorado River Basin. Many questions remain unsettled on this issue. Water rights remain unadjudicated for many river basins in the West—particularly in the Northwest; the total allocation of water within many basins remains unknown, and conflicts have not been settled for many multiple uses. Questions remain unresolved on the amount of water required for instream purposes; regional goals have not yet been backed by local goals and policies on land use, economic growth, and population growth, making uncertain the prediction of water requirements.

Indian tribes are increasingly exercising their water rights in the West. The present confusion over Indian water rights may be detrimental to the planning of water and related land resources in general, and to energy development in the West in particular. The key to Indian water rights is a 1908 U.S. Supreme Court case that produced what is widely known as the "Winters Doctrine." The case involved a dispute between Indians of the Fort Belknap Reservation and non-Indian appropriators over waters of the Milk River, a nonnavigable Montana waterway. The Court held that, although it was not explicitly mentioned in the 1888 documents creating the Fort Belknap Reservation, there existed an implied reservation of rights to the use of waters that originate on, traverse, or border on the Indian land, with a priority dating from the time the treaty created the reservation. The Court's language has led

---
[76]Ibid.

to two interpretations of the source of the right. One line of reasoning argues that, with regard to Indian reservations created by treaty (or executive orders), water rights were retained by the Indians at the time of the treaty. Moreover, the document is silent on the question because the Indians did not intend to transfer the water rights. The alternative view holds that the water rights were in fact transferred, but that the Federal Government, under its own powers, "reserved" an amount of water from proximate streams to support an agricultural existence for the Indians. In *Arizona v. California*, the U.S. Supreme Court approved water allocations to various Colorado River Basin Indian reservations. Water quantities demanded and ultimately adjudicated for Indian reservations remain to be determined. Whatever the amounts, the water will come off the top of the available water in the Colorado River Basin, reducing the amounts remaining to the States for allocation to agricultural, energy development, municipal, recreational, and other uses.

Ground water currently supplies about 20 percent of all freshwater used in the United States. The estimated storage capacity of aquifers (underground reservoirs) is nearly 20 times the combined volume of all of the Nation's rivers, ponds, and other surface water. Ground water serves about 80 percent of municipal water systems and supplies 95 percent of rural needs; in all, it serves 50 percent of the Nation's population. Irrigation accounts for more than half of ground water use. In many parts of the country ground water mining (pumping water from an aquifer faster than the water in the aquifer is replenished) is occurring. The tendency to use saline water for energy development may add to the overall problem of soil subsidence, salt water intrusion, and the lowering of the water table in many aquifers. In many States, ground water law, like riparian surface water law, is inadequate to allocate the resource among competing users and is unresponsive to the problem of excessive use. The first defect results from vagueness of the rule of allocation ("reasonable use") and the second from failure of the legal system to perceive that the ground water is often a common-pool resource for which there is little incentive to save an exhaustible supply for use tomorrow. Any user who seeks to save it is subject to having his savings captured by another pumper from the same aquifer.

In virgin or undepleted conditions it is estimated that the average annual surface runoff or yield of the stream systems of the 11 Western States totaled 427 million acre-feet, including 50 million acre-feet of inflow from Canada. Today, owing to the cumulative activities of man, this virgin water supply has been depleted by 83 million acre-feet of consumptive water use annually, leaving 344 million acre-feet or about 81 percent of the virgin yield still unconsumed. Within this western region are large areas such as the Upper Rio Grande and the Gila River Basins, where the total annual surface water supply, for all practical purposes, is completely consumed. In the Colorado River Basin this condition will be reached when the Central Arizona project is completed. The Missouri River and its tributaries in Montana and northern Wyoming appear to have sufficient unused water supplies to meet needs for the foreseeable future. The Platte River tributaries in Wyoming, and particularly in Colorado, are approaching the saturation point in water use.

Present and projected water demands for 1985 and 2000 indicate, on one hand, severe regional water shortages and problems, and on the other hand adequate water availability to meet energy needs on a nationwide basis. According to the 1975 National Assessment of the U.S. Water Resources Council, the Nation's freshwater withdrawals in 1975 from ground and surface water sources for all purposes average about 404 million acre-feet/year (mafy). Of this amount, 125 mafy were consumed through evaporation or incorporation into products, and the remainder was returned to surface water sources for possible reuse in downstream locations. By 2000, total withdrawals are projected to be about 348 mafy, with about 151 mafy being consumed — a 14-

percent reduction in withdrawal, but a 20-percent increase in the amount consumed.

The production of energy and the extraction of fuels from which energy is produced constituted 27 percent of the total U.S. water withdrawals and 3 percent of total consumption in 1975. By 2000, mining and energy production (excluding synthetic fuels) will constitute 27 percent of the U.S. withdrawals and 10 percent of total consumption.

The geographic and temporal distribution of the Nation's surface waters is so variable as to pose substantial problems for energy development where billion dollar localities depend on a continuous supply of water. Rainfall varies widely from region to region, from season to season, and from year to year. Similar variations occur in runoff and streamflows. For example, within a normal year, the ratio of maximum flow to minimum flow may be 500 to 1 or greater. Year to year variations in the average flow also are substantial. Even in areas of high precipitation and runoff, a series of dry years may occur, resulting in serious drought problems such as those in the Northwest from 1961 to 1966 and in the Pacific Northwest in 1976 and 1977.

The United States appears to have enough water available to supply its most urgent needs, but there are numerous legal, institutional, and political constraints on its use, and many areas may have severe water shortages unless concrete physical measures — restraints on development, water conservation requirements, construction of additional storage, etc. are taken. Some critical elements of legal/institutional change include:

- resolving the problems of Indian water rights;
- making the price of water commensurate with its cost while recognizing the non-price value to a locale of alternative types of development; and
- developing a cooperative, instead of competitive, relationship between energy and agriculture.

# CAPITAL AVAILABILITY

The costs of coal production and conversion facilities have increased rapidly because of hyperinflation in the equipment market and the addition of environmental protection controls. Coupled with the rapid growth expected, much more capital will be needed. The constraints and uncertainties discussed in other sections will make financing more difficult by increasing overall production costs. Nevertheless, if demand for coal actually develops, financing the new mines probably will not raise insurmountable problems. As discussed previously in the *Industry Profile* section, changes legislated in the industry structure will have an unpredictable effect in this area. Utilities may have more difficulty raising capital for a variety of reasons, which this report has not studied in detail, but most analysts feel the problems will not be insurmountable. If necessary, a number of different incentives could be considered to make all the necessary financing available. These could include tax incentives, loan guarantees, guaranteed purchase prices for products, investment tax credits, and a more uniform application of regulatory policy.

# EQUIPMENT

Most studies of the potential availability of mining equipment to meet expanded coal demands have concluded that equipment can be supplied as needed. The one exception is the long leadtime for delivery on large draglines used to strip mine Western coal. However, these conclusions were based on the expectation that much of the expansion of production would come from Western coals. If there is a shift toward more production from Eastern underground mines, the dragline delays may not be serious but deliveries of underground equipment could be delayed.

Little information is available on the condition of existing mining capacity and how much upgrading will be required to meet new underground production targets. There are few mining equipment suppliers and most are small; a sudden increase in demand for new equipment would strain their ability to supply the industry's needs and they may have difficulty raising the capital to expand their production to meet the new demand.

# TRANSPORTATION

On a national scale, coal development probably will not be constrained by the availability of transportation for the period under study. Minor capacity expansions will be required by 1985, and more significant ones will be needed thereafter. However, transportation facilities can be planned and constructed within the same amount of time as powerplants or large mines, so that with adequate investment and planning, improvements will take place as needed.

Rather than reduce the total level of national coal production and use, transportation bottlenecks that do develop will tend to alter the pattern of mining, shipment, and consumption. Routes and mode choices will be adjusted to alleviate capacity limitations, for example, and mining and power generation will tend to be expanded only where transportation service is adequate.

At the local level, however, availability of the means of transporting coal, or the electric power derived from it, will be significant. The following limiting factors may serve to inhibit coal growth in specific regions.

## Rail

Some northeastern and midwestern railroads are characterized by poor financial performance and deteriorated track conditions. Without either public or private investment in right-of-way improvements, particularly in Appalachia, growth in mining in this region may be inhibited, truck use increased, and mine-mouth power generation expanded.

Western coal-producing areas are served by railroads in somewhat better financial condition, but there rail lines must be extended to serve new mines.

## Highway

Investment in Appalachian coal roads has also not kept up with increases in truck traffic. Deteriorated roads, like deteriorated rail lines, may eventually inhibit coal production in affected localities.

## Waterway

Barge transportation is ultimately limited by the capacity of channels and locks, which are constructed and maintained by the Army Corps of Engineers. Major extensions and improvements to the waterway system are authorized by Congress; failure to undertake projects as capacity limitations are reached

also will affect regional production and market patterns and will tend to divert traffic to railroads.

## Slurry Pipelines

Slurry pipelines become an alternative to railroads only where rights-of-way and water rights can be acquired. Slurry pipeline enterprises do not enjoy the power of eminent domain in many States, and opposition by railroads and other landowners can impede development of this type of facility. Also western coal-producing States are characterized by scarcity of water, which may not be made available by local authorities for slurry pipelines.

## High-Voltage Transmission

Installation of long-distance transmission lines generally requires Federal approval, which provides an opportunity for opponents to intervene, particularly on environmental grounds. Siting powerplants to take strategic advantage of this means of transporting power may also be difficult, particularly in western coal-producing areas: Regulations to prevent significant deterioration of pristine air quality under the Clean Air Act and local opposition to water use and other environmental impacts serve to inhibit mine-mouth power generation, for which long-distance transmission generally is suited.

# REGULATORY RESTRICTIONS ON COMBUSTION

Federal agencies indirectly regulate the siting of large new coal combustion facilites and directly regulate the construction and operation of these facilities as well as the disposal of combustion wastes. The principal constraints are imposed by the Clean Air Act; additional considerations include the Clean Water Act, the National Environmental Policy Act (NEPA), the Resource Conservation and Recovery Act (RCRA), the Endangered Species Act, and regulations governing the use of Federal lands and of navigable waters. The legal framework of most of these provisions is

discussed in chapter VII. This section addresses the constraints imposed by these regulations on coal users; the next section discusses the options available for operating within those constraints.

## Facility Siting

Facility siting is affected primarily by the Clean Air Act; other considerations may arise under the Clean Water Act, NEPA, the Endangered Species Act, and regulations related to the use of Federal lands.

The Clean Air Act is structured around National Ambient Air Quality Standards (NAAQS) (see table 18), which are implemented through a variety of regulatory programs designed to limit emissions of airborne pollutants from stationary and mobile sources. Programs applicable to stationary sources primarily use control technology requirements and numerical emission limitations to regulate pollution from new and existing facilities, from facilities located in clean and dirty air areas, and from facilities located near scenic areas such as national parks.

**Table 18.—National Ambient Air Quality Standards**

| Pollutant | Averaging time | Primary standard | Secondary standard |
|---|---|---|---|
| Particulate matter.... | Annual (geometric mean) | 75 $\mu$g/m³ | 60 $\mu$g/m³ |
| | 24-hour | 260 $\mu$g/m³ | 150 $\mu$g/m³ |
| Sulfur dioxide ... | 3-hour | — | — |
| | Annual (arithmetic mean) | 80 $\mu$g/m³ | — |
| | 24-hour | 365 $\mu$g/m³ | — |
| | 3-hour | — | 1,300 $\mu$g/m³ |
| Nitrogen dioxide ... | Annual (arithmetic mean) | 100 $\mu$g/m³ | 100 $\mu$g/m³ |
| | 24-hour | — | — |
| | 3-hour | — | — |

As of 1977, 116 of 247 air quality control regions (AQCRs) reported violations of the primary annual particulate standard while 108 reported violations of the 24-hour standard. Similarly, 12 AQCRs reported violations of the primary annual SO₂ standard while 37 reported violations of the 24-hour standard. These regions are designated nonattainment areas for these pollutants, as shown in figures 17 and 18. Many of the AQCRs that are shown as nonattainment areas may only have localized violations with the remainder of the AQCR in compliance. The States are responsible for developing control strategies for areas showing violations that will provide for the attainment of the primary NAAQS as expeditiously as practicable, but not later than December 31, 1982. The State strategy must require significant annual incremental emission reductions from existing stationary sources in areas that show violations, as well as permits for the construction and operation of new or modified sources that would exacerbate existing NAAQS violations.

Nonattainment programs will impose severe constraints not only on increased coal use but on all growth in these areas. As mentioned above, most of the AQCRs have not attained the primary particulate standards, while 15 percent have not met the SO₂ standards. Wherever a new source would exacerbate an existing NAAQS violation, the permit applicant must apply the lowest achievable emission rate (LAER) and must secure from existing sources in the area emission reductions that more than offset the emissions from the proposed facility. The cost of meeting these two requirements is high, and securing the offsetting emission reductions is difficult. Consequently, new sources are more likely to be located in rural areas where the NAAQS have been achieved. In addition, major modifications to existing sources in nonattainment areas probably would be rejected in favor of new sources in clean air areas.

However, the Clean Air Act also includes comprehensive provisions to prevent the significant deterioration (PSD) of air quality in areas where the air is cleaner than the NAAQS require. The PSD increments for Class I areas (usually parks or monuments, or wilderness areas) allow the lowest increase in ambient concentrations over the baseline, and thus the fewest new stationary sources, while the Class III increments allow the greatest increase (see table 19). In no event may a new source located in a clean air area cause the concentration of any pollutant to exceed either the national primary or secondary ambient air quality standard, whichever concentration is lower. At present the PSD regulations apply only to emissions of particulate matter and SO₂. Regulations for NO₂, hydrocarbons, photochemical oxidants, and carbon monoxide are to be promulgated by August 1979.

To obtain a permit to locate a new source in an area subject to the PSD regulations, the applicant must demonstrate that the source will meet all applicable emission limitations under the State implementation plan (SIP) as well as performance standards for new sources and emission standards for hazardous pollutants, and that the source will apply the best available control technology (BACT). BACT is

**Figure 17.—AQCRs Status of Compliance With Ambient Air Quality Standards for Suspended Particulates**

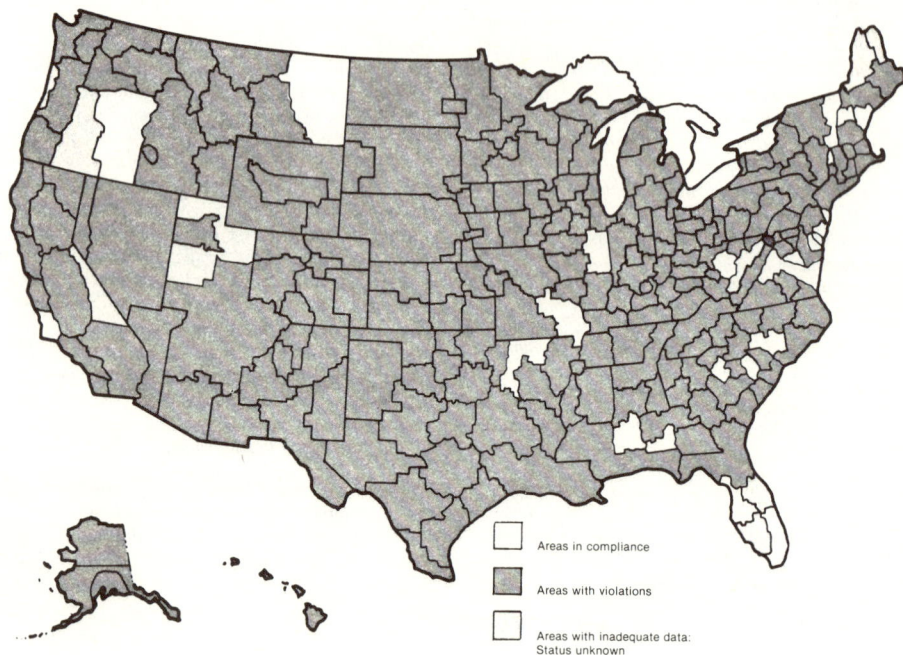

Areas in compliance

Areas with violations

Areas with inadequate data: Status unknown

SOURCE: U.S. Environmental Protection Agency. 1976a.

determined on a case-by-case basis, taking into account energy, environmental, and economic impacts and other costs. In addition, the applicant must demonstrate, based on air quality monitoring data and modeling techniques, that allowable emissions from the source will not cause or contribute to air pollution in violation of the NAAQS or PSD increments. The permit applicant also must provide an analysis of the source's air-quality-related impacts on visibility, soils, vegetation, and anticipated induced industrial, commercial, and residential growth.

Under the 1977 Clean Air Act amendments, PSD provisions apply to a source in any of 28 categories (including fossil-fuel-fired steam

electric generating units of more than 250 million Btu/hr heat input and coal-cleaning plants with thermal dryers) with uncontrolled emissions of 100 tons per year, or to any source with uncontrolled emissions of 250 tons per year. Previous regulations applied only to sources in 19 specified categories. However, under the new regulations, only those sources with controlled emissions of 50 tons/yr or greater or that would affect a Class I area or an area where the increment is known to be violated, will be subject to full PSD review. Sources with controlled emissions of less than 50 tons per year need only demonstrate that they will meet all applicable emission limitations. Each SIP must include a program to assess periodically whether emissions from

**Figure 18.—AQCRs Status of Compliance With Ambient Air Quality Standards for Sulfur Dioxide**

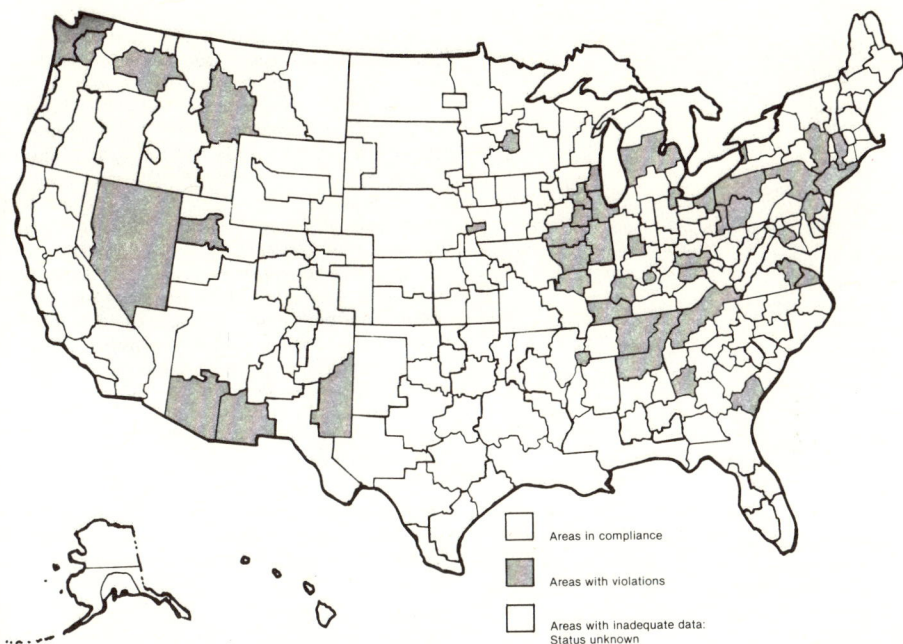

☐ Areas in compliance

▨ Areas with violations

☐ Areas with inadequate data:
Status unknown

SOURCE: U.S. Environmental Protection Agency, 1976a.

these small, exempt sources, and from any other sources that may be unreviewed because of their date of construction, are endangering PSD increments.

Because the 1977 PSD provisions apply to a wider range of sources, they are expected to increase the costs of facility siting significantly. EPA estimates that 1,600 sources of all types will be subject to the permit requirements each year (as compared to 164 per year under the previous regulations). In addition, there are two situations in which facility siting probably will be constrained. First, PSD permits will not be available for large new sources in areas where the difference between the baseline concentration and the NAAQS already is less than the allowable increment. Second, where there are several sources that are exempt from the PSD requirements because of their size or date of construction, these exempt and unreviewed sources may capture the available increments and foreclose siting for larger new facilities, as mentioned above, States are required to assess periodically the emissions from these exempt and unreviewed sources.

Finally, provisions in the Clean Air Act designed to protect visibility in areas primarily important for scenic values, such as national parks, are expected to affect the siting of coal-fired facilities. EPA's regulations, to be promulgated not later than August 1979, are to require SIP revisions in order to achieve visibility

**Table 19.—Ambient Air Increments**

| Pollutant | Maximum allowable increase ($\mu g/m^3$) |
|---|---|
| **Class I** | |
| Particulate matter | |
| Annual geometric mean | 5 |
| 24-hr maximum | 10 |
| Sulfur dioxide | |
| Annual arithmetic mean | 2 |
| 24-hr maximum | 5 |
| 3-hr maximum | 25 |
| **Class II** | |
| Particulate matter | |
| Annual geometric mean | 19 |
| 24-hr maximum | 37 |
| Sulfur dioxide | |
| Annual Arithmetic mean | 20 |
| 24-hr maximum | 91 |
| 3-hr maximum | 512 |
| **Class III** | |
| Particulate matter | |
| Annual geometric mean | 37 |
| 24-hr maximum | 75 |
| Sulfur dioxide | |
| Annual arithmetic mean | 40 |
| 24-hr maximum | 182 |
| 3-hr maximum | 700 |

improvement by retrofitting plants in existence for less than 15 years as well as by adopting a long-term strategy for progress toward a national visibility goal. Proposed fossil-fuel-fired powerplants with a design capacity of more than 750 MW must demonstrate that they will not cause or contribute to the significant impairment of visibility in any of the specified areas. However, until the visibility regulations have been promulgated, it is not possible to determine their impact on the costs of, or site selection for, coal-fired facilities.

Additional, less significant constraints on siting are imposed by the permit requirements of the Clean Water Act, the Army Corps of Engineers, and agencies having jurisdiction over Federal lands, as well as by the general requirements of NEPA, the National Historic Preservation Act (NHPA), the Fish and Wildlife Coordination Act (FWCA), and the Endangered Species Act. The Clean Water Act is structured around the quality of water necessary for a variety of uses, including public water supplies, propagation of fish and wildlife, recreational, agricultural, industrial and other purposes, and navigation. New facilities must obtain a permit under the National Pollution Discharge Elimination System (NPDES). The permit incorporates all applicable effluent limitations and water quality standards promulgated under the Clean Water Act, and an applicant must demonstrate that these limitations and standards will be met. In addition, if the plans for a combustion facility require any structure to be built in navigable waters (such as a cooling-water intake structure or barge unloading facility), a permit must be obtained from the Army Corps of Engineers. Corps regulations stipulate that no such permit may be issued until the applicant demonstrates that all other necessary Federal, State, and local permits, certifications, or other authorizations have been obtained. Finally, if the coal combustion unit or any of its support facilities (such as transmission lines) are to be located on Federal land, a permit must be obtained from the agency having jurisdiction over that land.

Most federally issued permits for coal combustion facilities will trigger the environmental impact statement (EIS) requirements of NEPA. Although the EIS is prepared by the agency issuing the permit, it is based on analyses submitted by the applicant, and the length of time required to prepare the EIS depends on the quality and completeness of those analyses. In addition, before issuing a permit an agency must obtain a certification from the Secretary of the Interior that the facility will not jeopardize the continued existence of an endangered species. Under the FWCA, when a federally permitted project would result in the modification of any water body (for example, reduction of water flow through consumption by cooling towers), the permitting agency must consult with the Fish and Wildlife Service and with the State agency having supervisory authority over fish and wildlife prior to issuing the permit. Issuance of the permit may be enjoined until consultation and coordination has occurred, and serious consideration must be given to recommendations for mitigation of impacts to fish and wildlife. Finally, regulations promulgated under the NHPA require all permitting agencies to determine whether there are historic, archeological, architectural, or cultural resources affected by the proposed action that are listed in the National Register of Historic Places or are eligible for listing. If any of these resources may be affected, the permitting agency must obtain comments from the Advisory Council on Historic Preservation. In most cases, however, historic and other sites must only be studied, not necessarily preserved.

Of all these requirements, the Clean Air Act will have the most far-reaching consequences in terms of the number of sites foreclosed to coal combustion facilities. However, the cumulative effect of all provisions, each with extensive interagency and public participation requirements, will be to lengthen significantly the time necessary for site approval. When numerous State and local permits and other requirements are added in, this leadtime can become costly.

## Combustion

As with siting, the Clean Air Act imposes the most significant constraints on increased coal combustion. Other limitations include the Clean Water Act and the RCRA. The Clean Air Act affects coal combustion through standards of performance for new sources and through the provisions related to nonattainment areas and the prevention of significant deterioration.

Standards of performance for new or substantially modified facilities establish allowable emission limitations for those facilities and require the achievement of a percentage reduction in emissions from those that would have resulted from the use of untreated fuels. Under the 1977 amendments to the Clean Air Act, these standards must reflect the degree of emission limitation and the percentage reduction achievable through application of the best technological system of continuous emission reduction that has been demonstrated adequately (taking into consideration the cost of achieving the reduction and any health and environmental impacts and energy requirements

not related to air quality). That is, New Source Performance Standards (NSPS) must be met through the use of a technological method of control, such as a scrubber or precombustion treatment of the fuel, rather than through alternative fuels such as oil or low-sulfur coal.

The principal coal combustion facilities for which NSPS have been promulgated include steam electric-generating plants of more than 250 million Btu/hr heat input, large industrial boilers, and coal preparation plants (including any facility that prepares coal by breaking, crushing, screening, wet or dry cleaning, and/or thermal drying). The standards for these sources are shown in table 20. EPA plans to announce new standards for large industrial boilers in 1980.

Each SIP must include a procedure for preconstruction review of new sources to ensure that NSPS will be met. In addition, sources that undergo a modification that will increase the kind or amount of pollutants emitted must undergo a similar NSPS review. However, the provisions relating to source modifications do not apply to facilities subject to a coal conversion order under the National Energy Act.

Although NSPS restrictions will eventually apply to most boilers that could emit at least 100 tons of a pollutant per year, it will be several years before these regulations are promulgated. Meanwhile, small boilers are regulated only by State and local authorities and do not require a Federal permit. A utility could be concerned that a number of small units could start up between the time of a powerplant's permit application and the time of permit award and make it impossible for the plant to operate without violating air quality restrictions. For example, six 250-million Btu/hr boilers, each burning 50,000 tons of 3 percent sulfur coal (without control) can release the same amount of $SO_2$ as a large powerplant burning 2 million tons of the same coal with 85-percent control. Because small coal plants can be ordered as package units with short delivery schedules, this scenario may be plausible. The Clean Air Act requires the States to monitor small sources to prevent them from

**Table 20.—New Source Performance Standards**

| | Pollutant | | | | | |
| | Particulate matter | | Sulfur dioxide | | Nitrogen dioxide | |
| Source | Emission limitation | Emission reduction | Emission limitation | Emission reduction | Emission limitation | Emission reduction |
| --- | --- | --- | --- | --- | --- | --- |
| Coal-fired steam generator | 0.03 lb/10⁶ Btu and 20-percent opacity | 99 percent | 1.2 lb/10⁶ Btu | 85 percent | 0.60 lb/10⁶ Btu | 65 percent |
| **Coal Preparation:** | | | | | | |
| Thermal dryer gases | 0.031 gr/dscf and 20-percent opacity | | | | | |
| Pneumatic coal-cleaning equipment gases | 0.018 gr/dscf and 10-percent opacity | | | | | |
| Other | 20-percent opacity | | | | | |

"using up" available air quality increments, but State enforcement programs may, in some cases, have difficulty complying with these requirements.

EPA's proposed regulations under the 1977 NSPS provisions have become controversial. The principal issue is whether a plant burning low-sulfur coal should be required to achieve the same percentage reduction in potential $SO_2$ emissions as one burning higher sulfur coal. EPA's proposed regulations for steam electric-generating units set forth a full or uniform control requirement regardless of the sulfur content of the fuel, but with a 3-day-per-month exemption from the percentage reduction to accommodate high-sulfur Midwestern coals. Alternative proposals considered by EPA include a sliding-scale standard under which the required percentage reduction declines proportional to the sulfur content of the coal, and monthly, rather than daily, averaging of the percentage reduction requirement. In addition, the General Accounting Office (GAO) has recommended that EPA continue to allow supplementary controls until mandated studies of $SO_2$, sulfates, and fine particulates are completed in late 1980.[77]

A variety of considerations will affect the

[77]Sixteen Air and Water Pollution Issues Facing the Nation (Washington, D.C.: General Accounting Office, U.S. Comptroller General, October 1978).

final NSPS regulations. Under most circumstances, the proposed full-control requirement would achieve the greatest reduction in $SO_2$ emissions, but would increase the amount of scrubber sludge from approximately 12 million metric tons dry basis under the previous NSPS to around 55 million tons in 1985. A full-control requirement would promote the use of locally available, higher sulfur coals, especially in the Midwest, and discourage the use of more expensive low-sulfur coals. Partial scrubbing would reduce flue-gas desulfurization (FGD) costs and permit the bypassing of a portion of the flue gas and thus alleviate the need for plume reheat and associated energy costs. Full scrubbing would delay the construction of new plants causing existing coal- and oil-fired plants to be utilized more than they would have been without the proposed standards, thus causing an increase in emissions from existing plants in the short-term future that would partially offset reductions achieved by new plants.

The provisions of the Clean Air Act related to nonattainment areas and the prevention of significant deterioration (as described above) also affect emissions from existing facilities. The offset policy included in the requirements for nonattainment areas and the availability of PSD increments will put pressure on existing sources to install costly pollution control technology. In addition, each SIP must include a control strategy for existing sources in order to achieve and maintain the national standards.

The Clean Water Act imposes effluent limitations on quantities, rates, and concentrations of chemical, physical, biological, and other constituents that are discharged from coal combustion facilities. Limitations have been established for total suspended solids, oil and grease, copper, iron, hydrogen-ion concentration (pH), free available chlorine, and materials used for corrosion inhibition including zinc, chromium, and phosphorus. These limitations are implemented primarily through NPDES. As mentioned above, a facility may be issued an NPDES permit for a discharge on the condition that it will meet all applicable water quality requirements. However, these limitations are not as difficult to achieve as the air quality standard, and the necessary controls do not add significantly to the cost of a coal combustion facility.

Finally, the Federal Government regulates disposal of combustion byproducts—ash and scrubber sludge—under RCRA. The Act establishes guidelines for the identification, transportation, and disposal of hazardous and nonhazardous solid wastes. The extent to which RCRA will constrain increased coal combustion is unclear until all final regulations have been promulgated. However, a preliminary industry analysis indicates that both ash and sludge meet at least one criterion for the "hazardous" designation.[78] If either were listed as hazardous it would have to be disposed of in accordance with a State plan, and the generator of the waste would be subject to extensive recordkeeping provisions. If both ash and sludge were listed as hazardous, RCRA could prohibit the use of sludge ponds, increase the cost of waste disposal by as much as 84 percent, and foreclose the sale and/or use of the wastes either directly or by making them noncompetitive with raw materials.

# COMPLYING WITH THE CLEAN AIR ACT

The three major strategies for complying with the provisions of the Clean Air Act are appropriate siting, pollution controls, and new combustion technologies. The range of pollution controls and combustion technologies available to minimize emissions from coal conversion are summarized in table 21.

Theoretically, complying with Federal emission restrictions should not be a problem. EPA is supposed to set the restrictions for new plants on the basis of demonstrated BACT. BACT for control of particulates and $NO_2$ emissions is relatively noncontroversial, although more stringent control measures could be enacted in the future. $SO_x$ emission control, however, is a matter of acrimonious dispute. The proposed NSPS for $SO_x$ requires all new coal-fired powerplants begun after September 1978 to use continuous technological controls. For the immediate future, this is in essence a requirement for FGD. This technology is described in chapter III. The controversy surrounding its technological adequacy is analyzed here along with alternatives under devel-

opment. The environmental costs and benefits of applying it are discussed in chapter V.

## The FGD Controversy

The recent history of FGD has been one of substantial disagreement between EPA and the utility industry over reliability, secondary effects, and costs and benefits of scrubbers. Most, though not all, of the operating FGD systems have experienced rather severe operating difficulties such as scaling of calcium sulfate on scrubber surfaces, corrosion of operating parts, erosion of stack liners, and acid fallout around the powerplants. Lime/limestone systems, the technologies that have been ordered by most powerplants, produce large quantities of calcium sulfite sludge that, unless specially treated, has poor structural strength (and thus does not provide a stable foundation) and represents a potential water pollution problem. Finally, the forced installation of FGD will cost

[78]*The Impact of RCRA (Public Law 94-580) on Utility Solid Wastes* (Electric Power Research Institute, EPRI FP-878, TPS 78-779, August 1978).

#### Table 21.—Applicability and Status of Pollution Control Technologies

| Pollutants and control technology | Pollutant reduction efficiency (%) | Applicability | | | Cost |
|---|---|---|---|---|---|
| | | Utility | Industrial | Residential and commercial | |
| **SO₂** | | | | | |
| Mechanical beneficiation | 20–40 | Current | Current | Current | Low |
| Low-sulfur coal | Varies | Current | Current | Current | Varies[a] |
| Flue-gas desulfurization (FGD) | 85–95 | Current | Current (large installations) | Not suitable for small boilers. | High |
| Fluid bed combustion (FBC) | 80–90 | 1985[b] | Very near term or current | Not evaluated | Low to moderate |
| Chemical coal cleaning | 40–60 | Post-1980 | Post-1980 | Post-1980 | Moderate to high |
| Solvent-refined coal | 70–90 | Post-1980 | Post-1980 | Post-1980 | Moderate to high |
| Coal gasification | | | | | |
| Low BTU | 90–95 | Post-1980 (preferable for new units) | Post-1980 (larger units) or industrial parks. | Probably not applicable except in commercial centers | Moderate to high |
| High BTU | 90–95 | Post-1980 | Post-1980 (highly applicable) | Post-1980 (highly applicable) | High to very high |
| Coal liquefaction | 90–95 | Post-1985 | Post-1985 | Post-1985 | High |
| Coal-oil slurry | varies[c] | 1980 | 1980 | Probably not applicable | Low |
| Magnetohydrodynamics (MHD) | 90 | Post 1990 | Not applicable | Not applicable | |
| FGD combined with mechanical beneficiation | 85–95 | Current | Current (large installations) | Not suitable for small boilers | Moderate to high |
| **NOx** | | | | | |
| Combustion modification | 20–80 | Current for some units | Current | Partially applicable | Low |
| Flue-gas denitrification | 60–95 | Post-1985 | Post-1985 | Not suitable for small boilers | High |
| **Particulates** | | | | | |
| Inertial devices | 90 | Current | Current | | Low |
| Electrostatic precipitators | > 99 | Current | Current | | Low |
| Fabric filters (bag houses) | > 99 | Current | Current | | Moderate to high |
| Scrubbers | | Current | Current | | Moderate to high |

a When Western low-sulfur coal is used at the source, the cost is low.
b TVA is presently planning for a FBC powerplant which is expected to come on line in 1985.
c Strictly speaking this is not a sulfur oxide control technology, but rather a technology to utilize coal in boilers designed for oil with minimal changes in boiler design. If a high or medium sulfur coal is used, then the $SO_2$ emissions will be higher than if oil alone were burned.

billions of dollars over what would have been required under the old NSPS, as well as several percentage points in lost electrical conversion efficiency. Industry will acknowledge no significant benefits from such an investment. On the other hand, EPA can point to smoothly running FGD units in Japan as well as a steady improvement in scrubber reliability in this country as counterweights to the industry's reliability arguments. In addition, EPA can point to the role of $SO_2$ as a precursor of acid rain, which is a serious ecological problem in the Northeast. Finally, EPA claims that the issue of sludge disposal is more of an enforcement and cost problem than a technological one; that techniques for stabilizing the sludge have been demonstrated; and that the volume, although extremely large in absolute terms, is not disproportional to the volume of fly ash that the industry has been handling for many years. Environmentalists who either support EPA's position or who want stricter standards, point to the possibility, strongly disputed by the industry and by many in the health community, that further controls on $SO_2$ will lower mortality rates (see *Health Effects*, chapter V).

The level of controversy over these issues has become so intense that it was felt necessary to add a discussion of the technological arguments over the problems of FGD. The health and ecological effects of $SO_x$ are discussed in chapter V, as are the mechanics of disposal and the potential environmental problems associated with the sludge. The various FGD systems are described in chapter III.

## FGD Status

The first FGD installation was made in England in the 1930's (the Battersea A powerplant, a 228-MW unit using alkaline Thames River water as the scrubbing medium). In 1968 the Union Electric Meremac Unit became the first U.S. commercial installation, using dry limestone injection in the boiler followed by wet scrubbing. Between 1968 and 1972, five utilities had installed this type of system, but all encountered operating difficulties in the form of plugging of boiler tubes, low removal efficiency, and interference with particulate collection efficiency. This technology is no longer sold, and the five original systems are either shut down or being converted to other systems. This record, along with operational difficulties with newer scrubber systems, has created the basis for years of arguments and controversy between EPA and the utility industry.

As of March 1978 there were 34 operating FGD systems, totaling 11,508 MW, 42 systems under construction (17,741 MW), and plans for installation of 56 other units (27,230 MW). Of the systems installed since 1968, EPA counts 15 (all "demonstration units") as having been terminated.[79] With only a few exceptions, utilities have chosen direct scrubbing with lime or limestone as the scrubbing medium. Most of the new systems have been retrofitted to existing boilers, although this situation should change within the next few years with the construction of new plants, which must comply with the percentage removal requirements stipulated by the 1977 Clean Air Act amendments.

Although EPA and the utility industry would agree that most of the existing scrubber systems have encountered operational problems, their viewpoints then diverge widely. What follows is a simplified, abbreviated summary of the major EPA and industry arguments about scrubbers. Note, however, that the utility industry is hardly a monolithic structure, and thus the summary "industry argument" should be interpreted accordingly. Some segments of the industry have found scrubbers to be the best alternative for their plants and are trying to make them work well.

## The Industry Viewpoint

As electric power demand grows, constraints on capacity growth will demand high levels of reliability for individual units. The industry contends that scrubbers are unreliable and their use endangers the system reliability of the powerplants they serve. They take the view that virtually every scrubber that has been installed on a major unit in the United States has had severe operating difficulties and major shutdowns. In many cases, these shutdowns would have forced reductions in boiler capacity except for the utility's ability to bypass the scrubber. New EPA regulations, if approved, will eliminate these bypasses. EPA points to a few smoothly operating scrubber systems, but the industry counters that in every case these systems have unique properties that avoid problems that most plants have to face. In some cases, this "smooth" operation is due to almost continuous maintenance, such as nightly cleaning of the unit, which the industry does not feel is practical or reasonable for the average plant.

Case studies of individual plants illuminate these problems (for more detailed descriptions of actual operating experience at a number of plants from an industry perspective, see appendix V of volume II). For instance, the Southwest plant (Springfield City Utilities), with 200

---

[79]"Summary of the Operability of FGD Technology" (Washington, D.C.: U.S. Environmental Protection Agency, Office of Energy, Minerals, and Industry, Sept. 14, 1977).

MW of scrubber capacity, started up in April 1977. Operation has been poor; only one of the two scrubbers has been kept in service. The operating scrubber is only 60-percent reliable, with problems such as: (1) mist eliminator plugging, (2) corrosion of dampers, (3) failure of the lining materials in the scrubber duct work and stack, and (4) damage to nozzles and pumps from foreign material in the limestone. Another problem is that, due to a lack of reheating for the cooled gas coming from the scrubber, acid fallout around the plant has occurred. Another plant, the 830-MW Mansfield powerplant (Penn Power), ran well for its first year of operation because of the ability to cut boilers back and thus maintain the scrubbers, but it was reduced to half capacity after failure of the stack liner, a problem that has occurred elsewhere. Operation of this plant has been said to be simplified because its ponded wastes are being discharged to the river rather than recycled. This has been said to reduce scaling problems because the concentration of dissolved solids in the scrubber circuit is kept low.

As noted above, some of the better experiences with scrubbers are felt by the industry to be unique and inapplicable to general experience. For example, Paddy's Run (Louisville Gas & Electric) has received special attention because it operates reliably and is quite effective in $SO_2$ removal. However, critics point to a number of factors that raise questions about the validity of extrapolating the Paddy's Run design to larger installations (the units are only 35 MW each). The operation is said to be atypical because the plant has a low load factor. Moreover, unlike other installations, the unusually low-chloride content of the coal reduces problems associated with chloride build-up in the scrubber. Finally, the lime employed by the system is a byproduct that is not generally available and appears to have unique properties that provide better operation than standard lime.

Some of the major scrubber problems identified by the industry are:

- Achievement of high $SO_2$ removal efficiency: Although high $SO_2$ removal effi-

ciency can be achieved, the problem of achieving high levels without incurring excessive costs becomes very complicated for the high removal levels that may be required by EPA. The major complicating factors include the wide variety of scrubber types and selection of type and amount of absorbent.

- Wet-dry interface deposition: Drops or slugs of slurry become detached, splash back into the gas stream, and stick to surfaces.
- Plugging and scaling of mist eliminators.
- Gas reheat: The requirement to reheat the scrubber gas to increase its buoyancy requires several percent of the total plant output.
- Corrosion/erosion: Especially due to the chloride concentrations in the scrubber liquids. Failure of the stack lining is a common problem, as are plugging, erosion and corrosion of the reheater.

In short, industry feels that a commitment to FGD is premature because the technology creates problems that are substantially more severe than those utilities have historically had to face, the units require maintenance and system debugging of a kind and intensity the industry has heretofore been spared, and the reliability of the units is seen as unacceptably low or unproven.

Beyond technological problems, however, the industry has severe economic problems with scrubbers. They are the most capital-intensive controls it has ever been asked to install—in an industry that is currently undergoing capital shortages. The industry is being asked to increase its power costs by as much as 15 to 20 percent even when no ambient air quality standards or PSD increments are being threatened. Moreover, where control of existing plants is required, the industry is almost forced by the nature of its regulatory system to prefer increases in fuel costs (i.e., switching to low-sulfur coal) to installing capital-intensive equipment. While general rate increases usually involve a lengthy hearing procedure, fuel adjustment clauses in many States allow the utilities to recoup immediately the cost of a fuel switch.

## The EPA Viewpoint

EPA justifies its very strong support for FGD systems despite the severe operating problems by the following points:

- The industry, in its description of "problems," is said to ignore the progressive improvements both in debugging existing scrubbers and designing new ones. EPA concludes that the performance of utility FGD systems has consistently improved over the last 5 years, that many of the problems described in the literature either are essentially solved or are more in the nature of startup problems requiring differing degrees of "fine-tuning" at each installation. Features of the newer systems (improved mist eliminators and reheaters, a trend toward open scrubber types, high scrubber liquid to gas ratios, improved materials) and newer systems (double alkali, magnesium oxide) should prevent many of the operational problems of the past. EPA asks that the viability of its FGD requirements be judged on the basis of what designers can achieve today and in the future, rather than on the basis of units that were designed several years ago and that were often improperly operated and maintained.
- The Japanese experience with FGD is frequently cited as an indication that FGD in this country could achieve far higher reliability if more optimum designs and more intensive maintenance programs were utilized. Japan has five exemplary coal-fired facilities operating at greater than 90-percent removal efficiencies and 95-percent operabilities. All are greater than 150 MW in size, and four burn 2 to 2.5 percent sulfur coal. A recent interagency report on these scrubbers[80] notes their extremely high operability (degree to which the scrubber is used when the boiler is operating) and performance, while also concluding that their operating conditions are not at all dissimilar to those applicable in

the United States. In fact, some of the systems were designed and installed by U.S. vendors. EPA can point to the Japanese experience as one where basically the same physical problem existed, but where a completely dissimilar industry attitude (one of commitment to scrubber success), recognition of the need for careful maintenance and chemical expertise in operation, and a conservative attitude towards design, materials, and contractor accountability led to an extremely successful scrubber-based control system.

- Utility industry shortcomings (EPA contends) are a major cause of FGD problems:
  - Utilities' tendency to select the lowest bid regardless of vendor experience or system design.
  - Inexperience in dealing with complex chemical processes, and unwillingness or inability to hire trained operating and maintenance personnel.
- EPA notes that FGD units can build in redundancy, allowing a more intensive maintenance program and compensating for any unexpected failures. This translates a reliability problem into an economic one.

Although EPA is concerned about the substantial economic impact of scrubber requirements, EPA's viewpoint is different from that of the industry. EPA is much more likely to consider the public at large rather than utility shareholders as its major "constituents," and thus will not automatically prefer a strategy that will raise operating costs over capital costs as the industry does. Second, EPA must respond to the wording of the 1977 Clean Air Act amendments, which stipulates the use of technological controls for power production.

### Interpretation

The substantial improvements in operating experience of scrubbers in the United States, the availability of new designs and new and improved systems, and the excellent operability of the scrubbing systems on Japanese coal-fired powerplants represent substantial evidence that FGD is both a perfectable technol-

---

[80]M. A. Maxwell, et al., *Sulfur Oxides Control Technology in Japan,* report prepared for Henry M. Jackson, Chairman, Senate Committee on Energy and Natural Resources, June 1978.

ogy and one that U.S. powerplants can install with reasonable confidence that high levels of reliability can be obtained.

The experience at the La Cygne (Kansas City Power & Light) powerplant illustrates the improvement that has occurred with experience in scrubber systems. The scrubber, which is designed to remove 80 percent of the $SO_2$ and most of the particulate matter from a very "dirty" coal (5 percent sulfur, 30 percent ash), was installed in June 1973 and initially experienced a very low 31-percent reliability. This reliability has been gradually increased over the past several years with experience in proper operation and the appropriate materials to use in critical components. The current reliability is 92 to 93 percent, and the manager of the plant sees no reason why this level should not be maintained or improved in the future.[81] This experience also is being applied to scrubbers now being designed. The significance of this experience is that many of the problems described by the industry are important only in a historical context, or in the context of understanding the problems that can develop if utmost care is not taken in equipment design and maintenance.

The Japanese experience, as EPA claims, appears to indicate that the major FGD problems are avoidable. Although many in the utility industry continue to challenge the applicability of this experience to the U.S. situation, the evidence indicates that the two most visible criticisms—that the Japanese success has been with oil-fired plants, and that their scrubbers run in an open loop mode that is inapplicable to U.S. utilities—are, respectively, no longer true and incorrect. Although the highest sulfur coal used in the Japanese facilities is only the equivalent of a 3 percent sulfur Midwest or Eastern coal, the La Cygne experience indicates that the problems of scrubbing very high-sulfur coal are not insurmountable.

The EPA proposal for an NSPS for $SO_x$ control of utility boilers asks for an 85-percent reduction in $SO_2$ emissions determined on a 24-hour daily basis for most coals. A reduction in

control efficiency to 75-percent removal is allowed for a maximum of 3 days per month. Although all of the existing Japanese coal-fired plants meet these requirements, recently designed U.S plants have not been required to meet them and have not met them. It would be naive to expect that the problems associated with scrubbers will evaporate even when (and if) U.S. powerplants are asked to satisfy the EPA standards. U.S. coal characteristics, utility operating practices, and plant conditions vary enough to ensure that each scrubber will require carefully tuned design and operation to attain satisfactory performance. U.S. utility operators must emphasize conservative design, extremely careful and constant maintenance, use of trained operating and maintenance personnel, and pressure on FGD vendors—just as the Japanese and the successful U.S. operators have. To the extent that utilities do not take these requirements seriously, they invite scrubber breakdowns and consequent loss of system reliability. At the same time, EPA should share the responsibility for ensuring scrubber success by acting in a watchdog capacity over vendor design and installation and utility operations. EPA must take an extremely vigorous position in disseminating its substantial experience; at a minimum, it should sponsor a series of courses or seminars for utility personnel who are responsible for ordering FGD systems, to assist them in selecting systems appropriate to their needs.

In conclusion, existing evidence points to FGD as a viable technology, albeit one with remaining problems. In most situations, scrubbers should be able to obtain sufficiently high levels of both control efficiency and reliability to satisfy EPA and utility requirements. Some doubt remains about satisfying EPA requirements for very high-sulfur coals, and both DOE and the Utility Air Regulatory Group (UARG) have expressed concerns about the need for scrubber bypass capability for plants using these coals.

Given the technological viability of FGD, there remains the critical question of its costs. Present estimates range from $80 to $120/kW out of a total plant cost of about $800/kW. Some utilities report higher FGD capital costs

[81]Personal communication with Mr. Cliff McDaniel, Dec. 4, 1978.

will be necessary to achieve adequate reliability. DOE and EPA estimate the cost of the EPA NSPS proposal by 1990 to be on the order of $22 billion over and above the cost of the previous standard. The accuracy of these and other estimates is very much in doubt, given the rapidly changing state of the art of scrubber design, the substantial degree of uncertainty as to the degree of scrubber module redundancy that will be needed to maintain the required reliability, the uncertainty as to how risk-averse the utilities will be in their scrubber purchases and operating programs, and the extent to which systems other than lime/limestone are used. Nevertheless, the order of magnitude of the estimate is certainly correct, and it demonstrates how much is at stake in the current argument over NSPS. Both DOE and UARG have proposed alternative standards that would reduce costs substantially with small increases in emissions. However, none of the actors in the regulatory process appears to have analyzed the probable air quality effects of any of these alternatives. The existing analyses are based on gross emissions, which are an inadequate measure of benefit. EPA, in turn, has satisfied the requirements of section III of the Clean Air Act by analyzing the "nonair quality health and environmental impact" of the proposed standard, but it has not attempted an analysis of the actual environmental benefits of the $SO_x$ reductions it proposes to achieve. Thus, neither DOE nor EPA has evaluated the relative costs and benefits of the alternative NSPS options. This raises some interesting questions about the adequacy and policy orientation of the environmental research programs sponsored by these agencies, as well as their attitude toward the need to attempt to balance costs and benefits.

## Other $SO_x$ Control Options

In the immediate future the only significant variant to the use of FGD as the total $SO_x$ control mechanism will be the use of physical desulfurization in conjunction with the scrubber. Any sulfur removed in cleaning counts toward the continuous removal required by the Clean Air Act. Very few coals can undergo

sufficient sulfur removal to affect FGD designs significantly. If 30 percent is removed in cleaning, FGD still has to remove 80 percent to meet an 85-percent standard. However, despite its failure to reduce significantly FGD removal requirements, coal cleaning does allow a significant reduction in limestone and sludge handling requirements and an improved operating environment for the scrubber. FGD units are sensitive to the variability of the sulfur in raw coal and the presence of other contaminants that may be partially removed by cleaning. For some new plants, this combination of front- and tail-end cleaning can be economically advantageous. EPA[82] has shown that, under some limited conditions, physically cleaning the coal before scrubbing lowers the total cost of $SO_x$ control.

Current FGD technology may be inappropriate for control of smaller, intermittently operated industrial boilers. Unit capital costs rise sharply with decreasing size, and if a boiler is used only 20 percent of the time, capital costs alone make FGD prohibitively expensive. The emissions standards for industrial boilers have not been promulgated yet. If they are similar to utility limitations, only the largest industrial units will be able to comply; until new coal combustion technologies are developed, smaller units simply will not burn coal. If these facilities and the even smaller residential/commercial-size equipment are to burn coal cleanly in the short term, it will be possible only by feeding them "clean" coal and accepting a higher level of pollution per ton than from utilities.

The list of options should lengthen considerably in the future. Fluidized bed combustion (FBC) is an efficient method of burning coal while simultaneously controlling $SO_x$ emissions. FBC furnaces can be smaller than conventional ones; this should lower capital costs. The residue of ash and sulfur compound is dry, simplifying disposal. FBC currently is developed to the stage that some industrial-size units are offered commercially. Utility-size units still pose substantial design prob-

[82]*Coal Cleaning With Scrubbing for Sulfur Control* (Washington, D.C.: U.S. Environmental Protection Agency, August 1977), EPA-600/9-77-017.

lems and may never offer any cost advantages over FGD. Although the technology is expected to be able to meet all new emission limitations, the performance has not been demonstrated on a large scale. If FGD is required as an add-on to FBC, as could theoretically happen, there would be little incentive to undertake FBC. The initial (and possibly sole) application would appear to be in industrial units. If the units now on order work out as expected (and the British experience indicates they will) a very rapid expansion could follow.

Low-Btu gasification is an intriguing near-term development concept for utilities and industry. A combined-cycle facility should prove more efficient than powerplants now in use, and the economics look attractive. Industry should find gasification more convenient than direct coal use.

Solvent-refined coal, chemical coal cleaning, coal gasification (both low- and high-Btu), and coal liquefaction are processes to produce coal-related fuels that are clean enough to meet emission limitations without the addition of FGD. "Clean" fuels assure that emissions limitations always are being met as long as the boiler is operating, regardless of load level. Those utilities that are least favored in the capital market are relieved of the capital costs burden of limiting emissions. Instead, the capital burden would be on a coal-refining industry that, like the oil-refining industry, can operate at constant load factor over the life of the equipment. Utilities are constrained to follow the load demands of their customers, and therefore have less flexibility even when output is reduced, which increases unit cost for pollution control. "Clean" ing" of fuel, therefore, is in some ways more attractive than FGD—but the economics still are questionable. "Clean" fuels also are applicable over a wider range of boiler sizes than FGD. They provide greater flexibility in siting coal-fired units. Existing units are more likely to be adaptable to "clean" fuels than to FGD.

All the processes are currently under development except the Fischer Tropsch coal liquefaction technology, which has been in continuous operation in the Union of South Africa.

The costs for U.S. construction and operation of this process, as well as most of the other "clean" coal processes, are the major uncertainty. Thus, none of the processes is expected to begin making much impact until the late-1980's. Chemical cleaning of coal is the only one of the "clean" coal processes that would require other control technologies in tandem since none of these methods removes sufficient sulfur to comply with new NSPS. This deficiency could limit its use.

## $NO_x$ Strategies

The control techniques now in use in the United States predominately involve combustion modifications. These techniques appear adequate to meet present standards without excessive economic penalty. Development of low-$NO_x$ burners is underway, and if successful, could reduce $NO_x$ emissions still further (to 85-percent control) at little cost.

The "clean" fuel technologies remove various degrees of nitrogen along with the sulfur. Coal gasification and liquefaction achieve almost complete removal. Chemical cleaning and solvent-refined coal are less effective in nitrogen removal as currently operated.

FBC produces lower $NO_x$ emissions than pulverized coal combustion because of its lower operating temperature and larger coal size. The first factor reduces atmospheric nitrogen reactions and the latter controls fuel bound nitrogen. This may prove to be a major attraction of the technology.

Flue gas denitrification processes are in various stages of development—mostly in Japan—and are designed for oil-fired units. Their commercial availability is not expected for coal-fired units in the United States until after 1985. These processes are considerably more expensive than combustion modifications.

## Particulates

Electrostatic precipitators (ESPs) are the major particulate control technology for large industrial and utility boilers. They are likely to remain the technology of choice for utilities burning high- and medium-sulfur coal unless a stringent standard for fine particulates is promulgated. Such a standard is an eventual possibility because fine particulates are suspected of playing a role in health and ecosystem damage that is disproportionate to their weight fraction of the aerosol complex.

Baghouses have recently been installed on a few utility units. They can be efficient collectors of fine particulates (ESPs are not) and they do not suffer performance degradations when low-sulfur coal is used, as do ESPs. Industry has extensive successful experience with baghouses, but this experience is not fully applicable to utility requirements. Thus a testing period is necessary, but there is no reason to believe that baghouses cannot be applied successfully to utility boilers.

## Control System/Fuel Compatibility

Coal-burning facilities must be designed as integrated systems. Thus, change in one part can affect others, and the switch of an existing plant to lower sulfur coal can cause a large loss to the efficiency of ESP's. This particular problem can be alleviated by use of flue-gas conditioning, but other changes in control may call for some extensive modifications. Existing facilities might find a conversion to baghouses from ESPs virtually impossible because of the increased space and pressure drop requirement.

New plants must consider the effect of the technology selected for control of each pollutant on all the other controls. Thus the use of low-sulfur coal may force the use of hot side precipitators at double the cost of a cold side ESP with high-sulfur coal.

# PUBLIC CONCERN

Expressions of public concern about energy development are relatively recent. Until the mid-1960's, active and organized opposition to energy projects arose primarily among property owners over mining or combustion methods that created nuisance-like conditions. In recent years, however, the increasing knowledge about the effects of energy development, as well as the growing distrust of large institutions of all kinds and the general concern for environmental quality and for future generations, has led to increasing opposition to energy-related projects. This opposition is not unique in American history; in many ways it echoes late-19th century populist revolts against the railroads and other large industries. In general, individuals appear to be increasingly unwilling to suffer personal and environmental risks, especially when they feel those risks have been forced on them with few or no counterbalancing benefits.

Initially, opposition to energy development focused on nuclear power and surface mining, but in recent years the trend has been spreading to coal-fired powerplants, transmission lines, and coal transportation systems. Opposition at first was limited to environmental groups, but recently such diverse groups as agriculture, labor, Indians, and local governments are acting to protect their interests when they perceive them to be threatened. In the past, this opposition has been constructive because it has focused national attention on the problems and has resulted in remedial legislation. For example, mine workers' protests about occupational hazards resulted in mine health and safety legislation and black lung benefits; public protests about the environmental degradation from strip mining brought the Surface Mining Control and Reclamation Act (SMCRA). However, even when concerns have been addressed by legislation, opposition to coal development can continue. For example, many Appalachians continue to oppose increased strip mining as well as mining not covered by SMCRA.

In order to devise more effective ways of addressing public concerns about coal growth, better ways must be devised to involve affected parties in the decisionmaking process. Federal, State, and local agencies that regulate energy development already have included in their regulations extensive interagency consultation and public participation procedures to ensure that all parties to development are identified and their interests heard. Most energy, natural resource, and environmental legislation provides for citizens' suits to ensure that the purposes of the legislation are attained. In many situations these mechanisms have adequately addressed public concerns about coal development. In others, however, the parties have found that they lack the resources to articulate or substantiate their concerns in public hearings or that their recourse lies with the legislature rather than the courts. In these cases, delaying tactics and civil disobedience have continued long after the conventional mechanisms for resolving disputes have been exhausted.

This section examines several cases in which public concerns have not been addressed adequately and discusses some alternate approaches to resolving energy development disputes. It should be noted that these cases were chosen because they do not fit the stereotype of zealous environmentalists blocking development to protect scenic areas far from their homes. Rather, these examples reflect disputes that emerged spontaneously among people concerned about their day-to-day quality of life and their long-term economic well-being.

One of the most dramatic examples of the failure of traditional citizen involvement mechanisms to resolve public concerns about coal-related development is the confict over a transmission line in Minnesota and North Dakota. The 470-mile, 400-kV direct current transmission line being constructed by two rural electric cooperatives (United Power Association and Cooperative Power Association) will be the largest of its kind in the country. Any high-voltage line on farmland will modify field work and irrigation patterns, limit future land use, disrupt drainage patterns and sometimes reduce the value of the land. The line's extra-high, direct-current voltage may involve health and environmental problems not previously encountered. Uncertainties about the potential health effects of the line's elec-

trostatic field make the farmers feel their families are being used as guinea pigs.

Much of the conflict surrounding the line stems from the planning process by which the route was selected. The line was routed under the 1973 Minnesota powerplant and powerline siting legislation, which authorizes the State Environmental Quality Council (EQC) to determine routes based on recommendations from utilities and citizens. The legislation precluded the use of parks and wildlife areas or highway and railroad rights-of-way, thus forcing the line onto farmland. Alternatives such as smaller decentralized powerplants built near the load center (rather than at the mine mouth) or underground transmission lines were not considered. The cooperatives misinformed EQC about the line's point of entry into Minnesota, limiting EQC's obligation to consider alternative corridors that might have crossed less productive, less populated lands.

Minnesota officials feel they have a siting process that protects individual rights and the environment while assuring timely and responsible energy development. The farmers were included in extensive public hearings during the siting process, and their lawsuits have been heard throughout the State court system. The Governor and church officials tried (separately) to negotiate settlements between the farmers and the cooperatives through mediation. The farmers received compensation for their easements, and the siting legislation has been amended to protect farmland in future route selections. Yet the farmers continued to oppose the line, often using civil disobedience tactics such as standing in front of surveyors' transits. Although they are pleased that farmland will be protected in future routings, that protection will not prevent this line from crossing their land. In addition, they object to the use of heavy-handed tactics and mis-information by the rural electric cooperatives to force acceptance of plans for energy development. To the energy industry such continued resistance seems not only irrational but selfish and irresponsible, given the role of northern Great Plains coal in the administration's energy plan. Thus, where an extensive public participation process was intended to produce consensus, instead the participants have emerged from the process even more firmly entrenched in their individual and opposing positions.

A second example of public concerns that have not been alleviated through the traditional mechanisms is the conflict over the community impacts of Western coal development. Concern about these impacts has risen in western towns such as Craig, Col.; Rock Spring, Gillette, and Wheatland, Wyo.; Colstrip, Mont.; Farmington, N. Mex.; Moab, Utah; and Page, Ariz. The rapid development occurring in these areas creates conflicts among long-term residents, newcomers, coal operators, and utilities, primarily over changes in quality of life as well as the nature of the growth and the responsibility for its adverse impacts. Long-term residents feel that their sense of community and continuity has been lost; coal miners and plant construction workers often are not included in community activities because they are perceived as transients. Those who will profit are in favor of the rapid growth; lower-income groups that will not share significantly in the community's increased wealth are opposed. Long-term residents and local government officials, who are aware of the historic cycle of booms and busts in the history of the West, are skeptical of rapid temporary growth but do not know how to control it. Coal developers, who could contribute to planned, orderly growth, tend to feel that the solution must come from government.

Although the impacts of rapid development of Western energy resources and the resulting patterns of conflict have received widespread publicity, little has been done to resolve them beyond conducting studies and holding public hearings. Rather, the early planning processes

**John and Bud Redding with a chunk of the sub-bituminous coal that underlies their land and some 25,000 square miles of Montana and Wyoming**

in more recently developed areas continue to repeat patterns that already have proved inadequate, such as minimal company support (sometimes in the form of company towns that provide housing but little more), ineffective or misdirected government assistance, and a lack of local money. Little effort has been directed toward determining the region's long-term comparative advantages (as opposed to relatively temporary, "boom and bust" growth) or toward accumulating resource tax monies that could be used to promote long-term improvement. Unless means of adequately addressing these concerns are found, active and organized opposition could develop in small western towns slated for coal development.

Opposition to Western coal development already has arisen among ranchers and farmers, who also are concerned about their quality of life and sense of continuity. Many ranchers are working land that has been in their families for three and four generations. They are proud to know that they were not pushed aside or bought out by a corporation. They also resent the extra taxes that have been imposed on their land and their cattle to pay for coal development in neighboring boomtowns. In addition, ranchers are concerned about the uncertainties associated with reclaiming arid and semiarid western land, and especially about potential disturbances to hydrologic systems vital to agriculture. Yet the ranchers are unable to counter effectively the influence exerted by large energy companies. Again, little has been done to resolve these conflicts beyond energy companies offering more and more money to buy out ranchers, ranchers expending large amounts of time and money to educate themselves about environmental and energy issues and to prepare for court battles, and local government continuing to increase cattle and property taxes rather than coal severance taxes.

In these and other instances of public opposition to coal and related energy development, the most common mechanisms for attempting to resolve disputes have been public relations campaigns, public meetings and hearings, studies, lawsuits, and legislation. However, as is seen in the above examples,

none of these has been entirely successful. Public relations campaigns present only one side of a conflict and do not contribute to its resolution. Although public meetings and hearings are designed to present all sides of an issue, those opposing development generally lack the resources to debate energy companies effectively. Studies can be designed to explore all facets of development, yet they cannot analyze changes in quality of life adequately, and they quickly become outdated. Lawsuits are costly and time-consuming and often merely serve to demonstrate to the parties that their relief actually rests with the legislature. Yet seeking new or amended legislation also can be time consuming and usually only affords relief in future conflicts. And while lawsuits or new legislation are being considered, uncertainties delay investments and increase economic risks.

Although each of these mechanisms may, to some extent, reach a result that favors what may be termed a common good or the majority view, present-day society increasingly seeks to protect the rights of minorities and increasingly questions the definition of "common" good. Yet an issue or conflict does not lend itself to a simple resolution that simultaneously pleases both the majority and all minorities, and no other traditional mechanisms are able to respond when the minority continues to rebel. This is especially true when the conflict arises over ethical questions such as whether a large corporation should be allowed to exploit energy resources for profit without paying local costs, as well as over questions related to national long-term priorities, such as energy versus agriculture.

However, some of the traditional mechanisms are amenable to modifications that could eliminate or mitigate some future disputes. For example, in the case of the Minnesota powerline, if farmers had been given an opportunity for public participation in the planning stages of the development—rather than at the permitting stage, when the cooperatives already had a vested interest in a particular route—the conflict might have been resolved. Similarly, in the case of the community impacts from Western coal development, tax revenues could be used to ensure that the affected States and localities receive an adequate share of the benefits of development to promote long-term economic improvement.

New methods for preventing or resolving conflicts must be devised. Two recent cases that show promise are greater local control over the manner in which development occurs, as typified by the Navajo experience, and mediation, such as the compromises negotiated during the National Coal Policy project.

The Navajo Nation owns an estimated 20 percent of U.S. strippable coal reserves. The Black Mesa coal mining complex, one of the largest in the world, is located on Navajo-Hopi lands, as are major existing and planned coal-fired powerplants. Yet Navajo per capita income remains at about one-third of the U.S. average, their unemployment rate is about 40 percent, and they are becoming increasingly concerned about the air quality and other environmental effects of coal use. The Navajos have indicated that in the future they will demand more favorable leasing arrangements as well as needed social and environmental benefits such as jobs, management training, and pollution controls as prerequisites to energy development on tribal lands. They already have won the right to impose a possessory interest tax equivalent to a property tax on powerplants and other energy development, as well as a business activities tax. They also are attempting to institute an emissions charge system in order to resolve their concerns about the air quality effects of coal combustion. In effect, the Navajos are ensuring their participation early in the planning stages of energy development as well as ensuring that revenue will be available for needed social and economic benefits.

An even more promising mechanism is mediation of public concerns outside the courts, legislatures, and bureaucracies in which developers and parties-at-interest negotiate a compromise. For example, a group based at the University of Washington has successfully mediated a controversy over the route and size of a major freeway. Recently a foundation-supported nonprofit corporation called RESOLVE

was formed to help settle environmental disputes at the national level. In the energy area, the National Coal Policy project was intended to produce agreements on how coal can be mined and burned without unacceptable externalities. In a series of meetings, leading conservationists and executives from coal-mining and coal-consuming industries agreed in principle but not necessarily in practice on a variety of public concerns. For example, industry accepted the principle that environmentally sensitive areas should be closed to mining; the environmentalists agreed to back off from their insistence that surface miners always must level off high walls. The environmentalists also agreed that their standard delaying tactics in powerplant siting and licensing procedures are counterproductive and concurred in a recommendation for one-step licensing. Industry representatives agreed that the public should be notified in advance of license applications and assented to the principle that qualified public interest groups participating in mine and powerplant hearings should receive public financial support. The two sides also agreed in principle that powerplants should be sited near the area where the bulk of the power would be sold, rather than at the mine mouth or in a remote rural area. Some of these and other agreements may require new or amended legislation, but many could be implemented privately.

The National Coal Policy project has been criticized because it did not include some of the parties at interest and because the participants failed to agree on all the issues. However, its example of conciliatory behavior provides a model for speedier and wiser alleviation of public concerns about energy development. This and other models for constructive public participation in both short- and long-term energy planning must become more common. Yet it must be recognized that some public concerns about coal and related development reflect basic value conflicts rather than objections to specific projects. Where these value conflicts exist, some people will continue to believe their rights are not protected. Thus, in some instances, even when energy companies make every effort to anticipate and address legitimate concerns and to demonstrate that all alternatives have been considered, opposition in the form of lawsuits and civil disobedience will continue.

# CONCLUSIONS

Many of the factors affecting coal production discussed in this chapter may limit the pace of coal expansion in the future. Some manifest themselves as temporary impediments to production—e.g., strikes and siting disputes—while others have become ground rules within which the industry must work. A number of these factors enable coal to be produced, but become constraints in certain circumstances. For example, coal cannot be mined without labor, but prolonged strikes by miners can disrupt coal supplies. Rail transport carries the bulk of the Nation's coal each year, but bottlenecks and car shortages slow down the tonnage carried from mine to market.

The most important constraint on coal production has been lack of demand; it is likely to continue to be the most important for at least the next decade despite the rapid growth in demand projected in chapter II. The Clean Air Act has been said to be a major constraint on meeting national goals for coal development, in part because of the requisite emission control equipment. This report does not concur. The now mandatory $SO_x$ controls on new coal-fired powerplants obviously increase the costs of burning coal, but the evidence indictates that the new standards can be met. This evidence, it must be noted, is based on a relatively few plants operating for only a few years. Hence some utilities may experience difficulties with their control equipment, especially if they have not made the necessary commitment of capital and manpower.

Despite the burden of regulations, labor-management unrest, transportation breakdowns, and other constraints, most operators report they can mine as much coal as they can sell.[83] If coal demand grows faster than the scenarios outlined in chapter II suggest, supply constraints may become significant. If these constraints do materialize, they are likely to be found among the factors analyzed in this chapter.

No insurmountable supply constraints now exist. However, Federal leasing will have to resume in the 1980's for Western coal production to meet expected demand in the 1990's. Coal transportation systems need upgrading and expansion, as is being planned. Existing data are insufficient to determine whether industry structure has been or will be a constraint on production. In recent years, poor labor-management relations have cut into coal production. However, as more and more coal is mined in the West and from non-UMWA operations, this factor should become less important nationally. Even during the 3½-month UMWA-BCOA strike in the winter of 1977-78, few power shortages occurred and only 25,500 workers were laid off at the height of the shutdown. Further, wildcat strikes slacked off in 1978. Labor-management relations may either be improving or each side may be regrouping.

The highest growth scenario would require almost all supply and demand factors to work out well. No new complex environmental control strategies could be accommodated in all probability. All the major resources required for production and use would have to be available. This situation is plausible, but it probably will not occur without additional Federal policy actions that promote the use of coal.

---

[83]See testimony of Stonie Barker, Jr., E. B. Leisenring, Jr., and Robert H. Quenon (coal operators) before the President's Commission on Coal, Oct. 20, 1978.

Chapter V
# ENVIRONMENTAL IMPACTS

# Chapter V.—ENVIRONMENTAL IMPACTS

# Chapter V
# ENVIRONMENTAL IMPACTS

## INTRODUCTION

Memories of the era when coal dominated the U.S. energy supply are blackened by visions of soot-laden cities, discolored streams, and scarred landscapes. Since the heyday of coal, Americans have become not only "spoiled" by the cleaner fossil fuels—oil and natural gas—but also more protective of their natural environment. They want to breathe clean air and drink pure water, and they insist that, after its resources have been removed, the land remain capable of supporting both natural ecologies and a thriving agricultural economy. Thus, if coal is to return to prominence in U.S. energy plans, its comeback must be staged under environmentally sound methods of extraction, cleaning, transportation, and combustion.

The "residuals"—unintended products that may affect the environment—from each operation involved in handling coal are illustrated in figure 19. The avenues of disturbances to the environment are abundant, and the intensity of even the minor disturbances will be magnified by the large quantities of coal forecast by most current energy projections. The most vociferous objections to coal use have stemmed from concerns about the pollutants released to the air during combustion and the variety of damages to land and water that result from mining. Most of these impacts are now addressed by Federal legislation, as indicated in figure 20. In general the legislation establishes either ambient air, land, and water quality standards, or minimum pollution control performance standards that can be achieved economically and effectively by the best state-of-the-art technology. In response to the passage of this large body of environmental legislation, public attention is shifting to questions of whether the new laws and technology are adequate, whether they are effectively enforced, and whether they are themselves the cause of environmental problems—such as the pollution of one environmental medium by the wastes collected to protect another.

Figure 19 summarizes only the byproducts of coal-related activities and not the resulting effect on public health or on the ecosystem at large. These effects may vary in their severity—from the annoyance and property damage of blasting to a possible tragic increase in ill-ness or mortality from air pollution. The effects also vary in their temporal and regional extent—from the short-term creation of dust at a local minesite to the possible long-term global alteration of climate by carbon dioxide ($CO_2$). The environmental effects of coal also differ from region to region, not only because of the unequal concentrations of coal extraction and combustion activities but also because of the difference in geologic, demographic, topographic, and climatic factors. Thus, for example, surface mining may be more benign in the lignite fields of Texas than on the steep slopes of Kentucky. As another example, the most significant impact of a coal-fired powerplant in New England might be to worsen the urban air breathed by millions, while that of a utility in Utah might be to degrade the quality and visibility of the clean air that is a major resource in that region.

Figure 19 emphasizes that each environmental impact may be envisioned as primarily affecting one of the three media—air, land, and water. This viewpoint is adopted as the organizational structure for this chapter, although it should not obscure the very extensive interactions among the three media. Each of the next three sections examines the impacts of coal use on one of the three media, describes the resulting effects on health or ecosystems, and assesses the ability of relevant legislation and control technology to mitigate these effects.

This chapter draws selectively on analysis of future energy impacts available in the liter-

## Figure 19.—Environmental Disturbances From Coal-Related Activities

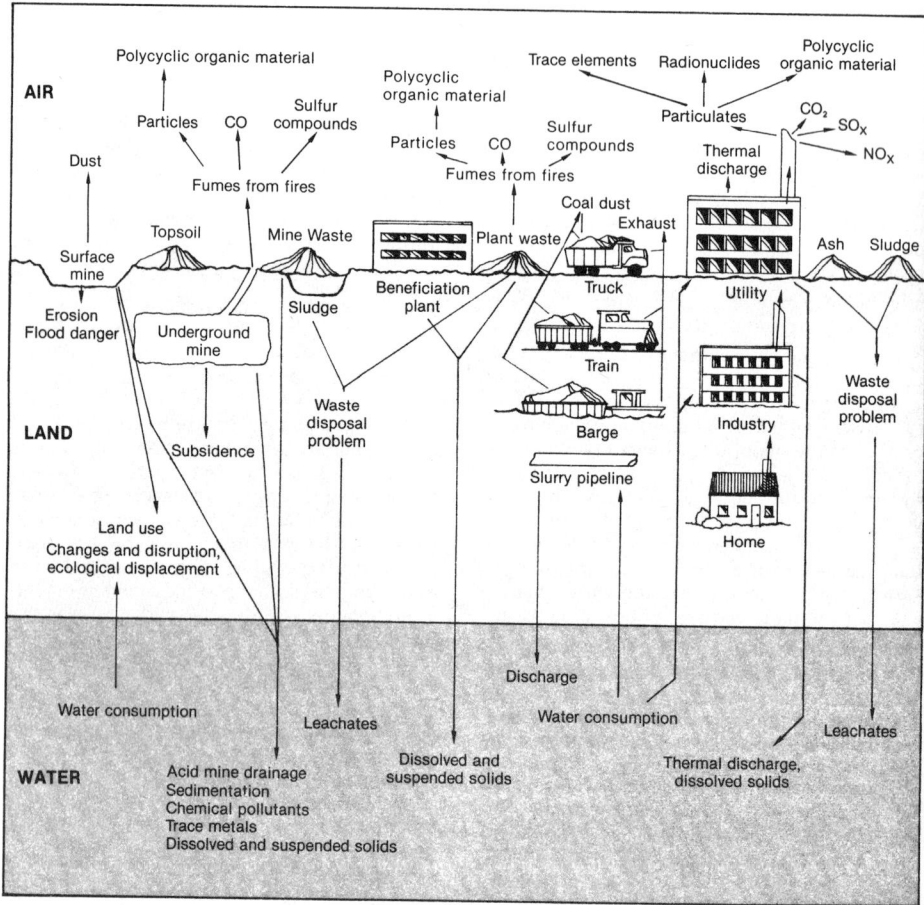

ature—and especially on the 1978 National Environmental Forecast No. 1 (NEF #1)[1] —to illustrate the level of residuals that would be generated by future increases in coal development. The forecast uses a "business as usual" scenario developed by the Department of Energy's (DOE) Energy Information Administration,

[1]*1978 National Environmental Forecast No. 1,* Division of Environmental Impacts, U.S. Department of Energy, September 1978 (draft).

which assumes medium economic growth, medium energy supply expansion, and constant (corrected for inflation) world oil prices. The scenario assumes an electricity growth of 4.8 percent *per year,* with total energy demand growing from 70.6 quadrillion Btu (Quads) in 1975 to 94.6 Quads in 1985 and 108.5 Quads in 1990 (see table 22 for a more complete breakdown of the projections). Coal use is assumed to double by 1990 to 25.4 Quads, from 12.8 Quads in 1975. Table 23 presents the environmental control assumptions used to predict future emissions of the three principal air pol-

lutants. This projection is similar to scenario B in chapter II.

**Figure 20.—Jurisdiction of Federal Control Legislation**

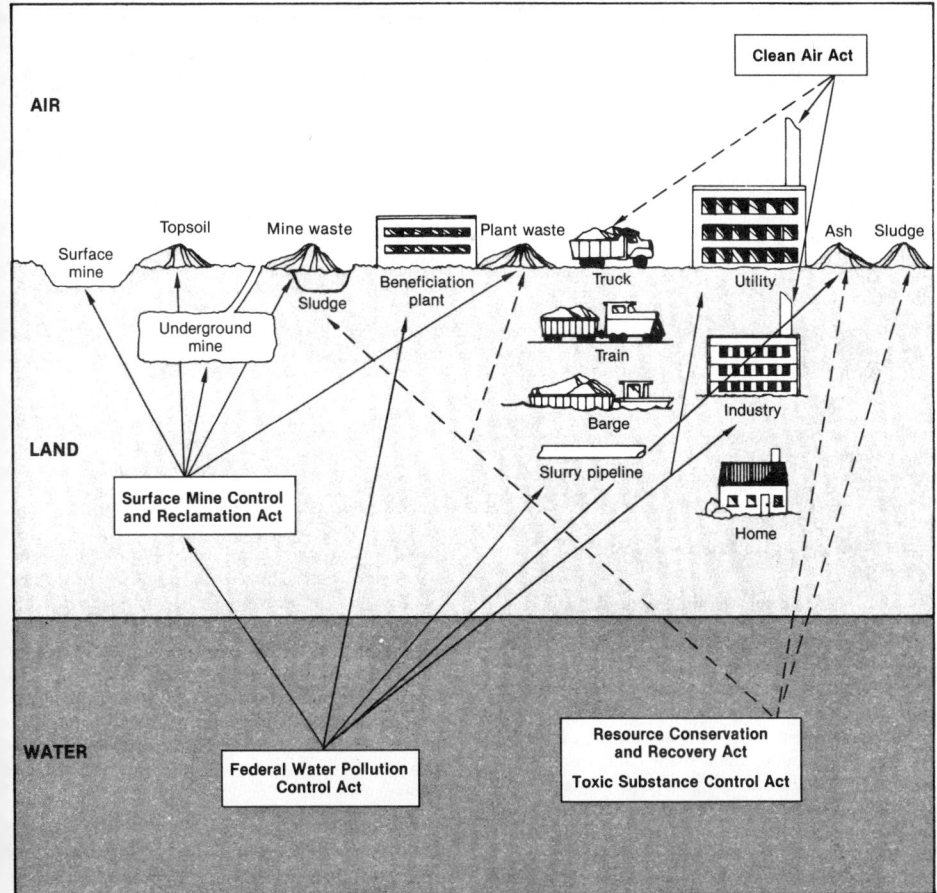

Table 22.—Energy Demand Projection Used to Develop DOE'S "National Environmental Forecast No. 1"
(quadrillion Btu per year)

|  | 1975 | 1985 | 1990 |
|---|---|---|---|
| Domestic Consumption |  |  |  |
| Oil | 32.8 | 43.9 | 48.5 |
| Natural gas | 20.0 | 19.1 | 19.3 |
| Coal | 12.8 | 21.2 | 25.4 |
| Nuclear | 1.8 | 6.2 | 10.3 |
| Hydro & geothermal | 3.2 | 4.2 | 5.0 |
| Total domestic consumption | 70.6 | 94.6 | 108.5 |
| Exports |  |  |  |
| Coal | 1.8 | 1.9 | 2.1 |
| Refinery loss | .2 | .4 | .3 |
| Total consumption & exports | 72.6 | 96.9 | 110.9 |
| Domestic consumption by sector |  |  |  |
| Residential | 14.7 | 19.0 | 21.2 |
| Commercial | 11.3 | 13.5 | 15.0 |
| Industrial | 26.0 | 40.7 | 49.0 |
| Transportation | 18.6 | 21.4 | 23.3 |
| Total domestic consumption | 70.6 | 94.6 | 108.5 |

Table 23.—Environmental Control Assumptions Used to Develop DOE'S "National Environmental Forecast No. 1"

| | | Revised NSPS | |
|---|---|---|---|
| Pollutant | Electric Utilities | Industrial boilers 25 MWe or greater | Industrial boilers under 25 MWe |
| TSP | Revised NSPS .05 lbs/MM Btu | Revised NSPS .05 lbs/MM Btu | Revised NSPS .05 lbs/MM Btu |
| SO$_2$ | Revised NSPS (scrubber efficiency 89 percent) | Revised NSPS (scrubber efficiency 89 percent) | Revised NSPS 1.5 lbs SO$_2$/MM Btu |
| NO$_x$ | NSPS (.7 lbs/ MM Btu) Schedule: For revised NSPS, all new utilities on line in 1984 and after. | NSPS (.7 lbs/ MM Btu) Schedule: For revised NSPS, all new boilers on line in 1981 and after. | None Schedule: For revised NSPS, all new boilers on line in 1981 and after. |

# AIR POLLUTION IMPACTS

## Pollutant Emissions

### Coal Activities Other Than Combustion

Most concern about air emissions from the coal fuel cycle centers on coal combustion, the emissions of which can affect natural ecosystems as well as human health and welfare over a broad region. The other portions of the fuel cycle also have important impacts on man and nature, although these impacts tend to be confined to a more local area. Mining impacts involve primarily the fugitive dust from surface and underground mines. After combustion, the dry disposal of ash or waste from flue-gas treatment produces particulate loadings that can be, on the site, in excess of Federal standards.[2] This dust problem is naturally of

[2]Report of the Committee on Health and Environmental Effects of Increased Coal Utilization ("Rall Report"), F.R., vol. 43, no. 10, Jan. 16, 1978; "App. on Coal Ash and FGD Sludges," U.S. Environmental Protection Agency.

greater intensity in arid climates. Othe sources of particulates are the construction ac tivities at each stage of the fuel cycle, the transportation of coal in open hopper cars o on barges, the storage of coal or coal wastes ir piles, and coal processing and cleaning. Al particulate emissions that include either coa dust or ash contain quantities of trace ele ments and radionuclides that may exacerbate the health and environmental impact of these emissions.

Additional releases can result from uncon trolled fires at mines or within mine piles. Suc fires, not uncommon, may smolder for years In January 1973, a survey counted 59 uncon trolled fires in abandoned mines in Appalachia and 185 uncontrolled fires in unmined coa seams in the West.[3] Another survey by the

[3]Underground Coal Mining: An Assessment of Technol ogy, Electrical Power Research Institute, AF-219, July 1976.

Bureau of Mines in 1968 registered 292 fires in coal refuse banks, principally in Appalachia.[4] Because such fires burn under oxygen-deficient conditions, they generate a somewhat different spectrum of air pollutants than coal-fired boilers. Table 24 presents a crude estimate of the emissions of carbon monoxide (CO), hydrocarbons, sulfur oxides ($SO_x$), nitrogen oxides ($NO_x$), and fine particulates due to such fires. Other releases include hydrogen sulfide and ammonia. It is reasonable to speculate that a significant portion of the hydrocarbon emissions could be polycyclic organic matter, which is particularly dangerous; however, appropriate data are not available.

**Table 24.—Gases Emitted From Burning Coal Mine Refuse Banks in 1968**

| | Emissions — $10^6$ tons | Percent of total nationwide emissions |
|---|---|---|
| Carbon monoxide | 1.2 | 1.2 |
| Sulfur oxides | 0.6 | 1.8 |
| Hydrocarbons | 0.2 | 0.6 |
| Nitrogen oxides | 0.2 | 1.0 |
| Fine particulates | 0.4 | 1.4 |

SOURCE: Nationwide Inventory of Air Pollutant Emissions, 1968, National Air Pollution Control Administration, AP-73, August 1970.

Development of controls for these air emissions has not received the attention given to air emissions from combustion. Dust from mining operations is not always completely controlled, and may be a serious occupational health problem. Coal dust from trains can be avoided by covering the hopper cars if spontaneous combustion of the coal can be avoided. The dust from waste disposal sites can be kept to a minimum by wetting agents and its impact reduced by a buffer zone around the site. Dust from coal hauling usually is controlled by watering the roadbeds. Unfortunately, most of these controls are not required by Federal regulations, and State and local requirements vary widely and often are not strictly enforced.

The general techniques for prevention of mine or refuse pile fires are to reduce air cir-

[4]*Coal Refuse Fires, An Environmental Hazard* (Washington, D.C.: U.S. Department of the Interior, Bureau of Mines, 1971) I.C. 8515.

culation to the combustible material, minimize the concentration of combustible material, and promote cooling. For abandoned mines these techniques may not always be possible to apply. The best means of prevention is to remove the excess combustible material and seal holes where feasible. The Surface Mining Control and Reclamation Act of 1977 requires that combustible materials exposed, used, or produced in underground mining (including exposed coal seams) be treated, if necessary, and covered. The Act also requires that all openings to the surface be capped, sealed, or backfilled when no longer needed for mining. Presumably, enforcement of these provisions will substantially decrease the incidence of new-mine fires.

Combustion

The air pollutants released in large quantities by combustion units include the oxides of sulfur, nitrogen, and carbon, as well as particles of ash that become entrained in the hot flue gases. Smaller quantities of trace inorganic elements, radionuclides, and hydrocarbons are also emitted; these are often adsorbed on the surface of the ash particles. Once emitted to the atmosphere, most of these substances may be transformed by a variety of chemical reactions, and all will be transported and deposited under a variety of different meteorological conditions. The pollutants may react with other pollutants from the same or different source or with natural atmospheric components. The nature and extent of resulting human health hazards and environmental damage depend on the types and rates of reactions in conjunction with the atmospheric transport and deposition, as well as the nature of the "receptor"—the ecosystem and/or human population. This discussion examines the emissions of each coal stack component. The following sections discuss the transport and transformation of each component. Understanding of these factors is crucial to predicting and mitigating the health or environmental impacts of coal combustion.

**Sulfur oxides** receive more attention than any other emission from the combustion of coal, primarily because of the large quantity of emissions, the diversity of and controversy surrounding the potential impacts (human health effects, acid rain, crop damage, etc.), and the great expense involved in $SO_x$ controls.

Coal combustion is a major source of man-caused sulfur emissions, and all of these sources may now exceed the emission of sulfur compounds from natural processes worldwide.[5] Estimates of the fraction of man-caused sulfur emissions range from about one-third to about two-thirds.[6] In industrialized parts of the United States the man-caused emissions may be at least an order of magnitude greater than natural emissions.[7]

Coal burning contributes more than half of man-caused sulfur emissions and 80 percent of the sulfur emitted from stationary fossil fuel combustion. All stationary combustion sources together release 70 percent of man-caused sulfur, with industrial processes running a distant second.[8]

The regional sulfur dioxide ($SO_2$) "emissions densities" (emissions per unit area) shown in figure 21 exhibit a wide geographic variation. The peaks in emission density do not necessarily correspond to peaks in ambient sulfate concentrations shown in figure 22, because of complex transport phenomena that may carry pollutants far from their source.

Total emissions of $SO_x$ may be expected to decline slightly in the next decade if current State implementation plans (SIPs) for air pollution control are strictly enforced, because the effect of existing sources coming into compliance with local regulations in the next few

years should more than balance the addition of new sources. NEF #1 computes the following total $SO_x$ emissions for its moderate coal use scenario:

1975 . . . . . . . . . . . . . . . 29.9 million tons
1985 . . . . . . . . . . . . . . . 28.8 million tons
1990 . . . . . . . . . . . . . . . 30.6 million tons

After full compliance has been achieved, $SO_x$ emissions will begin to creep upwards. Coal burning—and especially that from the electric utility industry—plays a major part of this trend. Electric utilities and industrial boilers burning coal are projected to remain at their present 70 percent of the total emissions through 1990. Existing coal-fired powerplants had 1975 emissions of 18.7 million tons; these will decline to about 15 million tons if full compliance is achieved, then drop only as existing plants are decommissioned. Utilities commissioned after 1975 would have a maximum (no control) emission of 15 million tons by 1990 under moderate growth conditions; given the strong controls listed in table 23, actual net emissions will be about 3 million tons. This imbalance between old and new source emissions is a critical point, because it illustrates the need to consider existing sources in devising a national $SO_x$ control strategy. Proposed New Source Performance Standards (NSPS) for steam electric utility boilers will cost tens of billions of dollars over the next few decades but will reduce total emissions less than 10 percent under emissions produced under the current standards (as projected by the Environmental Protection Agency (EPA) and DOE in a joint, ongoing analysis). Meanwhile, the major sources of $SO_2$ emissions for the next several decades—coal-fired power-plants already in existence—are allowed to operate with far less control—and sometimes none.

Industrial coal combustion also plays an important role in $SO_x$ emissions. Because industrial use of coal is expected to grow substantially in the next few decades, the NSPS will affect a higher percentage of the total emissions than is the case with powerplants. Existing industrial plants emit about 2 million tons of $SO_x$/year . By 1990, new plants will emit about half this amount if maximum control is as-

[5]R. B. Husar, et al., "Report on the International Symposium on Sulfur in the Atmosphere," Third National Conference on the Interagency Energy/Environment R&D Program, 1978, p. 77.
[6]E. Robinson and R. C. Robbins, "Gaseous Sulfur Pollutants From Urban and Natural Sources," J. Air Pollut. Assn. 20, 1970; Kellog, et al., 1972, "The Sulfur Cycle," Science, 175; and J. P. Friend, "The Global Sulfur Cycle" in Chemistry of the Lower Atmosphere, S. I. Rapeol (ed.) (New York: Plenum Press, 1973).
[7]R. B. Husar, personal communication.
[8]Preliminary Emissions Assessment of Conventional Stationary Combustion Sources: Volume II-Final Report, GCA Corporation, GCA/Technology Division, EPA-600 2-76-046b, U.S. Environmental Protection Agency, March 1976.

**Figure 21.—SO₂ Emission Contours—Emission Densities in g/m²/year**

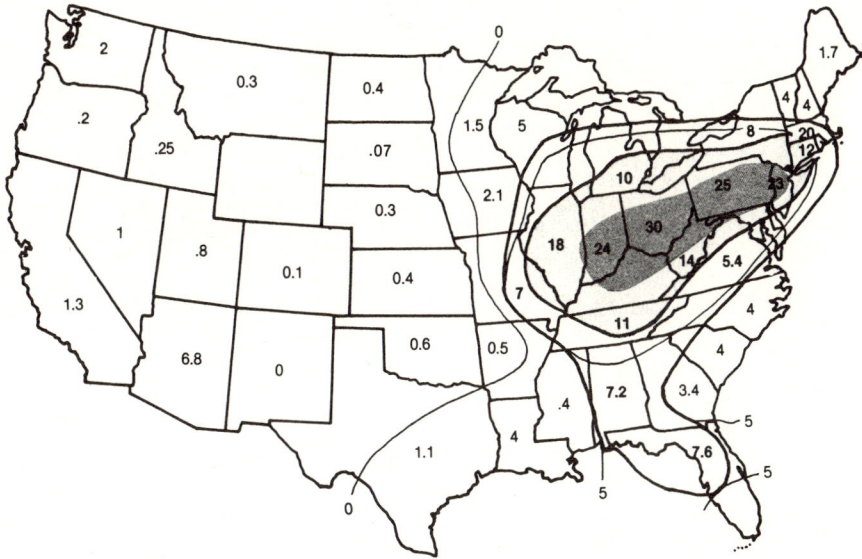

sumed. However, final emission standards for industrial sources have not been established.

More than two-thirds of the $SO_x$ emissions occur in the industrialized Middle and South Atlantic States and in the Midwest, both in 1975 and 1990, although emissions will decline in the Middle Atlantic and Midwest if existing sources comply with local standards according to schedule. Substantial emission increases should occur in the Southwest and Northwest as coal begins to replace natural gas as the dominant fuel.

**Nitrogen oxides** from coal combustion are associated with some plant damage, possible health effects, acid rain, and production of deleterious secondary pollutants. Unlike the sulfur compounds, however, man-caused sources produce only 10 percent of total $NO_x$ emissions (50 million tons per year compared to 500 million tons per year from natural sources). $NO_x$ in the atmosphere consist primarily of nitric oxide (NO), nitrogen dioxide ($NO_2$), and small amounts of nitrous oxides.

About 95 percent of the $NO_x$ of manmade origin is emitted from fossil fuel combustion. The direct product is primarily NO, which is readily converted in the atmosphere to the more toxic $NO_2$ and corrosive nitric acid ($HNO_3$).

The amount of NO produced is determined partly by the composition of the fuel itself. Coal has considerable fuel-bound nitrogen, and most of the NO produced by a normal boiler is created by the reaction of this nitrogen with the oxygen in the combustion air. NO is also produced by the high-temperature reaction between the oxygen and nitrogen in air; this reaction is the sole NO source from the combustion of nitrogen-free fuels (such as natural gas). Coal's high nitrogen content explains the major $NO_x$ control strategy for coal boilers, which is to burn the coal initially in a

**Figure 22.—Yearly Average Sulfate Concentrations, µg/m³**

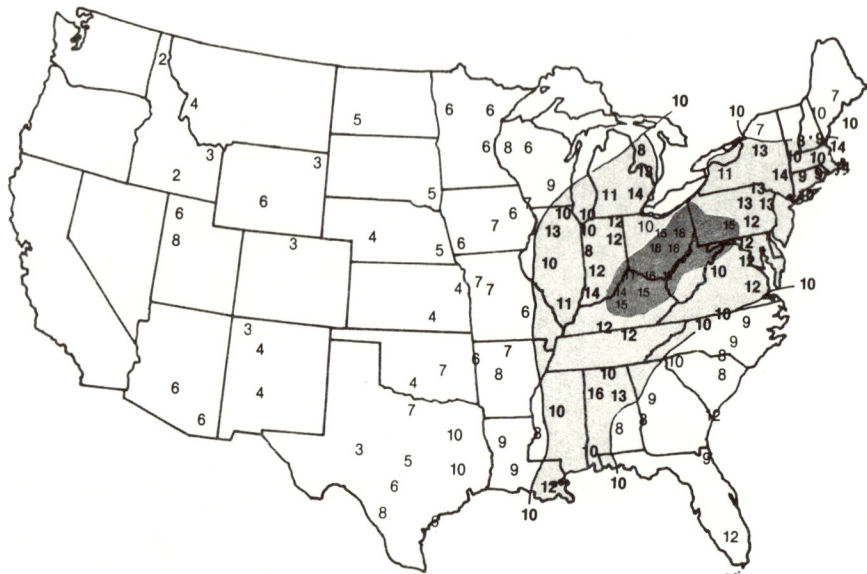

SOURCE: Report on the International Symposium on Sulfates in the Atmosphere, in EPA-600/9-78-022, Energy/Environment III, U.S. Environmental Protection Agency, Office of Research & Development, October 1978.

fuel-rich condition that promotes reduction of NO to nitrogen gas ($N_2$), and then to allow complete combustion after the fuel-bound nitrogen has been depleted.

Combustion of coal accounts for 24 percent of all $NO_x$ produced by man, and about 49 percent of $NO_x$ from stationary fossil fuel combustion. Mobile and stationary sources contribute about equally to the $NO_x$ national emissions. Although manmade sources are only 10 percent of all $NO_x$ sources globally, they can result in high local concentrations. Thus, although the natural background for $NO_x$ is about 8µg/m³, the national standard is 100µg/m³, on an annual average, and short-term local concentrations can be much higher. Figure 23 shows the 1973 national emissions densities for $NO_x$.[9] Spatial variability of $NO_x$

[9] *National Emissions Report 1973: National Emissions Data System (NDES) of the Aerometric and Emissions Reporting Systems (AEROS)* (Washington, D.C.: U.S. Environmental Protection Agency, 1976), EPA-450/2-76-007.

levels and composition is related to the quantities emitted, the availability of ultraviolet light to affect chemical transformation, meteorological conditions, and the presence of other contaminants in the atmosphere. Because of the very large amounts of $NO_x$ emitted by stationary and mobile sources, urban areas tend to have the highest $NO_x$ levels. Figure 24 shows the ambient $NO_3$ concentrations on a nationwide basis, averaged by State.

Levels of $NO_x$ are subject to temporal as well as spatial variation, because of the complex relationship of NO and $NO_2$ to light intensities as well as to the variation in time of emission levels from industrial and automotive sources (strong time-of-day variations have been observed and some evidence exists for seasonal variations).

Although $NO_x$ emissions from automobiles are supposed to decline in the next few decades, the lack of effective controls on station-

Figure 23.—Nationwide Ambient NO₃ Levels, $\mu g/m^3$

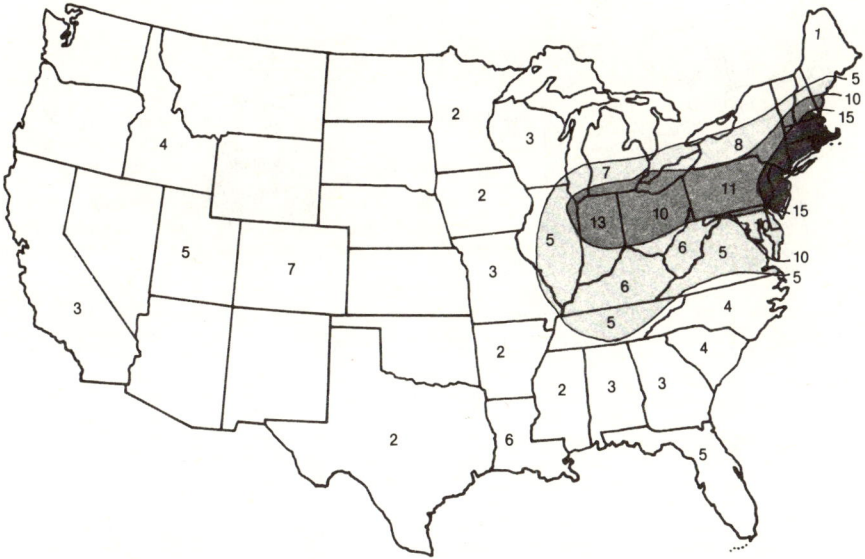

SOURCE: "Air Quality Data for Non-Metallic Inorganic Loss 1971-1974 from the National Air Surveillence Networks," Gerald G. Akland, EPA-600/4-77-003, Research Triangle Park, NC

ary sources will lead to a substantial increase in total emissions. NEF #1 projects national emissions to increase by 21 percent by 1990, from 19.4 million to 23.5 million tons, reflecting both increased energy production and a shift from gas to coal (coal combustion produces more $NO_x$ than natural gas combustion; although the ratio varies, it is about 3:1 for powerplants). Coal combustion by electric utilities and industrial boilers will play an increasing role in these emissions. In 1975, these sources produced 4.6 million tons of $NO_x$, or 24 percent of total emissions; with a moderate growth in coal use they will produce 8.7 million tons or 37 percent of the total by 1990. The shift in emissions away from automobiles and toward powerplants and industry should be accompanied by an increase in $NO_x$ concentrations in rural areas and possibly a slight decrease in urban areas.

Coal combustion produces **particulates** either by direct emission of primary particles

(fly ash) or by release of gases that are "precursors" to the formation of secondary particles. Secondary particulates include sulfates, nitrates, and particulate hydrocarbons.

Although particulate emissions from fuel combustion sources are generally well controlled, the most common control mechanisms are less efficient in controlling fine particulates (those less than 3 $\mu m$ in diameter). Thus, proportionately more particles in this size range escape capture and are emitted. In addition to this direct emission, most secondary particles (particles formed by chemical transformation of gaseous pollutants) are also in this size range.

Fine particulates have a greater impact on visibility and are more likely than coarse particulates to become lodged deep in the lungs and to travel great distances. In addition, directly emitted fine particulates have a larger surface area than an equal weight of larger

**Figure 24.—National NO$_x$ Emission Densities, g/m²/year**

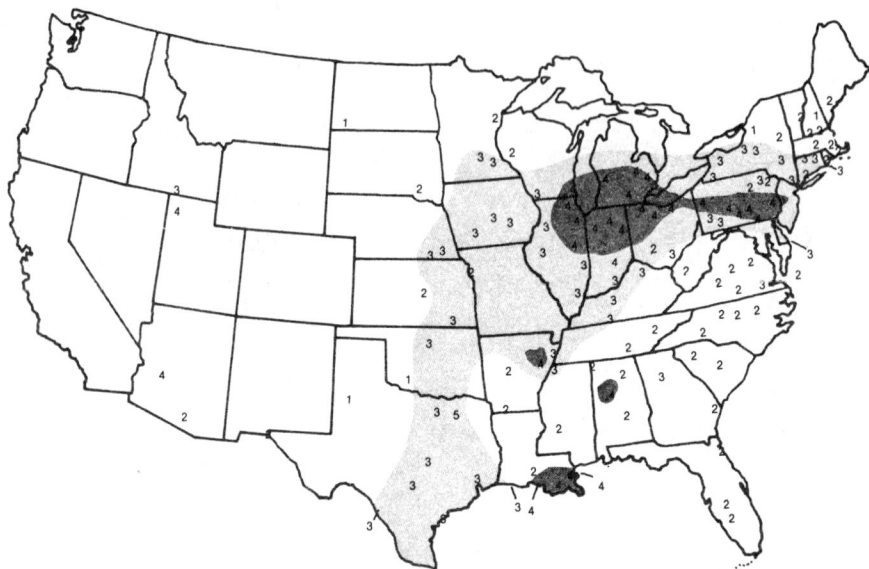

SOURCE: "National Emissions Report 1973," National Emissions Data System of the Aerometric and Emissions Reporting System, EPA-450/2-76-007, Research Triangle Park, NC

particles and thus tend to adsorb on their surfaces a disproportionate share of the toxic elements in the flue gas.

Present regulations and measurement techniques are based on total concentrations of particles, which may have little relevance to their chemical or biological behavior. In recognition of this, EPA has recommended setting a standard for particulate matter of less than 15 μm, and examination of the effects of particles smaller than 2 μm. Information gathered by this examination may provide the basis for a new "inhalable particulate matter" standard.[10]

Natural sources of particulates far outweigh manmade sources but are difficult to measure. Coal combustion yields 33 percent of the man-

made particulates and 90 percent of those from stationary fossil fuel combustion sources. All stationary combustion sources account for 36 percent of U.S. emissions, the second largest source after industrial processes, which constitute 49 percent (NEF #1).

Because compliance with all SIPs is far from complete today, total particulate emissions may decline quite significantly despite continued economic growth if compliance is achieved according to schedule. NEF #1 projects a decrease in the total emissions from 14.5 million tons in 1975 to 8.2 million tons in 1985, with a rise (to 9.1 million tons in 1990) thereafter that is characteristic of the assumption that future controls on both existing and new emission sources will not change. Clearly, at some point controls on particulates and other pollutants will have to become stricter if continued growth without unacceptable en-

---

[10]"Health Effects Considerations for Establishing a Standard for Inhaled Particulate Matter," EPA-HERL, internal publication, January 1978.

vironmental degradation is to occur. Also, none of these projections is likely to hold for fine particulates, where emissions could increase sharply unless control strategies shift to devices that are more efficient in collecting particles in this size range.

The role of utilities and industrial boilers will shrink from its present one-third contribution of total national emissions to one-fourth by 1990 despite the substantial shift to coal. The major portion of these particulates is contributed by coal combustion.

These trends differ substantially by region. Slight increases in particulate emissions are expected in New England and the Southwest, while substantial decreases should occur in those regions that now have many noncomplying sources—the Middle Atlantic and Midwest.

**Hydrocarbons** from coal combustion may be divided into two main groupings, corresponding to the way they are emitted. First, there is a volatile fraction composed mostly of low molecular weight species that are emitted as vapors. Second, there is a fraction of higher molecular weight species that are emitted as fine particulates or are preferentially adsorbed on the surfaces of fine particulates.[11] Although EPA has recently increased its efforts to characterize these emissions, little reliable information exists on their composition. Available data indicate that for moderate to large boilers (greater than about 25 MW), the volatile fraction (measured as methane) and particulate fraction (measured as benzene soluble organics) are emitted in amounts of the order of only a few milligrams per cubic meter in the flue gas. Emissions of polycyclic organic matter (POM), species of which are known carcinogens, are emitted in quantities considerably smaller. The available data indicate, however, that both the total hydrocarbons and the proportion of POM to the total may increase substantially as boiler or furnace size decreases. For example, hydrocarbon concentrations in the flue gas of boilers less than 1 MW in capacity were found to be more than 60

times as great as those from utility boilers more than 500 MW, while POM concentrations were several thousand times as great (but still less than 1 percent of the total hydrocarbon emissions).[12] Recent EPA-sponsored tests on a 10-MW industrial boiler, on the other hand, indicated that POM emissions were either zero or below the limits of detection of the available equipment.[13] There are now 12 million tons of coal burned each year in the residential/commercial sector, presumably in small boilers and furnaces in populated areas. The potential of these units, and additional units in the future, to generate quantities of the more toxic hydrocarbons is uncertain but a source of concern.

Coal combustion plays only a minor role in total hydrocarbon emission and will continue to do so in the future. In 1975, coal-fired utility and industrial boilers emitted about 70,000 tons of hydrocarbons, only one-half of 1 percent of national emissions. By 1990 these boilers may double their hydrocarbon emissions and increase their share of the total to 1.4 percent. During this time, strict automotive controls should (if the standards are successfully enforced) cause a sharp decline in national emissions from about 14 million tons in 1975 to 10 million in 1990. A notable exception to this sharp decline will occur in those areas slated for new petroleum and organic chemicals development. Thus, hydrocarbon emissions in Louisiana will increase somewhat in the next decade, while emissions in Texas and Delaware, sites of expected new refineries, will remain relatively stable.

**Inorganic trace elements and radionuclides** are emitted during coal combustion in particulate form or as vapors that can condense or be adsorbed on the particulates in the flue gas as it cools. These elements can be divided into three classes:

1. those that are not volatilized and thus are either left in the slag and bottom ash or become entrained in the flue gas as particulates,

---

[11]N. Dean Smith, *Organic Emissions From Conventional Stationary Combustion Sources* (Research Triangle Park, N.C.,: U.S. Environmental Protection Agency, August 1977).

[12]Ibid.
[13]Personal communication with Wade Ponder, U.S. Environmental Protection Agency, Research Triangle Park, N.C.

2. those that are volatilized and are absorbed on the fly ash as the flue gas cools, and

3. those that are volatilized and remain in vapor form (see table 25).

Those elements that volatilize and become adsorbed are (as are the volatile hydrocarbons) preferentially concentrated on smaller particles because, as noted above, fine particulates offer greater surface area than an equal weight of larger particles for condensation and adsorption to occur. This presents a serious problem in controlling trace element (and hydrocarbon) emissions, as conventional particulate control equipment grows progressively less efficient as particle size decreases.

**Table 25.—Distribution of Trace Elements in Coal Combustion Residues**

GROUP I—Equally distributed in bottom ash and fly ash

| Barium | (Ba) | Manganese | (Mn) |
|---|---|---|---|
| Chromium | (Cr) | Rubidium | (Rb) |
| Cerium | (Ce) | Scandium | (Sc) |
| Cobalt | (Co) | Samarium | (Sm) |
| Europium | (Eu) | Strontium | (Sr) |
| Hafnium | (Hf) | Tantalum | (Ta) |
| Lanthanum | (La) | Thorium | (Th) |

GROUP II—Preferentially concentrated in fly ash

| Arsenic | (As) | Lead | (Pb) |
|---|---|---|---|
| Beryllium | (Be) | Antimony | (Sb) |
| Cadmium | (Cd) | Selenium | (Se) |
| Copper | (Cu) | Uranium | (U) |
| Gallium | (Ga) | Vanadium | (V) |
| Molybdenum | (Mo) | Zinc | (Zn) |
| Nickel | (Ni) | | |

GROUP III—Discharged as vapors

| Chlorine | (Cl) | Mercury | (Hg) |
|---|---|---|---|
| Fluorine | (F) | | |

Concentrations of both trace elements and radionuclides vary considerably among different coals. For instance, thorium concentrations are quite high in Pennsylvania anthracite and low in interior bituminous coal (4.7 versus 1.6 parts per million (ppm), mean concentrations) while uranium varies from a high (mean) concentration of 2.4 ppm in Gulf lignite to a low of 0.7 ppm in northern Great Plains coals.[14] Looking at different coal qualities or

---

[14]R. I. van Hook, "Appendix on Trace Elements and Radionuclides, Oak Ridge National Laboratory, *Rall Report.*

"ranks," lignite has the highest concentrations of both radionuclides. This is unfortunate because lignite is the lowest rank coal and requires the most tonnage to generate a unit amount of electricity. Substantial variation also exists between different coal deposits of equivalent rank in the same region; some Wyoming and New Mexico coals are so high in uranium (up to 6,200 ppm) that the coal is essentially uranium ore.

Total emissions of most trace elements are expected to decline between 1975 and 1990, mainly because of increased particulate controls. The importance of such controls is demonstrated by an examination of arsenic emissions; electric powerplants produce more than half the gross (uncontrolled) national emissions but less than one-sixth of the total actually emitted (NEF #1), owing to the high particulate control levels in powerplants.

NEF #1 projects future trace metal emission increases only in arsenic, cadmium, fluorine, and selenium (however, it is not clear that these projections account for increases in fine-particulate emissions); of the four, utilities and industrial combustion are the major source only of selenium. Selenium emissions are projected to increase from 1,270 tons in 1975 to 2,900 tons in 1990, nearly all from coal combustion. Although coal combustion is a substantial source of other trace elements, its emissions are unlikely to decline much in view of the future growth of coal, controls on other sources— such as smelters—will drive down total emissions.

**Carbon dioxide** extracted during photosynthesis many millions of years ago is returned to the atmosphere by combustion of fossil fuels. Although $CO_2$ is naturally recycled through absorption by plants and by the oceans, the rapidly growing worldwide consumption of fossil fuels could lead to increased concentrations of $CO_2$ in the atmosphere. In fact, measurements in Hawaii have shown approximately a 5-percent increase in $CO_2$ levels since 1958. Fossil fuel combustion facilities may be the largest manmade source of this emission with coal yielding 11 percent more $CO_2$ than oil and 67 percent more than natural gas for the same energy value.[15] By projecting worldwide uses of fossil fuels and by modeling $CO_2$ exchange processes between the atmosphere and the oceans and the biosphere, a number of studies have developed estimates of atmospheric $CO_2$ levels over the next several centuries. Most show peak levels in the 21st and 22nd centuries from two to eight times higher than today. However, these estimates and their projected effects are subject to considerable uncertainty. (See $CO_2$ under *Ecological Effects.)*

### Sensitivity of Emission Estimates to Assumptions

Predictions of emissions are sensitive to assumptions about rates of growth, energy technologies used, environmental controls installed, quality of fuels burned, and other factors. Changes in the following assumptions are especially important:

- **State Implementation Plan compliance.—** Virtually all forecasts of future emissions assume full compliance by 1985 with existing State standards. There is substantial pressure in many States to revise emission standards upwards, especially for powerplants. The Mitchell and Kammer plants in West Virginia recently won adjustments in their $SO_x$ emissions reduction requirements, and the requirements for several large plants, including Cliftie Creek in Indiana and Labadie near St. Louis, Mo., are under review. Also, assumptions of compliance for other sources—especially for automobiles— may be too optimistic.
- **Growth rate.—**The growth rates of electricity and coal demand are critical factors in determining emissions, yet are highly uncertain. As an example of the effect of varying growth rates, emissions from a "moderate growth" scenario (5.8-percent electricity demand growth for 1975-85, 3.4-percent growth for 1986 to 2000, 1 billion tons of utility coal demand by 2000) and a "high growth" scenario (5.8 percent for 1975-85, 5.5 percent for 1986

to 2000, 1,900 million tons demand by 2000) are compared for the year 2000.

**Utility Emissions, Millions of Tons/Year[16]**

|  | Moderate | High |
|---|---|---|
| $SO_2$ .. | 14.0 | 17.1 (assuming 90-percent control for new plants). |
|  | 15.6 | 20.7 (assuming 80-percent control). |
| $NO_2$ .. | 9.5 | 17.5 |

Higher growth rates will have a maximum effect on those pollutants that are not efficiently controlled by available equipment. Thus, $NO_x$ and fine-particulate emissions are of particular concern in a high-growth situation.

- **Environmental controls.—**Most forecasts examine different versions of soon-to-be-enacted New Source Performance Standards (NSPS) for steam-electric powerplants but assume constant control levels after enactment. This is unlikely, as EPA must periodically review these standards. The very high levels of control achieved in Japan for scrubbers imply that future $SO_2$ emission standards may become more stringent, thus lowering national $SO_2$ emissions below expected levels. Other control variations that would affect emissions levels are:
  - *"Full scrubbing" versus "sliding scale."* Although national utility $SO_2$ emissions are not significantly different (in 1990, 19.6 million tons versus 20.2 million tons) for the EPA proposal (for a uniform percentage removal requirement) and the DOE proposal (to allow utilities using low-sulfur coal to apply only partial scrubbing), $SO_2$ emissions would be significantly higher in Western clean air areas under the DOE proposal.
  - *$NO_x$ controls.* Since $NO_2$ emissions are expected to increase substantially from expanded coal use, pressure for further controls is likely to increase. The probability of such additional control depends primarily on the success of EPA's

[15]*Analysis of the Proposed National Energy Plan* (Washington, D.C.: U.S. Congress, Office of Technology Assessment, 1977), p. 161.

[16]*Review of New Source Performance Standards for Coal-Fired Utility Boilers,* vol. 1 (Washington, D.C.: U.S. Environmental Protection Agency, August 1978), EPA-600/7-78-155A.

$NO_x$ control program. A particularly promising control technology appears to be a low-$NO_x$ burner that has achieved 85-percent emission reductions on a single full-size burner in the laboratory. Preliminary indications are that there may be little or no cost penalty for using these burners in a new facility. However, it is unlikely that sufficient full-size demonstrations can be completed and an NSPS promulgated before the mid-1980's.

—*Particulate standards.* Electrostatic precipitators (ESPs) are less effective at capturing fine particulates than are baghouses (fabric filters) but are generally less expensive and can achieve required emission reductions except when used with low-sulfur coals. Thus, the effect of $SO_2$ standards on low-sulfur coal use will also affect particulate control choices and, consequently, fine-particulate emissions.

It is clear that certain future conditions could lead to levels of emissions substantially greater than those projected. For example, a greater than expected level of coal development coupled with a significant relaxation of SIP control requirements could lead to significant increases in emissions of $SO_x$ and fine particulates. $NO_x$ emissions would be significantly increased if energy demand increased at a more rapid rate than projected.

## Chemical and Physical Transformation and Removal

As soon as each of the pollutants considered here is released to the atmosphere, it is free to participate in a complex series of chemical and/or physical processes that alter its form and location. Before the specifics of atmospheric chemistry are addressed, the following general relationships and conditions should be understood:

• Individual pollutants in the atmosphere are not static. That is, they react with other pollutants to give more or less harmful products, they decay, and they are ultimately deposited to varying degrees on soil, waterbodies, forests, farms, residential developments, and all other areas of the Earth's surface.

• On a large spatial scale, pollutants in the atmosphere are not well mixed. Consequently, many air pollution problems are localized over cities and large metropolitan regions. Further, climatological phenomena (such as inversions), topographic features (such as the Los Angeles Basin and the Appalachian Chain), and development patterns (such as the concentration of powerplants along the Ohio River) can influence what pollutants are present, in what quantities, and what types of reactions are taking place.

• The nature of the atmospheric chemical reactions is exceedingly complex and, at present, incompletely understood. Important factors that determine the rate at which the reactions occur are:

— *The form of the reactants* (gases or aerosols).

— *The presence of catalysts* (metals, light, and other reactants). Some of these may be present in the same emission as the reacting substance, or may come from other emission sources (including natural sources); some may exist only at certain times of the day or in certain seasons.

— *The concentration of reactants and type of reaction.* Doubling or halving the amount of a particular pollutant present in the atmosphere will not necessarily double or halve the rate of formation of reaction products; reactions often are nonlinear with respect to concentrations. This same nonlinearity applies to the eventual atmospheric concentration of the reaction products; thus, reductions in "precursor" pollutants could, depending on the circumstance, produce proportionately greater or lesser reductions in ambient concentrations of the secondary pollutants.

• Considerable uncertainty exists as to which chemical reactions predominate under various atmospheric conditions. This uncertainty results in severe difficulties in predicting the effects of new control strategies on concentrations of secondary pollutants.

*Reactions involving SO$_x$* have been the most intensely studied to date. The predominant set of reactions involves the conversion of SO$_2$ to some form of sulfate. The reactions believed to be of major importance are summarized in table 26. The importance of understanding the mechanism of production is quite obvious from the table. For example, indirect photooxidation requires free radicals often present in a polluted atmosphere, and indicates that conversion to sulfates may then proceed more rapidly in urban centers, especially in daylight or summer conditions. As another example, the fourth reaction requires ammonia to convert sulfuric acid to a sulfate salt. This buffering action could be important where acid rain was an important problem; in such circumstances, controlling atmospheric ammonia could be undesirable.

Removal of sulfur pollutants from the atmosphere takes place by gaseous, dry, and wet

deposition. Dry deposition (particles settling out or colliding with surfaces) occurs continually, and is more important close to the SO$_x$ source. Wet deposition (scrubbing of the atmosphere by precipitation) is sporadic and may be more important in regions remote from the source. The relative importance of the two mechanisms depends on the climate and the form of the sulfur pollutant. However, from observed data in temperate climates, wet and dry removal of all SO$_x$ is believed to be of roughly equal importance.[17] The relative importance of each deposition mechanism is different for each sulfur species, however; wet deposition is 10 times more important for sulfates than for SO$_2$. Thus, keeping track of precipitation is crucial in considering long-range transport of sulfates.

The mean residence time in the atmosphere for SO$_2$ is about 1 day. For sulfates it is 3 to 5

---

[17]R. B. Husar, et al., see note 5.

## Table 26.—Reactions in the Atmosphere

| Reaction type | Reaction | Catalyst/special conditions | Comments |
|---|---|---|---|
| Indirect photooxidation | $SO_2 \xrightarrow{\text{Free radicals}} H_2SO_4$ | Smog, sunlight, water vapor, Ho·, Ho·$_2$, CH$_3$O$_2^-$ radicals | Important reaction rates up to 5 percent per hour giving half-life of SO$_2$ of 1/2 $\longrightarrow$ 2 days. Rate depends on SO$_2$, hydrocarbons, NO$_x$ levels, amount of sunlight. |
| Heterogeneous catalytic oxidation | $SO_2 \xrightarrow{O_2} SO_4 =$ | Liquid water; metal ions | Because of dependency upon catalyst concentration, probably important in plumes and polluted urban atmospheres; probably minor in rural atmospheres. Virtually zeroth order in SO$_2$. |
| Heterogeneous oxidation by strong oxidants | $SO_2 \xrightarrow[H_2O_2]{O_3,} SO_4 =$ | Water droplets | Importance in dispute, rate estimates vary by a factor of 100. |
| Heterogeneous oxidation in the presence of ammonia | $SO_2 \xrightarrow{H_2O} H_2SO_4$   $NH_3 + H_2SO_4$ $\rightarrow NH_4^+ + SO_4 =$ | Water droplets | Rates unknown, dependent upon availability of ammonia and pH of droplet. |
| Surface catalyzed reactions | | Soot | Soot has been shown in the laboratory to catalyze oxidation of SO$_2$. Importance unknown. |

days and for particulate S, about 2 to 3 days. At a speed of 500 km/day, which is typical in the Midwest, half the particles or molecules would travel over 500 km, 1,500 to 2,500 km, and 1,000 to 1,500 km respectively.[18] In temperate climates, 20 to 50 percent of $SO_2$ is converted to the sulfate ion before removal. Overall rates of conversion ordinarily range from about 1 to 4 percent per hour during the day and less than 0.5 percent per hour at night for an average daily rate of 1 to 2 percent per hour during the summer.[19] This gives a summertime $SO_2$ half-life of from about 1½ to 3 days. Figure 25 shows, in schematic form, sulfur removal mechanisms and the relative contribution of wet and dry deposition.

The transformation of NO from coal combustion is less well understood than that of $SO_x$. Although a variety of reactions involving $NO_x$ are known, no reaction rates for actual atmospheric conditions can yet be assigned to them. Important inorganic reactions involve transformations between the oxides NO and $NO_2$, formation of $N_2O_5$, and reaction with water to form $HNO_3$. Organic reactions of $NO_x$

[18]Ibid.
[19]Ibid.

are particularly important. These result in various complex organic nitrogen species, including peroxyacyl nitrate (PAN) and peroxybenzoyl nitrate (PBN). PAN and PBN are extremely damaging to plants and irritating to animal tissues.

Gaseous, wet, and dry deposition are the sinks for atmospheric $NO_x$, with the first two of roughly equal importance and the dry deposition somewhat less important.

The chemistry of formation of **polycyclic hydrocarbons** is both complex and incompletely understood. This subject requires more research, especially with respect to formation of benzo(a)pyrene and other species of POM from advanced coal technologies.

### Atmospheric Transport

Superimposed on the chemical interactions just discussed are the largely physical mechanisms that govern not only the degree and duration of the mixing among pollutants but also the direction and distance to which the reaction products are transported. These physical factors may be divided roughly into mechanisms that primarily affect shorter range transport and those that determine the extent of long-range transport.

Figure 25.—Flow Diagram of Sulfur Transmission Through the Atmosphere. Over Half of the SO₂ is Removed or Transformed to Sulfate During the First Day of Its Atmospheric Residence

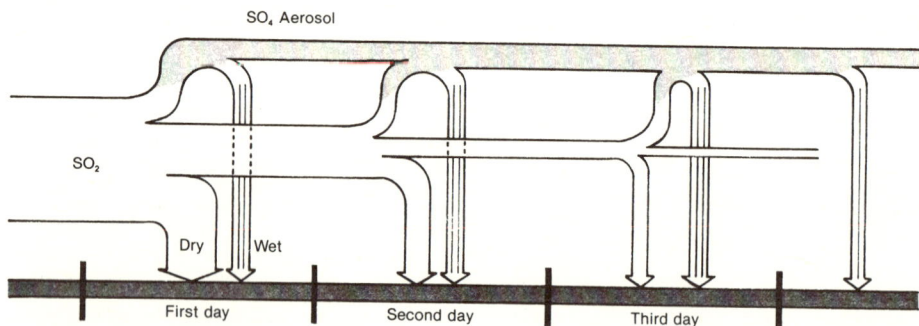

The short-range geometry of the plume is critical to both short- and long-range transport. Under some conditions, the plume spreads quickly to the ground level ("fumigation") and high concentrations of primary pollutants may result. Under other conditions, the plume remains isolated in a narrow vertical layer that disperses little and travels far ("fanning"), carrying high levels of secondary products far from the source. A variety of other possibilities falls between these two. Each type of plume behavior is associated with different atmospheric conditions.

An important factor in the short-term dispersion of pollutants is the vertical temperature variation with altitude. In daylight hours of intense sunlight, air that is heated by the warm surface continually rises, causing the atmosphere to be unstable and well mixed. These conditions promote the local deposition of pollutants, especially when persistent surface winds are light. Under the conditions of a temperature inversion (warmer temperatures at higher altitudes)—whether it is caused by nighttime cooling of the surface faster than the air or by a cold air mass sliding under a warm air mass—the air layers are stable and do not mix. At these times, effluents from moderate stack heights may be trapped below the inversion, while those from tall stacks may become embedded in the layers above, and hence subject to longer range transport. Nighttime conditions often lead to fanning plumes that remain in a narrow region and may be transported especially rapidly by a nighttime jet stream.

Other local factors determine the extent of vertical mixing of a plume. One is the wind turbulence caused by rough terrain, surface obstacles, or variation of wind speed with altitude. Another is the local wind pattern, such as the urban-rural wind or the land-sea breeze. If the local wind system becomes a temporary closed loop, pollutants can be sloshed back and forth with no real dispersal, and production of secondary products can further deteriorate the air quality until the cycle is broken.

Once the plume type has been determined by short-range transport mechanisms, several long-range factors govern the way the pollut-

ants travel. Before 1973 there was little documented evidence to suggest that air pollutants were transported in significant quantities to regions remote from their origin.

Numerous studies since then have confirmed that this phenomenon does indeed occur.

One such example is the plume of the Labadie Power Plant, outside of St. Louis, Mo., which has been extensively tracked by the EPA's Project MISTT. This plume has been followed for more than 180 miles, and at that distance its width was not much greater than 15 miles. Other examples include the Electric Power Research Institute's (EPRI) SURE program's observations that sulfate episodes occurred only when the air mass over the area had first traveled over high-emission regions; satellite tracking of "hazy blobs," masses of polluted air that have been observed over the Midwestern and Eastern United States, lasting for more than a week; and observations and analyses of acid rain in Europe, where Scandinavian rivers and lakes have apparently been degraded by long-range transport of $NO_x$ and $SO_x$ from the United Kingdom and other Western European countries.[20]

One of the four factors important in long-range transport is the blend of local factors, discussed above, that determine dispersal of pollutants from a given source. Local conditions promoting long-range transport are tall stacks, nighttime or cloudy conditions, and relative constancy of wind speed and direction with altitude.

A second factor is the persistence of surface winds. These have been studied in great detail at many stations in recent years. It has been found that many regions are characterized by surface winds that persist for 6 hours or more in a given direction (or, to be more accurate, within a narrow angular sector). They help determine the dominant direction(s) of transport.

A third factor is the type of winds prevailing at the altitude typical of plant stack heights. These, too, are characterized by dominant directions that often may differ only slightly

[20]Altshuller, et al, "Appendix on Transport and Fate," *Rall Report.*

Conditions leading to long-range transport of a plume from a coal-fired powerplant: persistent winds, and a stable atmosphere

from that of the persistent surface winds. Thus, the two often couple positively to carry pollutants still longer distances from their point of generation.

These last two factors can contribute to the additive effect of pollutants from several sources. Because the winds tend to blow more frequently in some directions than others, any emission sources lined up in the same direction will contribute to intensification of pollution concentrations. For example, powerplants clustered along the Ohio River are also in line with prevailing winds blowing to the northeast; occurrence of these winds has been correlated with high pollutant concentrations in Pennsylvania and other States to the northeast.

A final factor is the large-scale meteorological conditions resulting from the relative movements of low- and high-pressure systems.

These conditions can create large regional areas of stagnation.

Several research organizations are developing long-range transport models and are currently using such models primarily in an experimental mode. Models in current use have not undergone extensive peer review and thus are not likely to be immediately acceptable for regulatory purposes. This situation could change drastically within a short time—perhaps a year—and conceivably lead to greater regulatory attention to long-range transport problems.

Although long-range transport models are still not fully verified, regions of the United States can be surveyed for some of the key factors in long-range transport mentioned above. Data compiled by Teknekron for EPA on regional air quality and aerometric parameters yield the following brief survey.[21]

One potential air quality problem lies in the West, especially in the northern Great Plains and southeastern Utah, an area slated for considerable future powerplant development. Many areas nearby currently are designated as Class I (i.e., areas where air quality is to be strictly protected, with very little development allowed), and many of these lie in the direction of extremely persistent winds coming from

---
[21]"An Integrated Assessment of Electric Utility Energy Systems Briefing Materials," prepared for EPA/NASA Meeting on Cooperative Program on Synoptic Regional Air Pollution Problems, Sept. 11-12, 1978, by Teknekron Inc., Berkeley, Calif., and Waltham, Mass./pt. 2, Aerometric Data Compilation and Analyses for Regional Sulfate Modeling.

areas with a large number of current and projected urban and utility emissions sources. In addition, the prevailing winds tend to be from the south and may entrain elevated levels of sulfates from smelters in the Southwestern United States.

The gulf coast is expected to increase its utilization of coal, partly in response to the growing development of lignite fields. These fields lie primarily in eastern Texas and are alined approximately parallel to the prevailing wind direction (from southwest to northeast). If powerplants develop along the same line, intensification of concentrations may result.

The region that has been the most heavily studied for air pollution transport is the Ohio River Basin. This region not only contains a large cluster of coal-fired powerplants but also

*Photo credit: EPA—Documerica*

**Episodes of poor visibility in the northeast are influenced by the long-range transport of pollution generated hundreds of miles to the southwest**

exemplifies some of the meteorological factors that contribute to long-range transport. Extremely persistent winds in the Basin lead to intensification of pollution levels from a string of sources. Prevailing winds in the Upper Basin blow predominantly east and north, carrying polluted air as far east as Pennsylvania and beyond, while winds in the Lower Basin carry pollutants north into Canada.

Several regions in the Eastern United States have been found to be susceptible to prolonged atmospheric stagnation. Many of those cases are centered over a line from southern West Virginia to Georgia. Although emissions are generally low along this region, they are expected to grow.

## Health Effects[22]

### Introduction

It is known that very high levels of air pollution cause illness and death. Present ambient air standards are well below these levels, and it is unlikely that episodes of high concentrations of "criteria pollutants" similar to the notorious catastrophes that occurred in Donora, Pa., in 1948 and London in 1952 will be repeated in this country. However, occasional episodes of adverse meteorological conditions such as those that occurred in the Northeast during November 1966 could lead to high (and possibly dangerous) levels of currently unregulated pollutants. Also, elimination of acute, high-concentration episodes does not preclude the possibility of chronic damage and life shortening as a result of exposure to low levels of pollution. The existence of these types of effects is quite difficult to establish because the damages may take the form of common illnesses (e.g., bronchitis, asthma, cancer) that are known to have other causes and would not necessarily be attributed to air pollution. The effects, if any occur, would also tend to be masked by the very widespread dis-

tribution of air pollution and the large number of different pollutants usually present.

Increased coal combustion will eventually increase the ambient concentrations of $SO_x$, $NO_x$, fine particulates, and other pollutants, as discussed in the previous section. The population exposed to the increased concentrations will be large because pollutants travel long distances and, due to prevailing wind patterns, will be concentrated along the heavily populated Eastern seaboard. The present regulatory structure fails to address this problem directly.

If present and projected levels of ambient pollutant concentrations are shown to cause severe chronic effects, the implications for coal use and the degree of control to which it is likely to be subjected will be profound.

### Evaluating Health Effects

The determination of the potential of a substance for causing adverse health effects can be accomplished by combining the results of: 1) short-term tests, 2) animal bioassays, 3) chamber exposures to human beings, and 4) epidemiologic studies on human populations.

**Short-term tests** expose tissue, cellular, and microbial cultures to controlled levels of pollution to measure toxic effects, such as cellular or bacterial death, or mutagenic changes reflecting alterations in genetic structure.

**Animal bioassays** are used to measure acute and certain chronic effects. Tests for acute toxicity may measure the dose that causes death or impairment of target organs such as the lungs, kidneys, liver, reproductive system, etc. Chronic effects might include gradual impairment of target organs (e.g., chronic bronchitis through long-term insult to the lungs), teratogenic effects, or carcinogenesis.

**Exposure studies using human beings** are restricted, for ethical and legal reasons, to (usually) short durations and exposure levels no greater than might be expected in the normal working or outdoor environment. Further, "health effects" induced by controlled human exposures are limited to those responses that are known to be reversible, such as increased carboxy-hemoglobin levels in the blood or

---

[22]This section summarizes a detailed and extensively referenced report on health effects to be published separately as an appendix to this report. Full referencing of source material was considered impractical for this summary.

changes in airway resistance. Controlled studies of the short-term effects of air pollutant exposures to human beings provide a vital link between toxicologic animal studies and large population epidemiologic studies.

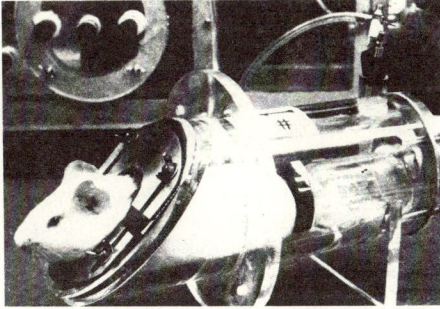

**Tests on mice are used to measure acute and chronic health effects**

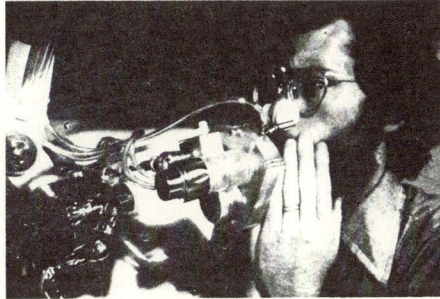

**Clinical studies on humans are limited to pollution levels that result in treatable health problems**

**Epidemiologic studies** are statistical examinations of illness and death rates, air pollution levels, and other variables—studies that seek to discover whether there is a direct correlation between a suspect air pollutant and adverse health effects. As such, they extend beyond the function of short-term tests and animal bioassays that are used to identify the mechanism of damage and the dose response function of the substance under study, and attempt to quantify those effects in human beings.

All of these methods of assessing health effects are necessary to establish the damage potential of a substance. Animal tests are insufficient by themselves because a biological effect in any living system cannot always be applied to a biological effect in others. Although the cellular or subcellular effect may be similar, the clinical effect may be quite different. The chamber tests can show human sensitivity to the substance, but it is impossible to infer from studies exposing 5 to 20 subjects to low levels of the substance for relatively short times what the chronic morbidity or mortality effects on the entire population may be. Finally, epidemiologic studies can establish statistical associations between a substance and damage to human beings, but they cannot prove cause and effect; the short-term and animal tests, by establishing the existence of damage mechanisms, and the chamber tests, by demonstrating human sensitivity, provide credibility to the cause-and-effect argument.

The epidemiologic studies have proven to be a weak link in the analytical chain needed to establish air pollution standards that will guard against chronic health effects. The studies that have attempted to examine the effects of low levels of pollution on the health of the general population have run into several problems:

- heterogeneous populations with different thresholds to particular toxic substances and different socioeconomic characteristics;
- very poor measurements of pollution levels—in most cases, the limited number of monitoring sites and the failure to incorporate levels of pollution indoors lead to poor determination of the pollution levels that the population is actually exposed to;
- outright failure to measure some potentially important pollutants (such as fine particulates);
- lack of data on variables (such as smoking habits) that are important determinants of health;
- wide variations of pollution levels with time, raising the question whether any observed health effects are caused by the mean levels or the occasional peaks; and

- presence of multiple pollutants that might act in combination.

Some of these problems are discussed in more detail later in the context of epidemiologic studies of sulfate damage to human health.

### Effects of Specific Combustion Pollutants

Coal combustion releases a mixture of gases and particles into the atmosphere. These include sulfur compounds, nitrogen compounds, fly-ash particulates, CO, $CO_2$, hydrocarbon vapors and particulates, radionuclides, and many trace metals. These emissions may lead to the formation of secondary pollutants, including photo-oxidants and particulates. Human beings are exposed to a complicated mixture that may act additively, competitively, or synergistically. However, clinical studies have seldom examined more than two pollutants in combination. Even epidemiologic studies, in which the populations are exposed to the entire gas-aerosol mixture, rarely attempt to sort the effects of more than two of the pollutants of the mix. This adds to the uncertainty of predicting health effects from future coal use.

The following discussion summarizes the health effects of specific pollutants from combustion. Table 27 presents the acute and chronic effects of specific gases, elements, and particulate compounds that are emitted from coal combustion. Little is known about the actual magnitude of the effects attributable to increased coal use. The ambient concentrations of radionuclides, organic compounds, and various trace metals due to increased coal combustion are projected to be very low. However, these pollutants are still a concern because of multiple exposures from other sources. Health effects from these pollutants are likely to appear first in the population working in coal conversion or related technologies because exposures will be higher.

**Total suspended particulates (TSP)** are composed of many different substances. The ambient aerosol (suspension of solid and liquid particles in air) includes trace metals, radionuclides, and organic compounds, as well as silica and other minerals. TSP can be roughly divided into fine (usually defined as less than 2 or 3 $\mu$m in diameter) and large particulates. This differentiation is important because:

1. Fine particulates are not explicitly regulated and are not as efficiently controlled by existing technologies as the large particles.
2. The most significant route of exposure to air pollutants is through inhalation. Fine particulates can pass through the filtering mechanisms of the upper respiratory tract and penetrate deeply into the lungs. Several studies using zinc ammonium sulfate and sulfuric acid aerosols have demonstrated that small particles result in greater functional and structural changes in the lungs than result from larger particles of the same substance.
3. Toxic substances, either gases or recondensed metal vapors, can be transported into lung tissues by inhaled fine particles. A unit mass of fine particulates has a greater total surface area than the same mass of large particulates, hence can transport more toxic material on its surface.

Thus, although there is not much explicit evidence defining their impacts on health, fine-particulate emissions are a primary concern in considering the impacts of expanded coal combustion.

Both chemically inert and active aerosols have been shown to alter the mechanical behavior of the lungs. The principal response of different aerosols as measured by animal and human controlled studies is to increase air flow resistance in the lungs. This indicates that lung airways are in some way altered. Several physical mechanisms are known to cause this symptom: reflex constriction of the trachea and bronchi, excessive secretions of mucus, edema, and local enzyme stimulation causing smooth muscle constriction. In general the deeper that relatively insoluble particles are deposited in the lungs, the longer the time required for their clearance. Some particles, such as asbestos fibers, penetrate the lung lining and lodge in the interstitial space or are carried by the lymphatic and circulatory systems to other body sites. In these cases, the manifest health effects may not be respiratory, but rather carcinogenic or systemic. The toxic elements contained in insoluble aerosols may ultimately affect extrapulmonary tissue irrespective of where they enter the body.

Besides changing airway resistance or causing systemic changes, aerosols can cause harm by altering lung clearance, transporting dissolved gases that may damage lung tissue, and affecting the compliance (tissue elasticity) of the lung (the importance of the transport of dissolved gases makes it very difficult to consider the health effects of specific chemical compounds or particulates in general without considering gas-aerosol interactions). A number of coal-related aerosols have been studied in animals and human beings. These include several sulfates, chemically inert dusts, metal oxides, and fly ash. Surprisingly, the effects of nitrate aerosols have not been reported.

Table 27.—Some Gaseous and Particulate Substances From Coal Combustion[a]

| Substance[b] | Toxicity | | Sources of Pollution | Environmental standards | Comments |
|---|---|---|---|---|---|
| | Acute | Chronic | | | |
| **Particulates** | | | | | |
| Total suspended particulates | With $SO_2$ in episode conditions contributes to mortality and morbidity. | Pulmonary irritation, chronic obstructive and restrictive lung disease. | Soil erosion, natural volcanos and fires, industrial activity, fossil fuel combustion: coal and oil, secondary atmospheric conversion of gaseous compounds. | 260 $\mu/m^3$24 hour max. 75 $\mu g/m^3$- annual ave. | A very broad class, undifferentiated by particle size or chemical composition. |
| Sulfates | Increased respiratory disease; breathing difficulty in asthmatics. | Respiratory disease and increased mortality suspected. | Conversion of $SO_2$ to sulfates in the atmosphere, therefore primary sources are $SO_2$ emissions from coal and oil combustion. Smelters, kraft paper mills, sulfuric acid plants also produce sulfates. Natural sources—$H_2S$ emissions, volcanos, sea salt. | | |
| Nitrates, nitrites | Increases infant susceptibility to lower respiratory infection due to conversion of nitrates to nitrites. | May combine with amines to form carcinogenic nitrosamines, also mutagenic and teratogenic. Nitrites a direct animal carcinogen. | Conversion of $NO_2$ to nitrates and nitrites in the atmosphere; therefore primary sources are NO emissions from fossil fuel combustion, fertilizer production, munition production, chemical plants, auto and industrial emissions. | | |
| Organic matter | Unknown for many compounds. Specific toxicity for others. | Long-term is potentially carcinogenic and mutagenic. | Fossil fuel direct and indirect use—combustion, refining, plastics, tars, coking, chemical production. | | Higher concentrations likely to be associated with nondirect combustion of fuels. |

**Table 27.—Some Gaseous and Particulate Substances From Coal Combustion[a]—Continued**

| Substance[b] | Toxicity Acute | Chronic | Sources of Pollution | Environmental standards | Comments |
|---|---|---|---|---|---|
| Arsenic (oxide forms) | Effects large to small depending on form and route of exposure; rarely seen. | Carcinogen, and teratogenic cumulative poison. | Weathering; mining, and smelting; coal combustion; pesticides; detergents. | 0.05 mg/l drinking water. | |
| Beryllium | Short-term poison at high concentrations, especially toxic by inhalation. | Long-term systemic poison at low concentrations; carcinogenic in experimental animals | Industrial; combustion of coal, rocket fuels. | 0.01 μg/m³ hazardous air pollutant | |
| Cadmium | Very toxic at high concentrations; to animals and aquatic life. Toxic by all routes of exposures. | Possible carcinogen, cumulative poison; associated with hypertension, cardiovascular disease, kidney damage | Weathering; mining and smelting, especially of zinc; iron and steel industry; coal combustion; urban runoff; phosphate fertilizers. | 0.010 mg/l drinking water 40 μg/l/day proposed effluent standard (withdrawn) | Chronic cadmium poisoning resulting in illness and death has occurred in Japan, where cadmium mobilized by mining contaminated daily diet. Margin of safety—measured levels of cadmium in renal cortex compared to threshold for renal dysfunction—is low: 4 to 12.5. |
| Chromium | Hexavalent form most harmful; skin and respiratory tract irritant | Carcinogenic; workers engaged in manufacture of chromium chemicals have incidence of lung cancer, no evidence of risk in nonoccupational exposure. | No chromium now mined in U.S. Emissions from industrial processes, including electroplating, tanning, dyes; coal combustion. | 0.05 mg/l drinking water | |
| Mercury | Methyl mercury and mercury fumes very toxic; other forms of variable toxicity | Methyl mercury very toxic, cumulative poison; affects central nervous system | Weathering; volcanoes; mining and smelting; industrial; pharmaceuticals; coal combustion; sewage sludge; urban runoff; fungicides. | 0.002 mg/l drinking water; maximum of 2,300 grams mercury in emissions from stationary sources; 20 μg/l/day proposed effluent standard (withdrawn) | Environmental pollution leading to contamination of fish and shell fish caused illness and death in Japan; contamination of fish in U.S. has caused closure of waters to commercial fishing. |
| Selenium | Soluble compounds are highly toxic | Probable carcinogen; also essential for life. | Natural; mining and smelting; industrial process; coal combustion. | 0.01 mg/l drinking water | Interacts with other metals, increasing or decreasing toxicity. |

**Table 27.—Some Gaseous and Particulate Substances From Coal Combustion[a] —Continued**

| Substance[b] | Toxicity | | Sources of Pollution | Environmental standards | Comments |
|---|---|---|---|---|---|
| | Acute | Chronic | | | |
| **Gases** | | | | | |
| Sulfur dioxide | Increased respiratory impairment— morbidity and mortality—in combination with particulates. | Increased respiratory disease and decreased respiratory function with particulates. | Sulfur contained in fossil fuels, smelters, volcanoes | 365 $\mu$g/m³-24 hour max 80 $\mu$g/m³-annual mean | Coal combustion presently represents between 60% and 70% of U.S. SO₂ emissions. |
| Nitrogen dioxide | Increase respiratory infections | Changes suspected in lung function; emphysema. | Nitrogen fixation in high temperature combustion, and from nitrogen contained in fossil fuels: coal, oil, gasoline combustion. | 100 $\mu$g/m³- annual mean | Organically bound fuel nitrogen is a more important component for coal NO emissions than for the other fossil fuels. |
| Carbon monoxide | Behavior changes, nausea, drowsiness, headaches, coma, death | Increase risk of coronary heart disease—arterial sclerosis suspected | Incomplete combustion of fossil fuels: coal, oil, gas, gasoline combustion. | 40 mg/m³ 1 hr. max. 10 mg/m³ 8 hr. max | |
| Ozone | Increased respiratory infection, eye irritation, headaches, chest pain, impaired pulmonary function | Unknown | Photochemical reactions involving hydrocarbons, nitrogen oxide and other compounds in lower atmosphere; reaction of atomic oxygen and oxygen in upper atmosphere. | 160 $\mu$g/m³ 1 hr. max. | |
| Aromatic hydrocarbons | Fatigue, weakness, skin paresthesias (> 100 ppm) | Irritation, leukopenia and anemia. Certain compounds are mutagens and carcinogens. | A broad class of compounds naturally evolved from organic material, and from the evaporation and combustion of fossil fuels and other organic industrial chemicals. | non-methane HC 160 $\mu$g/m³ 3 hour | Higher concentrations likely from less efficient and smaller boiler operation. Higher concentrations possible proximate to coal conversion facilities. Standard designed for photochemical oxidant control. |

[a] This table is provided merely to indicate some of the substances which are potential environmental hazards along with some information on each regarding toxicity, sources, standards, etc. It is not to be interpreted as definitive.
[b] The substance listed is not necessarily the form in which it becomes a potential environmental threat. In some cases the oxide or some metabolite, rather than the substance itself, is the culprit.

Controlled exposures to sulfates represent the vast majority of aerosol-health effects exposure studies. Overall, the aerosol studies report "health effects" at levels many times higher than common urban levels. There is evidence that chronic exposures below these levels adversely affect health by increasing respiratory disease (although a recent review of health effects literature concluded that "research on the toxicity of aerosols has not infrequently led to an overinterpretation of technically weak or insufficient data."[23] Some studies suggest that particulate pollution,

[23] *Airborne Particles*, National Academy of Sciences, National Research Council (Baltimore: University Park Press, 1978).

much of which is sulfur compounds, can initiate the development and progression of bronchitis and emphysema. Ambient particulate pollution has been associated with the development of lung cancer; the benzo(a)pyrene component of soot has been advanced as the causal factor. Although benzo(a)pyrene is a known carcinogenic material, it may be serving as a surrogate for a class of hazardous organic compounds found in urban aerosol samples.

Polycyclic organic matter (POM) is a fractional constituent of particulate matter emitted during coal combustion. POM emissions appear to vary inversely with combustion efficiency. There is a factor of approximately 10,000 difference between POM emission factors (emissions per pound of coal) for hand-stoked coal stoves and large new utility coal boilers.[24] Therefore, the major health risk of POM is not from the large sources but with the older, less efficient utility and industrial boilers near populated areas. Depending on dispersion characteristics, POM from these coal sources may add to POM concentrations from other sources such as refuse burning, coke production, and motor vehicles. Also, any substantial increase in residential and other small-source coal combustion could aggravate problems with POM emissions.

**Nitrogen dioxide** health effects are normally associated with exposures close to the pollution source, because $NO_2$ is reactive and takes part in complex reactions within the atmosphere that produce ozone, PAN, and other complex organic nitrogen species.

The Clean Air Act Amendments of 1977 require EPA to promulgate a short-term $NO_2$ standard (not more than a 3-hour averaging time) by 1980. An extensive review of $NO_2$ is reported in the EPA Office of Research and Development's draft document "Health Effects for Short-Term Exposures to Nitrogen Dioxide," March 1978. Although this draft document does not contain a synthesis, it does contain extensive review with some critical interpretations and serves as a draft criteria document. EPA has requested public comment

[24]N. Dean Smith, see note 11.

on a 1-hour $NO_2$ concentration in the range of 500 to 1,000 $\mu g/m^3$. It is expected that the proposed standard will most likely be at the lower end of the range.

$NO_2$ oxidizes molecules in cell membranes, attacks connective tissues, and interferes in metabolic reactions. Several small mammal species are affected by 1,000 $\mu g/m^3$ of $NO_2$ exposure repeated or continuous over the course of a week or more. Epithelial cells are damaged and biochemical changes occur in a manner that might be expected in human tissues. Furthermore, defenses of normal mice against massive bacterial infection are reduced at that concentration, perhaps mimicking the response of a severely impaired animal to a normal exposure to bacteria.

The airway resistance of normal subjects is not highly sensitive to the irritant effect of $NO_2$; bronchitics and others with chronic obstructive lung disease are sensitive; asthmatics are very sensitive. If the lower limits of the ranges of concentrations of $NO_2$ producing effects in human beings are correct, there would be respiratory resistance in sensitive individuals at concentrations above 200 $\mu g/m^3$. Thus, the short-term concentration patterns of $NO_2$ in the immediate vicinity of coal-burning facilities may represent an increased public health concern. However, the evidence from in vitro studies and from controlled $NO_2$ exposure studies with asthmatics at these relatively low concentrations is not universally accepted. In fact, the World Health Organization (WHO) has recommended a short-term (1 hour) $NO_2$ standard of 320 $\mu g/m^3$.

Table 28 summarizes the reported effects of low-level, short-duration exposure to $NO_2$, the levels reported in U.S. cities, and the proposed WHO and EPA standards.

**Ozone** may act by itself or in synergistic combination with other pollutants to damage human health. Our ability to quantify the health impacts of ozone attributed to increased coal combustion is severely limited by an inadequate understanding of $NO_x$ transformation processes and the long-range transportation characteristics of ozone and its precursor pollutants. Table 29 summarizes some of the existing data on ozone health damage.

**Table 28.—Nitrogen Dioxide: Levels and Effects**

| Comments | Concentrations | Reported Effects (exposure time) |
|---|---|---|
| | µg/m³, ppm | |
| | 2000 ⌐ 1 | • infectivity, mouse (3 hr. interpolated, 17 hr.) |
| | 1800 | |
| | 1600 | |
| | 1400 | |
| | 1200 | |
| peak level — — — — in Los Angeles | –1000 .5 ⌐ | • multiple biochemical changes, guinea pig (8 hr/day, 1 week) • increase in protein uptake by lung, guinea pig (4 hr/day, 8 days) • infectivity, mouse (1 week) • tracheal mucosa & cilia, autoimmune response, mouse (one mo.) • cilia, clara cell, and alveolar edema, mouse (10 days) |
| peak levels — — — — in four cities | – 800 .4 | • acid phosphates and serum proteins enter lung, guinea pig (1 wk.) |
| EPA lowest suggested 1 hr std — — — — — — | – 600 .3 ⌐ | • detected in blood, monkey (9 minutes) • human dark adaptation impairment (immediate) • collagen, rabbit (20 hr/wk for 24 days) |
| peak levels — — — — in many cities | 400 .2 | |
| WHO highest (320) lowest (190) suggested 1 hr std | 200 .1 ⌐ | • human asthmatics provoked resistance, (1 hr) • bronchial epithelial cells, alveolar macrophages (1-2 hr, in vitro) |
| | 0 ⌐ | |

SOURCE: Health Effects Appendix

Although there is extensive literature on experiments relating health effects of ozone and other oxidants, assessment of a safe standard is difficult. The problems parallel similar difficulties with interpreting health effects from other pollutants. First, questions are raised by the specificity and accuracy of ambient monitoring used in epidemiological studies. Second, there are uncertainties in extrapolating from cell and organ studies, as well as from whole animal studies, to estimates of human damage. Finally, field studies measure only the ozone component of total oxidant levels but are actually observing the effects of all oxidants. The evidence is consistent that nose, throat, and eye irritation occurs in the range of 200 to 294 µg/m³ (0.1 to 0.15 ppm) of measured ozone. Breathing difficulties in exercising adults and chest pain in children exposed to

ozone have been reported in the range of 200 to 400 µg/m³ (0.1 to 0.2 ppm). Decreased running performance has been associated with similar ozone levels. More severe symptoms of chest discomfort and coughing have been reported in healthy young adults at measured ozone levels of about 725 µg/m³ (0.36 ppm) and at 490 µg/m³ (0.25 ppm) in hypersensitive individuals. Levels high enough to cause these more severe symptoms occur in Washington, D.C., Houston, Tex., Philadelphia, Pa., and Los Angeles, Calif., with some regularity.

Japanese investigators have observed that the frequencies of several symptoms in school children, including sore throat, headache, cough, and dyspnea, are higher on days when maximum hourly oxidant concentrations equaled or exceeded 0.15 ppm than on days

**Table 29.—Ozone (O₃): Levels and Effects**

| (µg/m³) (ppm) | Effects |
|---|---|
| Peak levels in Los Angeles Area | |

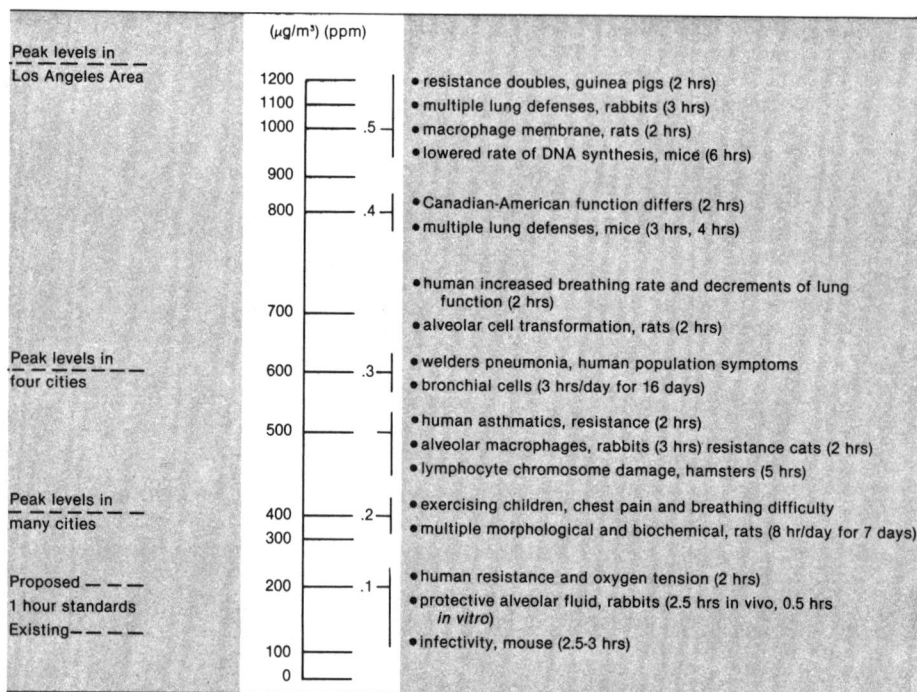

Peak levels in Los Angeles Area

| 1200 | • resistance doubles, guinea pigs (2 hrs) |
| 1100 | • multiple lung defenses, rabbits (3 hrs) |
| 1000 — .5 | • macrophage membrane, rats (2 hrs) |
| | • lowered rate of DNA synthesis, mice (6 hrs) |
| 900 | |
| 800 — .4 | • Canadian-American function differs (2 hrs) |
| | • multiple lung defenses, mice (3 hrs, 4 hrs) |
| 700 | • human increased breathing rate and decrements of lung function (2 hrs) |
| | • alveolar cell transformation, rats (2 hrs) |

Peak levels in four cities

| 600 — .3 | • welders pneumonia, human population symptoms |
| | • bronchial cells (3 hrs/day for 16 days) |
| 500 | • human asthmatics, resistance (2 hrs) |
| | • alveolar macrophages, rabbits (3 hrs) resistance cats (2 hrs) |
| | • lymphocyte chromosome damage, hamsters (5 hrs) |

Peak levels in many cities

| 400 — .2 | • exercising children, chest pain and breathing difficulty |
| 300 | • multiple morphological and biochemical, rats (8 hr/day for 7 days) |

Proposed 1 hour standards
Existing

| 200 — .1 | • human resistance and oxygen tension (2 hrs) |
| | • protective alveolar fluid, rabbits (2.5 hrs in vivo, 0.5 hrs in vitro) |
| 100 | • infectivity, mouse (2.5-3 hrs) |
| 0 | |

SOURCE: Health Effects Appendix.

when corresponding concentrations were below 0.10 ppm.[25] In Japanese studies, the number of individuals experiencing symptoms has generally been more strongly associated with ozone concentrations than with total oxidant concentrations. The Japanese studies also suggest that people with allergic tendencies are more susceptible to short-term photochemical oxidant exposure than are other segments of the population.

Japanese studies have generally shown oxidant- or ozone-associated effects at lower oxidant or ozone concentration than have Ameri-

can studies. The consistent difference between Japanese and American findings raises several questions, including whether the components of oxidant pollution in Japan are different from those in Los Angeles, or indeed, from those in any U.S. location. It is possible, for instance, that oxidants in Japan are accompanied by higher $SO_x$ levels than in the United States. In any case, the Japanese results underscore the point that epidemiological results gathered in one area can be generalized to other areas only with the greatest caution. The Japanese results also demonstrate the need to gather comprehensive data in U.S. locations other than Los Angeles.

Because the discrepancy among these results is along national lines, it is of particular

[25]J. Kagawa and T. Toyama, "Photochemical Air Pollution: Its Affects on Respiratory Function of Elementary Schoolchildren," *Archives of Environmental Health*, 30: 117-122, 1975.

interest to note the results of a study on nine new arrivals to the Los Angeles area.[26] They showed substantial decreases in pulmonary function for ozone exposures of 800 $\mu g/m^3$ for 2 hours, whereas six local residents did not show measurable effects. In a further study,[27] four southern Californians and four Canadians were exposed to 700 $\mu g/m^3$ for 2 hours with light intermittent exercise. The Canadians exhibited discernible decreases in lung function, but the Californians did not. Thus, either a possible adaptation or a "selective migration" effect is inferred. ("Selective migration" occurs when people moving into or out of an area show a sensitivity to a pollutant that is markedly different from the average). Any adaptation would indicate an altered lung surface due to air pollution, which is an effect that should arouse considerable concern. This observation is consistent with the observation that at low ozone levels smokers are generally less reactive than nonsmokers. An alteration in the lung would also be consistent with deactivated surface cells in the bronchi, which may lead to long-term respiratory problems.

In his comprehensive review of health effects from exposure to low levels of air pollution, Ferris concludes:

> Because of the highly reactive character of $O_3$, prudence dictates that we should abide by the present standard until more valid data have been collected. Chamber studies will be useful to define acute or short-term effects but cannot identify subtle long-term effects. For these we shall have to turn to epidemiologic studies which have many confounding variables. To date there is no evidence that health effects result from chronic exposure to levels of oxidants that currently occur in Los Angeles, but careful studies to confirm this have yet to be done.[28]

---

[26]J. D. Hackney, et al., "Studies in Adaptation to Ambient Oxidant Air Pollution: Effects of Ozone Exposures in Los Angeles Residents vs. New Arrivals," in *Environmental Health Perspectives*, 18: 141-146, 1976.

[27]J. D. Hackney, et al., "Effects of Ozone Exposure in Canadians and Southern Californians," *Archives of Environmental Health*, 32: 110-116, 1977.

[28]B. G. Ferris, Jr., "Health Effects of Exposure to Low Levels of Regulated Air Pollutants, A Critical Review," *J. Air Poll. Control Assoc.*, 28:482 (1978).

As yet, no convincing association has been shown between short-term oxidant exposures and rates of mortality due to any cause. Although a positive association has been observed in one study, a re-analysis appears to negate this conclusion. Interpretation of mortality studies has been hampered by limitations in statistical methodology. It has not been possible to fully separate the effects on mortality of oxidants, other pollutants, and meteorologic factors, most notably temperature.

**Sulfur dioxide.**—Although there have been many studies of $SO_2$, there is little evidence of acute adverse effects at ambient levels (usually below 0.4 ppm). The gas is soluble and removed almost entirely by the upper airways. At levels much higher than the ambient standard of 0.14 ppm for 24 hours, controlled exposure studies have demonstrated increased airways resistance in subjects. These and other reversible effects occurred at 1 ppm (2- to 3-hour exposures) and higher. Long-term animal exposures in the range of 340 to 15,000 $\mu g/m^3$ (0.13 to 5.72 ppm) show no functional or structural lung damage. All long-term exposure studies on human beings have involved a simultaneous exposure to the gas-aerosol mixture. However, it would appear on the basis of a lack of cumulative functional or structural impairment of the lungs from high-$SO_2$ animal exposures that $SO_2$ alone at ambient levels is not hazardous for human beings. The observed effects are shown in table 30.

**Sulfur dioxide and particulates.**—Exposure to $SO_2$ virtually always occurs in the presence of particulates and other gases. Generally, there is a heightened response from the dual exposure that is not observed from single exposures. A popular explanation of this synergism is that the aerosol acts as a carrier for the $SO_2$ (which would otherwise be screened by the lung's defenses), thereby delivering a higher fraction of the gas to the deep recesses of the lungs. An alternative explanation is that chemical reactions can occur between the gas and particle, forming new compounds (such as sulfuric acid) that are more toxic. Both explanations are supported by laboratory animal

## Table 30.—Health Effects of Sulfur Dioxide

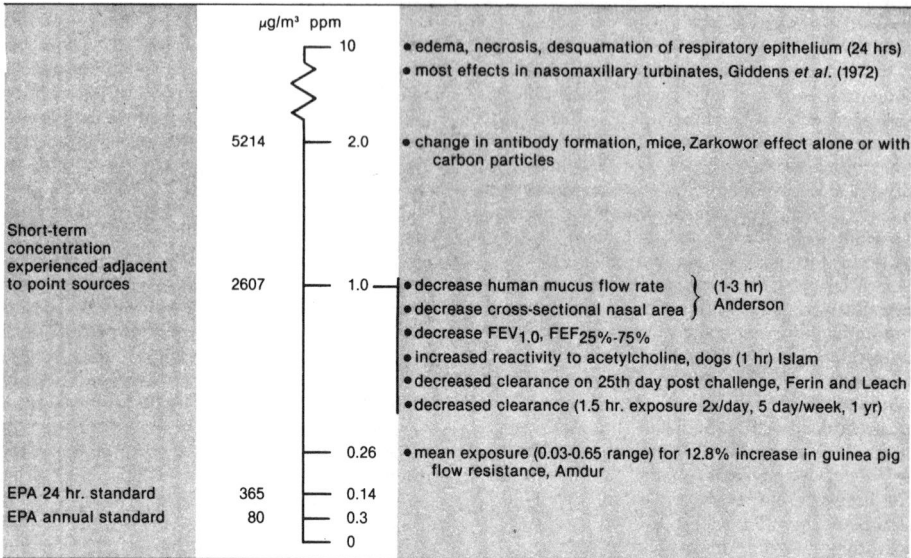

| | μg/m³ | ppm | |
|---|---|---|---|
| | | 10 | • edema, necrosis, desquamation of respiratory epithelium (24 hrs)<br>• most effects in nasomaxillary turbinates, Giddens *et al.* (1972) |
| | 5214 | 2.0 | • change in antibody formation, mice, Zarkowor effect alone or with carbon particles |
| Short-term concentration experienced adjacent to point sources | 2607 | 1.0 | • decrease human mucus flow rate    } (1-3 hr)<br>• decrease cross-sectional nasal area  } Anderson<br>• decrease $FEV_{1.0}$, $FEF_{25\%-75\%}$<br>• increased reactivity to acetylcholine, dogs (1 hr) Islam<br>• decreased clearance on 25th day post challenge, Ferin and Leach<br>• decreased clearance (1.5 hr. exposure 2x/day, 5 day/week, 1 yr) |
| | | 0.26 | • mean exposure (0.03-0.65 range) for 12.8% increase in guinea pig flow resistance, Amdur |
| EPA 24 hr. standard | 365 | 0.14 | |
| EPA annual standard | 80 | 0.3 | |
| | | 0 | |

SOURCE: Health Effects Appendix.

studies using combined exposure of $SO_2$ and various particles.

Short-term health effects from simultaneous exposure to high concentrations of $SO_2$ and particulates have been investigated in many European and American cities. Respiratory symptoms, including general lung function impairments and increased asthma attacks, have been associated with 24 hour 200 to 300 μg/m³ concentrations of $SO_2$ and 150 to 350 μg/m³ concentrations of particulates. Similar lung effects have been reported for long-term (yearly averages) levels of about 100 μg/m³ $SO_2$ and 150 μg/m³ particulates.

Gas-aerosol interactions have been documented in the laboratory and observed in population studies at levels slightly higher than the present standards. Given that these potentially harmful reactions are quite complex, perhaps impossible to understand completely, violations of National Ambient Air

Quality Standards (NAAQS) by coal combustion may be damaging to human health.

**Sulfate particulates.**—Over the past decade emissions of $SO_2$ have remained fairly stable, but their distribution has changed substantially. Urban $SO_2$ levels have decreased considerably as powerplants and other sources have either moved away or come into emissions compliance. At the same time, however, sulfate levels in most cities have not declined. This can be attributed to the formation of sulfate from the $SO_2$ emitted by new rural sources, and its long-range transport to the urban areas. Sulfate particulates can travel more than 1,000 km, with atmospheric residence time of between 2 and 4 days. The available data indicate that large regional reductions in total $SO_2$ emissions would be required to reduce sulfates. Thus, current policies designed to reduce ambient pollutant concentrations are not fully successful because they inadequately deal with long-distance transport of pollutants across jurisdictional and international boundaries.

Review of the available human experiments and epidemiologic studies indicates that high levels of some sulfates irritate eyes and the respiratory tract, decrease lung function, aggravate emphysema, asthma, and chronic bronchitis, and may increase mortality.

Research seems to indicate that sulfates function independently of $SO_2$ in producing health and ecological effects. Sulfates exist as several chemical species. Sulfuric acid mist and several species of sulfate particles have been demonstrated to be more toxic than $SO_2$, as measured by flow resistance in pulmonary airways in animals, while at least two sulfate species appear less toxic than the $SO_2$ gas. It has also been demonstrated that for a given concentration of a sulfate compound, toxicity increases with decreasing particle size. Atmospheric sampling has repeatedly shown that more than half of the mass concentration of sulfate is found in the aerosol fraction of particles less than 1.0 μm in diameter.

The most extensive epidemiological study of $SO_2$ and sulfate is the Community Health and Environmental Surveillance System (CHESS) study conducted by EPA.[29] As a consequence of many valid criticisms, it is generally felt that insufficient information exists in the CHESS results to propose a standard for sulfate.[30] Upon re-analysis, the CHESS results do not substantiate the original conclusions that ambient sulfate levels are associated with respiratory symptoms in asthmatic adults and acute lower respiratory illness in children. However, other CHESS results are not inconsistent with other findings and represent a large body of epidemiological data.[31]

Laboratory studies provide mixed evidence on the adverse health effects of sulfates on both human beings and experimental animals. Airway resistance has been measured in animals at sulfate levels higher than those experienced in ambient exposures. Impaired lung clearance in animals has been reported in the upper range of ambient exposures. Some human studies show effects at levels below 100 μg/m³ for short-term exposures. However, these observations are not confirmed in other studies. Table 31 summarizes the data available over the range of ambient exposures.

In summary, animal and human respiratory studies indicate health effects for sulfates by themselves and in combination with other gaseous pollutants at levels higher than current ambient exposure. Further, these studies distinguish among sulfate compounds. Ammonium sulfate, the principle constituent of airborne sulfates, is less hazardous than other sulfate compounds. Sulfuric acid aerosol, also detected in urban atmospheres, is the most hazardous. At times, urban and rural sulfate levels are almost exclusively comprised of sulfuric acid.

## Effects of Air Pollution on Death Rates

Several epidemiologic studies of mortality (death) rates have shown an association between sulfate and TSP levels and community mortality rates. Most of these studies are "cross-sectional;" that is, they examined differences in mortality, air pollution, and other variables among geographical areas over one time period. Increased levels of mortality are observed at or near air pollution levels that prevail in urban regions of the United States. Although different studies show a considerable range of pollution/death rate associations, the majority of these studies derived coefficients of air pollution-induced mortality (deaths per unit population exposed per unit of pollutant concentration, or annual number of deaths that would be expected to occur if 100,000 persons were exposed to an average of 1 μg/m³ of pollution) that indicate that several tens of thousands of premature deaths annually are associated with current pollution levels. Owing in part to limitations in the pollutant

[29]*Health Consequences of Sulfur Oxides: A Report From CHESS, 1970-71* (Washington D.C.: U.S. Environmental Protection Agency, May 1974), EPA-650/1-74-004.
[30]B. G. Ferris, Jr., see note 23.
[31]R. S. Chapman, discussion in: "Health Effects of Exposure to Low Levels of Regulated Air Pollutants: Discussion Papers," *Journal of the Air Pollution Control Association*, vol. 28, no. 9, September 1978, pp. 887-889.

**Table 31.—Health Effects of Sulfates**

| Exposure | Conc. (µg/m ÷) | Health Effects |
|---|---|---|
| | 500 | |
| | | • Increase in breathing resistance in animals |
| 24 hour levels frequently occurring, summer, N.E. U.S. | 100 | • Reported decrease in clearing mechanism in healthy adults |
| | 25 | • Increased daily mortality (24 hour)* <br> • Aggravation of heart and lung disease in elderly (24 hour)* |
| | 10 | • Increase risk, chronic bronchitis* <br> • Increase acute respiratory disease in children (yr)* |
| Non-urban levels N.E. U.S. | 5 | • Aggravation of asthma (24 hour)* <br> • Increased risk chronic bronchitis in nonsmokers* |
| | 0 | |

measurements (several important pollutants were unmeasured), it is impossible to determine from these studies whether sulfates and suspended particulates are the true causes of the statistically observed deaths, or whether they serve as a measure of some other pollutant or combination of pollutants. In the absence of a plausible alternative hypotheses, however, any other causative pollutants appear likely to have been the products of the same (fossil fuel combustion) sources that produced the sulfates and TSP.

These studies have been widely criticized in the health community for deficiencies similar to those described earlier on pages 202 to 204, and their results have been rejected by many. Statements of the more frequent arguments against the studies and discussions of the arguments' validity are presented below:

1. These studies typically use data from a single monitoring station to represent the exposure to the residents of a metropolitan region. Measurements taken at a single point are a poor measure of the mean exposure of the population.

   **Discussion:** The effect of using poor data is, if the error is random, to bias downward the estimated coefficient of the independent variable being measured. It is likely that the error introduced by using central station data as measures of population exposure would make it less likely that the analyst would find a positive association between air pollution and mortality. Thus, this is not a logical argument against a positive finding.

2. The single monitoring station is typically located in the center of the urban area, which often coincides with the areas of highest pollution exposure. This biases results.

   **Discussion:** The bias produced by location of monitors in areas of relatively high pollution may be either positive or negative, depending upon the relative gradients of exposure among the cities under consideration.

3. It is unlikely that current air pollution exposure is an adequate exposure parameter. Past exposure or cumulative history of exposure is undoubtedly more relevant.

   **Discussion:** It can be demonstrated that regardless of whether air pollution levels are increasing or decreasing, it is possible that this bias could operate in either direction. To the extent that present exposures are less than perfectly colinear with past exposures, there will be a tendency to underestimate the true magnitude of the relationship. Also, efforts to include measures of cumulative exposure have generally failed to increase the explanatory power of air pollution mortality models.

4. The studies have been confounded by the similarity of patterns of certain critical socioeconomic variables with patterns of air pollution. Whereas the researchers believe that they are seeing the relationship of death rates to pollution, they are actually seeing the relationship of death rates to occupation and other factors that happen to be somewhat colinear with pollution in the United States. If these variables were taken into account, the pollution/mortality association would disappear.

**Discussion:** Confounding is possible in any epidemiologic study. This can influence results in two ways: either an important variable is not included or two or more seemingly independent variables are really measures of the same thing. In statistical studies on mortality and air pollution, smoking and occupation are frequently cited as confounding variables left out of regression analyses. The effect of confounding would be the production of seemingly valid but spurious association. For confounding to occur, the variable must be a cause of disease, and must be highly correlated with the variable of interest, air pollution.

a. *Smoking.* It is certainly reasonable to argue that smoking may be confounding the results. Data on the smoking habits of metropolitan populations are virtually nonexistent. It is generally accepted that smoking can induce lung cancer, induce and/or exacerbate bronchitis and emphysema, and reduce life expectancy. Therefore, if there were substantial differences in the amount of smoking across metropolitan areas—differences that varied similarly to air pollution—then smoking may be the real cause of mortality now attributed to air pollution. There is mixed evidence on this issue. Several investigators have independently examined the mortality experience of men and women

in relation to air pollution exposure. In most of this work, air pollution exposure appears to have a stronger effect on men than women. Combining this with the knowledge that historically men in the United States smoke more heavily (and more men smoke) than women, two hypotheses become possible:

i. smoking is confounding the association;

ii. smokers have more respiratory diseases, and thus are more susceptible to air pollution.

Because of a lack of data on community smoking habits, cross-sectional studies can do little to resolve these two hypotheses. However, it is important to note that a very high correlation between air pollution and the smoking habits of entire urban populations would be required to account for the mortality differences now attributed to air pollution. It is unlikely that this is true.

b. *Occupational exposure.* It is argued that the regional differences in heavy manufacturing and other industrial activities vary similarly to the air pollution concentrations and that the differences in mortality reflect the hazards of working rather than breathing outdoor air.

When occupational variables such as the percentage of the work force employed in each of the following categories—agriculture, construction, manufacturing, transportation, trade, finance, education, white collar, and public administration—are added to regressions, the amount of mortality associated with air pollution is reduced. This indicates that adequate accounting for the occupational variable could reduce the effect attributed to air pollution. Lave and Seskin[32] suggest an alternative interpretation. They argue that occupational variables are surrogates for unmeasured air pollutants that affect the general population. If differences in occupation patterns are correlated with air pollution levels, this

[32]L. B. Lave and E. T. Seskin, *Air Pollution and Human Health*, John Hopkins U. Press, Baltimore and London, 1977

represents a limitation of the air pollution mortality associations derived from cross-sectional studies. Currently available data do not permit resolution of this issue.

c. *Differential migration.* Occasionally it is argued that people have been moving out of the heavily polluted Northeast and East Central United States to the less polluted Southeast, South, Mountain States, and West. It is well established that those who move tend to be relatively young and healthy. Thus it is possible that selective outmigration from the polluted areas might result in a spurious association between air pollution and mortality. Lave and Seskin have approached this problem by splitting their data base into areas of "high" and "low" migration and performing independent analyses on these two groups. The coefficients of air-pollution-induced mortality derived from the low-migration data set were about half as large as those from the high-migration data. Differential migration may be partially confounding the results, but a positive association remains for air pollution.

These arguments are the most important of many criticisms of cross-sectional studies of air-pollution-induced mortality. Others are considered in volume II, appendix IX. Although the findings and interpretations of these cross-sectional studies are still being debated, a positive relationship between air pollution and mortality cannot be rejected.

## Quantitative Assessments of Health Damage From Future Pollution Levels

The coefficients of air-pollution-induced mortality derived from the studies discussed above can be used to project the deaths that might result from future coal development. In conducting such a projection, the analyst must develop a "scenario" that describes the development, compute its air quality effects, and overlay those effects on the population.

The Biomedical and Environmental Assessment and the Atmospheric Sciences Divisions of Brookhaven National Laboratory (BNL) have developed a methodology for quantifying the

mortality associated with fossil-fuel-derived air pollution.[33] The Atmospheric Sciences Division has applied a long-distance trajectory model to predict the ground-level concentrations of a number of pollutants emitted from coal combustion. The model is capable of incorporating alternative siting and emission patterns to show the air quality implications of various scenarios. The modeling effort:

1. projects pollutant emission from 1,088 coal-fired sources (711 industrial point sources and 377 utility point sources) using 1985 and 1990 emissions projections, and assuming 80-percent $SO_x$ control for existing sources and 85-percent SO control for new sources;
2. computes the resulting ground-level air quality concentrations with a model which accounts for long-range transport and chemical conversion (i.e., $SO_2$ and $SO_x$ concentrations are calculated by the BNL model); and
3. calculates the annual population weighted sulfate exposure with 32 x 32 km grids for the entire country.

Figure 26 displays the calculated sulfate annual concentrations for the 1990, 25.7 Quads coal-use scenario. The pattern reflects very high sulfate concentrations in the Northeast with peak levels in the region of eastern Ohio, western Pennsylvania, and West Virginia. After obtaining population sulfate exposure, BNL applied to them a mortality coefficient of 3.25 deaths per year for every 100,000 people exposed to a 1 $\mu g/m^3$ annual averaged sulfate level. This coefficient is a "most likely" coefficient of mortality derived from BNL's examination of the available cross-sectional mortality studies and their data bases. In the BNL work the mortality coefficient is expressed in a probability function that ranges from no health effects (actually a negative coefficient) to four times this "most likely value." In this way the

[33]For example, see L. D. Hamilton, "Alternative Sources and Health," in *CRC Forum on Energy*, ed. Robert. T. Budnitz, by CRC Press, Cleveland, 1977 and M. G. Morgan, et al, "A Probabilistic Methodology for Estimating Air Pollution Health Effects from Coal-Fired Power Plants," in *Energy Systems and Policy*, vol. 2, No. 3, Crane Russak & Co., Inc., 1978

**Figure 26.—Annual Concentrations of Sulfate for a 1990 25.7 Quads Coal-Use Scenario
(Brookhaven National Laboratory, 1978)**

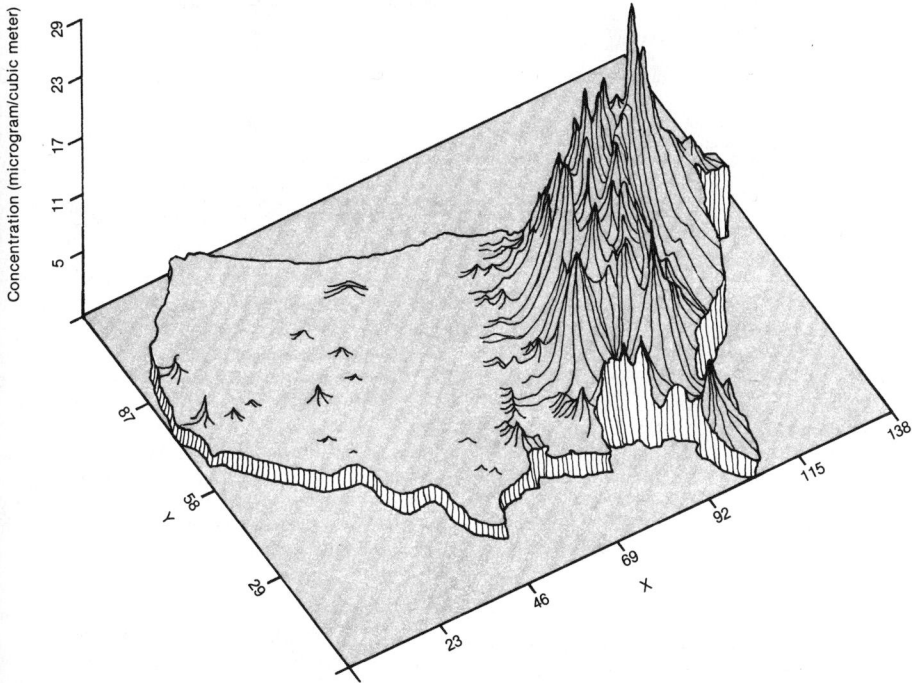

BNL excess mortality, as shown in table 32, is approximately the 50th percentile. That is, there is a 50-percent chance that the actual expected mortality is greater than the estimates presented, based on the BNL probability function for the mortality coefficient.

The implications of the estimates shown in table 32 are startling. First, they imply that coal emissions are responsible for approximately 2.5 percent (48,000 deaths) of the total annual mortality in this country. The second implication is that an increase in coal use could raise the air-pollution-caused annual mortality by approximately 7,000 in 1990. In comparison to the more than doubling in coal use, the projected annual air pollution mortality would increase by less than 20 percent over

the current air pollution mortality estimates. This increase is approximately 0.4 percent of the annual total U.S. mortality.

The credibility of estimates of this type has been severely challenged by many in the health community. As noted above, there is widespread skepticism that any air pollutants at current levels of concentration are killing people in large numbers. Furthermore, even the supporters of the air pollution/mortality associations used in the analysis generally agree that sulfates may not be the "guilty" pollutant. Other problems with the analysis include potential inaccuracies in the sulfate projections and the use of a simple linear model for health effects. However, this work is the first comprehensive attempt to formulate a

**Table 32.—Annual Projected Mortality From Coal Combustion**

| Year | Quads of coal use | Expected mortality |
|---|---|---|
| 1975............................. | 11.3 | 48,120 |
| 1985............................. | 22.5 | 49,543 |
| 1990............................. | 25.7 | 55,835 |

NOTE: Proportional Mortality Model—3.25 deaths/yr/100,000 populations/(1 $\mu g/m^3$—annual average for sulfate).
SOURCE: Based on population exposures from: Meyers, R. E., Cedarwall, R. T., Kleinman, L. I., Schwartz, S. E., and McCoy M., "Constraints on coal utilization with respect to air pollution production and transport over long distances: summary," Oct. 2, 1978 (draft report), Brookhaven National Laboratory.

quantitative methodology for assessing air pollution health effects. Even if the damage coefficient is not accepted, the model can identify population exposures to sulfates and related pollutants and can identify the relationships between regions of high exposures and regions with multiple pollution sources; this information can be used to evaluate the relative effects of alternative siting patterns.

### Research Needs

DOE, EPA, the National Institutes of Health (NIH), and other agencies support air pollution health effects research in areas of bioscreening for hazardous substances, cellular toxicology, animal toxicology, human exposure studies, and risk analyses. This research is funded through the Federal Interagency Energy Environmental Research and Development Program, as well as the basic contract and grant programs of various agencies. Health effects research programs related to fossil fuels are reviewed in volume II, appendix IX.

Quantification of mortality and morbidity from fossil fuel air pollutants is currently limited because:

1. Population exposures to pollutants are poorly defined.

2. The mechanisms of biological injury induced by specific pollutants are not adequately understood.

3. Insufficient epidemiologic evidence exists to relate mortality and/or morbidity to ambient levels of air pollution.

Significant improvement in the ability to quantify the effects of coal combustion is contingent upon continued development in exposure assessment, health effects research, and quantitative applications.

**Exposure.**—Comprehensive pollutant monitoring is fundamental to epidemiologic studies. Present monitoring assesses outdoor concentrations of relatively few pollutants. Particulates, for instance, are determined on a weight basis only and are not differentiated by size or chemical composition. Encouragingly, EPA is conducting a nationwide fine-particulate monitoring program. The relationship of outdoor air pollution to indoor exposures and, in fact, the total integrated dose of pollutants to the individual are not understood. Integrated exposure assessments that measure the human body burden from multiple exposures to pollutants are urgently needed. Moreover, to facilitate epidemiologic work, the exposure of entire urban populations must be better characterized by implementation of more "dense" monitoring networks, including personal monitors.

**Damages.**—The specific mechanisms causing biological injury from many atmospheric pollutants are unknown. Knowledge of these mechanisms will provide a firmer scientific basis for epidemiologic investigation of specific health damages and pollutants and, consequently, a foundation for effective pollutant-specific control strategies.

Ideally, understanding of the biochemical and metabolic fate of pollutants is desired. However, the precise fate of pollutants is often impossible to determine. When this is the case, research must be focused on understanding the grosser, aggregate effects of pollutant insults. For example, pollutant effects may manifest themselves as increased susceptibility to infection, or as inhibition of lung-clearing mechanisms. In pursuit of a better understanding of these effects, new techniques and traditional research for assessing defense mechanism inhibition should be expanded. New techniques are currently being developed for non-invasive methods for mapping particulate deposition and particle retention in the lungs. These techniques would permit rapid surveying of pollutant loadings in human lungs; they may allow the detection of health effects in human populations before the onset of clinical symptoms.

As the result of suggestive evidence on synergistic interactions of multiple pollutants, additional studies examining the combined effects of two or more pollutants should be begun. Few clinical studies of this type have been done. Federal health organizations as well as EPRI have recently recognized this fact. A few animal and human studies of multiple pollutant exposures have started; however, more such studies should be encouraged. In addition, there have been too few controlled exposure studies on subjects under stress. Stress may be induced by exercise or with medication. Stress-pollutant studies are important in determining biomedical effects in sensitive individuals.

**Epidemiology.**—Epidemiologic studies will be necessary for monitoring possible changes in morbidity and mortality as the total popula-

tion exposure to coal-related air pollution increases. Such studies must receive long-term funding commitment to comprehensively examine multiple health indicators while properly assessing pollutant exposures.

Until direct information from controlled toxicologic and epidemiologic studies is available, quantitative assessments of health impacts from fossil fuel combustion must rely on cross-sectional statistical studies. These studies can be improved by a more accurate measurement of socioeconomic variables. For example, as noted above, differences in regional smoking habits have not been adequately accounted for. This deficiency could be corrected if social, economic, and geographic smoking patterns were assessed as part of the National Census.

In order to resolve the question of whether long-term health effects are caused by peak exposure or are the cumulative effect of mean exposures, studies should be undertaken of the effects of periodic exposures to sulfate, nitrate, ozone, and fine particulates. Summertime "episodes," where these and other pollutants are elevated for hours to days over large areas of the Northeast, have been documented in ongoing field monitoring studies. Selected populations must be closely monitored for respiratory and other physiological changes before, during, and after such episodes.

Conclusions

The discussions in this section and in appendix IX indicate that there are grounds for speculation (for example, our knowledge of the ability of fine particulates to penetrate deeply into the lungs) and some positive but not conclusive statistical evidence that current and expected future concentrations of coal-related pollutants may be dangerous to human health. A significant segment of the health community considers the statistical evidence to be meaningless. Several of the objections they raise are valid but do not appear to justify rejection of the evidence.

Policymakers are faced with a difficult tradeoff. More stringent regulation of air pollution can involve very large costs and can even incur new environmental damages. For example, an insistence on full scrubbing will cost several tens of billions of dollars by 1990 and create a substantial disposal requirement for millions of tons of scrubber sludge. On the other hand, a failure to control incurs the risk that significant health damages may occur. The tradeoff is complicated by the current inability to definitely associate the (statistically observed) health effects with one specific pollutant; although sulfates have been implicated by most of the statistical analyses, they may be an indicator for another pollutant or mixture of pollutants.

Uncertainty in environmental policymaking is inevitable. However, the level of uncertainty and the stakes in coal policy are very high. There is thus a high premium on reducing the uncertainty by improving predictions of future coal-related health effects. The research needs identified above (increased monitoring, improvement in the capability to conduct cross-sectional and time series statistical studies, and expanded investigations of the mechanisms of pollution-caused biological injury) are critical elements in this process.

## Ecological Effects

### Sulfur Oxides

As emphasized in the discussion of pollutant transformation and transport, $SO_2$ may retain its chemical identity or may convert in the atmosphere to sulfuric acid or to a variety of sulfate salts. Within about 40 km of a coal-fired powerplant, the form of sulfur most likely to be deposited is $SO_2$. To place experimental evidence in proper perspective, one must compare levels at which damage is observed both to levels mandated by the current air quality standards and to those actually prevailing near and far from powerplants. The secondary NAAQS $SO_2$ standard is 0.5 ppm (3-hour average) and the primary standards are 0.14 ppm (24-hour average) and 0.03 ppm (annual average). In the United States as a whole in 1975,

the annual average ambient levels were under 0.01 ppm.[34] Levels near a plant vary greatly depending on the pollution controls, combustion conditions, weather, and topography. The $SO_2$ level calculated for an area within a few kilometers of an "ideal" 1,000-MW plant (meeting NSPS, having a 500-foot stack) is 0.04 ppm (24-hour average), with the concentration dropping by a factor of 10 within 5 to 10 km of the plant.[35]

Experiments have shown that levels as low as 0.02 to 0.05 ppm of $SO_2$, even at relatively brief exposures, can cause significant physiological disorders in important crops such as Eastern white pine,[36] ryegrass,[37] alfalfa,[38] and some others. Still more evidence exists for significant plant damage at doses that are higher than 0.05 ppm but lower than the current secondary standard.

From an economic standpoint, long-term yield losses are equally as important as the more visible signs of injury to vegetation. Yield loss may occur even without the manifestation of visible symptoms, especially in forest tree species and alfalfa (see appendix X, volume II). Usually, however, yield loss in crops is accompanied by the visible leaf damage that has come to be associated with the effects of $SO_2$ pollution.

Lower plant species—mosses, ferns, lichens, and liverworts—are even more sensitive to $SO_2$ than are agricultural and forest crops. Field studies indicate that many such species are sensitive to levels as low as 0.03 ppm. Numerous lichen species have already become extinct downwind of large cities in Europe, North America, and Asia.

[34]*Energy/Environment Fact Book* (Washington, D.C.: U.S. Environmental Protection Agency, 1978), EPA-600/9-77-041, p. 10.
[35]*An Analysis of the Impact on the Electric Utility Industry of Alternative Approaches to Significant Deterioration: Volume II* (Washington, D.C.: U.S. Environmental Protection Agency and Federal Energy Administration, October 1975).
[36]C. R. Berry, "Age of Pine Seedlings with Primary Needles Affects Sensitivity to Ozone and Sulfur Dioxide," *Phytopathology Vol. 64* (1971), pp. 207-209.
[37]Bleasdale, *Nature* 169 (1952), pp. 376-377.
[38]J. H. Bennett and A. C. Hill, "Acute Inhibition of Apparent Photosynthesis by Phytotoxic Air Pollutants," *Am. Chem. Doc. Sym.* series No. 3 (1974), pp. 115-127.

Although ambient levels near powerplants usually are below the current secondary standards, much higher values can occur during plume touchdowns. The $SO_2$ levels during a touchdown, as well as the frequency of touchdowns, strongly depends on local factors surrounding a given powerplant. Measurements of concentrations during these episodes are scarce, although measurements 3.6 miles from an Indiana powerplant showed nearly 100 episodes per year of 1-hour $SO_2$ concentrations above 100 $\mu g/m^3$ (.037 ppm) in the direction of persistent winds.[39] Better documentation exists for the actual damage done by plume touchdowns, or fumigations, from coal-fired powerplants and industrialized sources. These fumigations have caused acute foliage burning to commercial soybeans,[40] ponderosa and lodgepole pine forests, Douglas fir forests,[41] and to other plants.

Damage caused by mixtures of pollutants have been shown in the field and in chamber studies over a wide range of environmental conditions, including variations in soil type, temperature, light, and humidity. Combined effects appear to take place through several mechanisms. The most common seems to be that the combination of pollutants reduces the plant's threshold of injury; frequently the effect of the more toxic of two pollutants is enhanced.

A wide variety of field, vegetable, fruit, nut, forage, and forest crops have been shown to be sensitive to $SO_2$—even at levels below the Federal secondary standard—in mixtures with common pollutants (ozone, nitrous oxides, hydrofluoric acid, and peroxyacetyl nitrate). Such pollutant mixtures disrupt critical plant functions, including growth, yield, root/shoot ratios, photosynthesis, and respiration. The

physical appearance of produce can also be injured by the combined impact of different air pollutants. As a result, the marketability of vegetables and ornamentals may be reduced.

Damage to animals occurs at far higher doses of $SO_2$ than those levels at which plants are affected. Rather than being affected directly by $SO_2$, animals may be more seriously endangered by its indirect effect; sulfur compounds can destroy or alter their natural habitat.[42]

$SO_2$ is transformed in the atmosphere to various forms of sulfate that are damaging to plantlife. Plants are directly injured when excessive sulfates accumulate in their tissues. The rate at which this occurs varies according to the plant species and according to the amount of sulfur already available in the soil (some possibly stemming from chemical fertilization). The California Air Resources Board has reported growth and yield reductions from sulfate accumulation in cotton, oats, rye, tomatoes, barley, and in forests in New Hampshire. Other studies have shown damage to spruce needles from sulfate accumulation over periods of time as brief as 10 to 100 days.

Some sulfates, notably ammonium sulfate, can act as plant fertilizers. Nevertheless, even apparently beneficial sulfate compounds may be deposited at times not conducive to good farming practices, and the deposition of such sulfates may interfere with the individual farmer's schedule of chemical fertilization.

As discussed previously, the emission of $SO_x$ from all sources may be relatively stable in the next 10 years (at moderate coal use growth rates), although emissions may begin to grow again unless controls are placed on older powerplants. DOE projections show that the emissions in the eastern portion of the United States will generally decline during this period. This is fortunate because that is where the brunt of $SO_x$ impact is now borne. However, some impacts—for example, the acidifying effects of acid rain—may be somewhat cumulative and thus could become more intense unless changes in control strategies are forth-

[39]"Results of Multiscale Air Quality Impact Assessment for the Ohio River Basin Energy Study," Teknekron, Inc., Berkeley, Calif., and Waltham, Mass., 1978.
[40]C. C. Gordon and P. C. Tourangeau, "Biological Effects of Coal-Fired Power Plants," *Fort Union Coal Field Symposium*, University of Montana, pp. 509-530.
[41]C. E. Carlson, M. D. McGregor, and N. M. Davis, "Sulfur Damage to Douglas Fir Near a Pulp and Paper Mill in Western Montana," U.S. Department of Agriculture, Forest Service, Northern Region, Division of State and Private Forestry, Report No. 74-13.
[42]N. R. Glass, ed., "Appendix on Ecological Effects," U.S. Environmental Protection Agency, p. 6., *Rall Report*.

coming. An intensification of $SO_x$ controls on existing combustion sources is the only practical alternative available, as the proposed NSPS is near the present limits of control technology and increased requirements on new plant facilities would have only minor effects in this period.

## Nitrogen Oxides

The direct action of man-caused $NO_x$ on the ecosystem is difficult to analyze because nitrogen is an element essential to all biological systems and because natural sources of $NO_x$ (especially anaerobic bacterial action and chemical decomposition of nitrate) account for more than 90 percent of global levels.[43] The limited evidence suggests that $NO_2$ is more toxic than NO and that the most susceptible receptors in the terrestrial ecosystem are the higher plants, among them oats, barley, tobacco, apples, and lettuce. Still, the threshold for almost all observed effects on ecosystems is well above the $NO_x$ levels in most U.S. cities. $NO_x$ is not nearly as obviously damaging to vegetation as either $SO_x$ or photochemical oxidants at similar levels of concentration. Also, very high $NO_2$ concentrations are required to produce morbidity or mortality in most animals.[44]

However, there are some poorly understood areas of ecosystem effects that may be a cause for concern. For example, soils absorb NO and $NO_2$ quite readily and change them to nitrate, decreasing soil pH and possibly affecting the micro-organisms in the soil. The effects of chronic low-level or intermittent $NO_2$ exposures on vegetation also deserve attention.

The role of $NO_x$ in the formation of photochemical oxidants is of greater concern than $NO_x$ by itself, for oxidants are the most damaging air pollutants affecting agriculture and forestry in the United States.[45] $NO_2$ is the one primary "criteria pollutant" (pollutant regulated by a NAAQS under the Clean Air Act) that is likely to increase significantly in the next few decades. The 21-percent increase in $NO_2$ projected for 1990 by DOE (NEF #1) may

lead to an increase in concentrations of ozone and peroxyacyl nitrates. However, the degree of increase is complicated by a certain degree of quenching of oxidants by $NO_x$ and by an expected decline in hydrocarbon emissions, which are also oxidant precursors and may be the limiting component for oxidant production.

Any increase in oxidants appears particularly worrisome when it is recognized that elevated ozone concentrations caused by long-range transport of pollutants from urban areas have become a severe regional problem throughout the United States. "Repeated episodes during 4 to 6 months of the year . . . cause obvious foliage injury, impaired photosynthesis, and growth reductions of sensitive species populations, in both agricultural and forest ecosystems."[46] Oxidants have caused well-documented damage to agricultural crops, forests, and native vegetation in southern California as well as widespread damage to crops in the East, and have become a new stress on ecosystems in the Southwest. Although animals do not appear to be sensitive to existing levels of oxidants, they are affected by the loss of cover and food accompanying such levels.

$NO_x$, like $SO_x$, has the potential for synergistic effects. It may facilitate the creation of nitrosamines, which are known carcinogens, and it also seems to have more adverse effects when present together with $SO_x$.

Finally, the most serious effect of $NO_x$ may be its involvement with $SO_x$ in the formation of acid rain, which is discussed separately below.

## Acid Rain

Both $NO_x$ and $SO_x$ are implicated in the occurrence of acid rainfall at many locations around the world. "Normal" precipitation is expected to have a pH of about 5.7, based on theoretical calculations that postulate an equilibrium with $CO_2$.[47] Precipitation samples from various points in the United States indicate that in large portions of the Eastern

[43]Ibid., p. 10.
[44]Ibid., p. 13.
[45]Ibid., p. 19.
[46]Ibid.
[47]J. N. Galloway, G. E. Likens, and E. S. Edgerton, Science, 194, p. 722, 1976.

United States, the pH falls between 3 and 5[48] (see figure 27). In northwestern Europe, where acid rain has been more intensely studied, a network of monitoring stations yields evidence that the pH of the precipitation has dropped over the past 20 years.[49] It has been widely stated that the region of acid precipitation in the United States is extending westward and southwestward.[50] The data available are strongly suggestive of this statement, but are neither systematic nor extensive enough to be considered conclusive. EPA and the Forest Service are currently establishing networks of monitoring stations that will provide better data in the future.

One effect of acid precipitation is to lower the pH of bodies of water into which it falls. If the pH dips below 5.0, fish such as trout or salmon may perish, for they are very sensitive to pH. Fish kills have in fact been reported in areas such as New York's Adirondack Mountains, where 51 percent of the lakes now have a pH below 5.0, and where 90 percent of these support no fish populations.[51] A valuable recreational trout industry has thus been endangered. In Norway the losses to the salmon fisheries attributed to acid rain are estimated in the tens of millions of dollars.[52] Other aquatic effects include reduction in algal communi-

---

[48]*Energy/Environment Fact Book,* p. 70.
[49]L. S. Dochinger, paper presented at the *Third Annual Energy/Environment R&D Conference,* June 1978.
[50]See, for example, G. E. Likens, "Acid Precipitation," *Chemical and Engineering News,* Nov. 22, 1976.

[51]G. E. Likens, *Science* 22, p. 720, 1976.
[52]L. S. Dochinger and T. A. Seliga, "Acid Precipitation and the Forest Ecosystem," *Journal of the Air Pollution Control Association* 25, p. 1104, 1975.

## Figure 27.—Regional Impact of Acid Rainfall

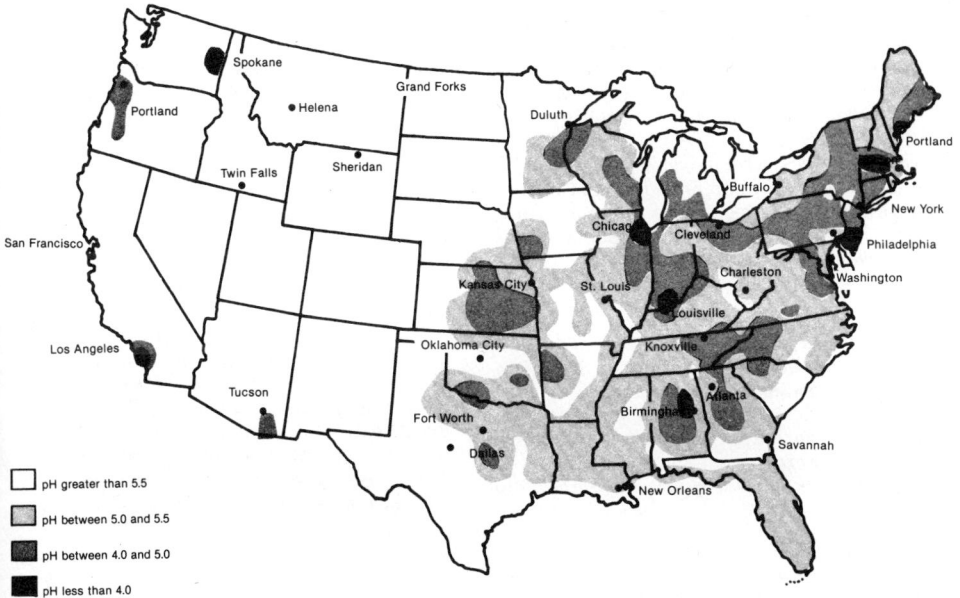

□ pH greater than 5.5

pH between 5.0 and 5.5

pH between 4.0 and 5.0

pH less than 4.0

SOURCE: Environmental Challenges of the President's Energy Plan: Implications for Research and Development. Report of the Subcommittee on Environment and the Atmosphere, Committee on Science and Technology, U.S. House of Representatives, 95th Congress, October 1977.

ties, growth of thick mats of benthic moss, and hindrance of nutrient recycling.

The evidence for damage to terrestrial vegetation is somewhat more elusive. An association has been postulated between acid precipitation and the decline in forest growth observed in both the Northeastern United States and in southern Sweden, where growth declined between 2 and 7 percent from 1950 to 1965.[53] The ecosystem is unfortunately too complex for the cause to be laid so unequivocally at the door of just one of many environmental stresses. Nevertheless, numerous experimental studies have shown that acid rain can harm terrestrial systems in several ways: it may "damage foliage; accelerate cuticular erosion; alter responses to associated pathogens, symbionts, and saprophytes; affect the germination of conifer and hardwood seeds and the establishment of seedlings; affect the availability of nitrogen in the soil; decrease soil respiration; and increase leaching of nutrient ions from the soil."[54] Acid rains have been strongly implicated in causing extensive damage to Christmas tree farms in West Virginia downwind of the Kyger Creek coal-fired powerplant in Ohio. The disease afflicting the trees has been duplicated in at least two independent laboratory studies using sulfuric acid mists. In other studies, several important food crops and trees were adversely affected by simulated acid rain and mist. It has also been shown that increasing acidity of rain significantly increases the uptake of cadmium and other toxic substances by plants that may subsequently be included in human and animal diet.[55]

The effects of acid precipitation depend largely on the buffering capacity of the region on which the rain falls. The damage may be quite dramatic in Scandinavia, for example, because 80 percent of the forest land in Sweden is acid podzolic soil and thousands of its lakes are underlaid with granitic bedrock.[56] By contrast, in the Western United States, where indications are that rainfall may be growing more acidic and may continue to do so, the effects may never become serious because the soils and lakes are so alkaline. More studies must be initiated to identify sensitive regions with low buffering capacity and to understand the processes that affect the buffering capacity of soils, such as acidification by natural processes and by chemical fertilization.

Any increase in soil acidity may alter the delicate balance of soil calcium, magnesium, and potassium to aluminum. Essential nutrients, including nitrogen, potassium, and phosphorus, can form compounds that are not useful to plants, or they may be leached from the soil altogether. The living organisms of the soil are very sensitive to pH changes in their environment, and these include not only important lower plant forms such as algae, lichens, and fungal associates of higher plants, but also the nitrogen-fixing bacteria, which are essential if airborne nitrogen is to be made available for use as a nutrient by plants. Current high nutrient erosion rates from soils subjected to acid precipitation of 4.0 pH or less in remote forested areas of the Adirondack Mountains[57] and elsewhere are evidence that forest soils in the Northeastern United States are becoming acidic at an accelerated rate.

Fossil-fueled powerplants may be major contributors to the occurrence of acid rain through their generation of $NO_x$ and $SO_x$. Researchers in Scandinavia attribute its acid rain to the long-range transport of sulfates from the heavily industrialized areas of northern Europe and England. Data from New York and New England indicate that about 60 to 70 percent of the acidity in precipitation there is caused by sulfuric acid, with the contribution from nitric acid becoming increasingly important.[58] Because the background of air pollutants from stationary and mobile sources varies with locale, these figures may not universally represent the composition of acid rain.

[53]Ibid., p. 1105.

[54]L. S. Dochinger, see note 49.

[55]I. Shen-Miller, M. B. Hunter, and J. Miller, "Simulated Acid Rain on Growth and Cadmium Uptake in Soybeans," Plant Physiology, vol. 57 (1976), p. 50.

[56]L. S. Dochinger, T. A. Seliga, p. 1104.

[57]C. W. Cronan, R. C. Reynolds, Jr., and G. E. Lang, "Forest Floor Leaching: Contributions From Mineral, Organic, and Carbonic Acids in New Hampshire Subalpine Forests," Science, vol. 200 (1978), pp. 309-11.

[58]G. D. Likens, "Acid Precipitation," Chemical and Engineering News 22, 1976.

Future increases in energy development, and in coal use, may seriously aggravate acid rain problems through two mechanisms. First, although tall-stack emissions of $SO_2$ will not increase rapidly if best available controls are applied, serious increases in $NO_2$ emissions will be inevitable if Federal emission standards — which require only about 50-percent control — are continued. As noted earlier, success of EPA's low-$NO_x$ burner will be a primary determinant of this required control level. Second, there is a possibility that the fraction of sulfates and nitrates that are in the strong acid form — currently thought to be about one quarter — may rise because of exhaustion of the neutralizing capacity of the atmosphere. If exhaustion occurs, a 1-percent increase in emissions would subsequently lead to a 4-percent increase in acid rain.[59] Very little is known about the atmosphere's neutralization capacity. Finally, as some of the impact of acid rain may be cumulative, even a stable level of acidity may lead to an increase in acid rain damages.

## Visibility Reduction

$SO_x$ emissions from coal combustion have been strongly implicated as a major factor in the occurrence of visibility ' reductions throughout the United States. The evidence consists of the following.

- Sulfate particles formed as the result of the atmospheric transformation of $SO_2$ emissions tend to be in the size range of fine particles that are most effective in scattering light and hence in producing haze.
- Field studies by EPA indicate that powerplant $SO_2$ emissions are converted to sulfates at significant rates.
- Sulfates account for about half of the particles in the "maximum light scattering" size range in industrial regions.[60]
- Visibility reductions and sulfate concentrations have been strongly correlated; for example, large-scale polluted air masses characterized by decreased visibility, or "hazy blobs," as they are sometimes

called, have been tracked over periods longer than a week as they moved over large portions of the United States. These "blobs" are correlated closely with sulfate levels.[61]
- Visibility substantially increased during the 1967-68 Southwest copper strike and shutdown of $SO_2$-producing copper smelters.[62]

Visibility degradation has recently become a major concern in the West, where clear blue skies are still a common and valued sight. In fact, some studies indicate that visibility in the Southwest appears already to have deteriorated significantly over the past two decades, although it is still quite good.[63]

Although the prevention of significant deterioration (PSD) restrictions in the Clean Air Act amendments are designed in part to protect western visibility, the present inability to compute the long-range impacts on visibility of new coal-fired powerplants may considerably complicate the enforcement of these restrictions until acceptable long-range transport and visibility models are available to enforcement agencies.

Visibility degradation will present a special problem in those areas of the West that are expected to sustain considerable increases in $SO_x$ emissions. The largest increase is expected in the Southwest (New Mexico, Texas, and Oklahoma), with an increase of 67 percent between 1975 and 1990. The expected upturn in national $SO_x$ emissions by the late 1980's is an indicator that visibility problems could substantially increase on a nationwide basis by 2000.

## Particulates and Associated Trace Elements

Because of the widespread use of controls that remove large particles with high efficiency, the major ecosystem impact of particulate emissions from coal-fired powerplants will be

---

[59]"Appendix on Transport and Fate," *Rall Report.*
[60]"Appendix on Ecological Effects," *Rall Report.*

[61]Husar, R. B., et al, "Ozone in Hazy Air Masses," *International Conference on Photochemical Oxidant Pollution and its Control,* EPA-600/3-77-0010, 1977.
[62]"Visibility in the Southwest," U.S. Environmental Protection Agency, EPA-600/3-78-039.
[63]Ibid.

caused almost exclusively by fine particulates and the associated trace elements and hydrocarbons that are preferentially adsorbed on their surfaces. Although total particulate loadings will be substantially reduced in the future as the result of strict controls on new plants and progress in obtaining conformance with State regulations from existing plants, emissions of fine particulates are unlikely to decline unless controls that are effective on fine particulates (such as baghouses) are retrofitted on older plants as well as being installed on new ones.

Fine-particulate effects on ecosystems may be more of a regional than a local problem. Glass[64] notes that more than 95 percent of the particulate emissions from a large powerplant with an ESP travel beyond a 20 km radius, and most studies of trace element contamination near powerplants have shown few significant effects.[65]

Soils are the principal repository for fine particulates, and there is some concern that the heavy metals contained within or on the surface of the particles could eventually alter the normal soil processes associated with nutrient recycling and with soil micro-organisms. Plants are also an important sink for fine particulates, which are captured by the fine hairs on plant surfaces. Plants are generally not considered to be endangered by such pollution; Van Hook[66] notes that "there is no reason to expect acute effects on plants . . . the potential chronic toxicity is relatively low except in local areas already enriched with a particular element." However, more subtle effects, such as reduction in photosynthesis due to reduced light or interference with stomatal function, and foliar lesions have been attributed to fly ash.[67]

Both inorganic trace elements and trace hydrocarbons present a potential danger to animals in water and on land. Biomagnification, whereby trace materials can be concentrated as they move up the food chain, is a par-

ticular problem in aquatic ecosystems (see table 33). Another source of concern is the extent to which trace element impacts are cumulative; logic argues that many will be cumulative, although the ambient levels caused by coal combustion are not likely to be high in most instances. Although the uncertainty surrounding these effects is very great, concentrations of polycyclic aromatic hydrocarbons that have been demonstrated to be damaging to fish have been observed in Norwegian rainwater.[68] However, the potential for significant impacts on either terrestrial or aquatic ecosystems is clearly a very speculative area that requires considerably more evidence before firm conclusions can be drawn.

Carbon Dioxide

$CO_2$ concentrations measured at Mauna Loa, Hawaii, and elsewhere have increased at a rate that appears to be tied to fossil fuel combustion.[69] The combustion of coal yields 11 percent more $CO_2$ than oil and 67 percent more than natural gas.[70] The confluence of these two facts has led to concern about the role that increased coal development might play in altering the Earth's climate.

$CO_2$ is relatively transparent to short-wave radiation (sunlight), but absorbs long-wave (infrared or heat) radiation. Thus, atmospheric $CO_2$ admits incoming sunlight that warms the Earth, but absorbs a portion of the infrared heat radiation coming from the surface of the warmed Earth and re-radiates part of it back to the surface. The heat is "trapped" in the atmosphere in the same way that heat is trapped inside a greenhouse, or an automobile with its windows closed; hence the "greenhouse effect." (See figure 28). Although no effects on global temperatures have yet been measured, most scientists believe that a continued increase in $CO_2$ levels must eventually affect the Earth's temperature.

[64]"Appendix on Ecological Effects," Rall Report.
[65]"Appendix on Trace Elements," Rall Report.
[66]Ibid.
[67]"Appendix on Ecological Effects," Rall Report.
[68]Ibid.
[69]G. Marland, and R. M. Rotty, "The Question Mark Over Coal; Pollution, Politics and CO," Futures, February 1978.
[70]Analysis of the Proposed National Energy Plan, 1977.

**Table 33.—Biological Concentration Factors for Selected Trace Elements in Aquatic and Terrestrial Environments**

| | Biological Concentration Factor[a] | | | | |
| | As | Cd | Hg | Pb | Zn |
|---|---|---|---|---|---|
| Aquatic | | | | | |
| Water ............................................................... | 1 | 1 | 1 | 1 | 1 |
| Plants ............................................................... | 170 | 1,000 | 1,000 | 200 | 1,000 |
| Invertebrates ................................................... | 330 | 2,000 | 100,000 | 100 | 10,000 |
| Fish ................................................................. | 330 | 200 | 1,000 | 300 | 1,000 |
| | | | | | |
| Terrestrial | | | | | |
| Soil ................................................................. | 1 | 1 | 1 | 1 | 1 |
| Plants ............................................................... | 0.01 | 0.3 | 0.4 | 0.07 | 0.4 |
| Invertebrates ................................................... | 0.01 | 17 | | 0.02 | 8 |
| Mammals .......................................................... | 0.001 | 0.008 | 5 | 0.001 | 0.6 |
| Birds ................................................................ | 0.001 | | 50 | 0.001 | |

[a] Ratio of concentration in organism to concentration in substrate.

SOURCE: Van Hook, R. I., "Transport and transformation pathways of hazardous chemicals from solid waste disposal," *Environ. Health Perspectives* (in press).

**Figure 28.—Heat Radiation Emitted to Space by CO₂**

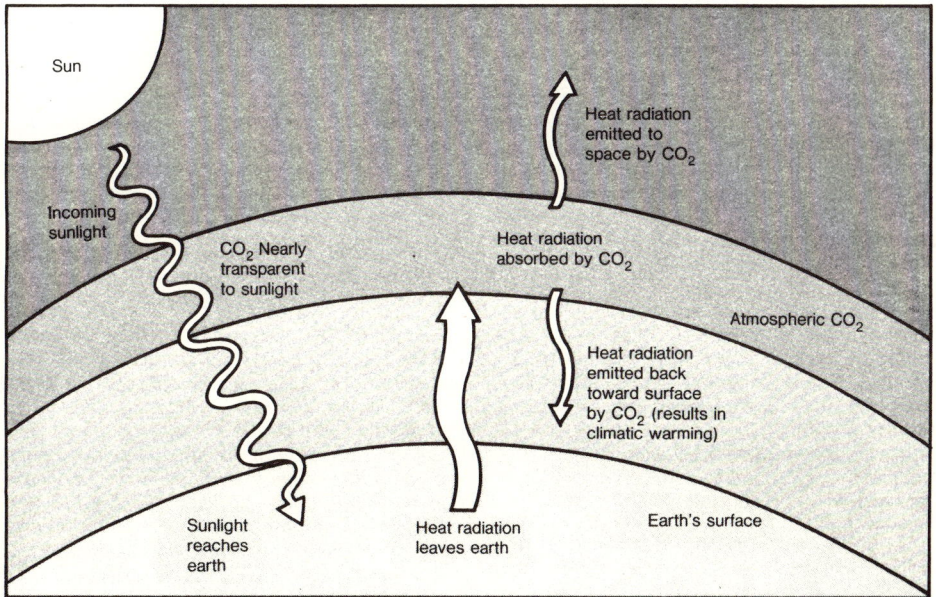

SOURCE: Mitchell, J. Murray, Jr., "Carbon Dioxide and Future Climate," EDS, March, 1977 cited in Norwine, J., "A Question of Climate," Environment. Vol. 19, No. 8, November 1977.

The legitimacy of environmental concern over the climatic influence of coal production hinges on the answers to two critical questions:

- How will future atmospheric $CO_2$ concentrations be affected by fossil fuel combustion and other activities such as forest clearing?
- What effect will increased $CO_2$ in the atmosphere have on global climate?

Considerable uncertainty clouds both questions.

**$CO_2$ Concentrations.**—There is general agreement about the amounts of carbon currently being released as $CO_2$ through fossil fuel combustion (about 5 billion kilograms per year) and the increase in the amount of carbon in the atmosphere (about 2.3 billion kilograms per year).[71] Until recently, scientists thought they could readily explain where the remaining 2.7 billion kilograms per year went; a maximum of about 2.5 billion kilograms per year was thought to be absorbed by the oceans, and the remainder by the biota-forests, grasslands, etc. A number of researchers have now challenged this view based on evidence that the largest and most intensive pool of carbon in the biosphere—the forests (both the trees themselves and their humus floors)—are shrinking significantly in area because of wood harvesting and clearing for agriculture and other development. This is especially true of the tropical rain forests such as those in the Amazon Basin, which are said[72] to play the major role in carbon storage in the biosphere. Thus, a net carbon release rather than a net absorption now is theorized. Woodwell[73] estimates this release from the biosphere to be about 8 billion kilograms, although he estimates the range due to uncertainty to be 2 billion to 18 billion kilograms per year. Other researchers have developed lower estimates, ranging from about one-tenth to one-third of the fossil fuel contribution.[74]

Accepting even the lower value still leaves an enormous gap in the existing models of carbon balance; it is not known where all this excess carbon is going. However, if Woodwell is correct that the biosphere is a $CO_2$ source rather than a "sink," this means that an unknown mechanism, capable of removing and storing atmospheric $CO_2$ at a substantial rate, must exist and might be available to remove excess $CO_2$ from fossil fuel combustion if the shrinkage of the forests could be halted

Uncertainty also is added by a lack of full understanding of the various known mechanisms involved in oceanic uptake of $CO_2$. For instance, carbonate ions in the surface layers of the ocean react with and remove atmospheric $CO_2$. As the pressure of $CO_2$ increases and the carbonate ions are used up faster than they can be replaced by mixing, the ocean's absorption of $CO_2$ from the atmosphere could slow and the atmospheric $CO_2$ buildup could accelerate. On the other hand, increased acidity of surface waters caused by greater dissolved $CO_2$ concentrations could increase the dissolution of calcium carbonate and thus increase the supply of carbonate ions. The extent of these potential changes presently is unknown. Uncertainty also is caused by the lack of understanding of the effects of increased $CO_2$ on ecosystems and the corresponding effect on their photosynthetic production and carbon storage capacities.

Any changes in these factors would alter the capacity of the biosphere to act as a sink—or source—of atmospheric $CO_2$.

Despite these uncertainties, however, probably few scientists would disagree with the statement that continued reliance on fossil fuels, in the absence of drastic measures to change worldwide development patterns that are radically altering the biosphere, will result in substantially increased levels of $CO_2$ in the atmosphere. Kellogg's[75] estimate of a 400-ppm $CO_2$ concentration by 2000 and a doubling of that level by 2040 (if fossil energy trends continue) should be taken quite seriously.

[71]G. M. Woodwell, "The Carbon Dioxide Question," *Scientific American*, January 1978, vol. 238, No. 1.
[72]Ibid.
[73]Ibid.
[74]R. A. Kerr, "Carbon Dioxide and Climate: Carbon Budget Still Unbalanced," *Science*, vol. 197, pp. 1352-1353, Sept. 30, 1977.

[75]W. W. Kellogg, "Global Influence of Mankind on the Climate," in *Climate Change*, J. Gibbin, ed., Cambridge University Press, 1978.

**Climatic Effects.**—Even if precise forecasts of future atmospheric $CO_2$ concentrations were possible, the effects of these concentrations on the climate and on man would be uncertain owing to several confounding factors. One factor is that the climate, with or without man's influence, may be changing as part of a natural cycle. Scientists disagree as to the nature of these changes. Secondly, the "anthropogenic" (man-caused) effect on global temperatures includes several effects aside from the "greenhouse effect." Their role complicates the calculation of the temperature effect of man's activities. Finally, the translation of more precise knowledge of the "greenhouse effect" and the other temperature-changing factors into a global climatic effect is a modeling problem that requires considerably more knowledge and sophistication than is currently available. It is the distribution of temperature and other effects, rather than the global average temperature effect, that is critical to understanding the changes that will occur in actual living conditions.

There are three major schools of thought on the nature of future climatic cycles.[76] The first is that the recent climatic record is not sufficient to allow accurate forecasts of future climate. The second is that the Earth is due for a short period of cooling followed in the early 21st century by mild warming, and that man's activities will not affect these trends. The third and most widely accepted is that global warming will occur sometime in the next century or so as a result of man's activities, and that any manifestation of natural climatic cycles will eventually be overwhelmed by this induced warming.

In addition to the "greenhouse effect," the other man-created effects that may affect global temperatures[77] include:

- Change in reflectance of the Earth: continued injection of particles into the upper atmosphere will lead to increased reflection of incoming solar energy and, consequently, a cooling effect.
- Changes in cloud cover: the increasing level of fine particulates (which act as condensation nuclei), the increasing areas of irrigation projects, the expanded use of cooling towers, and any warming trend (leading to increased evaporation rates) would all tend to increase the cloud cover of the Earth and increase its reflectance. This provides a "negative feedback" to compensate for the greenhouse effect.
- Thermal pollution of the atmosphere: by the year 2050, the total heat added to the Earth's atmosphere by energy production will approach $10^{19}$ Btu per year, which is on the order of one-half of 1 percent of the solar energy reaching the Earth's surface. A 1-percent increase in this energy "flux" would generate about a 1° F increase in global temperatures.

A reliable computation of the climatic (rather than the mean temperature) effect requires "a model that includes a fully coupled and interactive atmosphere and ocean, and an improved simulation of ice and snow."[78] No accepted model of this sort now exists. However, a simpler general circulation model predicts that a doubling of atmospheric $CO_2$ concentration will yield temperature increases of 2° to 3° C at the equator and 10° C at the poles.[79] This geographic variation in warming is ominous from the standpoint of the survival of the polar ice caps.

The concept of massive changes in the Earth's climate—in essence, a retrogression to a warmer climate characteristic of the Earth several thousand years ago—sounds more like a science fiction theme than a plausible projection for the near future. Realistic projections are hampered by the complexities of climatic modeling, the substantial uncertainties involved in the potential feedback effects of the atmosphere, the incompletely understood system of sources and sinks of $CO_2$ and the substantial differences in alternative energy projections. Nevertheless, the scientific

---

[76]J. Norwine, "A Question of Climate," *Environment*, vol. 19, no. 8, November 1977.

[77]R. M. Rotty, "Energy and the Climate," Institute for Energy Analysis, Oak Ridge Associated Universities, September 1976.

[78]J. Williams, "Global Energy Strategies; The Implications of $CO_2$,"*Futures*, August 1978.

[79]Ibid.

community generally agrees that the probability of significant global warming and other climatic changes is sufficiently high to warrant exceptional attention in form of expanded research and monitoring, and caution in weighing any decisions that might tend to tie us irrevocably to a fossil-based energy economy.

It must be stressed that the "scenarios" postulated by the more pessimistic researchers would involve massive disruption of the world economy and well-being. A 2° to 3° C change in mean world temperatures could lead to melting of the polar icecaps sufficient to cause flooding of coastal regions. Another critical effect would be significant shifts in the viability of the world's food-producing areas owing to massive shifts in rainfall distribution. In the long run, such shifts might not lead to a net decrease in potential worldwide productivity. However, the transient effects would be severe. Given appropriate climatic conditions, overall agricultural viability depends upon existing land use, ownership patterns, transportation, water, and other resource availability, etc. These factors represent a massive physical and societal infrastructure that would have to be virtually torn down and rebuilt to make a successful transition to a warmer climate. Problems with such a transition process could mean food shortages and social and economic disruption on a massive role.

## Effects on Materials

Although human health and ecological effects are the most serious problems associated with air pollution from the coal fuel cycle, damage to nonliving materials from this pollution is substantial today and may remain a potent policy issue because of memories of coal's past. The soiling problem commonly associated with coal combustion has been all but eliminated by extensive controls on particulate emissions and the phasing out of most of the small coal-fired furnaces. However, problems that remain include the deterioration of building materials and works of art, fading of dyes, weathering of textiles, and corrosion of metals.

A 1970 estimate of total damage to materials from air pollutants (not limited to emissions

from powerplants or coal combustion sources) is $2.2 billion per year. The relative contributions of the four major pollutant components are: $SO_x$—41 percent, particulates—27 percent, ozone—26 percent, and $NO_x$—6 percent.[80]

There are five major causes of damage:

- abrasion by high-velocity and/or sharp-edged particles,
- direct chemical attack,
- indirect chemical attack—where the pollutant becomes harmful only after a transformation when in contact with the material,
- electrochemical corrosion—where pollution increases conductivity of thin surface films of moisture and helps to destroy protective films on material surfaces, and
- discoloration by particulates.

Each mechanism depends on a variety of independent variables, including moisture, temperature, sunlight, etc. If estimates of future pollution levels are a guide, one might expect that acid rain and ozone damage caused by powerplant $NO_x$ emissions will rise in the future, that $SO_x$-caused damage will not undergo radical changes if best available control technology (scrubbers) is enforced, that particulate soiling damage should decrease with substantial decreases in particulate emissions as all new powerplants control to better than 99-percent efficiency and existing particulate emitters comply with State and local control requirements, and that chemical damage due to particulates might increase because of increases in emissions of fine particulates.

## Regulations

Air pollution from coal combustion is regulated under the Clean Air Act, which sets ambient air quality standards and establishes limits on the emissions from a variety of sources in order to achieve and maintain those standards. The legal framework of the Clean Air Act is discussed in chapter VII; this section

[80]T. E. Waddell, "The Economic Damages of Air Pollution," U.S. Environmental Protection Agency, EPA-600/5-74-012, May 1974.

summarizes the way the Act seeks to regulate the health and ecological effects of the pollutants.

National Ambient Air Quality Standards (NAAQS) have been promulgated under the Clean Air Act for six pollutants: $SO_x$, particulate matter, $NO_2$, photochemical oxidants, hydrocarbons, and CO. There are two types of NAAQS: primary standards are those needed to protect public health, allowing an adequate margin of safety. The secondary standards are intended to protect the public welfare from pollution damage to soil, plants, animals, materials, and other factors not directly related to human health. The NAAQS are based on air quality criteria that reflect the latest scientific knowledge useful in indicating the kind and extent of identifiable effects of a pollutant on public health or welfare. These criteria must include information about variable factors, such as atmospheric conditions that may alter the effects of a pollutant, and about interactive effects among pollutants.

The Clean Air Act is implemented at the State level through a variety of regulatory programs designed to achieve and maintain the NAAQS. First, State Implementation Plans (SIPS) must include control strategies designed to reduce the contribution of existing sources to airborne pollutant loadings, generally through the application of reasonably available control methods. Second, the Act establishes performance standards for large new sources that mandate the use of the best technological system of continuous emission reduction as well as a percentage reduction in emissions over those from the uncontrolled combustion of untreated fuel. Third, in those areas where air quality now exceeds the NAAQS (nonattainment areas), new sources must meet the lowest achievable emission rate (LAER) by applying stringent technological controls and must secure emission reductions from existing sources in the area that will more than offset the emissions from the proposed facility. Fourth, in order to prevent significant deterioration of air quality in areas where the air currently is cleaner than the national standards, the Clean Air Act specifies maximum allowable increases in pollutant concentrations as well as ceilings for ambient concentrations, and requires large new sources to use the best available control technology. Finally, the Act requires the retrofitting of pollution control technology on existing large sources near areas primarily important for scenic values (such as national parks) in order to protect visibility.

There are problems associated with implementation of the Clean Air Act to protect public health and welfare. As of mid-1976, 178 of the 313 air quality control regions (AQCRs) or State portions thereof had not attained the primary particulate standard, while 46 had not achieved the primary $SO_2$ standard. Most conventional stationary sources (such as coal-fired powerplants) in these nonattainment areas are either in compliance with the standards or are on compliance schedules. The particulate violations tend to arise because of fugitive emissions, fugitive dust, and fine particulates, all of which are more difficult to monitor and control, and which are not now explicitly regulated under the Clean Air Act.

No deadlines have been set for attainment of the more stringent secondary standards. Rather, the States are given a reasonable time to achieve the secondary standards, where "reasonable time" depends on the degree of emission reduction needed to attain the standards and on the social, economic, and technological problems involved in implementing a control strategy adequate to achieve and maintain the standard. Thus the secondary standards are not being enforced, and soil, plants, animals, materials, and other environmental factors are unprotected to the extent they require greater protection then that afforded by the primary standard.

The air quality criteria on which the NAAQS are based were published in 1969-71 and reflect the scientific knowledge current at that time about the effects of the various pollutants. However, in 1969-71 the phenomena of pollutant transformations and long-range pollutant transport had not been documented. Consequently, the Clean Air Act and the regulatory programs that implement the Act generally treat $SO_2$ and particulate matter as static pollutants that do not convert. The transformation products, such as sulfates and acid precipitation, are not regulated. In addition, current EPA regulations limit the applicability of air quality modeling techniques to a downwind distance of not more than 50 km from the source (because of EPA's judgment that the models are not reliable beyond this distance). Although the 1977 amendments to the Clean Air Act include procedures for the prevention and abatement of interstate air pollution problems attributable to particular sources, longer range transport that may contribute to air quality problems farther than 50 km downwind currently is unregulated.

The 1977 amendments to the Clean Air Act include two provisions designed to remedy these problems. First, they require EPA to review the NAAQS and the criteria on which they are based, and revise them if necessary by 1980 and every 5 years thereafter. Thus more recent evidence on health effects and on pollutant transformations and transport should be reflected within a couple of years in the standards and in the programs that implement them. Second, the amendments require the States to revise their implementation plans by 1979 in order to provide for achievement of the primary standards by 1982. However, this requirement may be largely moot if the standards are made more stringent in 1980, and it does not address the problem of implementing the secondary standards.

# IMPACTS TO WATER SYSTEMS

The impurities in coal that are released to the air and land in all stages of the fuel cycle may wend their way eventually into the waterways. Thus water may become the ultimate repository for many of the byproducts of coal activities. The water system can transport these impurities both geographically—from the local source to distant sites—and biologically—from micro-organisms on up the food chain, often in ever-increasing concentration.

The uses for which our water systems are needed include—roughly in order of increasing demand for water purity—navigation, recreation, irrigation, industrial processing, watering of stock and wildlife, maintenance of aquatic life, and drinking water supply. A lowering of water quality may affect these uses and consequently the health, economy, or quality of life within a community.

The various activities of the coal fuel cycle affect not only water quality but also water quantity and hydrology. For example, mining can cause stream flows to change and water tables to decline, and in some instances may affect the intensity of floods. Of particular concern is the potential alteration of the ground water system.

## Sources, Effects, and Control of Water Pollution

### Coal Mining and Cleaning

**Surface Water Impacts.**—In mining, huge slices are made into the earth, and layers once deeply buried are brought to the surface to be deposited, often only temporarily. In surface mining, the top layers of overburden are placed in "spoil banks" until time for reclamation. In underground mining, the waste material is heaped at the surface, and the mine walls often remain exposed when the mine is abandoned. At the preparation plant, where impurities are removed from the coal, still more waste piles are generated. As long as these once-buried materials remain exposed and unstabilized, they are subject to runoff and erosion, and become the source of chemical and physical alteration of local streams. These impacts are addressed by a variety of State laws and, as discussed later, by Federal water pollution and surface mining laws.

Within the coal production waste piles are many soluble components that dissolve as rain water percolates through them. A consequence is an elevation of *total dissolved solids*

(TDS) within the local watershed. TDS is a parameter of water quality that very generally indicates the presence of soluble ions such as (in the case of coal waste runoff) sulfates, calcium, carbonates, and bicarbonates (hardness). Elevated concentrations of these ions make water largely unpotable by altering its taste but otherwise have only minor human health effects. TDS and hardness can also degrade the use of water for agricultural and industrial uses. Regions that will absorb substantial TDS as a result of expanded coal mining are the Middle Atlantic (sulfate and fluoride from underground mining), and the Midwest and North Central (sulfate from strip mining).

*Acid mine drainage* is a particularly severe byproduct of mining in those regions—Appalachia and parts of the interior mining region

(Indiana, Illinois, Western Kentucky)—where the coal seams are rich in pyrite (iron disulfide). The mining process exposes the pyrite to water and air, causing a reaction that forms sulfuric acid and iron. The lowered pH also tends to dissolve heavy metal compounds, adding metal ions (aluminum, manganese, zinc, nickel, etc.) to the acidic drainage water.

Acid drainage is a problem with active and inactive mines, and with both surface and underground mines. Recent data are not readily available, but in the 1960's underground mining accounted for fully three quarters of the problem (see table 34). About 10,000 miles of streams and rivers, principally in Appalachia, have been degraded by acid drainage. The effects are most severe in Pennsylvania and

*Photo credit: EPA—Documerica*

**Acid mine drainage from abandoned underground mines and other sources still affects thousands of miles of streams in Appalachia**

West Virginia, while less severe in southern Appalachia.[81]

**Table 34.—Acid Drainage in Mines**

| Source | Acidity 1,000 lb/day | Percent |
|---|---|---|
| Underground, active | 614 | 9 |
| Underground, inactive | 1,712 | 53 |
| Surface, active | 28 | 01 |
| Surface, inactive | 361 | 11 |
| Combined, active[a] | 60 | 02 |
| Combined, inactive[a] | 238 | 07 |
| Other | 245 | 07 |
| Total | 3,258 | 100 |

[a] Includes sources where underground could not be separated from surface.

The environmental effects of acid mine drainage are severe. The acid and heavy metals are directly toxic to aquatic life, and render the water unfit for domestic and industrial use. Zinc and nickel, often present in substantial quantities, are known carcinogens.[82] Other carcinogenic trace elements are present in smaller concentrations but may have additive effects.

Many of the metals found in acid mine drainage can become concentrated as they move up the food chain. Adsorption of these metals can occur on the surfaces of small suspended solids and sediments; thus, the largest reservoir of trace contaminants resides in the bottom sediment and is available to the rest of the ecosystem according to a complex web of factors. An additional ecosystem impact is the smothering of stream bottom-dwelling organisms by precipitated iron salts.

The acid content of the mine drainage also causes substantial material damage by eating away metal structures and destroying concrete.

A variety of measures exists to control acid mine drainage from active and inactive surface and underground mines. All involve preventing the formation of acid by controlling oxygen flow to the pyrites, preventing a dis-

charge by controlling water flow in the mine, or treating the discharge by neutralizing the acid (or by using reverse osmosis or ion exchange techniques, at much higher costs). Although neutralization treatment can increase hardness and dissolved solids content, it is the commonly used method for controlling acid mine drainage from active surface and underground mines and is considered an effective control. Surface mines also must utilize drainage control and special mining techniques to properly control acid drainage.

The only major problem facing successful control of acid damage from active surface mines is enforcement, which has been a long-standing problem with the smaller mines. This situation is essentially duplicated with the control of inactive surface mines: available reclamation techniques can control acid drainage from most inactive mines, but enforcement efforts on small mines must be strengthened to ensure adequate control. Also, a small percentage of mining situations still defy acid control attempts. For instance, a few mines in West Virginia have been found to produce acid seepage from horizontal water movement through buried spoils. However, strict enforcement by the new Office of Surface Mining and continued development of control techniques for the remaining problem situations should ensure an acceptable degree of control.

Unfortunately, available controls for inactive underground mines are not likely to be as successful. When an underground mine is being closed, the techniques for stopping acid drainage are:

• sealing the mine to prevent air infiltration;
• filling the mine with water;
• preventing water infiltration by grading, sealing, aquifer control, etc.;
• backfilling the mine; and
• removing mine pillars and allowing the mine to collapse.

Although these techniques can be effective in some circumstances, success in preventing a continuing significant flow of water has been intermittent; the technology available to control acid drainage from most inactive underground mines must be considered inadequate.

---

[81]"Appendix on Acid Mine Drainage and Subsidence," R. D. Hill and E. R. Bates, *Rall Report*, U.S. Environmental Protection Agency.

[82]Little, "Inorganic Chemical Pollution of Freshwater," U.S. Environmental Protection Agency, 1971.

An alternate approach to controlling drainage in new mines, both during active operations and after abandonment, is to mine "down dip"—i.e., in mining a sloping seam, to mine the coal from the top of the seam on down—rather than the usual "up dip." In this type of mining, water will accumulate at the "working face" during operations and must be pumped out. In a mine examined during an EPA study[83] acid levels in the drainage were extremely low. After deactivation, the mine can be flooded, shutting off oxidation of the pyrites, without the need for a watertight seal. The extent to which this technique is applicable has not been adequately demonstrated, but it is a promising approach to permanent acid drainage control.

As the shallower deep-minable coal reserves become depleted, underground mining will begin to shift to seams deep enough to be below the water table. Many of these mines can be flooded without the use of a watertight seal, allowing the seams to be shut off from air and preventing acid drainage problems after the mines are closed. Based on this expectation, and on the expectation that reclamation of abandoned mines (under provisions of the new surface mining legislation) will proceed as planned, projections of acid drainage to the year 2000 show a declining trend.

Regions with alkaline soils, such as the West, do not have problems of acid mine drainage but may have a problem with increased *salinity* caused by dissolution of soluble ions such as sulfates and chlorides. The impact depends on the elevation of TDS levels (for example, a 10- to 20-percent increase has little effect on aquatic biota) and on the uses to which the water is put. The impact of salinity in the northern Great Plains, where water is used for irrigating largely salt-resistant crops, may be less severe than in the Colorado River Basin, where water is needed for more salt-sensitive plants. In times of low water levels in the streams, TDS concentrations may rise.

The chemical pollution caused by coal preparation includes treatment water, often bearing organics used in the processing, as well as runoff from waste piles. The latter is similar in content but smaller in amount than the acid drainage from mines. Techniques are available to control these pollution sources. Federal standards for coal preparation plants require zero discharge of pollution, which should eliminate impacts from at least the larger plants. Enforcement of these standards on the smaller plants has been uneven in the past, and future performance clearly depends on State and EPA regional office resources.

The disturbances of land by mining results in physical as well as chemical pollution of streams. Erosion of unstabilized waste piles or of denuded land produces large quantities of *suspended solids* and subsequent heavy *siltation* of local streams. Erosion is greatest from surface mines in areas with rugged terrain and rainy climates, such as southern Appalachia. Research in Kentucky indicates that sediment yields from lands strip-mined for coal can be as much as 1,000 times that of undisturbed forests.[84] In more level and dry regions such as the Midwest or West, the problem is less serious. Sudden violent storms that occur in parts of the West, however, may wash large amounts of silt into local waterways.

One effect of the siltation is to alter the character of stream bottoms. It may, for example, interfere with the spawning of fish eggs. Siltation also causes material damage by shortening the life of reservoirs and increasing the need for channel dredging.

Coal mines are never the major cause of suspended solids pollution on a regional basis, but they have severe local impacts. At present, mining causes significant consequences in the Middle Atlantic (30,000 tons/yr, mostly underground mining) and Midwest (17,500 tons/yr, mostly surface mining) regions. Implementation of controls on both surface and underground mining will sharply reduce sediment loadings in most areas, and mining-caused suspended solids should de-

[83]John W. Mentz and Jamison B. Warg, "Up-Dip Versus Down-Dip Mining," EPA-670/2-75-047, June 1957.

[84]C. R. Collier, R. J. Pickering, and J. S. Musser, "Hydrologic Influence of Strip Mining," U.S. Geological Survey professional paper 427-C, 1970.

cline significantly in the future. For example, loadings in the Middle Atlantic are projected to decrease to one-third of 1975 levels by 1990. One notable exception to this trend may occur in the North-Central region, where sediment loadings may increase sharply by 1990 unless better controls are devised and implemented.

*Flooding potential* is affected by several factors related to surface mining:

• permeability of land surface,
• vegetation cover of surface, and
• carrying capacity of streams.

The mining and reclamation process may act to increase surface permeability and thus allow the ground to soak up more water in a given period than the premined surface could. The loss of vegetative cover that acted as a water retention mechanism will tend to counteract this effect. In addition, increased erosion and consequent siltation lead to decreased carrying capacity of streams and greater flooding potential. Studies of the effect of surface mining on flooding potential thus have tended to conflict, with one study of strip-mined watersheds in Kentucky showing increased flooding potential and studies in Indiana indicating the reverse.[85] An examination of these factors leads to the tentative conclusion that flooding potential may be decreased by area mines, while contour mines may cause increased flooding. Indeed, several studies now appear to support the association of surface mines with flooding in steep terrain. More studies of surface mine hydrology, soil mechanics, revegetation techniques, and the positive impacts of methods to stabilize benches are needed to settle this issue. However, the only practical control of this problem in some flood-prone areas may be a limit on surface mining.

A final group of impacts is the *alteration of the hydrology* of local water systems. Surface streams often are diverted around mines to lessen the degree of siltation or acid or alkaline drainage. Some surface water may have to be impounded for entrapment of silt or sludge

from treatment of acid mine drainage. More water also may be demanded for irrigation of revegetated surface mines, especially in the West where reclamation is more difficult. The quantities of water are comparatively small (see table 35) but may nevertheless become a public concern in arid regions where competition for water is becoming critical. However, the potential for water conservation in the West is great enough to offset any potential regional shortages if the legal and political barriers to conservation can be dealt with (see chapter IV, *Water Availability* ).

**Ground Water Impacts.**—Mining is likely to have significant effects on ground water as well as surface waters. Mining operations can both contaminate ground water and cause severe physical dislocations of aquifers.

A primary mechanism of ground water contamination is the seepage of acid mine drainage into aquifers via joints and fractures in the rock or via direct interception of the aquifer. Inadvertent connection of two aquifers also can lead to contamination; this is more of a problem in the West, where deep saline aquifers abound. Coal spoil banks also can be the source of this contamination through leaching and penetration of the soil. Investigations of ground water contamination typically have shown high local effects but extremely variable effects some distance from the mine.

Direct interception of aquifers and other physical disturbances can substantially affect the quantity of ground water as well as its quality. Where a mine is located below the water table, seepage of water into a mine pit can lower the water table and dry up wells in the vicinity. (Thus, the trend towards deeper mines, identified as providing a benefit in reducing acid drainage, can cause additional ground water problems.) Blasting may rupture the impermeable strata below an aquifer and cause water to seep to a lower level, or may inadvertently connect two aquifers. The process of mining and reclamation will change the permeability of the overlying soil, altering the rate of ground water recharge, and affecting flooding potentials (as discussed in the preceding section).

[85]"Appendix on Reclamation," Coal Extraction Land Reclamation Group," Argonne National Laboratory, *Rall Report.*

**Table 35.—Water Requirements for Coal System Activities**

| Coal activity | Size | Water consumed by energy facility | | Peak water requirements associated with population increases[a] | |
|---|---|---|---|---|---|
| | | | | Peak | |
| | | Acre-ft per year | Gallons/10⁶ Btu in product | Construction (acre-ft/yr) | Operation (acre/ft/yr) |
| Surface mine .............................................. | 12 MMtpy | 3,400 | 4 | 71 | 323 |
| Underground mine ...................................... | — | 0 | 0 | 275 | 1490 |
| Power generation ....................................... | 3,000 MWe | 29,000 | 54 (thermal) 157 (electric) | 853 | 260 |
| Lurgi gasification ...................................... | 250 MMscfd | 6,700 | 28 | 1570 | 350 |
| Synthane gasification ................................ | 250 MMscfd | 9,100 | 38 | 1570 | 350 |
| Synthol liquefaction .................................. | 100,000 Bbl/ day | 17,500 | 28 | 1750 | 1800 |
| Slurry pipeline .......................................... | 25 MMmtpy | 18,400 | 14 | Not considered | Not considered |

[a]Assumes 150 gallons per capita per day, a multiplier of 2 to account for added service personnel during construction, and a multiplier of 3.5 to account for families and service personnel during operation.

SOURCE: White et al., "Energy from the West", EPA-600/7-77-072a, July 1977.

## Transportation

Coal energy transportation systems—trucks, railroads, coal slurry pipelines, barges, and transmission lines—can directly or indirectly affect water quality. New clearing of rights-of-way and construction of facilities will increase sediment loadings. Where herbicides are used for clearing and maintenance, excess application can lead to both surface and ground water contamination. Both coal slurry pipelines and barge transportation on canals "consume" water, the former by exporting water (approximately 500 acre-feet per million tons of coal) out of arid regions, the latter by exposing more water surface to evaporation.

Transmission generally has insignificant direct effects on water quality. The major indirect effect is associated with powerplant siting. A mine-mouth plant with long-distance transmission lines may be sited in an arid region, where the impact on water availability is greater than if the coal were shipped to another site with abundant water.

## Coal Combustion

Coal combustion releases thermal as well as chemical effluents into water systems. The waste heat from a coal-fired boiler represents that percentage of the energy originally present in the coal (about 60 percent in a modern utility) that is not converted to electricity. This thermal energy is transferred to an adjacent river, lake, or cooling pond and to the local atmosphere, in proportions depending on the cooling system. The thermal pollution problem is not unique to coal-fired utilities but is shared to the same degree with oil and gas plants and to a greater degree by nuclear plants (which currently are less efficient than fossil-fuel-fired plants and thus produce more waste heat).

The large quantity of heat that emanates from a coal-fired utility boiler is bound to affect at least the local environment. If heat is released in an open cooling cycle (that is, water is withdrawn from a source such as a river, lake, or ocean, circulated within the plant's heat exchanger, and released at higher temperature, usually to the original source), many properties of the receiving waters are altered by the resulting temperature rise. Changes occur in such properties as salinity, vapor pressure, etc.; the amount of dissolved oxygen in the water will decrease owing to lowered solubility and re-aeration rates, while TDS will increase because of the increased solubility of many compounds with increasing temperature. These changes in turn alter the life functions of organisms within the water.

Higher temperatures also may elevate metabolic rates of organisms, thus influencing oxygen demand, total energy needs, foraging ability, reproduction, migration, and susceptibility to disease. Heated effluents accelerate the consumption of dissolved oxygen by organisms; when coupled with lower oxygen levels as a result of the physical mechanisms noted above, oxygen content can become critically low within the thermal plumes.

In addition to the thermal effect, physical impacts are caused by small organisms or even fish eggs and larva being drawn through the cooling system of the plant. This entrainment may damage these organisms and affect not only their own lifespans but those of large fish that feed on them.

These impacts can be mitigated by careful siting of powerplants and construction of cooling intakes and outflows. However, complete control can be obtained only with cooling systems that emit heat into the atmosphere or into a small artificial body of water. These are called "closed-cycle" cooling systems.

Current Federal regulations require that, with certain exceptions, closed-cycle cooling systems be used in large, steam electric-generating plants built after 1970 and certain smaller units built after 1974. Thus, closed-cycle systems will be essentially standard equipment on new powerplants, and open-cycle ("once through") systems will be used only in the limited cases where exemptions can be obtained (see chapter VII, section on *Present Federal Coal Policy*).

The two major water impacts of closed-cycle, or evaporative, cooling systems stem from water consumption and "blowdown" disposal. Although water consumption by evaporative cooling towers and cooling ponds depends on many site-specific and design factors, in general the evaporative systems associated with powerplants are the largest water consumers in the coal fuel cycle (see table 35). Their widespread use causes the issue of water use for energy to be a major concern nationwide, and especially in the arid Southwest (see *Water Availability*, chapter IV). However, the use of relatively water-"conserving" once-through (open-cycle) cooling, in those areas where a sufficient volume of water would be available, would still consume a large amount of water (50 to 60 percent of evaporative tower consumption) because of the evaporative cooling that takes place in the river or other body of water into which the heated water is discharged. This fact has not always been considered by critics of the current requirements for closed-cycle systems.

The environmental effect of massive consumption of water by powerplant cooling systems is twofold:

• The amount of water available to dilute downstream discharges is reduced, and water quality is thus degraded. This effect is especially evident in the Colorado River, where massive water consumption (by agriculture, in this case) coupled with substantial pollution loadings have elevated salt concentrations in the River to the point where desalinization plants are required to keep water quality within standards.

• In critical watersheds, the physical quantity of water remaining in the streams may fall below that needed to support the water ecosystem. The recognition that this "instream use" of water is a "beneficial use" within the legal system established for allocating water rights is a recent one, and thus protection of water ecosystems from destruction by sheer lack of water is a recent development.

The concentrating effect of evaporation increases the levels of mineral salts in the circulating water of an evaporative cooling system. This water must be continually bled from the system and replaced with freshwater to maintain system operation. This bleed water, which is extremely high in salt concentration, is called "blowdown." Although the majority of the "pollution" in the blowdown water is the same mix of salts as in the water source (with a small concentration of biocides to keep the tower clear of fungae and algae, and some dust scrubbed out of the air), this water can be damaging locally because of the high salt concentration levels, or when the water enters ground water or surface water other than the source. In some cases, the local effects of the biocides may be more damaging than those of the high salt concentration.

Photo credit: EPA—Documerica

**The enormous quantity of water that must be evaporated to cool a large generating facility
is graphically illustrated by these cooling tower "plumes", which overshadow the plume from the plant's stacks**

Uncertainty about the precise level of controls that will be required is considerable. Present EPA effluent guidelines forbid the discharge of biocides but allow the discharge of the salts originally present in the intake water. However, if the cooling process uses a significant portion of the water source, the blowdown could violate water quality standards by substantially increasing the overall salt concentration; in these cases, some level of treatment will be required. Also effluent standards for new powerplants in some parts of the West require totally closed-cycle operations, with blowdown being concentrated and either discharged to evaporation ponds or else undergoing further treatment and disposal as a sludge. Disposal problems are similar to those associated with disposal of fly ash and flue-gas desulfurization (FGD) sludge, except that trace elements should not be a problem.

The dissolved solids loading that electric utilities impose on the Nation's waterways as a result of blowdown (from both the cooling system and the boiler) is extremely large. Utilities produced nearly 6 million tons, or nearly 28 percent of the total national dissolved solids discharge in 1975 and may increase this to 10 million tons by 1990. The share of this burden contributed by coal-fired utilities was nearly 4 million tons in 1975 and will be around 6.6 million tons in 1990 with moderate coal growth. Increases in utility coal burning play important roles in the projected substantial growth in TDS by 1990 both nationally and in several regions: South Atlantic, Midwest, North-Central, and Central.

The large water requirements associated with operation of a coal-fired powerplant can be reduced substantially by the use of air-cooled or "dry" cooling systems, either by themselves or in combination with evaporative systems. In the combination or "wet/dry" system, the air-cooled tower may fulfill the total cooling requirements of the powerplant in winter and be supplemented by the wet tower when weather conditions lower the cooling capacity of the ambient air. Using either system, the cost of building and operating the powerplant increases substantially because of the large capital cost of the cooling systems, lessened efficiency of the plant due to a lower cooling capability, and the increased capital cost of turbines designed to operate in a higher pressure environment. In a strictly economic tradeoff, a decision to use these systems would be correct if the cost of water began to approach $700 to $1,000/acre-foot. In few circumstances would this tradeoff be made today, but these systems still might be utilized where water is unavailable for physical or political reasons.

In addition to the environmental impacts associated with their cooling systems, powerplants generate additional water pollutants in the form of a blowdown from the boiler and acid drainage from coal storage piles. However, these pollutant sources are relatively minor compared to their counterparts from the cooling and extraction portion, respectively, of the fuel cycle

## Wastes From Coal Combustion

Under current and future combustion control technologies, coal impurities are removed from the air—only to reappear in solid or slurried form. An air pollution problem is thereby transformed into a waste disposal problem. The concentrations of undesirable constituents within the wastes demand strict attention to disposal methods to prevent contamination of surface or ground waters.

The wastes generated by coal combustion facilities are the fly ash from particulate control devices, the bottom ash and slag from boilers, and the sludge from FGD systems. In the future these wastes may include spent sorbent from fluidized-bed combustion (FBC) units. By 1985, the quantities of ash and slag requiring disposal may total 80 million tons,

and that of sludge may be 19 million tons assuming moderate coal growth. These wastes represent approximately half of the Nation's total noncombustible solid waste and industrial sludges, respectively. By 1990, the quantities of coal combustion waste requiring disposal may grow to 90 million tons of ash and slag, and 35 million tons of scrubber sludge (NEF #1).

The major ash-producing areas are the Middle and South Atlantic and Midwestern States. By the end of the century, the Southwest is expected to join these regions as a major ash producer. With the exception of New England and the Northwest, the rest of the country should also experience substantial increases in ash production in the near future. Although sludge production is relatively minor now, by the end of the century the South Atlantic and Midwestern States should be major sludge producers, with the New York/New Jersey, Middle Atlantic, Southwest, and Central regions being major contributors.[86]

The potential disruption of water quality by these ash and FGD wastes depends largely on their chemical and physical structure. The ash is an alkaline material consisting primarily of insoluble inorganic elements (most of which are enriched in concentration compared to their natural abundance), as well as elevated levels of radionuclides. Up to 10 percent of the ash consists of water soluble species, principally calcium, magnesium, potassium, sulfate, and chloride. The presence of these materials creates a strong potential for leaching of ash components into surface and ground waters unless the ash is carefully disposed.[87]

The sludge from FGD systems varies with the scrubber design and operation but in general consists primarily of calcium sulfite or calcium sulfate with some carbonates, inert elements, and insoluble ions.[88] Calcium sulfite hemihydrate (the major species in the lime/limestone scrubbers) consists of small, thin crystals that settle very slowly, making the sludge difficult to dewater and contributing to its physical instability.

---

[86]"Appendix on Ash and Sludge," *Rall Report.*
[87]Ibid., p. 11.
[88]Ibid., p. 13.

**FGD sludge pond—Commonwealth Edison's Will County powerplant**

The concentration of trace elements in FGD sludge is much smaller than that in fly ash; those that are present tend to be elements that escaped the ESP in gaseous form or that came from the sorbent used. Often fly ash and FGD sludge are combined for disposal, and the mixture may be referred to as flue-gas cleaning (FGC) wastes. The ash helps increase the solids content of the sludge but also contributes a substantial load of trace elements to the mixture.

Whether these undesirable elements in FGC wastes enter surface or ground water depends on the disposal method used. The disposal methods traditionally used for ash and currently being considered for FGD are impoundments (ponds), landfill, and utilization, in decreasing order of common use for ash. Open-cycle ponds allow direct discharge of

ash transport water to receiving streams. Some of the solids and trace metals that are not in solution in the alkaline ash medium may dissolve as the pH is lowered in the receiving stream. If the ash transport water pH exceeds 9, it may have to be treated with sulfuric acid, causing the formation of a precipitate (floc) that can affect the receiving stream.

In addition to direct discharge to surface waters, ponded wastes may leach to ground water. In the case of sludge, the leaching comes both from the fluid (liquor) initially present and from the dissolution of sludge solids once the fluid is gone. The major soluble species in most FGC (combined) wastes are chlorides and sulfates (primarily in calcium, sodium, and magnesium salts).[89] These species

---

[89]Ibid., p. 11.

have been found in high levels in FGC leach-ates and would increase the TDS of ground and surface waters if leaching or surface discharges occurred. In addition, trace elements present in the ash may be leached out and added to the potential groundwater contamination.

The characteristics of FGC discharges that are most important from an environmental impact standpoint are the:

- presence of small particles,
- high trace element concentrations, and
- reducing capacity of sulfite-rich sludge.[90]

Trace element contamination of surface and ground water can lead to ingestion by aquatic organisms and contamination of soils (and, in turn, crops). Iron oxide floc and the FGC sludge itself can coat the bottom of streams and disrupt bottom communities. The oxygen demand of sulfite-rich sludge can cause serious stresses on aquatic life by depleting oxygen supplies. These and other impacts, coupled with the very substantial increases expected in FGC waste generation, focus attention on the importance of appropriate specification of required controls for waste disposal.

Virtually all of the above pollution impacts of waste disposal are controllable with available techniques, as discussed below, and both surface discharge and leaching to ground water fall under existing regulations (discussed in the following section). Although leaching of salts and trace elements into ground water generally is considered a major problem with FGC disposal, the actual impact is extremely dependent on site-specific factors. In some cases, it could be quite small because of the attenuation of pollutants as they pass through soil layers. Unfortunately, gaps in our knowledge of the pollutants present, the attenuative nature of different soils, and the actual physical characteristics of ground water systems complicate the specification of potential environmental effects.

As noted previously, the options for coal combustion waste disposal include impound-ment, landfill, and utilization. Considerable experience exists with the disposal of ash by all of these methods. Ponding has been the most common because it is the easiest and cheapest method; ash is sluiced to a lagoon near the plant, and the water is either returned or discharged. The imminent restrictions on ash transport water discharge and the declining availability of land near powerplants may discourage future use of ponds (at least open-cycle ponds) for ash disposal.

Ponds are less satisfactory for disposal of FGD sludge (either alone or mixed with ash) than they are for ash. The use of a pond for sludge disposal represents a long-term commitment of the land; ponding of untreated sludge (and perhaps even some treated sludge) does not result in the degree of settling that would be required for future development on the site.

The problems of leaching associated with ponding of wastes mandate attention to possible lining of the pond, selection of a pond site with highly impermeable soil, and treatment of the sludge itself to reduce its porosity and permeability. In addition, continual removal of the surface layer of water that develops in the pond from rainwater or other runoff is necessary.

These problems with ponding appear to be turning more plant operators in the direction of landfill for FGC wastes,[91] despite its higher cost. For landfill purposes the ash is collected dry and the sludge is dewatered and/or treated to improve its structural properties. In this form the wastes are more easily transportable than the wet sludge, allowing the landfill site to be further from the powerplant, perhaps in a less valuable location. However, the cost of transporting the dried and treated sludge is high, and onsite disposal will be favored where it is possible. Land impacts are decreased with landfill because of the smaller area and shorter retention time required before the land can be developed. As a tradeoff, the air pollution from fugitive dust (especially for a dry ash landfill in an arid climate) is greater during

[90]Ibid.

[91]J. Jones, "Disposal of Power Plant Wastes," presented at the Third National Interagency Energy/Environmental R&D Conference, Washington, D.C., June 1978.

transport and disposal. This effect is highly local and may be mitigated by use of a wetting agent or by a buffer zone around the site.

Before sludge may be used for landfill it must undergo some physical or chemical treatment to improve its structural stability and reduce its volume, permeability, and porosity. One method is to alter the operation of the scrubber itself to force the oxidation of sulfite to gypsum. Alternatively, one might exploit the finding that some modes of scrubber operation seem to produce forms of calcium sulfite that are easier to dewater.[92]

The FGC wastes may be dewatered by several physical techniques such as vacuum filtration, centrifugation, or gravity settling in order to increase the percentage of solids (65 to 70 percent is desirable). Adding ash to the sludge also increases the solids content and lends structural strength.

Several commercial chemical sludge treatment processes have been developed and are being used by coal-fired plants. Field tests of chemically treated wastes demonstrate significant structural improvement and a reduction in permeability, as well as a reduction in the concentration of dissolved solids in the leachate and a decrease in the actual volume of leachate.[93] An additional benefit of sludge treatment is the improvement in handling characteristics, which allows better sludge placement and thus additional control of leaching.

Another disposal method for FGC wastes is burial with the mine wastes. This technique is felt to be too expensive for underground mines but may be feasible for surface mines. The concept is made more appealing by the coal transportation link that connects a mine with a plant. Alternatively, the plant may be a mine-mouth plant with close ties, both geographically and economically, to the mine. The mine may welcome the wastes as additional fill to restore the original land contour and, in the East, as alkaline material to help neutralize some of the acid wastes. It is not known

whether the economics and institutional links between mine and plant are favorable enough on a large scale to promote widespread use of this disposal method.

Another disposal method for FGC wastes is to dump them in the ocean. The idea is appealing for northeastern utilities that are close to the coast but have little space to store wastes. The concept is further encouraged by the fact that the major chemical constitutents in FGC wastes are already in relatively high concentrations in seawater. Promising forms of disposal are concentration of treated FGC wastes in bricklike form or dispersion of treated or sulfate-rich FGC wastes in either deep ocean or on the Outer Continental Shelf.[94] The environmental impacts of ocean disposal are being investigated by EPA.

The utilization of wastes is more appealing from a resource conservation standpoint than is disposal, but it is not always as economically attractive. About 20 percent of all ash is now used for some purpose, and the National Ash Association is actively pursuing new uses for ash. The most promising use is in the manufacture of portland cement. Approximately one-third of the 1980 fly ash production could be used for this single application.[95] Other uses are as a light aggregate for construction, as a filler for asphalt, as an abrasive, and as a soil additive. In addition, extraction of valuable minerals from the ash may be economically feasible.

The potential uses for sludge are not as numerous. In Japan, scrubbers are operated so as to produce gypsum, which is then used for wallboard and portland cement. However, the economics are not favorable enough for FGC wastes to penetrate the market in this country.[96]

An alternative to sludge disposal is to avoid sludge generation altogether. Several "regenerable" systems have been developed for FGD. These typically produce elemental sulfur or sulfuric acid. Some wastes are emitted as a

[92]Ibid., p. 12.
[93]"Appendix on Ash and Sludge," *Rall Report.*
[94]J. Jones, "Disposal of Flue-Gas Cleaning Waste," *Chemical Engineering,* Feb. 15, 1977.
[95]Jones, *R&D Conf.,* p. 1
[96]Ibid., p. 2.

bleed stream, but the volume is far less than wastes from nonregenerable systems. Sulfur is unfortunately not a scarce element but may become valuable when present sources are depleted.

Only general statements can be made about the waste disposal problems associated with the spent sorbent from FBC, because no large-scale demonstration yet exists. The amount of sorbent waste generated will depend on the chemical balance needed to assure adequate removal of sulfur in the coal, but indications now are that the volume will be greater than that of sludge for a plant of comparable size burning the same coal. By contrast with sludge, however, the sorbent has a much higher solids content and greater structural stability. Among the possible uses being investigated is as an agricultural fertilizer.

A key to the actual waste disposal alternative selected will be the application of the Resource Conservation and Recovery Act (RCRA) described in the following section. Designation of sludge as a "hazardous" material under RCRA will result in extensive use of liners, chemical fixation, and landfill disposal of scrubber sludge. DOE has estimated that such a designation could increase sludge disposal costs by about 20 percent over costs incurred by a "nonhazardous" designation[97] to a total of about $600 million per year by 1990. Similarly, designation of ash as hazardous material is estimated to yield a 30-percent increase over the costs incurred with a nonhazardous designation, to a total of about $1 billion by 1990 for utilities. As even a nonhazardous designation will require an upgrading from present practice, the total cost of disposal for sludge and ash is estimated to increase up to 45 percent over present practice for a nonhazardous designation, and up to 84 percent if both are designated hazardous. The incremental cost of waste disposal on electric generation was estimated to range from 1.22 to 2.01 mils/kWh under RCRA; in comparison, continuation of current practice would cost approximately 1 mil/kWh (NEF#1). All of these cost estimates have been challenged by the utility industry as

[97]National Environmental Forecast No. 1.

being severe underestimates of the costs that will have to be borne under RCRA.

These costs must be balanced against the protection of ground waters and reduction in land use impacts provided by the increased controls. An unfortunate side effect of a hazardous designation for ash might be the elimination of the ash byproduct market, but this is clearly dependent upon the precise nature of EPA's regulatory stance.

## Regulations

Impacts to water systems are regulated under the Clean Water Act, the Surface Mining Control and Reclamation Act (SMCRA), and RCRA. These are discussed in detail in chapter VII and summarized below.

The Clean Water Act sets effluent limitations for chemical and physical discharges from steam electric-generating plants, coal preparation plants, and coal mines. For generating units, limitations have been established for discharges of: total suspended solids and oil and grease from low-volume waste sources, ash transport water, metal-cleaning wastes, and boiler blowdown; copper and iron from metal-cleaning wastes and boiler blowdown; free available chlorine from once-through cooling water and cooling tower blowdown; and materials added for corrosion inhibition, including zinc, chromium, and phosphorus, from cooling tower blowdown. In addition, discharges of total suspended solids from material storage and construction runoff are regulated.

For coal mines and coal preparation plants, limitations have been established for discharges of iron and total suspended solids from acidic and alkaline mine drainage, and for manganese from alkaline drainage. In addition, the pH of all discharges from generating units, mines, and preparation plants must be in the range of 6.0 to 9.0.

The Clean Water Act also limits thermal discharges from steam electric-generating plants. Standards of performance for new units provide that there shall be no discharge of heat from the main condensers; existing units (with some exceptions) must meet this limit by 1981. In addition, the Act requires that the location, design, construction, and capacity of cooling water intake structures reflect the best technology available for minimizing adverse environmental impacts such as impingement and entrainment of organisms.

Effluent sources must obtain a permit under the National Pollutant Discharge Elimination System (NPDES), which includes all applicable effluent limitations. Variances are available only for sources that can demonstrate that factors related to the equipment or facilities involved, the process applied, or nonwater quality impacts (such as economic factors) are fundamentally different from the factors considered by EPA in establishing the effluent limitations.

For the most part, the effluent limitations established under the Clean Water Act are intended to achieve a water quality goal that provides for the protection and propagation of fish, shellfish, and wildlife and for recreation in and on the water by 1983. Stricter limitations designed to eliminate all pollutant discharges will not be enforced until at least the late 1980's.

No Federal regulations deal with water consumption by coal combustion facilities. Water rights traditionally have been regulated by the States, and the means of allocation vary widely by region. The 1977 amendments to the Clean Water Act reaffirmed this system by stating that the Act in no way impairs the authority of each State to allocate quantities of water within its jurisdiction. However, the Act does require that water resource management be considered in State water quality planning.

SMCRA sets performance standards for surface mines and for the surface operations of underground mines that are designed to prevent acid and other toxic drainage as well as runoff containing suspended solids, and to prevent disruptions in local water supplies. Mandated practices to control surface mine-related water pollution include stabilizing disturbed areas, diverting runoff, regulating the channel velocity of water, lining drainage channels, mulching, selectively placing and

sealing acid-forming and toxic-forming materials, selectively placing waste materials in backfill areas, achieving rapid revegetation and, where necessary, maintaining water treatment facilities. In addition, all surface drainage from disturbed areas must be passed through at least one sedimentation pond.

Additional requirements for underground mines include designing mines to prevent gravity drainage of acid waters and sealing and controlling subsidence. Similar provisions apply to support facilities such as coal loading and storage facilities, roads and other transportation facilities, and mine buildings.

SMCRA also requires mine operators to minimize hydrologic disturbances. Overland flows and stream channels may be diverted when necessary to comply with other performance standards in the Act and when the diversion meets specified engineering standards. In addition, mine reclamation must restore the premining capability of the area to transmit water to the ground water system. The mine operator also must replace water supplies used for domestic, agricultural, industrial, or other uses when they become contaminated, diminished, or interrupted as a result of mining activity.

In order to obtain a mining permit, the mine operator must demonstrate that these requirements will be met and that they will be adequate to comply with all applicable Federal and State water quality standards and effluent limitations.

The disposal of combustion byproducts is regulated under the Clean Water Act and RCRA. Where the wastes are discharged to receiving waters, they come under the Clean Water Act; an example, as discussed above, is ash transport water. Land disposal of the wastes is regulated under RCRA, which seeks to control open dumping of wastes under a system of State plans and permits for waste disposal. The Act distinguishes hazardous and nonhazardous wastes and places more stringent restrictions on the former. If either ash or scrubber sludge is determined to be hazardous, the disposal options for these substances are constrained. For example, sludge ponds would have to be constructed in a manner that would prevent any leaching to the ground water. Similarly, burial of wastes or their use in landfills or in roadbeds would be prohibited where there is any possibility of leaching. In addition, disposal of hazardous wastes would be prohibited in wetlands or floodplains (where most large combustion facilities such as powerplants are located), adding substantial transportation costs to the waste disposal costs.

# IMPACTS TO THE LAND

The impacts from air pollution provide an interesting contrast to the direct impacts to the land caused by the use of coal. The air emissions from the coal fuel cycle stem primarily from coal combustion. Many of the impacts are subtle or are masked by other factors; others take a long time to become manifest.

By contrast, the land impacts of the use of coal stem primarily from the mining and waste disposal portions of the fuel cycle. These impacts are more site specific and often cause acute and drastic alterations of a local area. The land impacts of mine activities are more likely to damage plantlife, animals, or human beings by mechanical harm (such as that resulting from removal of habitat or from landslides and floods) than by chemically induced damage. Some land disruptions, such as blasting damage, may be short term. Others, such as subsidence, may linger for many years.

A variety of impacts to other media result indirectly from perturbation of the land by coal extraction and waste disposal. These include air pollution from coal mine fires and fugitive dust or water pollution from waste pile runoff. Although the specific details are treated in the sections concerned with the air and water media, the interaction of the three media should not be forgotten.

## Sources, Effects, and Control of Land Impacts

### Surface Mines

Although estimated future coal production levels, and therefore land use impacts, vary considerably with assumptions about environmental policy and economic conditions, most studies show a very substantial increase in sur-

face mining in the northern Great Plains and, to a lesser extent, in the Southwest. Surface production in the Midwest fluctuates widely in the long term with varying assumptions, although it is unlikely to increase much by 1985.

As shown in table 36, Western surface mining disturbs considerably less land per ton of coal extracted than do Eastern and Central surface mining. This occurs because many Western coal seams are over 50 feet thick, allowing extraction of more tons per mined acre.

The most obvious land impact of a surface mine, while it remains active, is the complete removal of the land from its normal uses. The

### Table 36.—Land Affected by Coal Utilization

| Facility | Acres over 30-year period |
|---|---|
| Mining: $10^{15}$ Btu's per year[a] | |
| Western strip | 15,000–96,000 |
| Central strip | 216,000 |
| Central room and pillar[b] | 525,000 |
| Eastern contour | 470,000 |
| Eastern room and pillar[b] | 560,000 |
| Gob and refuse disposal[c] | 15,000 |
| Combustion/Conversion:[c] Input of $10^{15}$ Btu's/year | |
| Powerplants | 13,600 |
| Lurgi gasification | 6,500 |
| Synthoil liquefaction | 6,300 |

[a] For a high heat value coal (e.g., Eastern coal with 12,000 Btu's/lb) this is equivalent to approximately 42 million tons per year. For a low heat value coal (e.g., Western coal with 8,500 Btu's/lb) this is equivalent to approximately 59 million tons per year. For a medium heat value central coal of 10,000 Btu's/lb, this is equivalent to 50 million tons pr year.

[b] Includes undermined land which is potentially subject to subsidence.

[c] A 1,000 MWe (megawatt-electric) powerplant would "consume" approximately 59 × $10^{12}$ Btu's/year; thus it would take nearly 17 such plants to consume $10^{15}$ Btu's/year. Similarly to use $10^{15}$ Btu's of coal per year would require about 8 Lurgi gasification plants of 250 million cubic feet per day and about 3 Synthoil liquefaction plants each producing 100,000 barrels per day.

SOURCE: Calculated from White et al., "Energy from the West", EPA—600/7-77-072a, July 1977.

surface covering, which may have been agricultural crops, rangeland, forests, or desert, is replaced by the sights of excavation and the sounds of blasting. Wildlife, deprived of food and cover, must migrate if possible to other areas. The unsightliness may degrade both property and scenic values, impacting a distance beyond the mine equal to the extent of the vista. Thus the surface mining of one mountain can alter the recreational quality of a large geographical area.

Once the coal resource has been extracted from a given area, the land generally can be recontoured and revegetated so that it can support former uses. There is sufficient reclamation experience with most of the coal-producing areas to suggest that successful reclamation will occur with the appropriate commitment of resources. Doubts about the success of reclamation procedures, however, have been fostered by the poor attention given to reclamation in the past. Some 420,000 acres of mine lands are unreclaimed. Although more stringent regulations and keener public awareness should produce a better record in the future, the past can serve as a reminder of the importance of constant monitoring. Doubts about reclamation also stem from lack of experience in reclaiming some of the more ecologically fragile regions to be mined in the future. In particular, experience with revegetating strip mines in the arid or semiarid regions of the West is not extensive. A key issue here is whether apparently successful reclamation efforts may degrade over time, especially in the face of periodic droughts.

The concern over reclamation centers on lands that play a unique vital role in a region, and many of these critical lands are protected by the 1977 SMCRA. Among these are the alluvial valleys, which often form the backbone of the ranching and farming economy in semiarid regions. Even though only 3 percent of the coal leases in an eight-State area of the West underlie such valleys, the shallow overburden there makes these lands most attractive for mining.[98] Some fear that even strip mining in the lands adjacent to alluvial valleys, where 10 percent of the strippable coal exists, may have

hydrological impacts that would affect the agricultural productivity of the valley floors. As discussed below, SMCRA strictly regulates mining on alluvial valley floors, but interpretation and enforcement of the law's provisions is critical to the actual degree of protection afforded these lands.

Another type of land where surface mining might threaten the resumption of a vital function is the agricultural farmland of the Midwest. Before passage of SMCRA, new strip mining would have encroached on prime agricultural lands. For example, in 1976, about 75 percent of the surface mine permits issued in Illinois were for lands classified as prime agricultural lands by the Soil Conservation Service.[99] The main concern surrounding agricultural lands is that the ecology of the soil system is too little understood for the success of reclamation to be guaranteed. Only recently have researchers begun to treat soil as an ecosystem rather than as a mineral.[100] It is not known whether the soil productivity deteriorates as the soil layers are removed, separated, piled, and stored.

Some concern also centers on stands of hardwood timber in southern Appalachia that require decades or even centuries of natural forest succession to become re-established. The Ponderosa pine forests in the West also are believed to be quite difficult to re-establish. A final concern is that some areas (southern Appalachia in particular) have such steep slopes that reclamation is unfeasible. The rapid erosion of soil prevents successful rooting of vegetation.

Surface mine reclamation usually proceeds in parallel with excavation; the overburden and topsoil from the active area are placed in the area of the previous cut. The overburden is backfilled, graded, and compacted. The topsoil is removed and replaced as a separate layer. The land is then replanted with some species of vegetation.

Besides merely replacing topsoil and overburden separately, these materials must be

---

[98]"Appendix on Reclamation," *Rall Report,* p. 27

[99]Department of Mines and Minerals, "1976 Annual Report—Surface Mine Land Conservation and Reclamation," Springfield, Ill., 1976.

[100]"Appendix on Reclamation," *Rall Report,* p. 35

Dragline operating at the Kayenta Mine of the Peabody Coal Co. on the Navajo Reservation, Black Mesa, Ariz.

replaced properly: toxic or acid-producing layers in the overburden must be identified and covered by sufficient depth of acceptable material to prevent contaminants from entering the aquatic environment; layers must be ordered to prevent fine topsoil particles from being swallowed by coarse particles. Finally, a thick layer of subsoil must underlie the topsoil. All plans for soil layer replacement must be based on thorough analysis of the overburden in the area.

In prime agricultural lands, the permanent provisions of SMCRA require the segregation of topsoil horizons so that the root zone can be reestablished. Before a permit is granted for a new surface mine in such areas, the operator must demonstrate that the original soil productivity will be restored. Some operators currently mining prime farmlands contend that they have indeed done this, but the regulations require proof in the form of scientifically valid experiments.

The provision to return land to its approximate original contour will help minimize any permanent disruption by surface mining in most locations. However, universal application of this requirement may be inappropriate. In the steep terrain of Appalachia, reclamation of slopes greater than 30° may be hindered if the land must be returned to its original slope, as the long, uninterrupted slopes will promote erosion. (The law does, however, allow the use of terracing and other erosion control measures to alleviate this problem.) In other areas, local governments and residents might prefer alternative configurations to allow new uses (or enhance old uses) of the land. For instance, land contour shaping can be used to trap water and allow growth of important wildlife cover in portions of the arid West.

## Underground Mines

Although less conspicuous than surface mines, deep mines nevertheless can have severe impacts on the land. Aside from their impacts on hydrology, underground mines have a potential for subsidence. This problem occurs when the support of the mine roof either shifts or collapses. While subsidence can occur during the active operation of a mine, it is more likely to be delayed for many years as the mine pillars slowly erode and collapse. The altered ground slopes can damage roads, water and gas lines, and buildings. The subsidence may further change natural drainage patterns and river flows, intercept aquifers and existing springs, or create new springs and seeps.

The extent, severity, and timing of the subsidence are complex functions of soil composition, overburden thickness, and mining method, to name only a few relevant factors. The difficulty in predicting and preventing subsidence complicates land use planning greatly. Just the potential for subsidence may reduce the land available for construction of homes and businesses. Even farming of the land may be discouraged by possible future unevenness of the surface.

Unfortunately, reliable surveys of subsidence do not exist, although a sizeable percent of the 8 million acres undermined to date have experienced some degree of subsidence, and more of this acreage may yet subside in years to come. Future subsidence impacts will depend heavily on the success of preplanning and control measures, and upon the degree to which underground mines will contribute to the future U.S. coal supply. A Bureau of Mines report estimates that a potential 1.5 million acres would be affected by subsidence by 2000.[101] A study of the environmental impact of the National Energy Plan assumed that 35 percent of coal production would come from underground mines by 2000. The study predicted that, in that year, underground mines would account for three-fifths of the land disturbance by total coal production. Most of the impact would be in the Middle and South Atlantic regions, where underground mines could affect up to a few percent of the total land in those regions.[102]

The amount of land potentially disturbed by underground mines is greater than that for surface mines of the same production capacity, as illustrated in table 36. This table compares the land affected by 30 years of various types of coal activity that produce or consume 1 Quad of energy. This comparison between the effects of surface and deep mines must be made cautiously. Surface mining disturbs all the land, but for a limited time. For deep mines, one can cite only the maximum amount of land that could be disturbed, with the actual acreage, duration, and degree of disruption subject to greater uncertainty.

Subsidence problems can be addressed by both preventive and corrective measures. The common preventive measure for subsidence is to leave a considerable portion of the coal itself—sometimes as much as half—in place as a roof support.

Another strategy is to allow subsidence to occur but in a controlled manner. In this scheme, called longwall mining (see chapter III) each section of the mine is collapsed after all the coal is extracted. This allows surface subsidence to occur sooner than with room and pillar mining, and the subsidence also occurs more evenly with major disruptions only at the perimeter of the mined area. Thus, future plans for the land may be made with greater certainty. Longwall mining is used extensively in Europe, but there has been less experience in the United States.

One of the few corrective measures that can be applied to existing mines is to backfill them with mine waste or other materials (including FGD sludge). At first glance this suggestion seems to solve the waste disposal and subsidence problems at the same time. More extensive studies, however, reveal that this method is quite expensive, involves considerable hazard to workers and may release contami-

[101]"Appendix on Acid Mine Drainage and Subsidence," *Rall Report*, p. 37.

[102]*Annual Environmental Analysis Report, Volume I Technical Summary* (Washington D.C.: U.S. Energy Research and Development Administration, Office Of Environment and Safety, September 1977).

nants into the ground water. In addition, some mine wastes and untreated sludge lack the structural strength to provide the needed support for the mine roof. An additional technique that is quite feasible for some of the shallower mines is "daylighting;" it involves surface mining to obtain the remaining coal and reclaiming the entire area. This technique is a cure for acid mine drainage as well as subsidence in abandoned mines.

## Regional Factors

The regional variation of the extent of land disturbed is largely a function of the thickness of the predominant coal seams. Western coal, although low in heating value (and thus at a certain disadvantage when comparing impacts on a unit energy basis), affects strikingly less acreage than either Eastern or Central coal because of the extreme thickness of the coal seams (seams greater than 50 feet in thickness are not uncommon). On the other hand, many Appalachian seams are only a few feet thick and thus require the disturbance of considerably greater acreage to extract the same coal energy. In fact, virtually all of the effects of mining vary greatly with geographical factors, so it is instructive to give an overview of the pertinent characteristics of each general province.

The major coal-producing area in the East—Appalachia—can be divided into two regions, based on differing concerns over mining. In northern Appalachia, stretching upwards from the northern coalfields of West Virginia, the topography is not rugged, and the soils contain substantial acid-forming materials. The region is densely settled, and the land is dotted with small farms and forested primarily with softwoods. By comparison, southern Appalachia has more steep, rugged terrain and smaller concentrations of acid-forming strata. The population is less dense and the forests are mainly hardwoods.

These characteristics make the impacts of acid drainage more severe in the northern region and necessitate careful attention to proper management of toxic spoil banks there. In addition, the subsidence problem is of greater impact in northern Appalachia, for more lands

are subject to residential or agricultural development. A considerable fraction of the subsidence problem in Pennsylvania has resulted from the underground mines in the urban areas of the anthracite region.

The problems of surface mining are more severe in southern Appalachia because of steep slopes and heavy rainfalls. The stands of hardwood forests require much longer times to reestablish than do the softwood forests of the northern region.

Surface mining in the Illinois Basin will be limited in those portions that are designated as prime farmland, as discussed earlier. In other regions, reclamation is facilitated by relatively flat topography and favorable rainfall. The impact of subsidence from underground mining in this region is relatively unknown, but subsidence may pose serious problems for agricultural lands.

Over the next several decades, surface mining is expected to dominate coal extraction in the northern Great Plains. The key problem in reclamation will be the scarce water supply and rainfall, coupled with the lack of experience with reclamation in this region. The alluvial valleys in these regions are protected by the SMCRA, as discussed earlier.

Surface mining also will be the dominant mining method in the gulf coast, where lignite reserves are concentrated in Texas. Except in the areas south of the Colorado River, rainfall appears adequate and potential land impacts minimal.

In the Rocky Mountains, most mines in western Wyoming, western Colorado, and Utah will be underground. The low population density and infrequent farms should minimize economic damage from subsidence. By contrast, the subbituminous coal deposits in New Mexico and Arizona will most likely be surface mined. The limited rainfall could hamper reclamation efforts, but some mines have been successfully revegetated when irrigated. Therefore, the surface mine impact there is uncertain.

## Mine Wastes

Both surface and underground mines generate large quantities of wastes that impact both water and land. The waste is of three types:

1. the solid waste from underground mines, which commonly is called "gob" and corresponds in composition to the overburden removed in surface mining;
2. the refuse from coal washing and preparation, which consists of coal waste and other impurities; and
3. the sludge resulting from treatment of acid mine drainage.

The total amount of solid waste generated by coal mines in 1975 was estimated to be almost 50 million tons.

Although the percentage of coal that is washed has decreased from about 65 to 41 percent between 1965 and 1975, the fraction of coal that is discarded has increased from about 20 percent to around 29 percent. One reason for the larger quantity of waste from coal cleaning is the greater use of automated equipment that digs less selectively than the hand-mining methods of the past. A second reason is that most coal is now being extracted from seams that are less rich and hence contain more impurities. More than 3 billion tons of waste lie in 3,000 to 5,000 refuse banks in eastern coalfields.

One of the most obvious impacts of these unsightly wastes is the degradation of local property values and destruction of esthetics. The land used for waste disposal, unless reclaimed, is no longer available for other uses. Furthermore, the gob, refuse, and sludge create dangers of many kinds if effective disposal techniques are not applied to them. First, the refuse piles contain flammable material that makes them susceptible to spontaneous combustion. They are also vulnerable to landslides and erosion because they usually are comprised of mixed, poorly graded, and uncompacted material. These wastes have been used to construct dams near the minesite for impoundment of water or slurry, a use to which they are poorly suited. The danger of this practice was dramatized in 1971 when such an impoundment, in Buffalo Creek, W. Va., broke

during a heavy rainstorm and flooded the valley below, causing more than 125 deaths and millions of dollars in property damage. A 1974 study of dams in the East, only 40 percent completed, found 30 similar water impoundments that were classified as imminent flood hazards and 176 additional structures that were classified as potential flood hazards.[103]

A major concern about coal gob and refuse piles — especially in the eastern regions — is their content of acid-producing materials. Exposure of these layers to air and water causes the leaching of coal impurities into surface and ground waters. Erosion of these wastes also contributes sediment and dissolved solids to local waterways, as was discussed in the section on water impacts.

## Transportation and Transmission

The key factors determining the land requirements for the various coal energy transportation modes are the right-of-way requirements, disturbances outside the right-of-way, the directness of the route, barriers created, other uses of the right-of-way, and other uses of the mode. Transmission and waterways generally require the widest paths, typically about 200 feet. Waterways may also produce flooding for a much wider distance, or require extensive areas for disposing of dredged material. Slurry pipelines may disturb large areas during construction. Rails, waterways, and highways may act as barriers to people or wildlife. These three modes can be useful for other purposes, while slurry and transmission lines are dedicated to one product. The rights-of-way of the latter, as well as those of waterways, however, may have multiple uses such as pathways or farming, which may provide a land benefit. In the case of very high-voltage transmission lines, multiple use may be compromised because of the possibility of electrical effects. The existence of these effects remains a controversial issue.

---

[103]*The National Strip Mine Study—Vol. 1, Summary Report*, (U.S. Department of the Army, Corp of Engineers, Washington, D.C.: 1974).

## Coal Combustion

As noted earlier, both the ash and the scrubber sludge produced by utility and industrial boilers represent a very sizable disposal problem. With particulate collectors on most large coal-fired boilers, 92 percent of all fly ash generated in the United States is now collected. This collection efficiency should increase with time because Federal standards now require greater than 99-percent collection for moderate and large boilers. Also, provisions for requiring technological controls for $SO_x$ will yield increasingly large amounts of scrubber sludge requiring disposal. As noted in the previous section, the total solid waste—fly ash, bottom ash, and slag—generated in 1985 is projected to be approximately 80 million tons per year, while scrubber sludge is expected to be about 19 million tons assuming moderate coal growth.[104] Although this weight of sludge is only about one-quarter of that projected for ash wastes, the two materials may necessitate comparable storage volumes because of the high water content of the sludge. For example, the sludge produced by a "typical" 1,000-MW powerplant (3.5 percent sulfur, 90 percent $SO_2$ removal, 12 percent ash, 6,400 hrs/yr, 20 percent excess lime, 50 percent oxidation, 0.88 lb coal/kWh) requires 630 acre-feet of pond volume if it is 40-percent solids and 270 acre-feet if the sludge is dewatered to 70-percent solids. The ash produced in 1 year would occupy about 140 acre-feet. Over a 30-year lifetime, the plant would require about 600 acres for waste disposal (assuming the ash and dewatered sludge are mixed and placed in a 20 feet thick storage area). By the year 2000 or so, the cumulative storage area devoted to these wastes could approach 100,000 acres, or well over 100 square miles (based on the same assumptions).

Although the land use requirements for sludge disposal are similar to those for ash disposal, the actual land impact attributed to the sludge could be considerably greater than that of the ash if ponding remains a predominant sludge storage mechanism; in that case, the land utilized will be unavailable for devel-

opment for a considerable period. The choice of disposal technique, if not dictated by RCRA, should depend largely on the value of land. Of the major sludge-producing regions, the more densely populated ones—the Midwestern, New York/New Jersey, and Middle Atlantic regions—might be expected to lean heavily toward landfill disposal. As the major problems associated with obtaining suitable land disposal sites may fall on industrial coal users rather than on utilities—because the industrial plants may be expected to locate closer to urban centers—industrial users may also be expected to lean more heavily toward landfills. However, other factors, such as the importance and vulnerability of ground water and the State and local regulatory climate, should play a major role in disposal decisions.

## Regulations

The land impacts of coal production and use are regulated primarily under SMCRA and RCRA, which are directed at mining and waste disposal practices that tend to cause land impacts. In addition, a variety of provisions cover the potential land impacts of transportation and transmission. These laws are summarized below and discussed in detail in chapter VII.

The primary purposes of SMCRA are:

• to ensure that surface coal mining operations are conducted in a manner that protects the environment,
• to ensure that adequate procedures are undertaken to reclaim surface areas as contemporaneously as possible with mining operations, and
• to strike a balance between protection of the environment and agricultural productivity and the Nation's need for coal as an essential source of energy.

To accomplish these ends, SMCRA mandates State permit systems in accordance with Federal guidelines that include comprehensive performance standards for surface mining operations and for the control of surface effects of underground mining.

[104]National Environmental Forecast No. 1.

For surface mining, the environmental performance standards regulate the removal, storage, and redistribution of topsoil; siting, erosion control, drainage, and restoration of coal haul roads; protection of water quality and the hydrologic balance; waste disposal; backfilling, grading, and revegetation, and postmining land uses. This section discusses those provisions of SMCRA that relate to areas of special concern, such as alluvial valleys and prime farmland. The overall legal framework of the Act is discussed in chapter VII.

Minimum performance and reclamation standards for alluvial valley floors in the arid and semiarid areas in the Western United States are intended to preserve existing or potential agricultural uses and productivity. An applicant for a mining permit in these areas must demonstrate that the operations will preserve the essential hydrologic functions of an alluvial valley floor by maintaining the geologic, hydrologic, and biologic characteristics that support those functions. These requirements apply both to disturbed and adjacent undisturbed areas. In addition, surface mining and reclamation operations may not interrupt, discontinue, or preclude farming on alluvial valley floors unless the premining land use is undeveloped rangeland or the area affected is small and production is negligible. The mine operator must conduct environmental monitoring to ensure that these requirements are met; where the monitoring shows they are being violated, mining must cease until approved remedial measures are taken. However, mines that were permitted or in production before August 1977 are exempt from these SMCRA requirements.

SMCRA also includes regulations designed to minimize the effects of surface mining on prime farmland so that the land will have equal productivity after mining and is not lost as a resource. Each soil horizon used in reconstruction of the soil must be removed before drilling, blasting, or mining in a manner that prevents mixing or contaminating the soil horizon with undesirable material. When replaced and reconstructed, the soil material must be at least 48 inches deep and must create a final root zone of comparable depth and quality to that which existed prior to mining. In addition, the surface soil horizon must be replaced in a manner that prevents excessive compaction and reduction of permeability and that protects against wind and water erosion. Finally, the operator must apply nutrients and soil amendments as needed to establish quick vegetative growth.

Protection of timber stands under SMCRA is not as rigorous as that afforded to alluvial valley floors and prime farmland. In general, mine operators must establish a diverse, effective, and permanent vegetative cover of species that are native to the area or that support the approved postmining land use. Introduced species may be substituted only if they are of equal or superior utility for the postmining land use and if they meet the requirements of State or Federal laws. Specific provisions for forest land in the Eastern United States, as measured 5 years after planting, require an average of 600 trees of commercial species of at least 3 years of age per acre. In the Western United States, where shelter belts, wildlife habitat, or commercial or other forest land is the approved postmining land use, the operator must conduct a premining and postmining inventory of vegetation. The density and number per unit area of trees and shrubs on the restored land must be equal to or greater than 90 percent of the premining density and number. The success of the reforestation, in terms of species, diversity, distribution, seasonal variety, vigor, and regenerative capacity of the vegetation, is evaluated on the basis of results that reasonably could be expected from the approved reclamation plan.

Finally, regulations promulgated under SMCRA provide special environmental performance standards for operations on steep slopes. Spoil and waste materials may not be placed on the downslope; the highwall must be completely covered with compacted soil and graded to the original contour; land above the highwall may not be disturbed; woody materials may not be buried in the backfill area, and unlined or unprotected drainage channels may not be constructed on backfills unless they are stable and not subject to erosion.

Photo credit: Union Pacific

COAL MINING.—This was a coal mine pit 3 years ago. Arch Mineral, operator of the Medicine Bow Coal Mine in Wyoming, follows rigorous reclamation, regrading worked over areas into gentle contours and reseeding with native grasses

Problems of subsidence from underground mining also are regulated under SMCRA. In general, the regulations require that underground mining be conducted so as to prevent subsidence from causing material damage to the surface. This may be accomplished by leaving adequate coal in place, backfilling or other measures to support the surface, or by mining in a manner that provides for planned and controlled subsidence. Underground mining is prohibited beneath or adjacent to streams or beneath impoundments of 20 acre-feet or more, beneath aquifers that serve as municipal water supplies, and beneath or in proximity to public buildings, unless the permitting agency determines that subsidence will not cause material damage to any of these factors. In addition, the mine operator is required to consult with the surface owner prior to conducting any underground mining activities, and the surface owner may request a comprehensive premining survey of dwellings, structures, and other physical factors that may be affected by subsidence. However, these measures primarily are designed to preserve the surface owner's rights under insurance and tort law; under SMCRA the mine operator is not directly responsible for subsidence damage and gives no warrantee to the surface owner.

The requirements of SMCRA are enforced through performance bonds that are conditioned on compliance with the terms of the Act. In general, liability under the bond continues for 5 years after reclamation is complete in areas with more than 26 inches of rain per year and for 10 years in areas with less than 26 inches. However, no long-term bonds are required for problems such as subsidence that may not occur until many years after mining is terminated.

As discussed in chapter VII, the success of SMCRA in correcting the abuses that have prevailed in the past will depend largely on the adequacy of enforcement under State programs. However, the history of compliance with previous State regulatory programs is not promising. A survey of 16 sites in Appalachia shows that only 9 of the sites are in complete compliance with the minimal grading requirements in effect during the period the sites were mined; 1 site is in partial compliance, while on 6 of the sites no grading at all has taken place. A similar study of 9 sites in the Eastern Interior region shows 5 had complied fully with the applicable grading requirements, while 3 sites are in partial compliance, and 1 in noncompliance. Of 13 additional sites mined before any regulations were in effect, 6 show voluntary compliance with subsequent grading requirements while 7 show no grading at all.[105]

The storage and disposal of mine wastes is regulated under SMCRA and under RCRA. In general, mine operators must maintain approved disposal sites within the mine permit area for coal processing wastes and for spoil not required to achieve the original contour. The waste disposal areas must be designed, constructed, and maintained in order to prevent combustion, adverse water quality impacts, and erosion. Mandated provisions in-

clude grading and compaction, then covering the waste with a minimum of 4 feet of the best available nontoxic material. Wastes may not be used in the construction of dams and embankments unless the mine operator demonstrates through appropriate engineering analyses that the waste will have no adverse effect on stability, and that the dam or embankment will be constructed, designed, and maintained in accordance with proper engineering principles.

RCRA seeks to prevent adverse effects on health or the environment from disposal of all types of solid wastes. The requirements of RCRA are primarily intended to prevent leaching into ground and surface waters, and are discussed in the section on impacts to water systems.

Finally, a number of Federal laws regulate transportation and transmission facilities. The construction of coal haul roads is regulated under SMCRA, which requires that erosion and siltation, air and water pollution, and damage to public or private property be minimized. In addition, roads must be removed and the land regraded and revegetated unless they are part of the approved postmining land use, in which case they must be maintained adequately. Permanent roads built with Federal funds are subject to extensive environmental protection and routing requirements under Department of Transportation regulations.

Rights-of-way for roads, transmission, and waterways across public lands must be permitted by the Federal agency having jurisdiction over the land. Federal right-of-way laws usually limit the width of the path. Dredging for waterways, or construction of any structure in the navigable waters must be permitted by the Army Corps of Engineers. Finally, most federally permitted activities that will have significant environmental effects will require an environmental impact statement under the National Environmental Policy Act.

[105]D. Willard, et al. "Land Use Changes Resulting from Strip Mining in the Ohio River Basin Energy Study Region" (Draft Report, unpublished) October 1978.

Chapter VI

# WORKPLACE AND COMMUNITY
# IMPLICATIONS

# Chapter VI.—WORKPLACE AND COMMUNITY IMPLICATIONS

Chapter VI
# WORKPLACE AND COMMUNITY IMPLICATIONS

## INTRODUCTION

This chapter discusses the occupational and social implications of increased coal production and use. It seeks to identify likely problem areas and enable policymakers to assess the need for remedial action. The trends in mine health and safety performance are traced and estimated future costs of increased coal production are presented. The socioeconomic impacts of more coal development differ in the East and West. Separate sections are devoted to the current experience in each region, and likely patterns of development are described.

## WORKPLACE HEALTH AND SAFETY

### Health

#### Background

Coal mines are inherently unhealthy places to work. Mortality studies of mineworkers who were employed in the 1940's, 1950's, and 1960's suggest that they died more often than others from respiratory diseases (influenza, emphysema, asthma, and tuberculosis), and lung cancer, and hypertension.[1] Pneumoconiosis and bronchitis were underlying causes of death in a significant number of fatalities attributed to nonmalignant respiratory diseases. Since 1969, Federal officials, mine operators, and coal workers have sought to minimize exposure to coal mine dusts and noise, and they have succeeded to a degree. If current dust standards are met on a daily basis, coal workers' pneumoconiosis (CWP) should disable far fewer workers 20 years hence than today.

Many occupational diseases require two or three decades to become manifest. Because about half of the total coal work force is under 34 years of age, the actual results of today's dust-control efforts will not be known until the 1990's and beyond. Efforts focused on CWP

after 1969 have lowered the amounts of respirable dust that miners now breathe on the job. But other hazards were overshadowed in the debate over CWP. This section discusses CWP and the programs to control the respirable dust that causes it, as well as other relevant health hazards.

The costs of occupational illness from coal mining have been immense. More than 420,000 coal workers ". . . who are totally disabled due to pneumoconiosis . . ."[2] or their widows were awarded Federal black lung compensation between January 1, 1970, and December 31, 1977. That number of beneficiaries equals the total number of coal workers employed industrywide in 1950. Occupational health officials at the United Mine Workers of America (UMWA) estimate that 4,000 Federal black lung beneficiaries have died from black lung disease each year since 1969. Other work-related illnesses — hearing loss, bronchitis, and breathing difficulties — are impossible to estimate.

Occupational illness usually affects older or retired workers, leaving them financially and psychologically strapped. For every worker disabled by CWP, two or three family members are also affected. Stress is created. Work schedules must be readjusted. Family income

[1]Howard Rockette, *Mortality Among Coal Miners Covered by the UMWA Health and Retirement Funds* (Morgantown, W. Va.: U.S. Department of Health, Education, and Welfare, National Institute for Occupational Safety and Health, 1977).

[2]Public Law 91-173, Federal Coal Mine Health and Safety Act of 1969, title IV, pt. A.

is reduced. Bills from doctors and hospitals rise.

Discussions of occupational disease often dwell on numbers, percentages, and sanitized scientific terminology. To the victim, black lung can be an overwhelming fact of life:

At work you [coal miners] are covered with dust. It's in your hair, your clothes, and your skin. The rims of your eyes are coated with it. It gets between your teeth and you swallow it. You suck so much of it in your lungs that until you die you never stop spitting up coal dust. Some of you cough so hard that you wonder if you have a lung left. Slowly you notice you are getting short of breath when you walk up a hill. On the job, you stop more often to catch your breath. Finally, just walking across the room at home is an effort because it makes you so short of breath.[3]

The cost of mineworker occupational illness are also borne by the public and, to a lesser extent, by the coal industry. Black lung compensation has been administered by the Social Security Administration and, since July 1973, by the Department of Labor. Table 37 summarizes the 8-year dollar cost of this program. Of the $5.569 billion paid to miners and their survivors from 1970 through 1977, the Federal Government paid $5.567 million, or almost 100 percent. An operator-paid tonnage tax was enacted by Congress in 1977 to provide for future benefits. Compensation is from funds contributed by the "last responsible operator." Labor Department spokesmen say coal companies contest many — if not most — determinations of their responsibility to pay. If the current trend continues, most new black lung claims may be paid from Federal resources. Recent legislation allows coal companies to deduct their contributions to the black lung trust fund from their Federal taxes. The deduction cannot exceed 38 percent of the company's payroll, which means, in practice, that many operators will be able to write off their black lung contribution.

Some health risks of mining are common to underground and surface miners, but they are

magnified many times for the deep miner. In an underground mine the dangers of dust, fumes, noise, and other contaminants are intensified by close quarters and artificial ventilation. Surface miners, too, face health hazards. The outdoor work environment of these heavy equipment operators, truck drivers, supply workers, welders, electricians, and mechanics exposes them to dusts, heat and cold, diesel and welding fumes, whole-body vibration, noise, and stress. Because the underground environment is so clearly more hazardous, comparatively little research has focused on surface miners.

**Table 37.—Eight-Year Summary of Black Lung Compensation (Jan. 1, 1970-Dec. 31, 1977)**

| | |
|---|---|
| Claims filed: | |
| Social Security | 586,400 |
| Labor | 124,200 |
| Total | 710,600 |
| Claims approved: | |
| Social Security | 415,200 |
| Labor | 5,744 |
| Total | 420,944 |
| Claims denied: | |
| Social Security | 171,200 |
| Labor | 68,062 |
| Total | 239,262 |
| Total federally paid benefits: | |
| Social Security | $5,513,712,000 |
| Labor | 54,000,000 |
| Total | $5,567,712,000 |
| Total operator-paid benefits: | |
| Social Security | -0-[a] |
| Labor (220 claims estimate) | $  1,500,000 |
| Total (estimated) | $  1,500,000 (estimated) |

[a] Coal operators were not required to pay for black lung benefits under the Social Security program.
SOURCE: Social Security Administration and Department of Labor, September 1978.

Preventive measures in the workplace are the most effective approach to controlling occupational disease. Prevention of all work-related diseases requires a holistic approach to analyzing hazards and controlling them. This approach recognizes that disability results from cumulative, total exposure. Unfortunately, most occupational disease research and

[3]Dr. Lorin Kerr, speech to the United Mine Workers of America at the 1968 UMWA Convention, reprinted in the *Congressional Record*, Sept. 25, 1968, p. 11446.

prevention programs dwell on the effects of single hazards. The cumulative and synergistic impacts of workplace hazards—that is, the way human beings experience them—have not been well researched.

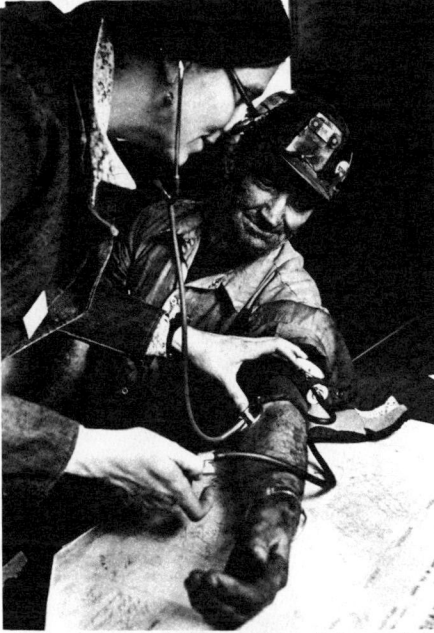

*Photo credit: Douglas Yarrow*

**Jan Chapin, a nurse with the Cabin Creek Medical Center, examines miner at minesite on Cabin Creek, W. Va., 1977**

### The Federal Role

Congressional concern for coal miners' health—as distinguished from safety—is relatively recent. The 1969 Coal Act established the Mining Enforcement and Safety Administration (MESA) to develop health standards and enforcement procedures. When the agency was shifted from the Interior to the Labor Department in the spring of 1978, it was renamed the Mine Safety and Health Administration (MSHA), seemingly promising a new emphasis on workplace health.

The 1969 Act regulated four health hazards—respirable coal mine dust, dust from rock drilling, respirable coal dust when quartz is present, and noise. Compensation for black lung was started. The 1969 Act failed to address other potentially significant hazards such as diesel emissions, heat, "nonrespirable" coal mine dust, and harmful fumes. As single factors, some of these may have little effect, but coupled with known hazards from respirable dust and noise, their total health consequences may be substantial. Enforcement of the standards since 1970 has been marked by litigation, technical problems, unsympathetic Federal administrators, and insufficient funds.[4]

MSHA's task is made more difficult by the lack of data on the health effects of coal mining. MSHA requires operators to report all occupational illnesses. The agency distributes form 6-347, listing seven categories of reportable occupational illness: occupational skin diseases, dust diseases of the lungs, respiratory diseases (toxic agents), poisoning, disorders (physical agents), disorders (repeated trauma), and all other occupational illnesses.[5] When a computer printout was sought from MSHA on disease experience for the 1972-76 period, data were available for only two categories—occupational skin diseases and all other occupational illnesses. MSHA statisticians said the other data were not available because the operators do not report them—and that even the available data were probably not very accurate. Without this information, policymakers are blocked from evaluating current prevention strategies.

---

[4]See Nancy Snyder and Mark Solomons, "Black Lung—A Study in Occupational Disease Compensation," Jan. 15, 1976 (Washington, D.C.: Department of Labor), reprinted in U.S. Senate, Black Lung Benefits Reform Act, 1976, p. 468.

[5]Bureau of Mines, Department of the Interior, title 30, pt. 80.1 (h), *Federal Register*, vol. 37, no. 55, Mar. 21, 1972. "Occupational illness" means any abnormal condition or disorder, other than one resulting from an occupational injury, caused by exposure to environmental factors associated with employment. It includes acute and chronic illness or diseases that may be caused by inhalation, absorption, ingestion, or direct contact, and which fall within the listing under the heading, "Occupational Illness" on form 6-347.

## Mortality

Disease recognition and prevention are hampered by the time required for many occupationally linked diseases to appear. The most recent mortality study surveyed more than 23,000 UMWA miners—of whom 7,628 had died—between 1959 and 1971. Rockette found coal miners died more often than the U.S. male population from respiratory disease (pneumoconiosis, influenza, emphysema, asthma, and tuberculosis), accidents, hypertension, and stomach and lung cancer.[6]

Rockette's data may be used cautiously to correlate coal mine employment with excess mortality from certain diseases. It is clear that death from accidents, pneumoconiosis, and nonmalignant respiratory ailments are associated with coal mining. Hypertension may be. Lung and stomach cancer may or may not be. It is conceivable that the combination of coal dust, trace-element exposure, noxious fumes and gases, and other irritants (including cigarette smoking and tobacco chewing) are related to excessive cancer among coal miners.[7]

### Dust—The Greatest Hazard

All mine dusts—of which coal dust is the most prominent—are classified as either respirable or nonrespirable according to the size of the dust particles. It is generally accepted that only the smallest particles (smaller than 5 $\mu$g in size) are respirable. They alone are retained in the alveoli—the gas-exchanging sacs of the lungs—and cause pneumoconiosis. Generally, the larger particles (nonrespirable dust) do not penetrate the alveoli and are not thought to cause CWP.

While the distinction between respirable and nonrespirable dust is scientifically valid, it is clear that both sizes can impair lung function when inhaled in quantity over time. The larger particles are probably linked to bronchitis among coal miners. Although these particles are generally not retained in the lung, continuous exposure to them during the normal work year produces a more or less constant irritation of the upper respiratory tract. Breathlessness has also been found to be sig-

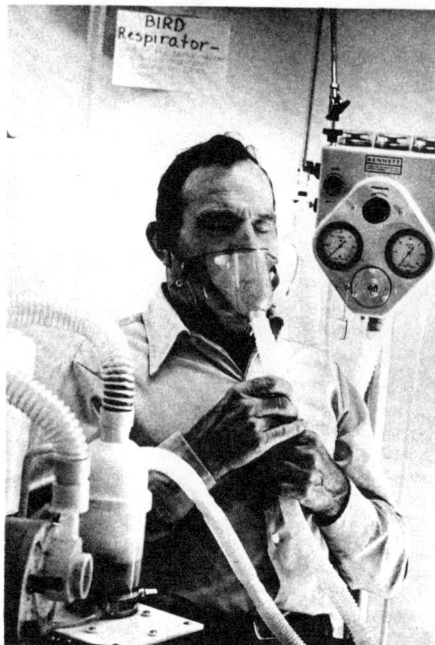

Photo credit: Douglas Yarrow

Black lung victim using vaporizer at the Cabin Creek Medical Center, Cabin Creek, W. Va., 1977

[6]Rockette, Mortality Among Coal Miners, p. 35ff.
[7]Rockette, Mortality Among Coal Miners. Also, W. Keith C. Morgan, "Coal Workers' Pneumoconiosis," Occupational Lung Diseases (Philadelphia, Pa.: W. B. Saunders Co., 1975), Morgan says: ". . . chronic bronchitis and emphysema account for much of the excess (mortality among coal miners), the SMR (Standard Mortality Ratio) for lung cancer and tuberculosis for miners is also increased above that of the general population," p. 160.

nificant among miners who do not show X-ray evidence of CWP. All researchers believe that breathlessness is related to chronic, nonspecific, obstructive bronchopulmonary disease. Some investigators, in addition to CWP and bronchopulmonary disease, have found a third, as yet unidentified disease process, that reduces the ability of the lungs to exchange gases.[8] Black lung disease has come to represent a broad definition of occupational respiratory disabilities in miners, of which CWP is one major component. The 1969 Act regulated respirable dust exposure but did not limit exposure to nonrespirable dust. Respirable dust, which is invisible to the unaided eye, accounts for less than 1 percent of the dust in a mine workplace. It is not clear how much nonrespirable dust is retained in the lungs when the Federal standard for respirable dust—2 mg/m³ of air—is being met.

Because compensable black lung is associated with other factors in addition to respirable coal mine dusts, prevention programs aimed exclusively at that hazard are likely to reduce CWP but may not reduce black lung disability proportionately.

## Coal Workers' Pneumoconiosis

The 1969 Act defined CWP as a "chronic dust disease of the lung arising out of employment in an underground coal mine."[9] It is caused by the inhalation and retention of respirable coal mine dust in the lower lungs. After a dozen years or so of exposure, a noticeable dose-response relationship may appear. CWP is classified by X-ray diagnosis into levels of ascending severity, from simple to complicated. If dust exposure continues, simple CWP can progress into more advanced stages. Miners with advanced CWP (characterized as progressive massive fibrosis) are usually totally disabled. Miners breathing high dust concentrations may be disabled after only a few years of exposure. Some miners seem more vulnerable than others, and this variability has yet to be explained. Cigarette smoking contributes to lung impairment among miners as in the population generally. Considerable controversy exists over the role smoking plays in miners' lung impairment compared to occupational factors.

The first major U.S. survey of CWP was done in the 1960's[10] by the Public Health Service. It collected evidence from 2,510 working coal miners and 1,190 nonworking miners. In this sample of 3,700 Appalachian, bituminous coal miners, almost 10 percent of the working miners and about 18 percent of the others had evidence of either simple or complicated CWP. About 5 percent of the working miners and about 7 percent of the nonworking miners were classified as having "suspect" CWP. Lainhart, et al., found "the degree of roentgenographic evidence of pneumoconiosis was not related to cigarette smoking [and] . . . the degree of pneumoconiosis was not related to the number of cigarettes smoked daily."[11] Cigarette smoking does not cause CWP, although it is associated with breathing impairment, which, in some cases, may pass as black lung. X-ray evidence of CWP was associated with increasing years of exposure. About 9 percent of working and nonworking miners had evidence of other chest illnesses—tuberculosis, cancer, pulmonary, and cardiac disease.

The most recent and comprehensive study is being done by the National Institute for Occupational Safety and Health (NIOSH). About 10,000 working miners in 31 mines were X-rayed in the early 1970's. An overall CWP prevalence of nearly 30 percent was found. Progressive massive fibrosis was diagnosed in 2.5 percent of the sample.[12] The NIOSH work—known as the National Study of Coal Workers' Pneumoconiosis—found that CWP increases with dust exposure and number of years worked underground. About 31 percent of Appalachian miners showed X-ray evidence of CWP, compared with 23.5 percent for midwestern miners and 10.5 percent for western miners. However, upon reinterpretation of the X-ray data, a CWP prevalence figure of between 10 and 15 percent seemed justified.

[8]D. L. Rasmussen, W. A. Laqeur, P. Futterman, H. D. Warren, and C. W. Nelson, "Pulmonary Impairment in Southern West Virginia Coal Miners," *American Review of Respiratory Disease*, vol. 98, 1968; and Donald L. Rasmussen, "Breathlessness in Southern Appalachian Coal Miners," *Respiratory Care*, March-April 1971.
[9]Public Law 91-173, 83, sec. 202. Surface miners are now eligible for black lung compensation.

[10]W. S. Lainhart, H. N. Doyle, P. E. Enterline, A. Henschel, and M. A. Kendrick, *Pneumoconiosis in Appalachian Bituminous Coal Miners* (Cincinnati, Ohio: DHEW, Public Health Service, 1969).
[11]Ibid., p. 53.
[12]W. K. C. Morgan, Dean B. Burgess, George Jacobson, Richard J. O'Brien, Eugene P. Pendergrass, Robert Reger, and Earl Shoub, "The Prevalence of Coal Workers' Pneumoconiosis in U.S. Coal Miners," *Archives of Environmental Health*, vol. 27, October 1973, p. 221.

The second round of this study in 1973-75 found CWP prevalence to be about 11.5 percent.[13] Of miners X-rayed in both rounds, about 6 percent showed disease progression. Some controversy exists over the reliability of these findings, however. Critics say the results should be considered with several caveats. First, different X-ray classification schemes were used in each round and side-by-side X-ray readings using one system were not performed. Second, although X-ray diagnosis is the accepted (and only) method for diagnosing CWP, it is not foolproof. The quality of X-rays is directly linked to the conditions in which they are taken. Critics say that differences in conditions were not weighed in the final evaluation. Third, only 40 to 50 percent of the miners X-rayed in the first round were X-rayed again in the second round. This has led some to argue that the second-round data reflect CWP prevalence for a "survivor population and the slightly lower prevalence rate between round 1 and round 2 reflects the "healthy worker" effect, that is, workers who are not healthy are forced out of the labor force leaving only healthy workers to be surveyed for CWP. The 50 to 60 percent of round 1 miners who were **not** X-rayed in round 2 may consequently have a considerably higher prevalence of CWP than those miners who continued working and were X-rayed in both rounds." (Many of the first-round CWP cases may not have been examined because they no longer worked in the mines or refused to be examined because they feared job discrimination if they had a "bad" X-ray.) Finally, dust exposures cannot always be correlated to X-ray results with a great deal of confidence, making it difficult to say anything conclusive about the safeness of the Federal 2-mg dust standard using these data.

Rough estimates of CWP prevalence for 1976, 1985, and 2000 appear in table 38. Actual prevalence of CWP in the future will differ from these projections according to the actual numbers of miners employed and actual prevalence rate. These estimates assume the existing Federal dust standard will significantly reduce

[13]Status of Mandated Programs, Receiving Center, National Institute for Occupational Safety and Health, unpublished reports, end of second round of examinations, 1975.

CWP prevalence in the future. Thus, each of the three scenarios forecasts steadily lower prevalence rates. However, even if prevalence declines, more workers will be exposed. Assuming 230,000 coal miners will be working in 1985, the number of likely CWP cases ranges from almost 13,700 to almost 18,800. Assuming about 411,000 miners will be employed by 2000, CWP cases might range from 10,600 to about 18,000.

Both the optimistic and pessimistic cases probably err on the side of optimism because they assume more or less full compliance with the 2-mg dust standard on a daily basis. If every mine strictly complies with this standard between 1975 and 2000, the prevalence rate may be 3-percent underground and 1-percent surface in that year. The current standard will not eliminate CWP, but it is likely to cut significantly the probability of its occurrence.

Along with CWP, coal miners will continue to experience other black lung diseases— bronchitis, severe dyspnea (shortness of breath), and airways obstruction. Much of this illness will be work-related and some of it will be caused or worsened by cigarette smoking. For every case of CWP, about three cases of bronchitis, one case of dyspnea, and two cases of airways obstruction may be found. These cases are not exclusive, and are often found in combination.

NIOSH calculated the health effects of increased coal production using current prevalence rates and higher estimates of the future work force. NIOSH estimated there were 19,400 cases of CWP in 1975, rising to about 28,500 in 1985, and 45,500 in 2000 (table 39). Cases of chronic bronchitis, dyspnea, and airways obstruction are estimated to increase proportionately. Both OTA and NIOSH estimates deal only with the illness found among working miners. Retired and disabled workers do not appear in these estimates. This group, which will be eligible for black lung compensation, is likely to be large.

Questions relevant to congressional policy include: 1) Are these estimated health costs of increased coal production unacceptably high? 2) If so, what can be done to reduce these

## Table 38.—Estimates of CWP Among Working Miners, 1976, 1985, and 2000

| | Optimistic | Prevalence Rates | Pessimistic | |
|---|---|---|---|---|
| | Underground | Surface[a] | Underground | Surface[a] |
| Working miners ........................................................ 1976 (208,000) | 10% | 4% | 15% | 4% |
| Underground (70%) 145,600 ......................................................... | 14,560 | | 21,840 | |
| Surface (30%) 62,400 ........................................................... | | 2,496 | | 2,496 |
| Total ........................................................... | 17,056 | | 24,336 | |
| Working miners[b] ................................................... 1985 (229,829) | 7% | 3% | 10% | 3% |
| Underground (74%) 169,922 ......................................................... | 11,895 | | 16,992 | |
| Surface (26%) 59,907 ........................................................... | | 1,797 | | 1,797 |
| Total ........................................................... | 13,692 | | 18,789 | |
| Working miners[b] ................................................... 2000 (410,893) | 3% | 1% | 5% | 2% |
| Underground (79%) 326,305 ......................................................... | 9,789 | | 16,315 | |
| Surface (21%) 84,588 ........................................................... | | 846 | | 1,692 |
| Total ........................................................... | 10,635 | | 18,007 | |

[a] Prevalence rates for 1976 are consistent with the range of current research findings. The 4 percent prevalence rate for surface miners comes from R. Paul Fairman, Richard J. O'Brien, Steve Swecker, Harlan Amandus, and Earle P. Shoub, "Respiratory Status of the U.S. Surface Coal Miners," unpublished manuscript done for ALOSH/NIOSH, 1976. The ALOSH study used X-ray evidence to determine the prevalence of CWP. Most of the surface miners had extensive experience in underground coal mining which contributed to the prevalence of CWP. Prevalence rates for 1985 and 2000 assume that compliance with the current dust standard will lower prevalence. In addition, increasing numbers of older miners who had years of exposure before 1969 will retire, thereby lowering prevalence among working miners. These prevalence rates have not been derived mathematically, and should be seen more as possibilities than as predictions.
[b] Workforce estimates are the average of OTA's low-case and high-case labor projections found in chapter II, table 6.

## Table 39.—Projected Health Effects of Increased Coal Production

| Year | Millions of tons produced (quads) | Number of employees | Cases[a] of CWP | Cases of[ab] chronic bronchitis | Cases of[ab] severe dyspnea | Cases of[ab] airways obstruction |
|---|---|---|---|---|---|---|
| **1975 estimates** | | | | | | |
| Underground ................................................. | 279 ( 7.3) | 139,500 | 18,100 | 41,800 | 11,200 | 41,800 |
| Strip auger ...................................................... | 332 ( 7.9) | 52,500 | 1,300 | 15,700 | 4,200 | 10,000 |
| Total ........................................................... | 611 (15.2) | 192,000 | 19,400 | 57,500 | 15,400 | 51,800 |
| **1985 estimates** | | | | | | |
| Underground ................................................. | 395 ( 9.8) | 197,500 | 25,600 | 59,200 | 15,900 | 59,200 |
| Strip auger ...................................................... | 735 (18.3) | 116,000 | 2,900 | 34,800 | 9,300 | 22,100 |
| Total ........................................................... | 1,130 (28.1) | 313,500 | 28,500 | 94,000 | 25,200 | 81,300 |
| **2000 estimates** | | | | | | |
| Underground ................................................. | 630 (15.7) | 315,000 | 40,900 | 94,400 | 25,300 | 94,400 |
| Strip auger ...................................................... | 1,170 (29.2) | 185,000 | 4,600 | 66,300 | 14,800 | 35,200 |
| Total ........................................................... | 1,800 (44.9) | 500,000 | 45,500 | 160,700 | 40,100 | 129,600 |

[a] Not mutually exclusive.
[b] Not necessarily due solely to coal dust inhalation.
SOURCE: National Institute for Occupational Safety and Health, "Occupational Safety and Health Implications of Increased Coal Utilization," computer printout and attachments (rev.) Nov. 4, 1977. Distributed at a NIOSH conference of the committee on health and ecological effects of increased coal utilization, Nov. 21, 1977.

costs? 3) Who should bear the costs of reducing dust levels? 4) How may tradeoffs be determined between miners' health and dollars spent on prevention?

### Enforcement of Federal Dust Standard

The current dust-control program has four components: 1) the federally established 2-mg/m³ dust standard, 2) dust suppression efforts in the mine, 3) monitoring dust levels in the work environment, and 4) enforcement strategies for compliance with the Federal standard. The Federal dust standard refers only to respirable coal mine dust.

The Federal Government plays an important role in each of these components. Ventilation standards are federally established. Federally sponsored research has helped develop water sprays and other dust-control techniques. Operators must have Federal approval of their dust-control plans. Monitoring the work environment is now a joint Federal and industry responsibility. Mine operators are required to submit dust samples to MSHA several times a year. These samples are weighed and the operator is told several weeks later whether the sample complied on the day it was taken. If the sample exceeds the standard, the operator must redo it until it complies. MSHA inspectors take their own samples two or three times a year. If the operator is out of compliance, he will be issued a notice of violation and may be assessed a civil penalty. The average assessed penalty was between $150 and $175 for a respirable dust violation in 1978, and the average amount collected was about $120.

The adequacy of the current 2-mg/m³ respirable dust standard is one key assumption that will determine the prevalence of respiratory disease among coal miners in the future. Although the U.S. standard is probably the strictest in the world, its inherent safeness is not entirely certain. When the 2-mg standard was mandated in the 1969 Coal Act, Congress assumed that full and comprehensive compliance would effect foolproof prevention, based on the best evidence available at the time. That evidence deserves renewed scrutiny.

Congress based the 2-mg standard on British research done in the 1960's.[14] More than 4,000 British miners were studied over a 10-year period. The study's methodology—how its data were collected and how they were used—may raise questions as to the soundness of its final conclusions.[15] Dust exposure of each miner was not monitored daily. Rather, a stratified random-sampling procedure was used, which took 50 to 250 samples annually at each of 20 mines. The dust sampler used in the study is no longer used. The exposure data were converted from particles to weight, but the conversion factors were developed experimentally rather than from in-mine measurement. Average 8-hour exposures for each job at the face were calculated from the stratified random sampling. Calculating a mean from a set of averages twice removes statistical results from reality. Probability curves were then constructed that related 10-year mean exposures to X-ray evidence of CWP. The probability curves suggest the current standard is reasonably safe (at 2 mg, 1 to 3 of every 100 miners would show X-ray evidence of CWP after 35 years exposure).[16] However, if the data collection and statistical manipulations were not totally sound, then the curves are flawed. The safeness of the 2-mg standard has never been confirmed by a long-term epidemiological study. A recent British update of the 1969 results showed that in 10 mines the "earlier predictions underestimated 35-year working-life risks [the probability of developing simple pneumoconiosis] by 1 to 2 percentage probability units."[17] Researchers could not explain

---

[14]M. Jacobsen, S. Rae, W. H. Walton, and J. M. Rogan, "New Dust Standards for British Coal Mines," Nature, vol. 227, Aug. 1, 1970; M. Jacobsen, "The Basis for the New Coal Dust Standards," The Mining Engineer, vol. 131, March 1972; M. Jacobsen, "Dust Exposure and Pneumoconiosis at 10 British Coal Mines," a paper presented at the 5th International Pneumoconiosis Conference, Caracas, Venezuela, October-November 1978.

[15]A detailed discussion of the British research may be found in the Coal Mine Health appendix in vol. II.

[16]No probability curves were constructed for bronchitis, emphysema, or dyspnea.

[17]M. Jacobsen, "Dust Exposure and Pneumoconiosis at 10 British Coal Mines," paper presented to the 5th International Pneumoconiosis Conference, Caracas, Venezuela, October-November 1978, pp. 5-6.

why CWP risks were five to six times higher than average at one mine, where the mean concentration of airborne dust to which the miners were exposed was 2.9 mg/m³, compared with another mine, where the CWP risks were about one-ninth the average but the mean dust concentration was 4.4 mg/m³. A 1979 draft of a NIOSH report, "Criteria for a Recommended Standard . . . Chest Roentgenographic Surveillance of Surface Coal Miners," on X-ray monitoring of surface miners said the "current U.S. permissible exposure limit for coal mine dust of 2 mg/m³ does not assure a zero risk for the development of either simple or complicated pneumoconiosis." The draft said the current standard is ". . . insufficient to eliminate the development of CWP in a small portion of American coal miners . . . ." The risk of developing more advanced stages of simple pneumoconiosis approaches zero at an average coal mine dust exposure concentration of 1 mg/m³ and below, NIOSH reported. Since the average dust sample submitted to MSHA in 1978 indicates that concentration levels of slightly over 1 mg/m³ are being achieved, Congress may wish to consider lowering the standard to attain zero CWP risk. In light of the large numbers of young miners joining the coal labor force in the 1970's, prudence suggests that new research confirm or change the 2-mg standard as soon as possible.

Assuming, for the moment, that 2 mg is low enough to prevent CWP, it is apparent that dust control depends on reliable monitoring of the work environment. MSHA says that more than 90 percent of the operator-submitted samples of mine sections have been in compliance since 1973. This seemingly encouraging record leads to the conclusion that CWP should be almost eliminated in the future. Yet the sampling system is so burdened by uncertainty and flaws that such a conclusion may prove to be unwarranted.

The dust-monitoring process has been criticized on a number of grounds. First, the sampling is performed only on perhaps 2 percent of the days worked each year. Second, the samplers used to monitor the dust are not always reliable. Third, control of the sampling program at each mine is exercised by the operator rather than the miners or MSHA. Serious mis-

use of this responsibility has been reported, including deliberate falsification of sample results. Miners express little faith in a sampling system managed by their employers—whose interest, they claim, is not served by submitting noncompliance samples to MSHA. Fourth, a lag of 2 to 6 weeks often occurs before the operator is apprised of the results of his sample. By that time, the work environment has changed considerably, for better or worse.

Behavioral factors influence the representativeness of the samples. Miners don't like to wear the samplers because they are noisy, awkward, and heavy. Both the company and the miner have incentives to take "good" samples (by tampering with the sampler or leaving it in a nondusty place). If a noncompliance sample is submitted to MSHA, the individual miner must take 10 more samples, which is considered a nuisance. The miner may not wear his sampler to minimize his inconvenience or possibly to appease an anxious supervisor. Management may try to distort the sampling, too. In one case, two company technicians at a mine of a major operator were found guilty—and then innocent—of deliberately falsifying samples once miners had turned them in at the end of their shift. The samplers themselves are far from perfect. MSHA stopped the use of the most common sampler in 1976 because it gained weight artificially. In the past, samplers were also found to lose weight for no apparent reason.

The General Accounting Office (GAO) found:

> . . . many weaknesses in the dust-sampling program which affected the accuracy and validity of the results and made it virtually impossible to determine how many mine sections were in compliance with statutorily established dust standards.[18]

GAO said factors that affected sample accuracy were: sampling practices used by operators and miners, dust-sampling equipment, weight loss of sampling cassettes, and weighing of cassettes by MESA and cassette manufacturers.[19]

[18]*Improvements Still Needed in Coal Mine Dust-Sampling Program and Penalty Assessments and Collections* (Washington, D.C.: U.S. Congress, General Accounting Office, Dec. 31, 1975), p. i.
[19]Ibid., p. 15

A National Bureau of Standards (NBS) study found:

> ... when the miners and mine operators perform and supervise the sampling and when the weighings are made in the normal manner ... the uncertainty [of accurate, actual measurements of dust concentrations] is estimated to be as large as 31 percent ...[20]

No followup study on the accuracy of the sampling program has been done since the GAO and NBS reports were released in December 1975. Both investigations noted that MSHA and NIOSH officials have made efforts to improve.

A recent study of underground mines in east Kentucky found that:

> Both the interviews and the Federal dust records suggest that many dust samples are being collected incorrectly. In some instances, extra dust has been added; but in substantially more cases, the samples are too low. One explanation for the inaccurate sampling, of course, lies in the fact that the mineowners control sampling in their own mines. Since the penalty for exceeding Federal standards may be to close the mine, coal operators have a strong incentive to err in the direction of samples showing less than the actual dust levels. The 1969 law attempts to circumvent this possibility by requiring that all samples be collected by actual miners on the assumption that, since their own health is at stake, coal miners will have a vested interest in insuring the accuracy of samples taken at their workplace. For various reasons, however, that assumption has proven wrong.[21]

Sharp found 27 percent of the 680 face and high-risk samples he studied were at the 0.1 and 0.2 mg levels, which are generally considered to be impossibly low. (That level represents dust levels in fresh, intake air.) About 23 percent of equipment operator's samples were also found to be inaccurate (low). Operator attitude toward the sampling program has an important effect on sampling accuracy, Sharp concluded.

[20]An Evaluation of the Accuracy of the Coal Mine Dust-Sampling Program Administered by the Department of the Interior, a Final Report to the Senate Committee on Labor and Public Welfare (Washington, D.C.: Department of Commerce, National Bureau of Standards, 1975), p. iii.

[21]Gerald Sharp, "Dust Monitoring and Control in the Underground Coal Mines of Eastern Kentucky," masters thesis, University of Kentucky, November 1968, p. 4.

Additional doubt is raised about the soundness of the sampling program when the operator-submitted samples are compared with MSHA-taken samples. While the operator-submitted samples indicate more than 90-percent compliance, MSHA inspectors cited 36 percent (of the almost 5,000 underground sections they sampled) for noncompliance in 1976. Further, MSHA issues an average of about 2,500 notices of noncompliance and 50 withdrawal orders annually for dust violations, increasing the doubt about the more than 90-percent compliance. Finally, sampling itself is meaningful only to the degree that the samples reflect conditions when sampling is not being done. When sampling does not occur, sections may or may not be consistently in compliance. Conceivably, a section might be in compliance on the sampling day, but be out of compliance until either the next official sampling day or when a MSHA inspector next samples.

MSHA and NIOSH recently proposed new dust-sampling procedures designed to improve their reliability.[22] MSHA proposed that dust samplers be placed in high-risk *areas* rather than worn by miners *individually*. Area sampling, as this technique is known, has some advantages, but its reliability is uncertain. MSHA has not been able to demonstrate that area sampling and personal sampling produce identical results. The proposed regulations have become a battleground between labor and industry over the question of who should control dust sampling to achieve data reliability. UMWA and some occupational health advocates support miner control of the monitoring. Industry replies that the current system is working and that miner control is an infringement on management's right to operate its mines. UMWA proposed to MSHA a combination area and personal sampling system that would be administered by a MSHA-certified, miner-elected dust person at each mine. If UMWA and its allies are successful in this demand, an important precedent will have been set for other workers in other industries. Another solution to dust control is a machine-mounted continuous sampling system that trig-

[22]See draft of pt. 70—Mandatory Health Standards, Underground Coal Mines—and pts. 11, 71, 75, and 90 of title 30, CFR. Available from the Mine Safety and Health Administration's Division of Coal Mine Health.

gers an automatic power cutoff when dust levels are too high. Such a system is being developed but is several years away from commercialization, the Bureau of Mines says. Longwall mining, in particular, needs more effective dust controls.[23]

## Other Mine Dusts

The 2-mg dust limit covers all respirable coal mine dust, not only coal dust. Coal mine dusts contain a wide range of noncoal constituents, including silica and hazardous substances such as benzenes, phenols, and napthalenes. Shultz, et al.,[24] found 13 polynuclear aromatic hydrocarbons in the respirable mine dusts they studied. Polynuclear aromatic hydrocarbons are listed among NIOSH's suspected carcinogens.[25] Arsenic, beryllium, cadmium, fluorine, lead, mercury, and selenium are all found in coal and all appear on the Environmental Protection Agency's (EPA) list of hazardous elements. At even one part per million, a mine producing 1 million tons of coal generates 1 ton of each element annually. As coal is cut from the face, some of each element will be liberated as dust or gas in the workplace.

Table 40 provides information on seven trace-element concentrations in various coals. The permissible levels of workplace air contaminants as developed by the Occupational Safety and Health Administration (OSHA) are also included in the table. The median concentration levels of these seven elements in uncleaned coal almost uniformly exceed OSHA's standards. Research has not determined how much of each element is actually liberated in the mining process or, further, how much of that is inhaled by miners. One sample of West Virginia miners found their lungs to contain an excess of beryllium, cobalt, copper, and other minerals, but was unable to determine the significance of these excesses.[26] An autopsy study found miners' lungs contain higher than normal concentrations of aluminum, barium, boron, chromium, germanium, iron, lead, magnesium, manganese, nickel, silver, tin, titanium, and vanadium—all common coal constituents.[27]

Little research has been done on the health effects of trace-element dust or trace-element compounds generated in coal extraction. Trace elements may have a role in producing black lung disability, either alone or synergistically. They may also play a role in the excess lung and stomach cancer found in miners.

## Harmful Fumes and Gases

Hazardous fumes and gases are often produced in underground mines under both normal and abnormal conditions. Common gases include nitrogen and its oxides, carbon dioxide ($CO_2$), methane and other hydrocarbons, sulfur dioxide ($SO_2$), and hydrogen sulfide. If ventila-

---

[23]Mine operators acknowledge and MSHA surveys have confirmed that longwall mining systems (nearly 100 are now operating) have not been able to comply consistently with the Federal respirable dust standard. Only 48 percent of the 57 longwall sections MSHA surveyed in a 1978 report were in compliance with the 2-mg respirable dust standard. Longwall plows and single-drum shears had much better compliance performance than double-drum shears (of which 22 out of 32 were out of compliance). Longwall shears using planned mining cycles were in compliance and averaged 790 tons per shift compared with noncomplying shears without planned mining cycles that produced an average of 546 tons per shift. If a significant portion of underground production in the future is to be mined by longwalls, better dust-control measures must be used. See Robert E. Nesbit, "Summary of a Technical Survey of All Longwall Sections for Compliance With Respirable Dust and Noise Standards," Mining Safety and Health Administration, Oct. 31, 1978.

[24]J. L. Shultz, R. A. Friedel, and A. G. Sharkey, Jr., *Detection of Organic Compounds in Respiratory Coal Dust by High-Resolution Mass Spectrometry*, Bureau of Mines Technical Progress Report 61 (Pittsburgh, Pa.: U.S. Department of the Interior, Bureau of Mines, 1972), p. 14.

[25]*Suspected Carcinogens* (2nd ed.; Washington, D.C.: DHEW, National Institute for Occupational Safety and Health), p. 189.

[26]J. V. Crable, R. G. Keenan, F. R. Wolowicz, M. J. Knott, J. L. Holtz, and C. H. Gorski, "The Mineral Content of Bituminous Coal Miners' Lung," *American Industrial Hygiene Association* 28, 8, 1967; and J. V. Crable, R. G. Keenan, R. E. Kinser, A. W. Smallwood, and P. A. Mauer, "Metal and Mineral Concentrations in Lungs of Bituminous Coal Miners," *American Industrial Hygiene Association Journal* 29, 106, 1968.

[27]Robert W. Freedman and Andrew B. Sharkey, Jr., "Recent Advances in the Analysis of Respirable Coal Dust for Free Silica, Trace Elements, and Organic Constituents," *Annals of the New York Academy of Science*, vol. 200 (New York: New York Academy of Science, 1972), p. 9.

## Table 40.—Median Trace Element Concentrations by Coal Type
### (parts per million dry weight, uncleaned basis)

| Type of coal (OSHA permissible level)[a] | As[b] (0.4) | Be (1.6) | Cd (.2) | F (.2) | Pb (.2) | Hg (.1) | Se (.2) |
|---|---|---|---|---|---|---|---|
| Northern Appalachian bituminous | 23.00 | 2.40 | 0.45 | 83.50 | 9.00 | 0.22 | 3.40 |
| Central Appalachian bituminous | 25.00 | 1.80 | 0.20 | 50.00 | 9.00 | 0.12 | 3.00 |
| Southern Appalachian bituminous | 31.00 | 1.80 | NA | 103.00 | 9.00 | 0.17 | NA |
| Interior Eastern bituminous | 26.50 | 2.50 | 2 90 | 75.00 | 11.00 | 0.13 | 2.00 |
| Interior Western bituminous | 26.50 | 1.10 | 11.00 | 92.50 | 4.00 | 0.19 | 2.90 |
| North Western Subbituminous | NA | 1.50 | 0.40 | 65.00 | 7.00 | 0.07 | 0.80 |
| Lignite | NA | 1.10 | 1.10 | NA | NA | 3.15 | neg. |
| Central Western bituminous | 1.00 | 0.90 | NA | 131.00 | 7.50 | 0.05 | 1.60 |
| South Western bituminous | 2.00 | 0.04 | NA | 130.00 | 7.50 | 0.05 | 1.85 |
| Texas lignite | 7.69 | 0.38 | NA | NA | NA | 3.08 | NA |

[a] Occupational Safety and Health Administration permissible levels were calculated from OSHA, *General Industry Safety Standards rev.* January 1976, pp. 206–509.
[b] As (arsenic), Be (beryllium), Cd (cadmium), F (Flourine), Pb (lead), Hg (mercury), and Se (selenium).

tion is maintained at required levels, these gases will be diluted and carried quickly from the workplace. Miners are often exposed to noxious or poisonous fumes from fires in machinery, insulated electric cables, conveyor belts, lubricating oils, and various synthetic materials. Friction from moving parts and electrical faults are common fire sources. Close observation of equipment by miners and supervisors and compliance with Federal equipment standards are well-known ways of preventing fires. Still, several hundred probably occur each year.

Hartstein and Forshey[28] analyzed coal mine combustion products from brattice cloth, ventilation tubing, belts, insulation, resins, foams, and oils—products commonly found in mines. Hazardous emissions identified were polyvinyl chlorides, neoprenes, and other chemical compounds plus hydrogen chloride, carbon monoxide (CO), $SO_2$, and hydrogen sulfide. When fires occur, these substances are liberated simultaneously in large doses. Harmful synergism with ambient dust may occur. No morbidity study has been done to assess the cumulative health impact of these substances.

## Noise

Noise is a proven hazard to both underground and surface miners. NIOSH believes

occupational noise has these actual, or possible, effects: temporary or permanent loss in hearing sensitivity, physical and psychological disorders, interference with speech communications or the reception of other wanted sounds, and disruption of job performance.[29] Excessive noise may also cause changes in cardiovascular, endocrine, neurologic, and other psychologic functions.[30]

NIOSH completed the most comprehensive survey of hearing loss in the underground coal mining industry in June 1976.[31] It found that "mining operations cause 20 to 30 percent of all miners to be exposed to noise that is potentially hazardous to their hearing."[32] However, disagreement was noted as to the proportion of miners suffering actual hearing loss as a consequence of exposure. Occupational hearing loss in the studies NIOSH surveyed ranged from 9 to 29 percent.[33] "The results of this study indicate that coal miners have measurably worse hearing than the national average,"[34] NIOSH concluded. MSHA has not con-

[28]Arthur M. Hartstein and David Forshey, *Coal Mine Combustion Products: Identification, and Analysis* (Washington, D.C.: U.S. Department of the Interior, Bureau of Mines, 1974).

[29]*Criteria for a Recommended Standard... Occupational Exposure to Noise* (Washington, D.C.: DHEW, National Institute for Occupational Safety and Health, 1972), p. IV-1.
[30]Ibid., p. IV-10.
[31]*Survey of Hearing Loss in the Coal Mining Industry* (Washington, D.C.: DHEW, National Institute for Occupational Safety and Health, 1976).
[32]Ibid., p. 2.
[33]Ibid., pp. 2-3.
[34]Ibid., p. 29.

ducted an equivalent noise survey for surface miners, although comparable data suggest similar hearing risk for truck operators and earth-moving equipment drivers. Data on active and retired miners covered by the UMWA Health and Retirement Funds in 1974 indicated that the Funds male population has a prevalence rate for hearing aids of 1.5 percent compared with a 1-percent rate for all U.S. males.

Operators of underground and surface equipment are usually exposed to high, often intermittent noise levels. In surface operations, where machinery is in continuous operation, noise is often excessive. Miners are rotated in and out of the workplace to limit exposure. The Federal noise level of 90 dBA average for an 8-hour period permits higher exposures for shorter periods of time. However, convincing evidence is not at hand to indicate the 90 dBA level is sufficiently safe. NIOSH has recommended that the permissible exposure level be lowered to 85 dBA; UMWA health advocates have urged a 75 dBA limit.

Operators are required to submit noise-sampling data to MSHA. Samples are taken only once or twice a year in most cases. The results indicate almost total compliance with the Federal standards. MSHA issues even fewer notices of noise violations than the small number of noncompliance samples warrant. MSHA has no way of systematically checking the reliability of the operator-submitted noise samples.

MSHA inspectors are required to sample noise during inspections. But the agency does not tabulate its inspectors' measurements. Thus no way exists to compare the operators' data with MSHA's.

Without reliable noise exposure data, it is impossible to predict the extent of hearing impairment miners will experience in the future. It is reasonably clear that a significant number of today's miners is exposed to and impaired by noise. Noise-related accidents are not uncommon. Lowering the Federal noise standard to 85 dBA, for example, and instituting a reliable monitoring and enforcement system would probably minimize future hearing impairment and accidents. Noise control at the 90 dBA level or lower requires careful engi-

neering of equipment and design of the work process. Exposure can be reduced by providing personal protective headgear, but this approach is usually less reliable than engineering control and may increase accidents. Machine modifications to achieve a lower noise standard would involve retrofit expenditures and perhaps increase the cost of new equipment. Concern for inflationary impact and engineering considerations have discouraged Federal safety agencies from promulgating the recommended 85 dBA standard.

## Diesels in Underground Coal Mines

Diesel-powered equipment is commonly found in surface coal mines throughout the United States. However, only 200 diesel units operate in underground coal mines. About 160 of those pieces of equipment are found in the West; most of the others are used in east Kentucky. Only 10 or 12 pieces are found in mines organized by UMWA.

Health hazards to both surface and underground coal miners from diesel emissions have not been studied definitively. In fact, no long-term epidemiological study of diesel exposure has been done of any work group. A preliminary survey covering a 6-year period that matched 772 diesel-exposed miners with an equal number of "controls" reported that the diesel miners reported significantly more symptoms of persistent cough, phlegm and exacerbations of cough and phlegm. Their pulmonary function performance was generally poorer, but they reported fewer symptoms of moderate to severe dyspnea and wheezing.[35] NIOSH expressed concern about the health hazards of diesels underground, but has not yet been able to justify a recommendation that diesels be prohibited. However, mine operators have been cautioned not to invest heavily in underground diesels pending further research. NIOSH is currently working up a criteria document, which surveys the literature and

[35]R. Reger and J. Hancock, "On Respiratory Health: Coal Miners Exposed to Diesel Exhaust Emissions (draft) (Morgantown, W. Va.: Appalachian Laboratory for Occupational Safety and Health, 1979).

recommends a standard, for diesel emissions. This should be issued within 2 years.

Diesel engines produce emissions that are known to be health hazards: CO, unburned hydrocarbons, oxides of nitrogen, particulates, polynuclear aromatic hydrocarbons, phenols, aldehydes, oxides of sulfur, trace metals, nitrogen compounds, smoke, and light hydrocarbons.[36]

With exhaust scrubbers, rigorous maintenance, and massive ventilation these emissions can be minimized. However, Federal researchers point out that diesel particulates (which include carcinogens and mutagens) and oxides of nitrogen cannot be trapped within the combustion process. The safeguard standard is 8,000 to 17,000 ft³ per minute of air per piece of diesel equipment. Normally, only 9,000 ft³ per minute is required in each section (and 3,000 ft³ per minute across each working face). The tremendous volumes of air required for diesel operation probably precludes their use in most older mines.

NIOSH's current assessment of the potential health effects of diesels underground is pessimistic.

    ... Although acute or subchronic health effects are important, chronic health effects are the central issue. The principal organ likely to be affected by diesel emission products is the lung. In addition to coal mine dust, many of the individual pollutants associated with diesel exhaust are known to cause chronic adverse respiratory effects after prolonged exposures. The introduction of carcinogenic materials into the mine environment through the use of diesel engines provides yet another insidious potential for chronic illness and death. These chronic effects may take years to manifest themselves. Thus, the real human impacts of introducing diesels into mines today may not be determinable for several decades. It must be emphasized that currrent or proposed standards for these individual pollutants offer no assurance of protection when the contaminants are found in combination with other toxic substances ...
    ... from a health protection posture it is obviously undesirable to use diesel engines in un-

derground coal mines. However, it is also recognized that safety and productivity considerations could potentially provide overall gains in protection of the miners' well-being. Unfortunately, none of the potentially positive aspects of diesel use have been scientifically and objectively documented at this time. Conversely, there are valid investigations which clearly demonstrate that toxic substances are introduced into underground coal mines from diesel engine exhaust ... such exposures would not be expected to produce a healthy working environment.[37]

## Shift Work

Several studies have linked shift work—especially rotating shifts—to a wide range of adverse health effects. Most U.S. mines operate on either two or three shifts,[38] and about one-third rotate shift times every week or two, according to BCOA. Larger and newer mines tend to use rotating shifts more than smaller and older mines. More than 30,000 Appalachian miners struck in protest for several weeks in 1973 over compulsory shift work. Miners' wives picketed at a number of northern West Virginia mines in 1978 over the family disruption caused by rotating shifts.

Akerstedt summarized the literature in 1975 and found:

    ... the studies ... seem to indicate that shift workers have:
    • An excess of sleep problems including short sleep time, difficulties of falling asleep and retaining sleep, not feeling refreshed after sleep, etc. The problems occur mainly in the nightwork period and often are associated with tiredness, bad mood, restlessness, digestive problems, etc.
    • An excess of minor nervous disturbance states.
    • An excess of gastrointestinal disturbances.[39]

[36]Gamble, et al., Health Implications, p. 5.

[37]Gamble, et al., Health Implications, p. 87.
[38]UMWA conventions in recent years have urged adoption of a 4-shift, 6-hour-per-shift pattern in order to increase job opportunities and safety. The fourth shift would be designated a safety and maintenance shift.
[39]Torbojorn Akerstedt, "Shift Work and Health—Interdisciplinary Aspects," Shift Work and Health (Washington, D.C.: DHEW, National Institute for Occupational Safety and Health), p. 180.

Akerstedt noted that "shift workers suffer from a host of social handicaps . . . that can be expected to affect physchological aspects of well-being, e.g., anxiety, tension, self-esteem, and similar parameters.[40]

The most recent NIOSH study by Tasto, et al., found that workers on a rotating-shift system experience significantly more accidents than those working either straight-day shift or a fixed-shift pattern.[41]

Several important questions about shift work remain to be studied. What are the biological costs of adaptation? What is the long-term cost of the "wear and tear" on the body as a consequence of repeated attempts to adjust or simply of being active at times when physiological preparedness for activity is at its lowest? What are the long-term consequences of being active during those portions of the 24 hours when the resistance to noxious agents is lowest? Do rotating shifts contribute to absenteeism and labor-management antagonisms?

## Health Effects of Job Stress

Job stress is produced by a miner's work environment. Working conditions that either make unattainable demands on the miner or do not fulfill his needs create stress. Strain is a psychological reaction to stress—that is, a deviation from normal responses. Although a recent NIOSH study found that miners do not *perceive* their work environments to be more stressful than the perceptions of other blue-collar workers, miners experienced more strain—irritation, anxiety, depression, and physical compaints—than their counterparts.[42] As measures of job stress are subjective (workers' perceptions of their own environment), it is less useful to ask "How stressful is mine work?" than to ask what possible stress-related health effects are found among miners.

The findings of the 1977 NIOSH study on stress ". . . provided support for the argument that strain produces illness."[43] The reported incidence of circulatory and gastrointestinal problems, nervous system disorders, urinary tract conditions, or musculoskeletal problems ". . . were associated with pyschological strain, and those miners with these disorders reported significantly higher strain than workers who did not experience these problems.[44] When high- and low-accident mines were compared, it was found that miners in high-accident mines reported more illness than those in low-accident mines.[45] Miners were found to experience higher morbidity and mortality from stress-related illnesses than expected.[46]

Negative health effects are likely to increase where newly initiated productivity plans increase job stress. Although some of the incentive plans tie cash bonuses to safety, none apparently tie them to illness. NIOSH concludes: ". . . it can be expected that there will be an increase in the number of health and safety problems related to behavioral factors if coal production is increased."[47]

## Conclusions

This catalog of occupational health hazards is long and discouraging. It is difficult to pinpoint the extent of occupational disease among working miners. MSHA's data are incomplete and unreliable. Epidemiological studies suggest an excess of respiratory disease, hypertension, lung and stomach cancer, and hearing impairment. But no one knows how many working miners die prematurely or suffer work-related illness. No one knows how many retired miners eventually die of work-related diseases. No one knows the extent of psychological and social stress a miner's illness inflicts on him and his family.

[40]Ibid., p. 191.
[41]B. Tasto, K. Cheaney, J. Isaacs, S. Jordan, S. Pally, and E. Skjer, *Health Consequences of Shiftwork*, Interim Contractor's Report 210-75-0072 (Washington, D.C.: DHEW, National Institute for Occupational Safety and Health, 1977).
[42]Ronald Althouse and Joseph Hurrell, Jr., *An Analysis of Job Stress in Coal Mining* (Washington, D.C.: DHEW, National Institute for Occupational Safety and Health, May 1977).

[43]Ibid., p. 63.
[44]Ibid.
[45]Ibid., pp. 124-125.
[46]C. M. Pfeifer, J. F. Stafanski, and C. B. Grether, *Psychological Job Stress and Health*, Contractor Report #CDC99-74-60 (Washington, D.C.: DHEW, National Institute for Occupational Safety and Health, 1975.)
[47]*Occupational Safety and Health Implications of Increased Coal Utilization* (Washington, D.C.: DHEW, National Institute for Occupational Safety and Health, Nov. 4, 1977), lines 848-850.

Estimates of the future number of CWP and black lung cases vary widely depending on assumptions about numbers of workers and disease prevalence. These estimates range from about 14,000 cases of CWP among working miners to 28,500 in 1985; from 10,600 CWP cases to 45,500 in 2000. A proportional range exists for bronchitis, severe dyspnea, and air obstruction. Thousands of additional cases of respiratory disability will be found among retired workers especially over the next decade (that is, among those who spent most of their work lives in uncontrolled dust conditions).

Other uncertainties must be added to this list. The safeness of the 2-mg dust standard is open to question. If 2 mg eventually proves to be too high, miners and compensators alike will pay a higher price for coal production. Further, the full health effects of coal-mine dusts are not clear. Lung and stomach cancer, skin disease, and gastrointestinal problems may be related to various noncoal constituents. While respirable dust is a known hazard, the degree of risk associated with nonrespirable dust has not been fully explored. Uncertainty also exists about whether (and to what extent) mine health hazards work synergistically. Do diesel emissions underground worsen the respiratory impact of coal mine dusts? Do large—but short—doses of mine-fire smoke increase respiratory disability?

This list of unknowns and uncertainties suggests a heavy research agenda before thousands of new miners are hired underground. Essentially, a list of research priorities can be grouped according to factors that need to be known and ways of doing things better.

### TO BE KNOWN
**Dust.—**

1. Safeness of the current 2-mg/m³ Federal standard for respirable dust.
2. Need for a standard for nonrespirable dust.
3. Need for a standard on trace-element and combustion-product exposure.
4. Possible relationship between mine dust both respirable and nonrespirable, coal and noncoal constituents with stomach and lung cancer.

5. Epidemiological study (including synergistic potentials) of underground coal miners exposed to diesel emissions.
6. Extent of work-related bronchitis and emphysema; causes and controls.
7. Epidemiological survey of respiratory diseases—both CWP and black lung—among surface miners and preparation plant workers.
8. Health effects of radioactive trace elements in underground coal miners.

**Others.—**

1. Health effects—synergistic and otherwise—of shift rotation in coal miners.
2. Epidemiological study of health effects of job stress.
3. Possible unique health effects of mining on women miners.

### TO BE DONE BETTER

1. Data collection and analysis of occupational disease.
2. Reorganization of the respirable dust-sampling program.
3. Noise sampling.
4. Training of miners and management in health hazard recognition and control.
5. Nonrespirable dust control.
6. Dust control on longwall mining units.[48]
7. Involvement of miners in occupational disease prevention programs such as hazard monitoring.
8. Engineering production systems and equipment with health-hazard control in mind.
9. Training MSHA inspectors in health matters.
10. Sampler design (both area and personal samplers should be able to give miners and management immediate feedback on dust levels).

Congress faces several obvious policy questions in the area of coal mine health. First, is

[48]In recent hearings on MSHA's dust-sampling procedures, industry health specialists said mine operators have yet to succeed in bringing long-wall units into compliance. To comply with Federal standards, mine operators are forced to rotate miners onto the mining machine for short periods to comply with the Federal 8-hour standard.

the current program of Federal standards and enforcement adequate? Second, are the projected levels of miner occupational disease acceptable? Third, if the estimated health costs of increased production are unacceptably high, what additional legislation, if any, may be needed to reduce workplace exposure to health hazards?

## Safety

Concern for the safety of American coal workers[49] has been expressed repeatedly by labor, coal operators, and public officials over the years. The reason is simple: mining is a dangerous job.

Mining is undoubtedly safer today than it was during the first three decades of this century. In 1926-30, for example, 11,175 miners were killed on the job compared with 715 fatalities in 1971-75. In 1977, 139 coal worker fatalities were recorded. Altogether, coal mining has claimed a recorded 110,833 lives since 1900 and more than 1½ million disabling injuries since 1930. In some respects, mining is safer today than it was 10 years ago but in other ways—the number and rate of disabling injuries, for instance—no improvement has been seen.

Congress passed five major pieces of mine safety legislation in this century, each one coming on the heels of a major mine disaster. Each law strengthened its predecessor. The 1969 Federal Coal Mine Health and Safety Act established the current legal framework following a 78-victim explosion at a Consolidation Coal mine in northern West Virginia in November 1968.[50] The 1969 Act mandated safety practices to prevent future methane explosions. The Act has reduced disaster-caused fatalities from 158 in 1965-70 to 37 in 1971-76.

But the 1969 Act did not address specifically the problem of injury prevention. In 1977, 229,000 coal workers experienced almost 15,000 disabling injuries and 10,000 nondisabling injuries (which did not result in an extra shift lost time). The actual number of injuries has increased each year since 1975. Injury frequency in underground mines has risen after a 2-year (1974-75) improvement. In 1977, the rate was 50.86 disabling injuries per million hours of exposure, about what it was throughout the 1950's and 1960's. About 60 coal workers experience a disabling injury every production day. Injuries rose in the 1970's because almost 100,000 additional coal workers were hired since 1969 and because the underground injury rate has not improved. If underground production in 1977 had accounted for the same share of total national output that it did in 1969 (62 percent), the industry's overall safety record would show little improvement at all.

The safety costs of producing coal are borne by mine operators, taxpayers, and miners and their families. The dollar costs alone are substantial. A recent study found that the average annual cost of each coal mine fatality was $125,000 (in 1974 dollars) and $4,000 for each disabling injury.[51] Mining companies paid about 41 percent of the total as compensation payments, lost production, and investigative costs. Wage losses to the injured miner and his family amounted to about 47 percent. Public agencies paid the remaining 12 percent as compensation and investigative costs. In 1977 dollars[52] each fatality would have cost $153,000 and each disabling injury $4,900. That means a cost of more than $73 million for 15,000 coal worker injuries in 1977 and more than $21 million for 139 fatalities. Added to these dollar costs (paid after the accident) are the expenditures for research, training, and equipment paid by operators, UMWA, and public agencies to prevent accidents from happening. The human costs of accidents—pain, mental anguish, lost opportunities, disorientation—cannot be calculated in dollars and cents.

---

[49]Coal worker refers to all workers in the coal industry, including miners (underground and surface) and workers in preparation plants, shops, and contract construction (shaft drillers, for example).

[50]Public Law 91-173.

[51]FMC Corp., *Accident Cost Indicator Model to Estimate Costs to Industry and Society From Work-Related Injuries and Deaths in Underground Coal Mining,* vol. 1, September 1976.

[52]1977 dollars calculated by assuming a 7-percent annual inflation rate in 1975, 1976, and 1977.

Mine safety can be measured in different ways. Most obvious is the actual number of recorded fatalities and injuries. This number depends on the number of workers in the labor force and the accident frequency rates. Rates for fatalities and injuries (both disabling and nondisabling) can be expressed as per million hours of worker exposure or per million tons of coal mined. Rates in exposure are more common and express accident risk from the workers' viewpoint. Accidents expressed in terms of output show directly the human costs incurred per unit of coal produced. Accidents are also measured in terms of severity—the average number of calendar days lost because of the injury. Finally, accidents are grouped according to mining method, underground or surface. Underground mining employs more workers per unit of production than surface mining and about three times as many industrywide.

Underground mining has resulted in more fatalities and higher fatal frequency rates

historically than surface mining (table 41). Between 1952 and 1970, both underground and surface fatality rates (measured in hours of exposure) did not change although actual fatalities declined because the work force was reduced by 70 percent. Since 1970, both rates have fallen: the 1977 underground fatality rate is about one-third the 1970 level; the surface rate about one-half. The gap between underground and surface fatality rates has narrowed as underground mining made proportionately greater improvement.

Post-1969 improvement in fatality reduction is not matched by a similar trend in disabling injuries. Underground mining has been consistently more than twice as hazardous as surface mining in the frequency of disabling injuries (table 42). Underground mines injure more than five times as many miners as surface operations each year. Except for 1974 and 1975, no improvement in underground disabl-

**Table 41.—Fatalities in U.S. Coal Mines, 1952-77**
**(hours of exposure)**

| Year | Underground Fatalities | Underground Rate[a] | Surface Fatalities | Surface Rate[a] | Surface fatalities as a percentage of underground |
|------|-----------|--------|------------|--------|------------|
| 1952 | 514 | .97 | 33 | .55 | 6% |
| 1953 | 440 | .96 | 21 | .39 | 5 |
| 1954 | 374 | 1.10 | 22 | .48 | 6 |
| 1955 | 391 | 1.06 | 25 | .51 | 6 |
| 1956 | 417 | 1.11 | 28 | .52 | 7 |
| 1957 | 451 | 1.30 | 22 | .42 | 5 |
| 1958 | 334 | 1.26 | 19 | .39 | 6 |
| 1959 | 268 | 1.11 | 20 | .43 | 7 |
| 1960 | 295 | 1.29 | 24 | .52 | 8 |
| 1961 | 275 | 1.34 | 17 | .39 | 6 |
| 1962 | 265 | 1.34 | 20 | .46 | 8 |
| 1963 | 257 | 1.28 | 22 | .49 | 9 |
| 1964 | 224 | 1.12 | 15 | .33 | 7 |
| 1965 | 240 | 1.21 | 18 | .40 | 8 |
| 1966 | 202 | 1.12 | 23 | .55 | 11 |
| 1967 | 186 | 1.05 | 22 | .52 | 12 |
| 1968 | 276 | 1.63 | 24 | .57 | 9 |
| 1969 | 163 | .95 | 28 | .63 | 17 |
| 1970 | 220 | 1.20 | 29 | .55 | 13 |
| 1971 | 149 | .86 | 23 | .39 | 15 |
| 1972 | 128 | .68 | 20 | .38 | 16 |
| 1973 | 105 | .56 | 16 | .28 | 15 |
| 1974 | 97 | .51 | 24 | .33 | 25 |
| 1975 | 111 | .47 | 32 | .33 | 29 |
| 1976 | 109 | .45 | 23 | .22 | 21 |
| 1977 | 100 | .43 | 27 | .23 | 27 |

[a]Expressed in million hours of exposure. Data do not include auger mines, culm banks, dredges, preparation, plants, shops, and contractors.
SOURCE: Mine Safety and Health Administration, 1978.

ing injury rates has occurred since 1952. The underground disabling rate and actual number of disabling injuries have risen each of the last 3 years. Modest improvement is seen in surface injury rates since 1970, although this trend seems to have bottomed out since 1973. The actual number of surface injuries has been rising as more miners are employed. In 1977, about 142,000 underground miners experienced 50 disabling injuries per million hours of exposure—that is, about 55 injuries every working day. In contrast, 65,000 surface miners experienced roughly 18 injuries per million hours of exposure, or about 10 injuries every working day.

Underground mining generally produces more severe disabling injuries than surface mining. In 1977, for example, the average severity of all temporary total disabling injuries in all underground mines was 73 calendar days per injury compared with 58 days for surface mines.[53] Although the average disabling injury idles a miner for 2 months or more, 98 percent are temporary total disabilities that do not produce permanent impairment. Of the 14,989 disabling injuries in 1977 for all coal production (mining, dredging, preparation, and shopwork), only 3 permanent totally disabling injuries were recorded and 225 permanent partial disabling injuries.

*Measured in terms of output rather than exposure,* a somewhat different perspective on safety is gained. When expressed in output, the productivity advantages of surface mining (more tons for less exposure) show it to be five times safer than underground mining in fatality frequency and almost nine times safer in injury frequency in 1977 (table 43). In 1977, underground mining showed 0.39 fatalities per million tons compared with 0.07 fatalities in surface operations. Surface mining showed a 42-percent decline in its *fatality rate* in 1967-77 compared with a 26-percent decline for underground mining. On the other hand, the *injury rate* in surface mining only registered a 10-percent increase in 1967-77 while underground

mining showed a 90-percent increase! In 1977, underground mining registered 46 injuries per million tons compared with 6 injuries for surface mining.

Safety data measured by output and exposure are compared in table 44. Fatality and injury rates are higher when expressed in exposure than in output. Rate comparisons for 1977 appear in table 45.

A central relationship in coal's future is that of productivity and safety. Productivity (tonnage mined per worker per shift) can be increased in different ways. Each produces a unique set of safety consequences. For example, the introduction of continuous-miner systems in the 1950's raised productivity by maintaining output and reducing the labor force. This had the effect of lowering the actual number of accidents (since fewer workers were exposed) but not accident rates.

A rough index relating safety and productivity for 1967-77 appears in table 46. Safety is expressed as fatalities and injuries per million tons of coal mined. Productivity is expressed in two distinct ways: first, as tonnage produced per worker per shift (labor productivity), and second, as tonnage produced per production day. By calculating and tracing the number of fatalities and injuries that occur every production day, safety can be associated with productivity.

The steady 43-percent decline in underground labor productivity since 1967 is matched by a 43-percent decline in the number of underground fatalities per production day. However, underground injuries per production day increased 47 percent, from 38 in 1967 to 55 in 1977. Surface productivity dropped 22 percent in this period, but surface fatalities per production day rose 20 percent from 0.10 to 0.12, and injuries rose 134 percent from 4 to 10 (table 46). If underground and surface mines are compared on a productivity-safety index, it is clear that surface mining is three times more productive, four times safer in fatalities, and almost six times safer in injuries per production day.

---

[53]Mine Safety and Health Administration, "Injury Experience at All Coal Mines in the United States, by General Work Location, 1977," table 7.

### Table 42.—Nonfatal Disabling Injuries in U.S. Coal Mines, 1952-77
(hours of exposure)

| | Underground | | Surface | |
|---|---|---|---|---|
| | Injuries | Rate[a] | Injuries | Rate[a] |
| 1952 | 28,353 | 53.26 | 1,698 | 28.11 |
| 1953 | 22,622 | 49.38 | 1,604 | 29.48 |
| 1954 | 16,360 | 47.98 | 1,342 | 29.01 |
| 1955 | 17,699 | 48.10 | 1,140 | 23.14 |
| 1956 | 18,342 | 48.88 | 1,318 | 24.31 |
| 1957 | 17,076 | 49.23 | 1,339 | 25.72 |
| 1958 | 12,743 | 47.94 | 1,150 | 23.86 |
| 1959 | 10,868 | 44.90 | 1,054 | 22.64 |
| 1960 | 10,520 | 46.09 | 1,125 | 24.45 |
| 1961 | 9,909 | 48.19 | 1,052 | 24.44 |
| 1962 | 9,700 | 48.88 | 1,027 | 23.50 |
| 1963 | 9,744 | 48.62 | 1,099 | 24.25 |
| 1964 | 9,692 | 48.35 | 1,116 | 24.75 |
| 1965 | 9,705 | 49.09 | 1,178 | 26.12 |
| 1966 | 8,766 | 48.78 | 1,043 | 25.00 |
| 1967 | 8,417 | 47.57 | 949 | 22.43 |
| 1968 | 7,972 | 46.99 | 1,039 | 24.59 |
| 1969 | 8,358 | 48.77 | 967 | 21.84 |
| 1970 | 9,531 | 51.79 | 1,346 | 25.67 |
| 1971 | 9,756 | 56.40 | 1,564 | 26.54 |
| 1972 | 10,375 | 55.32 | 1,305 | 24.84 |
| 1973 | 9,206 | 48.80 | 1,208 | 20.94 |
| 1974 | 6,689 | 34.95 | 1,229 | 16.83 |
| 1975 | 8,687 | 37.13 | 1,714 | 17.74 |
| 1976 | 11,390 | 47.09 | 2,071 | 20.04 |
| 1977 | 11,724 | 50.86 | 2,246 | 18.91 |

[a]Expressed in million hours of exposure. Data does not include auger mines, culm banks, dredges, preparation plants, shops, and contractors.
SOURCE: Mine Safety and Health Administration, 1978.

### Table 43.—Fatality and Disabling Injury Rates for U.S. Coal Mines, 1967-77[a]
(tons mined)

| | Underground | | | Surface[e] | | |
|---|---|---|---|---|---|---|
| | Tonnage[b] | Fatality rate[c] | Injury rate[d] | Tonnage | Fatality rate | Injury rate |
| 1967 | 349,133 | .53 | 24.02 | 203,494 | .12 | 4.99 |
| 1968 | 344,144 | .80 | 23.13 | 201,103 | .13 | 5.49 |
| 1969 | 347,132 | .47 | 24.04 | 213,369 | .14 | 4.79 |
| 1970 | 338,788 | .65 | 28.07 | 264,141 | .12 | 5.42 |
| 1971 | 275,887 | .54 | 35.09 | 276,303 | .09 | 5.91 |
| 1972 | 304,103 | .44 | 36.06 | 291,283 | .08 | 5.53 |
| 1973 | 299,353 | .37 | 32.20 | 292,385 | .06 | 4.63 |
| 1974 | 277,309 | .37 | 25.34 | 326,097 | .08 | 4.20 |
| 1975 | 292,826 | .40 | 31.18 | 355,612 | .10 | 5.28 |
| 1976 | 292,384 | .39 | 40.53 | 372,616 | .06 | 5.78 |
| 1977 | 255,385 | .39 | 45.54 | 407,326 | .07 | 5.53 |
| % Change 1967–1977 | − 27% | − 26% | + 90% | + 100% | − 42% | + 10% |

[a] Expressed in fatalities and injuries per million tons of coal mined.
[b] Thousands of tons.
[c]Fatality rates were calculated using fatality data from table 41.
[d]Injury rates were calculated using injury data from table 42.
[e]Surface data do not include the injury and fatality experience of auger mines, culm banks, dredges, mechanical cleaning plants, independent shops and yards, and contractors.

SOURCE: Mine Safety and Health Administration; National Coal Association, Coal Data, 1976, U.S. Bureau of Mines, Minerals Yearbooks.

**Table 44.—Accident Rate Comparisons for U.S. Coal Mines, 1967-77**

| | Fatality Rates | | | | Injury Rates | | | |
| | Underground | | Surface | | Underground | | Surface | |
| | Hours[a] | Tons[b] | Hours[a] | Tons[b] | Hours[a] | Tons[b] | Hours[a] | Tons[b] |
|---|---|---|---|---|---|---|---|---|
| 1967 | 1.05 | .53 | .52 | .12 | 47.57 | 24.02 | 22.43 | 4.99 |
| 1968 | 1.63 | .80 | .57 | .13 | 46.99 | 23.13 | 24.59 | 5.49 |
| 1969 | .95 | .47 | .63 | .14 | 48.77 | 24.04 | 21.84 | 4.79 |
| 1970 | 1.20 | .65 | .55 | .12 | 51.79 | 28.07 | 25.67 | 5.42 |
| 1971 | .86 | .54 | .39 | .09 | 56.40 | 35.09 | 26.54 | 5.91 |
| 1972 | .68 | .44 | .38 | .08 | 55.32 | 36.06 | 24.84 | 5.53 |
| 1973 | .56 | .37 | .28 | .06 | 48.80 | 32.20 | 20.94 | 4.63 |
| 1974 | .51 | .37 | .33 | .08 | 34.95 | 25.34 | 16.83 | 4.20 |
| 1975 | .47 | .40 | .33 | .10 | 37.13 | 31.18 | 17.74 | 5.28 |
| 1976 | .45 | .39 | .22 | .06 | 47.09 | 40.53 | 20.04 | 5.78 |
| 1977 | .43 | .39 | .23 | .07 | 50.86 | 45.54 | 18.91 | 5.53 |
| % Change 1967–1977 | – 59% | – 26% | – 56% | – 42% | + 7% | + 90% | – 16% | + 10% |

[a]Hours expressed as fatalities or injuries per million hours of work exposure.
[b]Tons expressed as fatalities or injuries per million tons of coal mined.
SOURCE: Mine Safety and Health Administration; National Coal Association, *Coal Data, 1976*; U.S. Bureau of Mines, *Minerals Yearbooks*.

**Table 45.—Accident Rate Comparison, 1977**

| | Hours | Tons |
|---|---|---|
| Fatality Rate[a] | | |
| Underground | 0.43 | 0.39 |
| Surface | 0.23 | 0.07 |
| Injury rate[a] | | |
| Underground | 50.86 | 45.54 |
| Surface | 18.91 | 5.53 |

[a]Fatalities and injuries per million units.
SOURCE: See table 44.

The decline in labor productivity coincides with higher accident rates in three out of four cases: 1) underground injuries per production day, 2) surface fatalities, and 3) surface injuries. Declining underground productivity coincides with fewer underground fatalities per production day. This tremendous reduction in underground fatalities is undoubtedly due to the safety practices required by the 1969 Act designed specifically to prevent multivictim disasters. It is a mistake, however, to conclude that higher accident rates are caused by declining productivity or its causes. Because Federal accident reporting requirements have been tightened twice since the late-1960's, officials believe more injuries are being reported today than 10 or 12 years ago. When many components in a system are changing rapidly, trend data measuring the change in only one must be used only with extreme caution to infer causality. Another consideration is that the number of production days was about the same throughout this period. This means that any rise in recorded injuries will show up dramatically as a rise in injuries per production day.

The decline in underground productivity was matched by a decline in underground production per day. The 23-percent reduction in underground production per day coincided with a 43-percent decline in underground fatalities per production day. But it also coincided with a 47-percent increase in underground injuries per production day. Surface mining showed a 111-percent increase in production per day that coincided with a 20-percent increase in fatalities per day and a 134-percent increase in injuries per day. The production-per-day (raw output) measure may be a better predictor of likely accident experience than labor productivity (efficiency of output) because of the great number of variables that affect labor productivity rates and the definitional problems in the concept itself (see chapter IV, *Productivity*).

Several ways of increasing labor productivity can be discussed in terms of safety. First, productivity can be increased by cutting the labor force without reducing current production. Since fewer miners are employed, the number of accidents will probably fall. But the rate of accidents—that is, the risk to the worker—may rise because of increased pro-

Table 46.—Accident and Productivity Experience, for U.S. Coal Mines, 1967-77
(tons mined)

| | Fatality[a] rate | Injury[a] rate | Productivity[b] | Employees | Estimated[d] production per day | Fatality[e] per production day | Injury[f] per production day |
|---|---|---|---|---|---|---|---|
| | | | **Underground** | | | | |
| 1967 | .53 | 24.02 | 15.07 | 103,993 | 1,567,175 | .83 | 37.64 |
| 1968 | .80 | 23.13 | 15.40 | 98,831 | 1,521,997 | 1.22 | 35.20 |
| 1969 | .47 | 24.04 | 15.61 | 97,395 | 1,520,336 | .71 | 36.55 |
| 1970 | .65 | 28.07 | 13.76 | 102,379 | 1,408,735 | .92 | 39.54 |
| 1971 | .54 | 35.09 | 12.03 | 97,740 | 1,175,812 | .63 | 41.26 |
| 1972 | .44 | 36.06 | 11.91 | 109,396 | 1,302,906 | .57 | 46.98 |
| 1973 | .37 | 32.20 | 11.66 | 100,843 | 1,175,829 | .44 | 37.86 |
| 1974 | .37 | 25.34 | 11.31 | 113,169 | 1,279,941 | .47 | 32.43 |
| 1975 | .40 | 31.18 | 9.54 | 137,060 | 1,307,552 | .52 | 40.77 |
| 1976 | .39 | 40.53 | 8.50 | 137,316 | 1,167,186 | .46 | 47.31 |
| 1977 | .39 | 45.54 | 8.58[c] | 141,411 | 1,213,306 | .47 | 55.25 |
| Percent change 1967–77 | −26% | +90% | −43% | +36% | −23% | −43% | +47% |
| | | | **Surface** | | | | |
| 1967 | .12 | 4.99 | 35.17 | 24,064 | 846,331 | .10 | 4.22 |
| 1968 | .13 | 5.49 | 34.24 | 24,400 | 835,456 | .11 | 4.59 |
| 1969 | .14 | 4.79 | 35.71 | 25,323 | 904,284 | .13 | 4.33 |
| 1970 | .12 | 5.42 | 36.26 | 31,103 | 1,127,795 | .14 | 6.11 |
| 1971 | .09 | 5.91 | 35.88 | 33,344 | 1,196,383 | .11 | 7.07 |
| 1972 | .08 | 5.53 | 36.33 | 35,364 | 1,284,774 | .10 | 7.10 |
| 1973 | .06 | 4.63 | 36.67 | 30,475 | 1,117,518 | .07 | 5.17 |
| 1974 | .08 | 4.20 | 33.16 | 44,491 | 1,475,322 | .12 | 6.20 |
| 1975 | .10 | 5.28 | 26.69 | 57,562 | 1,536,330 | .15 | 8.11 |
| 1976 | .06 | 5.78 | 25.50 | 55,993 | 1,427,822 | .09 | 8.25 |
| 1977 | .07 | 5.53 | 27.34[c] | 65,254 | 1,784,044 | .12 | 9.87 |
| Percent change 1967–77 | −42% | +10% | −22% | +171% | +111% | +20% | +134% |

[a]Expressed in fatalities and disabling injuries per million tons.
[b]Expressed in tons mined per worker per shift.
[c]Preliminary estimate.
[d]Productivity multiplied by number of employees.
[e]Fatality rate multiplied by production per day divided by $10^6$.
[f]Injury rate multiplied by production per day divided by $10^6$.
[g]Includes auger mine employment.

duction pressure. The net effect may be fewer fatalities and injuries, but higher accident frequencies. The health implications of this approach would be substantial because many of the laidoff workers would probably have lost jobs related to dust control.

Second, if production increases without any increase in employment, then productivity will increase. The introduction of a new mining technology may effect this or it may come from better work relations, new work processes, better job design, and more effective training. Incentive programs and increased production pressure may also bring about higher production and productivity from the existing work force. The safety implications of these alternative approaches vary and deserve study before being embraced.

To illustrate this point, it is useful to work out the possible safety costs of a hypothetical case of higher productivity leading to higher production. If 1977 underground productivity (8.58 tons per worker shift) had been what it was in 1967 (15.07 tons), production would have increased from 255 million to 449 million tons (assuming 210 production days and 141,752 miners). Similarly, if 1977 surface productivity (27.34 tons per worker shift) was the 1967 rate (35.17 tons), production would have increased from 407 million to 524 million tons (assuming 230 production days and 64,753 workers). Total national coal production would have increased from about 688 million to 973 million tons, a figure in line with current national goals for 1985. Underground fatalities would have risen from 0.48 per production day to 0.83 per production day; annually from 100 to 174.[54] Underground injuries would have risen from 55 per production day to 97, annually from 11,535 to 20,370. Similarly, surface fatalities would have gone from 28 (actually 29) to 37 annually, injuries from 2,146 to 2,760. This illustration, of course, is suggestive rather than definitive.

Productivity strategies that reduce the work force or increase production may prove to have substantial safety costs. However, other ways of improving productivity—technological innovation, redesign of the work process, or more efficient production systems—may actually improve safety performance either by making the production process safer or by reducing human exposure. The particular method of productivity enhancement will determine the level of safety costs. Analysis of safety and productivity is handicapped by the total lack of comparable mine-specific data that would lend itself to deriving hard-and-fast causal relationships between alternative productivity-enhancing strategies and safety. Such research is critically important to any national policy aimed at greater coal production and productivity.

Types of Accidents

Three major kinds of conditions are now associated with underground fatalities—roof falls, haulage, and machinery. One of the major pre-1969 causes of fatalities—mine disasters—has been sharply curtailed. The 1969 Act outlined some general roof-control require-

ments—prohibiting activity under an unsupported roof, requiring a roof-control plan, etc.—that have helped reduce roof-fall fatalities. Comparatively little legislative attention was paid to machinery and haulage, although some regulations—cabs and canopies, for example—have been promulgated since 1970. As a percentage of the total, roof-fall fatalities have declined from a 1956-60 average of 50 percent to a 1971-75 average of 34 percent. However, haulage and machinery fatalities are increasing as a share of total annual fatalities. Such accidents are often caused by speeding up production, taking short cuts, poor training, inadequate equipment maintenance, and not using caution as a regular part of the production process. Haulage and machinery are also the leading categories of surface mine fatalities.

*Underground disabling injuries* are most often associated with material handling, haulage, and machinery. Surface injuries are most often slips or falls (of persons), handling material, and machinery. Reducing injuries from these causes may be more related to altering production patterns, work routines, and workload/workpace pressures than to new equipment standards.[55]

On the whole, the larger the underground mine, the lower its fatality rate both in terms of exposure and output. Underground mines producing more than 500,000 tons a year generally had lower injury rates than smaller operations in 1975. Strip mines producing more than 1 million tons annually had the lowest fatality and injury rates of all (with one exception).

[54]This illustration assumes the 1977 underground fatality rate remains constant at 0.39 fatalities per million tons. Underground fatalities per production day are estimated by dividing annual tonnage by 210 days, which gives tonnage per production day. This number is multiplied by the 1977 fatality rate and divided by 1,000,000 leaving fatalities per production day. This method was used to calculate other accident-per-production-day estimates. In calculating surface-mine data, 230 days were used.

[55]A recent exception involved the requirement for cab and canopy installation on mobile equipment in low-coal. MSHA had required the installation of cabs and canopies to protect miners from roof and rib falls. However, when retrofitted on existing equipment, machine operators often found that cabs and canopies impaired their vision and forced them to assume uncomfortable and dangerous positions to do their work. Oversized equipment maximized productivity. But oversized machines cannot be safely retrofitted with cabs and canopies, although their inherent safeness has been documented. The tradeoff was between safety on one hand and productivity on the other. In this case, productivity won. Instead of requiring smaller machines fitted with cabs or canopies, MSHA rescinded its safety regulation for low-coal mines in 1977.

Auger mines—none of which produced more than 250,000 tons annually—had an extremely high fatality rate. The smallest underground and surface mines—producing less than 250,000 tons a year—generally had very high fatality and injury rates. Most coal production—both surface and underground—comes from mines producing 250,000 tons or more annually. Table 47 compares accident frequency rates for 1975 by mine size. At every size—with one exception—strip mines were safer than underground mines whether measured in exposure or output. In most cases, larger mines were safer than smaller mines, both underground and strip. Smaller mines are usually the most marginal economically. Their equipment is often older and they work more difficult seams. Large mines in the East are usually organized by UMWA, which has safety requirements negotiated into its contract.[56] Management attitude in small mines may affect safety as much as market economics, safety economies of scale, and worker participation.

### What Makes Mining Safe?

Certain economic and mine conditions appear to be associated with safe mining. Three economic variables—technology, price, and Federal regulations—are discussed below.

1. New extraction technology does not necessarily lower fatality and injury frequency. When continuous miners were phased in underground (1950-70), the fatality rate rose slightly and the disabling injury rate remained relatively constant. Longwall mining may reduce fatalities but not necessarily injuries.[57] Contour strip mining used in the Appalachian mountains is generally less safe than open-pit

mining found in the Midwest and West, but the difference is due more to terrain than to equipment used in each system.

Mining technology can be engineered for safety as well as productivity. Current underground mine technology—uninsulated electric powerlines, opaque ventilation curtains, belt haulage, electric-powered track haulage, cable-reel shuttle cars, and conventional mining machinery—deserve safety analysis. In surface mining, comparisons between different flatland excavating methods (power shovel vs. draglines) and mountain-stripping systems (auger vs. contour vs. mountaintop removal) would be instructive.

2. Coal price appears to be related to accident frequency. Both price and accident rates were relatively constant in the 1950's and 1960's. When prices rose after 1969, fatality rates, at least, declined. The coal industry's two lowest years in injury frequency were 1974 and 1975, the same years that coal prices hit record highs. Prices stabilized in 1976 and 1977, and so did fatality rates. But injury frequency—especially underground—went up. This pattern implies that operators are able to take extra steps to safen their mines when their profit margins are large and, conversely, do less when margins are squeezed.

3. Federal mine safety regulation has three components: setting standards, inspections, and civil penalties. Each component, however, does not appear to have had equal impact on safety performance.

Both miners and operators generally agree that most Federal regulations setting out electrical standards and minimum conditions and practices have safened mining. But many aspects of

[56]At UMWA mines, a miner-elected, health and safety committee exists with the power to inspect, consult with the company on mining plans, and withdraw workers from conditions of imminent danger. In addition, each individual UMWA miner has the contractually guaranteed right to refuse to work "under conditions he has reasonable grounds to believe to be abnormally and immediately dangerous to himself . . ."

[57]One study found that longwall mining had worse injury experience than either continuous or conventional systems. Leslie Boden, "Coal Mine Accidents and Government Enforcement of Safety Regulations," Ph. D. dissertation, Massachusetts Institute of Technology, 1977. Another found that no significant difference in injury trends among the various mining methods. See, D. P. Schlick, R. G. Peluso, and K. Thirumalai, "U.S. Coal Mining Accidents and Seam Thickness," The Aus. IMM Central Queensland Branch, Symposium on Thick Seam Mining by Underground Methods, September 1976.

**Table 47.—Injury Experience in Bituminous Coal Mines by Production Size, 1975**

| | Per million hours of exposure | | Per million tons of coal mined | |
|---|---|---|---|---|
| Annual tonnage | Fatal | Nonfatal disabling injury | Fatal | Nonfatal disabling injury |
| **Underground** | | | | |
| Less than 25,000 | .72 | 26.94 | 4.27 | 159.00 |
| 25,000 to 49,999 | 1.25 | 48.11 | .92 | 35.17 |
| 50,000 to 99,999 | .56 | 45.86 | .44 | 36.12 |
| 100,000 to 149,999 | .45 | 55.74 | .36 | 44.38 |
| 150,000 to 249,999 | .47 | 50.06 | .42 | 44.82 |
| 250,000 to 499,999 | .50 | 46.46 | .46 | 42.64 |
| 500,000 to 749,999 | .33 | 30.12 | .26 | 23.96 |
| 750,000 to 999,999 | .34 | 32.00 | .25 | 23.40 |
| 1,000,000 or more | .34 | 28.68 | .18 | 15.49 |
| Total average of all underground | .48 | 37.11 | .40 | 31.11 |
| **Strip[a]** | | | | |
| Less than 25,000 | .58 | 10.10 | .85 | 14.74 |
| 25,000 to 49,999 | .35 | 15.49 | .18 | 8.11 |
| 50,000 to 99,999 | .38 | 17.42 | .17 | 7.86 |
| 100,000 to 149,999 | .35 | 15.33 | .14 | 6.32 |
| 150,000 to 249,999 | .64 | 18.46 | .25 | 7.19 |
| 250,000 to 499,999 | .09 | 22.72 | .03 | 7.67 |
| 500,000 to 749,999 | .42 | 18.55 | .13 | 5.81 |
| 750,000 to 999,999 | .22 | 22.18 | .05 | 5.23 |
| 1,000,000 or more | .19 | 16.22 | .03 | 2.73 |
| Total average of all strip | .33 | 17.05 | .10 | 5.01 |
| Total all bituminous coal | .42 | 29.89 | .25 | 17.91 |

[a]Does not include auger mining and other miscellaneous surface mining.
SOURCE: Mine Safety and Health Administration, *Injury Experience in Coal Mining, 197,* p.77.

mining machinery have no standards set for them. Some in the industry challenge the inherent safeness of particular standards, such as underground lighting, temporary roof support, or cabs and canopies in thin seams. Often, a complaint against a standard is based on the costs of implementation rather than its intrinsic worth. Since compliance is basically a matter of management policy, it is reasonable to conclude that post-1969 improvement has resulted from operators complying with standards.

Federal inspection is the principal minesite enforcement tool. Several studies have found that safety improved with the frequency of Federal inspection. Boden's study of 539 bituminous underground mines producing more than 100,000 tons annually indicated that inspection leads to fewer injuries. A 50-percent increase in Federal inspection rates in this sample was predicted to lead to 11 fewer fatalities, 2,400 fewer disabling injuries, and 3,800 fewer

nondisabling injuries per year.[58] Another study estimated that one more Federal inspection per mine would decrease annual fatalities by 4 and nonfatal injuries by 52.[59]

It is difficult to discern the real effect of MSHA's civil penalty program on safety. The average assessed penalty in 1978 was about $180, of which about $150 was collected, according to MSHA. Both the average assessment and average collection in 1978 was about twice the level of earlier years. MSHA also has a "special" penalty program for serious violations—those involving negligence, and unwarrantable failure to comply, and viola-

[58]Boden, "Coal Mine Accidents."
[59]Louise Julian, "Output, Productivity, and Accidents and Fatalities Under the Coal Mine Health and Safety Act," mimeographed (University Park, Pa: The Pennsylvania State University, circa 1978).

tions leading to fatalities and injuries—that amount to $2,000 or $3,000 each. MSHA raised its assessed dollar costs on most of its violations in 1978 after an internal evaluation of its own performance found that operators said they did not alter their practices after being fined and that civil penalties were regarded as a cost of doing business—"a cheap nuisance."[60] However, while it may be true that fines themselves do not alter management practices, the threat of being fined appears to act as a deterrent. Section supervisors, for example, are evaluated according to how much coal their units produce and how many Federal violations are issued. Even the current dollar amount of civil penalties is probably not much of a deterrent, but it is unlikely that Congress or the administration would approve the necessary level of financial disincentive to create safer mines in this fashion. Further research, however, is warranted to determine the level of civil penalty that would serve as an incentive for compliance and safety.

In sum, each Federal safety component—standards, inspections, and civil penalties—has some effect on safety. But better safety performance probably will be found in emphasizing the first two and discriminating application of the third.

Many variables are related to good safety. Larger operations are generally safer than smaller ones. High output usually means geological advantages and high capital investment. Flatland surface mining is generally safer than mountain stripping. Large underground mines are often safer than contour strip mines. Thin seams (less than 48 inches) and very thick seams (over 8 feet) are the least safe underground conditions because of roof control problems. Noncompliance with Federal regulations is often associated with mine fatalities. High hourly production rates in underground mines of the same kind apparently lead to higher injury rates.[61] Miners inexperienced at a given job are more likely to be injured than task-experienced miners. Captive mines (steel-owned) are generally safer than non-captive mines, but this may have as much to do with reporting practices as safe operations.[62] No clear relationship is demonstrated between safety and unionization. Large, captive underground mines (almost all of which are organized by UMWA) are safer than small, noncaptive underground mines (many of which are non-UMWA), but it is not clear that the unionization variable makes the difference.

Several conclusions may be drawn from the available data. First, since surface mining is demonstrably safer than underground mining, additional research and enforcement should be directed at the latter. Research is needed to determine what combination of underground mine factors—equipment, management attitude, worker role, and so forth—safen operations. Since small underground mines are generally less safe than large mines, research and enforcement should be focused there.

Second, worsening underground injury experience needs to be addressed specifically. Injury-producing accidents appear to be linked to the interaction of three factors:

1. an incredibly narrow margin for error in the work environment,
2. inadequately trained workers who may not know how to—or are not permitted to—work safely, and
3. production-oriented supervisors who accelerate the normal pace of work and fail to maintain equipment properly.[63]

---

[60]*Report on Civil Penalty Effectiveness* (Washington, D.C.: U.S. Department of the Interior, Mining Enforcement and Safety Administration, June 1977).

[61]Boden, "Coal Mine Accidents."

[62]Theodore Barry and Associates, *Industrial Engineering Study of Hazards Associated With Underground Coal Mine Production,* 1971.

[63]Ibid. Management skill and management policy are variables of acknowledged importance in maintaining safety, but neither has received the necessary attention. One exception was the Barry study of mine accidents which found:

. . . almost every operator interviewed . . . cited the insufficient numbers of high-quality operating supervisors, and the resultant loss of supervisory effectiveness at the working face, as one of the major problems confronting the industry today in terms of both safety and production. (p. 259)

. . . too many foremen are not safety conscious
(continued)

Smaller mining machines that build in safety features, pneumatic, battery-powered, chain-type, or hydraulic face-to-portal haulage systems; packaging supplies in smaller, lighter units are all areas where research and development might reduce accidents. Behavioral factors also require examination. Job pressure, insufficient training, inexperienced supervision, inadequate hazard-recognition education, inappropriate worker and management attitudes—all may bear more heavily on injury-producing conditions than those more easily recognizable conditions that produce fatalities. Injury prevention may not be achievable by more rigorous enforcement of existing Federal standards if, as seems to be the case, current standards don't come to grips with the causes of mine injuries.

Third, mine safety may be improved considerably by changing certain psychological, behavioral, and organizational factors in the workplace. A recent report found significant differences in such areas between high- and low-accident mines.[64] Some of its major findings were:

- It appears that coal companies, especially high-accident mines, need to formalize safety as an organizational goal, and subsequently communicate to workers both verbally and through the behavior of management the relative importance of

[64]Westinghouse Behavioral Service Center, *Psychological, Behavioral, and Organizational Factors Affecting Coal Miner Safety and Health* (Washington, D.C.: DHEW, National Institute for Occupational Safety and Health, July 1976), pp. 125-134.

[63] (continued)
supervisors. In many cases they actively encourage unsafe practices or give instructions to perform unsafe practices. For example, the . . . the majority of fatal roof fall accidents involved unsafe deviations from roof control plans of other safety regulations; the foreman was usually aware of the deviation and had apparently given tacit approval to the unsafe procedures. (p. 261).

The Barry research is 7 years old, but it is not apparent that the industry has solved the problem of "quality" supervisors. Nor can it be proved that enough foremen have become sufficiently safety conscious. This is particularly true of the hundreds of section supervisors in their 20's and early 30's whose career advancement depends on strong production performance.

safety . . . . if coal companies truly do value safety over production, the message is apparently not reaching miners. If coal companies do not, in fact, value safety over production, then a more fundamental problem exists in terms of overall goals and orientation of the organization. Results indicate that in high-accident mines, problems stem from safety never having been adopted as an organizational goal.

- . . . a second area that deserves consideration in the coal industry is work organization and job design. It has been established in the literature that jobs which are designed with an awareness of worker needs for recognition, responsibility, and variety tend to contribute to higher worker satisfaction and improved work quality, including safety performance.

- Survey results substantiate the need for training by providing empirical evidence that good training is related to a low-accident record. Underground miners in low-accident mines indicated that their training in dealing with hazards such as gas, dust, and noise was significantly . . . better than the training in high-accident mines . . . . the fact that foremen could use additional training is evident from the fact that many foremen do not use safety equipment when necessary.

- In comparing high- with low-accident mines, two disorders were significantly more prevalent in miners working in high-accident mines—high blood pressure (29 percent versus 10 percent respectively) and 'nervous trouble' (16 percent versus 2 percent respectively) . . . it would seem that a program designed to reduce occupational stress could have a significant impact on the safety, health, and well-being of coal mining personnel.[65]

In the past, the cause of mine safety has been driven by headline-grabbing, multivictim disasters. In the future, this driving force is not likely to exist, yet fatalities and injuries are likely to rise steadily. While management reserves the right to determine the balance between production and safety, higher numbers

[65]Ibid.

of mine accidents may lead Congress to find it prudent to encourage modifications in management attitudes and priorities. A comprehensive injury-prevention program would equip miners with the knowledge, skills, and rights to protect themselves. Because so much of mining is a matter of individual miners adapting to ever-changing conditions, enfranchising them in this manner may prove to be an inexpensive and effective way of reducing accidents.

## Data Analysis

Coal mine injury and fatality data are imperfect and incomplete. A clear bias toward undercounting appears.[66] The actual human costs of coal production have been understated historically. Accurate data are crucial to wise policymaking. Valid projections of future human costs associated with increased coal production must be based on the most accurate data obtainable. It is probably reasonable to assume that fatalities have been undercounted by as much as 3 to 6 percent annually and disabling injuries by 25 percent. Both, admittedly, are "guesstimates." The *Mine Safety* appendix in volume II discusses the shortcomings in the reporting and counting of workplace accidents. That argument is highlighted here:

**Fatalities.**—Accidents involving employees of independent contractors, such as mine construction workers, were excluded from MSHA data until last year. Perhaps as many as 20,000 workers fall into this category. UMWA sources estimated that 32 coal construction workers died between 1970 and the end of 1976. A reasoned guess is that an average of four or five construction/contractor fatalities may not have been counted each year. MSHA is now trying to include contractor accidents in its

[66]Both the General Accounting Office and the Department of the Interior have reached this conclusion after studying MSHA's system. Every major study of occupational injury and disease reporting supports this finding. See for example, Jerome B. Gordon, Allan Akman, and Michael L. Brooks, *Industrial Safety Statistics: A Re-Examination, A Critical Report Prepared for the U.S. Department of Labor* (New York: Praeger, 1971), and Nicholas Ashford, *Crisis in the Workplace* (New York: Quadrangle, 1976).

data base, but it is too early to evaluate the effectiveness of the effort.

Other reporting loopholes exist. For example, persons who die from "an event at a mine" but who are not at the mine when the event occurs should be counted as mine fatalities. Usually, they are not. The 125 deaths that resulted from the collapse of a coal-waste impoundment on Buffalo Creek, W. Va., in 1972 were not reported in MSHA fatality data. Off-site fatalities that occur from boulders and debris launched by surface-mine blasting would not be usually reported. When a fatality from occupational illness occurs at a mine, it is usually not reported despite clear Federal regulations requiring the operator to do so. Determining whether such a fatality was caused by coal mine employment is medically difficult; consequently, an attempt is rarely made. MSHA data show no fatalities from occupational illness reported since 1972. For these reasons and others, puzzling discrepancies exist between Federal fatality data and those recorded by State worker-compensation programs. Finally, MSHA defined coal workers in such a way that when three of its own inspectors were killed at the Scotia mine in 1976 they were not counted as mine fatalities.

**Injuries.**—Until 1978, MSHA defined a disabling injury as "any work injury which does not result in death but which either results in any permanent impairment to the injured person or causes the injured person to lose 1 full day or more from work after the day of the injury."[67] A nondisabling injury did not result in a lost shift after the day of the injury. The distinction between disabling and nondisabling is important. Insurance premiums are linked to the former, but not the latter. Corporate safety performance is measured in the same fashion. Thus, employers had incentives to show as few disabling injuries as possible. Some mines and companies, therefore, adopted "light-duty" policies whereby an injured miner was encouraged to come to work on the day after an injury to be assigned light work. Miners call this "benchwarming" or the "bandaid brigaid." The injury was reported as

[67]Department of the Interior, Bureau of Mines, title 30, pt. 80.1(e)(f).

nondisabling rather than disabling. Miners often agreed to light-duty assignments because they feared job reprisals. It was also far less bother to go along with the company and collect an uninterrupted paycheck than it was to file for workers' compensation and await bene-

fits that were less than normal pay. But by not reporting the injury as disabling, the worker lost any claim to future compensation. In a case—such as a back injury—that may worsen over time, the miner could be severely penalized.

Photo credit: Earl Dotter

Disabled UMWA miner and family on strike in Harlan County, Ky., 1974

Injury data may also be affected by how individual companies choose to interpret their Federal accident-reporting obligations. Interior Department investigations in 1975 found extensive misclassification of accidents.[68] Interior found in its sample that only about 17 percent of all injuries were ever reported to MSHA's data collection center.[69] About 39 percent of disabling injuries were not reported, including: 1) finger amputated to first joint (8 lost workdays); 2) foot bruised with complication (13 lost workdays); and 3) knee bruised and infected, fractured fingers, and cracked rib. About 44 percent of nondisabling injuries were not reported. Underreporting resulted from oversight, misinterpretation, and deliberate circumvention of reporting regulations. An Interior update of the 1975 survey turned up similar findings.[70] In 1975 and 1976, nearly 65 percent of all active mines did not report a single injury. Underreporting was most suspect at larger mines and less suspect at smaller ones. Interior's "basic conclusion" was that MSHA's "accident/injury data submitted by mine operations cannot be relied on as reasonably accurate." Suspect reporting was the result of "corporate policy on reporting practices," Interior said. Recent survey-research of 30 underground mines found that "foremen in low-accident mines were reported to turn in inaccurate safety reports more often than foremen in high-accident mines."[71]

Within the last year, MSHA has tightened its definition of disabling injuries to reduce underreporting. MSHA believes that the new definitions of disabling injury has cut underreporting by two-thirds. Some MSHA officials say unofficially that more than 80 percent of all injuries are now being reported under the new system. Some observers argue that accidents were undercounted by as much as 60 percent under the old definitions.[72] The 1975 Interior audit found 39-percent undercounting of disabling injuries. A correction factor of 25 percent is used in table 48 to illustrate possible accident experience for 1967-77. Underreporting of tonnage and accidents lowers productivity, distorts accident data (thereby confusing policy analysis), understates future compensation costs, and lowers worker morale and respect for Federal safety efforts.

**Safety Projections.**—The safety implications of increased coal production are critically important to Congress. The dollar costs of fatalities ($153,000 each) and injuries ($4,900 each) are immense. Production and productivity are reduced when accidents occur. Coal workers and their families experience pain and suffering. For these reasons and others, Congress has expressed sustained interest in minimizing coal mine accidents. Unfortunately, fatalities and injuries will increase as more coal is mined unless steps are taken to reduce accident frequency.

The likely range of fatalities and injuries associated with higher production levels is shown in table 49. These estimates assume no change in certain variables—production pressure, productivity rates, accident reporting reliability, seam conditions, and work force and management characteristics. Those variables that have been considered are: number of coal workers and production levels.

Fatality and injury rates for 1977 were used to calculate the number of fatalities and injuries in 1985 and 2000. These rates may go down. If they are lowered sufficiently, they will offset the increasing number of workers exposed to safety hazards, thereby reducing estimated accident totals. However, if the rates improve only modestly, the actual num-

[68]"Review of Accident/Injury and Production/Manhour reporting under the Federal Coal Mine Health and Safety Act of 1969 administered by the Mining Enforcement and Safety Administration," U.S. Department of the Interior, Office of Audit and Investigation, memorandum of Nov. 19, 1975.

[69]Ibid., p. 5.

[70]Opportunities to Improve the Effectiveness of Mining Enforcement and Safety Administration Accident/Injury and Employment Production Data Information Systems (Washington, D.C.: U.S. Department of the Interior, Office of Audit and Investigation, 1977), pp. 6, 15.

[71]Westinghouse Behavioral Service Center, Psychological . . . Factors, p. 84.

[72]L. Thomas Galloway of the Center for Law and Social Policy argued that the MESA injury rates did not reflect up to 60 percent of all coal mine accidents because of underreporting. See Galloway testimony in U.S. Senate, Federal Mine Safety Health Amendments of 1976, hearings on S. 1302, 94th Cong., 2d sess., Mar. 24, 25, 30, and 31, 1976, p. 920.

Table 48.—All U.S. Coal Estimated Accident Experience, 1967-77
(hours of exposure)

| | Nonfatal disabling injuries recorded[a] | Rate as calculated by MSHA[b] | Unrecorded Nonfatal disabling injuries, additional[c] | Estimated adjusted rate[d] |
|---|---|---|---|---|
| 1967 | 10,115 | 41.84 | 2,529 | 51.82 |
| 1968 | 9,639 | 41.12 | 2,410 | 51.13 |
| 1969 | 9,917 | 41.76 | 2,479 | 54.70 |
| 1970 | 11,552 | 44.40 | 2,888 | 62.32 |
| 1971 | 11,916 | 46.89 | 2,979 | 53.70 |
| 1972 | 12,329 | 46.08 | 3,082 | 52.53 |
| 1973 | 11,220 | 40.41 | 2,805 | 36.58 |
| 1974 | 8,545 | 28.79 | 2,136 | 34.28 |
| 1975 | 11,107 | 30.17 | 2,777 | 45.51 |
| 1976 | 14,389 | 37.55 | 3,486 | 47.82 |
| 1977 | 14,989 | 38.37 est. | 3,698 | 45.68 est. |

[a]Nonfatal disabling injury data provided by Mine Safety and Health Administration. Includes all mining, culm banks, dredges, plants, and shops.
[b]Injury rate is expressed in injuries per million hours of exposure.
[c]Applies a 25 percent rate of undercounting.
[d]This rate is a rough estimate. It was calculated by multiplying the number of workers employed by 8 hours by the average number of days worked in each year (from National Coal Association, Coal Facts, 1974-1975 and Coal Data 1976). This sum was divided by one million and then divided into estimated total of accidents (column 1 plus column 3). Since the hours worked per worker and the number of days worked each year are very rough calculations these estimated accident rates should be seen as approximations.

Table 49.—Mine Worker Fatality and Injury Estimates

| | 1977 | 1985[b] Low | High | 2000 Low | High |
|---|---|---|---|---|---|
| **Surface** | | | | | |
| Fatalities | 29 | 25 | 30 | 32 | 46 |
| Injuries | 2,281 | 1,954 | 2,401 | 2,515 | 3,634 |
| **Underground** | | | | | |
| Fatalities | 100 | 118 | 140 | 203 | 291 |
| Injuries | 11,724 | 13,907 | 16,513 | 24,012 | 34,405 |
| **Other Coal workers[a]** | | | | | |
| Fatalities | 10 | 14 | 17 | 24 | 34 |
| Injuries | 984 | 1,586 | 1,891 | 2,653 | 3,804 |
| **Total** | | | | | |
| Fatalities | 139 | 157 | 187 | 259 | 371 |
| Injuries | 14,989 | 17,447 | 20,805 | 29,180 | 41,843 |

[a] 1977 data include workers in shops and cleaning plants, but not construction workers. Estimated data for 1985 and 2000 include all other coal workers and uses a 10 percent add-on to the total of underground and surface accidents.
[b]Low estimates are keyed to the workforce estimates found in the low-case growth assumptions (100 Quads) in chapter II, table 6. High estimates are keyed to the high-case growth assumptions (150 Quads).

ber of fatalities and injuries will still rise over current levels because of the larger number of workers employed. Accident frequency rates have not improved substantially for several years, and it may be that the 1969 Act will not have any additional impact on lowering them.

Between 157 and 187 coal workers are likely to be killed and between 17,400 and 20,800 injured in 1985. That represents a 13- to 35-percent increase in fatalities over 1977 and a 17- to 39-percent increase in injuries. By 2000, between 259 and 371 coal workers are estimated to be killed and between 29,200 and 41,800 injured. These estimates represent an 86- to 167-percent increase in fatalities over 1977 and a 95- to 180-percent increase in injuries. These calculations assume no underreporting and undercounting. The 25-year total (1976-2000) of mine fatalities may exceed 5,000 and injuries may exceed 500,000.

## Conclusion

These estimates of future fatality and injury costs are disquieting. No one likes to measure the cost of electricity in human life and limb. Yet it is clear that coal production two to three times higher than current levels will greatly increase the number of fatalities and injuries unless frequency rates are cut sharply or productivity rises spectacularly (which would mean fewer workers exposed). Neither seem likely to materialize in the next 10 years assuming "business as usual." Frequency rates have not improved since the mid-1970's. Productivity is likely to rise in the future, but very slowly.

Congress clearly stated its position in the 1969 Act, which provided for the "attainment of the highest degree of safety protection for miners." The Act recognized that "deaths and serious injuries from unsafe . . . conditions and practices in the coal mines cause grief and suffering to the miners and their families." They also cost money and coal.

Thousands of hours of production will be lost because of injuries. Productivity will be lowered. Hundreds of millions of dollars in worker compensation benefits will be paid. Coal workers and their allies may increase their resistance to management over safety issues. Absenteeism and wildcat strikes may erupt. Congress may again become the object of intense pressure for even more stringent mine safety legislation. For all of these reasons, the effort to step up coal production and productivity ought to be accompanied by increased efforts to minimize fatalities and injuries. So far, this has not occurred.

# COMMUNITY IMPACTS

## Introduction

Coal mining shapes the social, economic, and political life of the communities in which it occurs. The scope and intensity of coal's local impacts differ from region to region[73] depending on the nature of the precoal economy, the extent of local coal development, and the level and regularity of demand for the local product. Local and regional impacts were also determined by the kinds of mining methods employed, level of mechanization, size of the mines, extent of economic diversification, local sociopolitical structures, company ownership patterns (local or absentee), presence or absence of unionization, and topography.

Mining has usually occurred in remote areas, where industrialization and other forms of commerce were not well-developed. Often, coal became the only "cash crop." One-crop economies — be they cotton, cocoa, or coal — often mean many social and environmental costs are externalized. Where mining was the principal economic activity, its effect on community life was greatest. Where mining was part of a diversified economy, its impact was less broad and less deep.

Appalachia has been America's bituminous coal bucket for a century; it is there that coal has impacted most heavily. What strikes the observer of coalfield Appalachia is the near total absence of other primary economic enterprise. The land itself was not hospitable to much more than subsistence farming, so mining quickly replaced existing socioeconomic structures. In Ohio, Illinois, and western Kentucky the agricultural base was sufficient to resist being supplanted. Yet even there, other

---

[73]Coal production has been centered historically in three regional coalfields: the Appalachian fields ranging from Pennsylvania to Alabama; the Midwestern fields including Illinois, Indiana, west Kentucky, and portions of Kansas and Iowa; and the Western fields covering Colorado, Wyoming, Montana, Utah, New Mexico, and Arizona.

industrial activity did not generally follow the development of local coal resources.

Coal operators often had to build not only mining and transportation systems, but also whole towns, complete with commercial and professional services, housing, roads, cultural activities, and political systems. Each community was almost totally dependent on the marketplace success of its local operator. The public infrastructure in these communities — its institutions, services, and personnel—was an adjunctive activity for the operators, whose principal concern was the profitability of their coal business. Company town were often a profitmaking business for a coal company even in slack times. Many times the town's profits were the only ones an operator could show for being in the coal business. In some cases, operators saw community social investment as an unproductive and unprofitable use of scarce capital. When competition was intense and demand slack — as it was for most of the 20th century—support for community institutions was often the first economy operators effected.

The pattern of coal development in the Midwest and West differed less in manner than scope. There, too, the company-town model was used, but it was less frequent and less pervasive. More often, coal commercialization occurred in existing towns whose lifeblood was agriculture, which continued alongside coal production. Coal development in the Midwest did not snuff out traditional economies and social systems as it did in much of Appalachia. Economic diversity enabled midwestern communities to survive coal's lean times. Finally, because the demand for Midwestern and Western coals was generally less than for Appalachian coals, mining never dominated these regions as it did Appalachia.

Rocky Mountain coal development was even slower and less comprehensive than that of either the Midwest or Appalachia, although it impacted specific communities with equal intensity. Markets were generally limited to steam-grade coal for railroads and industrial activity. Mining towns and boomtowns existed in the West prior to coal development; coal continued the boomtown pattern. As the seams played out or demand evaporated, the jerry-built boomtowns tended to be abandoned. In contrast, Appalachians tended to stay where they were despite cyclical coal demand.

Coal development shaped the political structures and culture that evolved in coal towns. Coal entrepreneurs had economic reasons for developing harmonious relationships with the local officials who administered the legal apparatus, tax system, and public services. In many cases, coal operators had to create a political system because none existed, often tying the public sector directly into the administration of the mining enterprise. As a result, the political culture that took root was widely perceived by residents to be the public expression of the local coal company.

In terms of social structure, in much of the Eastern coalfields, coal's method of industrialization meant that communities were divided into two basic classes: management and labor. The antagonisms produced in the mine workplace never dissipated in the environs of the company town where the workplace division was maintained socially. Miners' grievances have usually been played out through on-the-job militancy rather than in electoral politics because miners believed their power was more effective in the workplace.

The two-class structure prevented a noncoal middle class from forming. In other parts of America where an independent middle class developed, civic reform was forced on entrenched political elites. Had an independent middle class existed, it is reasonable to suppose that coalfield politics would have been pluralized and more adaptive to cha... As it was, however, civic improvement was left to the discretion of individual coal companies. Some companies made good-faith efforts to supply their employees with decent housing, medical attention, education, and public services. The majority, however, could not or did not spare the money and effort. Yet even where operators made a good-faith effort, they did so on their terms.

For these reasons and others, public administration in many coalfield communities today—particularly in Appalachia—is often characterized by a lack of professional skills and institutions. Local politics is often tied to political machines and personalities. Corruption and inefficiency are not unknown. In the slack years of the 1950's and 1960's, public administration was heavily influenced by patronage considerations, as the public sector often replaced the private sector as the major local employer. Because the political system was so much an extension of local economics, public authority could do little to regulate the widespread cost externalizations of mining.

These historic patterns of underdevelopment have produced in much of the Eastern coalfields stunted private and public infrastructures that cannot respond to the needs of rapid coal expansion.

Because underground mining is the most common system in Appalachia, the community impacts there have been far greater in proportion to output than in the West where surface mining dominates and requires far fewer miners per unit of output. Even though most new coal output is scheduled to come from surface mines west of the Mississippi River, five times as many miners will work in the East than in the West in 1985 (chapter II, table 6). If the impact problems of increasing coal production were to be translated into one simple concept—more people—the socioeconomic impacts will be predictably greater in the East, and especially in central Appalachia, than in the West.

Coal mining brings benefits and costs alike to local communities. Both have a private and a public side. Further, both have a dollar side and a nondollar side. In developing a mine, however, the cost-benefit calculus lies entirely with the individual entrepreneur. He establishes the scope of the project, its methods, and its rate of development. It is his ratio of (private) dollar costs to dollar benefits that determines whether a project is begun. It is on the assumption of profitability (that is, benefits over costs) that he ventures capital. The costs in this case are those that the company pays (internalizes) to produce coal at a profit.

Public costs are generally not evaluated by the mine developer because they are not directly relevant to his balance sheet.

The local private sector bears a share of the costs of increased coal development. The private sector in this instance consists of three parties: the operator, his employees, and local private interests. Private costs can be divided into three categories: costs of production that are internalized by the operator, costs of production that are externalized and absorbed either by the workers or by local private interests, and opportunity costs. Internalized production costs are carefully projected and regulated by mine operators and are a normal part of doing business. Externalized costs and opportunity costs are not readily quantifiable but may be gauged by looking at occupational injuries and disease (including compensation programs), private health care utilization patterns and costs, lack of economic diversification, shortage of local investment capital, stress on community values and social structure, etc. Opportunity costs include all of the existing local business that ends as a result of coal mining, together with all noncoal business that might have occurred had mining not happened.

Coal production directly benefits its employees and their communities through wages and fringe benefits. The public sector also benefits from production. Economic growth helps a community prosper. Taxes may be levied on coal property, wages, sales, corporate income, and on the product itself to support public services. Presumably, a relationship can be drawn between the level of private-sector benefits on one hand with public-sector benefits on the other. It might reasonably be expected that where the private sector—both owners and employees—benefits, the public also gains.

But such a relationship has not always held in the Eastern coalfields throughout much of this century because of the volatility of coal demand, the boom-bust cycle, and other unique factors. In the first two decades, the level of public benefits was determined by the coal operator who owned the community. The public interest was defined by the private sector. The company-town model polarized the private sector into "company men" and employees; neither group was capable of advocating an independent public interest. Even today as increased spendable income from mine employment reflects "economic growth", the development of community services and institutions lags. Growth is not synonymous with— and may not lead to—development. Community polarization continued long after companies liquidated company towns. For these reasons, increasing private benefits have not been matched by increasing public benefits. Public benefits from coal production have also been circumscribed historically by protracted demand stagnation. Even as America's consumption of energy quadrupled from 1920 to 1970, coal's share dropped from about 78 percent of the total energy market to 19 percent.[74] Stagnating demand meant that coal towns were never able to levy reasonable taxes on their principal local business for fear of shutting it down. Thus they never had the money to provide adequate public services. Studies of taxation patterns in coal counties find historical patterns of underassessment of undeveloped coal property and undertaxation of that which is assessed. States passed severance taxes in the early 1970's with great difficulty, and only when demand and prices were rising. Low taxation functioned as a public subsidy to local operators. Whatever the merits of the subsidy, it resulted in the chronic underdevelopment of coalfield public services and institutions. The public sector—especially at the local level—was thus forced to bear disproportionately high costs while reaping disproportionately few benefits from coal development through the 1960's. On the other hand, whatever public benefits that existed in these areas came from coal. To that extent, coal development was a source of both community benefit and community deficit.

Similarly, local governments rarely regulated coal's social costs. Operators often sited

their mines outside of incorporated areas to avoid taxation and land-use restrictions. State legislatures and Governors rarely chose to do battle with politically powerful, economically imperiled coal operators. These economic realities and attitudes inhibited predictable and sustained development of public services in coal communities. Long-term planning was not undertaken. Eventually, coalfield expectations about public services became permanently undervalued. What people never had, they were encouraged never to expect. Doing without became the norm.

Congress has debated an energy-impact assistance bill in recent sessions. Funds would be provided from general Federal revenues rather than through a national severance tax. The Farmers Home Administration (FmHA) proposed guidelines to disperse $20 million to coal- and uranium-impacted communities in March 1979. The assistance would be used for growth-management plans, housing plans, housing, public facilities, and services. However, little of this aid will find its way to Appalachian communities because the FmHA growth requirement—8 percent or more in coal employment in the year following the base year—is too high to qualify the well-settled, high-employment areas there. Congress may increase appropriations for this purpose orconsider a coal-financed contribution.

Much of this history is discouraging and pessimistic. Perhaps sustained, increasing coal demand will correct these problems. If that is to happen, policymakers need a clear picture of current coalfield conditions and the reasons why things are as they are. The following sections describe these conditions in the East and West.

## Impacts on Eastern Communities

More than 90 percent of all of the coal ever mined in the United States has come from States east of the Mississippi River. Nearly 50 percent of the Nation's remaining demonstrated coal reserves are found there. Billions of dollars of coal have been—and will be—mined from under the Appalachian Mountains and the fields of the Illinois Basin.

The visitor to the Central Appalachian coalfields is struck by its contrasts. Trains of 100

---

[74]See National Coal Association, *Coal Facts, 1974-1975,* pp. 58-59. In 1920, about 592 million tons of bituminous and anthracite coal were consumed whereas in 1970, only 524 million tons were used domestically. Domestic consumption of bituminous coal was 509 million tons in 1920 and 516 million tons in 1970.

cars regularly haul hundreds of thousands of dollars worth of coal past grimy, railside shacks. Yet amid the dingy grimness of many old coalfield towns, the visitor will sometimes come upon a bustling county seat that has been turned into a multicounty, commercial center within the last 5 years. The natural beauty of the mountains contrasts starkly with the scars and blemishes they bear from past mining practices. Finally, the visitor will meet at least three distinct groups of people: the very poor, the middle-income working miner, and the well-to-do.

The typical Central Appalachian coalfield community is small, congested, and lies along a stream or small river. Mountains often circumscribe the town and define its growth. Here and there comfortable ranch-style brick houses appear. Most buildings are old and rarely renovated, except for an occasional new fast-food restaurant or quick-service grocery, a modern brick post office, or public building. The roads linking the outlying hollows to the towns (and the towns to each other) are often narrow, poorly engineered, dangerous, and falling apart. Mobile homes are wedged between mountain and highway wherever a flat place can be found or bulldozed. Yet the visitor will also see a meticulously tended garden beside each house. Well-kept churches and cemeteries are the rule. Evidence that this is coal country appears frequently: a tipple; a portal; an old strip mine highwall or slag heap; a bumper sticker that says, "We dig coal."

Within a typical community it is common to find severely inadequate water and sewage systems, a low level of most public services, an almost complete absence of public transportation, a shortage of adequate housing, undercapitalized financial institutions, limited education programs, and a general feeling that the quality of community life is not what it could be. Although coal miners make $17,000 or more in a normal year, many find it difficult to purchase quality goods and services that non-Appalachians with similar incomes take for granted. Thus coalfield residents often feel individually deprived and publicly disadvantaged. As a result, there is a general ambivalence about the coal industry. On one hand,

miners feel an intense pride in their profession and appreciate its wages and benefits. On the other, many believe they and their communities have been victimized by the ups and downs of coal demand and the cost-consciousness of their employers.

North into Pennsylvania and west through Ohio, Indiana, Illinois, and Kentucky, the coal towns change. The mountains flatten out. The visual impact of old strip mines is less severe. The towns are less cramped and seem more prosperous. Farming coexists with mining. In the Ohio River Basin, manufacturing and other businesses make coal less important. Public services appear to be closer to national norms. The towns are less coal towns than towns where coal mining occurs along with other businesses. The distinction is significant. In one, community life is totally dependent on the fortunes of a single industry. In the other, economic development is balanced and has been cushioned from the consequences of coal's quick booms and long busts. Finally, miners here appear to enjoy a better quality of community life and seem able to use their incomes more efficiently. One sees fewer very poor or very affluent.

Coal mining's impact on community life varies with conditions in the marketplace, terrain, the extent of economic diversification, and the kind of industrial socialization each community experienced when coal was developed. Generally, the social, political, and economic effects of coal mining have been most severe where communities were totally dependent on coal, where the terrain was inhospitable to other activity, and where mining was the principal socializing force in community life. When these factors were less dominant, the public benefits of mining were greater.

The extensive underdevelopment of much of the Appalachian coalfields is not difficult to explain. First, demand for coal has remained stagnant for most of the last 55 years. This placed tremendous cost-cutting pressure on coal companies. Economy was effected by holding down labor costs, mechanizing, and minimizing tax burdens through the exercise of political power at the local and State levels. Neither miners nor indigenous coal operators

Coalfield housing located in the Cityview area of Logan County, W. Va.

got very far ahead until the 1970's. That meant coalfield communities lacked investment capital, tax resources, and spendable income.

Second, mining companies were forced to externalize production costs in order to survive. Thus strip mines and slag heaps were not reclaimed; coal-haul roads were abused; streams were polluted. Most State and local efforts to internalize these costs through taxation and regulation were fought and generally defeated until the 1970's. Had such efforts been successful, some marginal companies would have gone under and others would have had a harder time.

Third, Appalachia has served America as a resource exporter. Coal mining did not stimulate investment in manufacturing, often limited development of other economic activity. Coal users usually did not locate their manufacturing and processing near coal mines. Again, a diversified economic base did not take hold.

Fourth, the biggest mining operations were usually owned by nonindigenous companies. Some of these companies developed adequate communities to service their mines; others did not. Mining profits—such as they were—were not deposited in coalfield banks, so mining's financial benefits to coal towns were limited principally to wage and salary income. Consequently, the private sector in many coalfield communities has been chronically short of cash. Native entrepreneurs usually engaged in the most marginal kinds of business, mining and otherwise, because capital was so hard to assemble. These two factors—initial inadequacies in community-building and subsequent capital shortages—left coal towns with chronically unmet needs and lacking the ability to solve their own problems.

Any analysis of the future socioeconomic impacts of increasing Eastern coal production must confront the social legacy of past coal development. The post-1973 Appalachian "coal boom" has not been based on increased production. Indeed, production has fallen—particularly underground tonnage. The boom was one of price and profit on management's

side and wages and jobs on labor's. That combination, lacking as it does the necessary premise of more output, cannot be and has not been sustained. Even as the promise of coal growth beckoned, 5,000 to 10,000 miners were laid off or placed on short work weeks in recent months as metallurgical demand dropped and because utilities chose to meet air pollution standards by using low-sulfur coals rather than through installing scrubbers or other sulfur control measures.

With these shifting developments, it is difficult to forecast a uniform set of socioeconomic implications from increased Appalachian coal mining. Two to three dozen counties may find it extremely difficult to manage either growth or stagnation, as neither condition is predictable. For other coal counties, production will expand steadily and a range of growth-related social and economic issues will demand attention. As a result, Federal policy will need to be flexible and address three coal-related issues of Appalachian community development:

1. The current residual deficit in facilities and services in both stagnating and expanding communities.
2. The problem of continued uneven coal demand affecting particular communities or sub-State areas.
3. The problems of rapid coal development.

The capacity of any community to benefit from coal expansion depends to a great extent on the seriousness of its existing underdevelopment. Where this deficit is greatest, expansion will produce the most problems; where it is least, growth will be accommodated and the ability of communities to cope with associated social problems will be greatest.

### Scope of Eastern Coal Expansion

Estimates of new Eastern coal production vary, rising or falling according to the optimism and assumptions of the estimator. One survey forecast 255 million tons of new capacity in 11 Eastern States by 1987, of which 197 million tons would be deep-mined and 58 million tons surface-mined.[75] This report estimates 212 million tons of planned capacity east of the Mississippi by 1985 (chapter II, table 5). More significantly, no net increase in Eastern coal production is foreseen through 1985.

Appalachia produced 390 million tons in 1977, (had there not been a 10-week wildcat strike that summer and a month-long shutdown in December, output would have been around 425 million tons). Appalachian production in 1985 should range from 355 million to 415 million tons and in 2000 from 510 million to 680 million tons (chapter II, table 5). The three remaining Eastern States—Illinois, Indiana, and Kentucky (west)—produced 133 million tons in 1977. Actual production should range between 128 million and 164 million tons in 1985. Substantial expansion of eastern production is likely to occur after 1985. From 510 million to 680 million tons of Appalachian production and 206 million to 299 million tons from Illinois, Indiana, and western Kentucky is expected by 2000.

These estimates suggest that the East as a coal-producing region has 6 years in which to plan for higher production. However, some communities will be heavily affected by coal-related growth within these 6 years. Others will not experience any expanded production but will probably be required to absorb additional coal-related employment.

Forecasting community impacts of increased coal production hinges on the amount of coal actually produced in the future (rather than theoretical capacity) and the proportion of new production mined by strip and deep methods. Underground mining now requires roughly 550 miners to produce 1 million tons; surface mining uses about 160 miners for the same amount at current productivity rates. Accordingly, between 144,000 and 167,000 miners will be employed in Appalachia in 1985. No increase in net Appalachian mine employment is forecast for the next 6 years, although in-

dividual counties may show net gains of miners. In Illinois, Indiana, and western Kentucky between 29,000 and 38,000 miners should be employed in 1985. As net production increases between 1985 and 2000, mine employment will probably increase proportionately. These projections assume no change in productivity and do not include nonminer coal workers, such as mine construction workers, and those who work in preparation plants, tipples, and shops.

For each new miner, it can be assumed that five other persons—spouse, children, and secondary-employees—will be added to local communities. The net population increase from coal production in the East is likely to range from zero to 100,000 persons by 1985. Specific counties will be most heavily affected when new deep mines are put in where the existing social deficit is greatest. The most impacted counties will probably be in southern West Virginia, eastern Kentucky, southwestern Pennsylvania, and some areas of the Illinois Basin.

### Existing Conditions

Social and economic conditions in Eastern coalfield counties differ widely. Compared with the United States as a whole, the 50 leading coal counties in the East showed gross deficits in income, educational attainment, and housing in the 1970 census.[76] Central Appa-

---

[75]George F. Nielsen, "Keystone Forecasts 765 Million Tons of New Coal Capacity by 1987," *Coal Age,* February 1978, pp. 113-134. The 11 States are Alabama, Georgia, Illinois, Indiana, Kentucky, Maryland, Ohio, Pennsylvania, Tennessee, Virginia, and West Virginia.

[76]Census data for 1970 require two caveats. First, they are dated. Socioeconomic improvement should be recorded in 1980. But the statistical improvement in income that will be seen must be weighed in light of inflation's impact on real income. While fewer coalfield residents will fall below national poverty criteria, this may reflect only statistical betterment. Gains measured in per capita or family income may not indicate any real relative improvement among the poorer sectors in the community. Per capita and family income data may also be somewhat misleading where a few residents have become very rich.

The second qualification of the poverty and income data has to do with who is poor in coal counties. Coal miners would have normally earned between $8,000 and $12,000 in 1970 (depending on wage grade and days worked), a level clearly above the poverty line. In 1977, this increased to between $16,000 and $23,000 in current dollars. But the coalfield poor in 1970 were not working
(continued)

lachian counties had the highest deficit, Illinois Basin counties the next highest, and northern Appalachian counties the least when compared with the other two. Measures of poverty have historically been higher in coal counties than in most noncoal counties. The more dependent a county has been on coal, the higher its poverty index. In the 1920's and 1930's, this was the result of the industry's low-wage policies and slumping demand. In the 1950's and 1960's, mineworker unemployment and lack of economic diversification caused poverty to increase. The 50-county coal region had a 43-percent greater incidence of poverty than the United States as a whole in 1970, while family poverty was 53-percent higher than the national average.[77] The comparatively strong showing of northern Appalachian coal counties is due to their diversified economies—especially high-wage industries such as iron and steel, chemicals, metal products, glass, transportation, and electric power. By contrast, much noncoal economic activity in the Illinois Basin is agricultural.

The range of social problems coinciding with poverty places extra burdens on communities whose tax resources are already limited. The 1970 census data show that much of the Eastern coalfields entered the mid-1970's "boom" with few financial resources. Despite the gains in mining wages, personal income in the Appalachian region was below the national

average in 1975. From Kentucky's low of 66 percent of the national average, with Virginia (73 percent), Ohio (79 percent), Alabama (82 percent), and West Virginia (84 percent) in between.

Similarly, the 50 coal counties had nearly one-fourth more than their share of persons with less than a high school education. About 59 percent of persons 25 years and older lacked a high school diploma. Northern Appalachian coal counties had the lowest incidence of poverty and the lowest percentage of those with less than a high school education. Central Appalachian counties had the highest incidence of both.

One reflection of educational quality in Appalachia is provided by comparing regional achievement test scores with National or State rankings (see table 50). The widest discrepancies between county performance and National/State reading and math standards occur in Central Appalachia, principally a coalfield area. There, 69 percent of those counties reporting were below National/State standards on reading achievement in 1976-77 and 94 percent were substandard on math. Students in northern and southern Appalachia performed better than National/State standards on reading, but did less well in math.

Rapid population expansion from coal development will add to problems of educational quality. In Raleigh County, W. Va., for example, even though a $20-million construction program was begun in 1973, the county superintendent admits that "in four or five cases we've had to put two teachers in one classroom with 50 kids." He anticipates "this sort of thing . . . will probably occur more often in the future." In Greene County, Pa., local school officials feel that school construction should be postponed until after the population expands, with portable classrooms used in the interim. When a large number of young, married coal miners settles near new coal mines, local school systems will face front-end financing problems in building facilities. Educational quality will be adversely affected by overcrowding, inadequate facilities, and teacher shortages.

---

[76] (continued)
coal miners. Many were unemployed miners, retired and disabled miners, and widows of miners. Some portion of the coalfield poor had nothing to do with the coal industry. Although miners' pensions have improved since 1974 and Federal black lung benefits now pump dollars into coal towns, there remains a thick slice of mining-related residents who live on modest fixed incomes. They are disadvantaged by coal-boom inflation. They do not contribute as much to local tax revenues on a per capita basis as active miners. Each coalfield community must absorb some of costs of providing services to this group. Even as coal growth brings income gains to working miners, a class of poor and lower income residents will remain. They will be economically and psychologically disadvantaged by the coal-based advancement of their neighbors.
[77] Calculated from *County and City Data Book, 1972* (Washington, D.C.: U.S. Department of Commerce, Bureau of the Census, 1973), tables 1 and 2. See vol. II, app. XII.

**Table 50.—County Rankings on Achievement Tests, 1976-77**
**(3rd or 4th grades)**

| Geographic region | Percent of Appalachian counties Reporting | Reading | | Math | |
|---|---|---|---|---|---|
| | | Percent of counties at or above National/State standard | Percent of counties below standard | Percent of counties at or above National/State standard | Percent of counties below standard |
| Appalachia ........................................ | 57% | 66% | 34% | 43% | 57% |
| Northern subregion .......................... | 42% | 63% | 37% | 58% | 42% |
| Central subregion ............................ | 19% | 31% | 69% | 6% | 94% |
| Southern subregion ......................... | 88% | 62% | 38% | 39% | 61% |

SOURCE: Appalachian Regional Commission, 1978. Assembled from data collected from State Departments of Education.

Housing is crucial to expanding coal production. Coalfield housing is of low quality compared with national norms. More than 31 percent of the Central Appalachian housing units surveyed in 1970 lacked plumbing facilities; 15.6 percent lacked them in the Illinois Basin; and 8.6 percent in northern Appalachia. But coalfield housing is more complicated than plumbing inadequacies, although they are a good index of housing quality. The most urgent issue is supply. Every study of coalfield housing finds the existing housing stock insufficient. It is not likely to expand fast enough to house adequately additional mining-related population. The current housing deficit is the product of the coal slump of the 1950-70 period when few units were built. It is compounded today by a lack of adequate water and sewage facilities, an absence of private builders, a shortage of mortgage capital, an unwillingness of coalfield banks to make housing loans, and, most fundamentally, the unavailability of land on which to build.

Water and sewage systems are expensive. They do not exist in many Appalachian coal towns. In Raleigh County, W. Va., only half the population is served by public sewage facilities. The rest use septic tanks, package plants, outhouses, or the nearest stream. The head of the local chamber of commerce says a total moratorium on new construction could be imposed by health agencies because of sewage problems. One local health official says the only thing preventing a typhoid epidemic from striking some sewerless hollows is the acid mine drainage that kills the bacteria in the

befouled creeks. A recent West Virginia State health department study found that 70 percent of the water from public supplies in the county was substandard. Raleigh County residents complain of tapwater the color of orange pop. In Mingo County, W. Va., primary sewage treatment facilities exist in only two communities. Drinking water for Williamson, the county seat, comes from the Tug River, which the county says is polluted by acid mine drainage, preparation discharges, and diesel-oil runoff from adjacent railroad yards. Williamson's treatment plant is old and overloaded. The city is seeking State and Federal funds to replace it. In Breathitt County, Ky., only Jackson, the county seat, has sewage service. The pervasive lack of water and sewage facilities makes coalfield housing development costly. It is especially difficult when State and Federal regulations require such services.

In many coal counties, most of the land is owned by a handful of corporations, which are reluctant to sell, preferring to hold the land for future development. The underlying coal makes the land worth considerably more than sale for housing would justify, especially when taxes are low. One study found that in the 14 major coal-producing counties in West Virginia, the top 25 landowners—all of whom are connected to the coal industry—owned as much as 2.1 million acres of land, or nearly 44 percent of all of the land in these counties.[78] The top 10 landowners who owned 31 percent

[78] J. Davitt McAteer, *Coal Mine Health and Safety: The Case of West Virginia* (New York: Praeger, 1973), p. 163.

of the land in West Virginia, were, by rank: Pocahontas Land Corporation (a wholly owned subsidiary of the N&W Railroad), Consolidation Coal, the Chessie System, Georgia-Pacific Corporation, Eastern Associated Coal Corporation, Island Creek Coal Company, Bethlehem Steel, Charleston National Bank, Berwind Corporation, and Union Carbide. Nine of the top 10 landowners in these 14 counties had headquarters outside of West Virginia.[79] This study reported that the top 25 landowners had their property assessed at about one-fifth of the 14-county total assessment. A second study in West Virginia found that "two dozen out-of-State corporations and land companies—all tied directly or indirectly to mineral industries—own a third of the State's 12 million privately held acres."[80] This report found that "in almost 50 percent of West Virginia counties, at least half the land is owned by out-of-State corporate interests . . . [that often pay] as little as $2 per acre in annual property taxes.[81] The most recent study found that nearly 80 percent of the land in Mingo County, W. Va., is owned by major corporations, most of which were related to coal mining and natural resources.[82] The Appalachian Regional Commission has recently funded a major landownership survey to facilitate housing and economic development.

Given the steep terrain and the flood-proneness of the valleys, flat land suitable for permanent housing is inherently limited. Mountainside housing is possible in many areas, but development costs (for water, sewage, and roads) are high. Individual homebuilders can rarely afford to build there. Federal programs have not adequately explored using this type of land for coalfield housing. Although the coalfields are rural, their population is congested along narrow creek bottoms. Without additional land, the population density of coalfield towns and hollows will continue to increase or miners will be forced to commute

long distances; this, in turn, leads to absenteeism at work.

A few landholders have begun to make some housing land available. The Beth-Elkhorn Corp., a subsidiary of Bethlehem Steel, is located in Jenkins, Ky. In 1976, Beth-Elkhorn provided title in fee simple to 92 acres of land on an abandoned strip mine bench for residential development to contain 48 single-family, 62-multifamily, and 33-clustered housing units. Unfortunately, the development costs of using surface-mined land make this apparently logical solution of limited feasibility in many places. Island Creek announced an even bigger housing development for southwestern Virginia. Another example is the Raleigh County-based Coalfield Housing Corporation (CHC), a joint industry-union effort begun in March 1977. CHC has 26 units under construction and 210 additional units scheduled. Land availability, financing, and lack of builders have been its three major obstacles. The CHC model has not been repeated elsewhere in the coalfields. Recently West Virginia Governor John D. Rockefeller IV initiated land condemnation proceedings against the Philadelphia-based Cotiga Land Co., which refused to sell land to the State at what the Governor considered a reasonable price for victims of the 1977 Tug River flood.

Housing and land shortages force miners to remodel old coal-camp houses or purchase mobile homes instead of permanent residences. These are bunched in towns or strung closely together along valley floors. Mobile homes may shortchange local tax coffers, as their owners use a full range of community services but often pay personal rather than real property taxes on their dwellings. Their owners often complain about construction quality, rapid deterioration, fire hazards, and low resale value. Some of the newer and more expensive models address these concerns. Many residents believe that coal companies prefer to have their workers housed in mobile homes so that they can be moved out of the way when a mine needs the land. On the other hand, some coal executives believe one solution to wildcat strikes is to have miners buy quality housing that comes with a long and

[79]Ibid., pp. 140-149.
[80]Tom D. Miller, "Who Owns West Virginia?," Huntington, W. Va., Huntington Herald-Dispatch, 1974, pp. 2, 23.
[81]Ibid.
[82]Study prepared by the staff of the Tug Valley Recovery Center and the Sandy New Era, published in the Sandy New Era, Feb. 1, 1979.

Flood damage in Wyoming County, W. Va., 1977

heavy mortgage. If significant production is expected from the eastern fields after 1985, land must be made available for single-family housing, and a housing construction industry must be created.

Coalfield underdevelopment is reflected in the quantity and quality of other public services. For example, the Appalachian Regional Commission funded an assessment of the impact of coal movement on Appalachian highways.[83] This study found that coal movement by truck affected more than 14,300 miles of Appalachian roadway in 1974. Of that total, 6,880 miles of coal-haul roads were inadequate to meet the present volume of coal-truck traffic. Between 897 and 1,103 bridges were inadequate. The cost of maintenance was placed at $66 million to $81 million annnually (1977 dollars), bridge replacement at $591 million to $726 million, and roadway reconstruction and rehabilitation at $4.1 to $4.9 billion.[84] Only a tiny fraction of these sums is now spent to maintain, replace, and reconstruct coal-haul roads. These numbers do not adequately convey the visible wreckage of the public road system in many Appalachian coal counties. Some portions have simply slid down a mountainside. Spillage and coal dust from overloaded trucks have caused problems for motorists and landowners. Major secondary lines must be traveled with a sharp eye for axle-breaking potholes. Taxpayers pay for the constant upkeep on coal-haul roads while consumers pay repair bills for their vehicles. The main cause of the problem — apart from the initial shortcomings in construction and upkeep — is the illegal overloading of coal trucks. Numerous studies of this phenomena have reached the same conclusion: coal trucks regularly exceed legal axle weights, offenders are rarely cited, and fines are so low as to be no deterrent at all. State officials are reluctant to tighten up on enforcement. Federal officials are reluctant to use Federal law — which requires States to enforce Federal gross and axle

weights on interstate highways in their boundaries or suffer a cutoff of Federal highway aid. Without strict enforcement of weight limits, road repair money will have little long-term value.

Coalfield health care is generally inadequate and extremely vulnerable to overloading from rapid population growth. The Appalachian Regional Commission reports that most of coalfield Appalachia is underserved by physicians, dentists, nurses, and hospital beds compared with the U.S. average. Many measures of community health — infant mortality, fluoridated water, prenatal care, per capita public health expenditures, immunizations, etc. — indicate that much of coalfield Appalachia falls below national standards. Many of these same problems were first brought to national attention in 1947 when the Department of the Interior published a comprehensive survey of coalfield health, under the leadership of Administrator Joel T. Boone.[85] Much improvement has been recorded since the Boone report, but current data reveal many unmet needs. In 1950, UMWA won an operator supported system of medical insurance and health services providing near comprehensive benefits to UMWA miners and their families. Hospitals and clinics were built. Preventive medicine and group practice were encouraged. The Fund brought dozens of doctors and hundreds of other health professionals into the coalfields. Today, part of this system is floundering, and some of the rest is uncertain. Severe health care curtailments have already been recorded at about four dozen Funds-dependent clinics. The cutback in clinic services resulted from the Funds ending first-dollar coverage for miners along with replacing retainer payments to the clinics with fee-for-service reimbursement. The West Virginia University Department of Community Medicine found that total doctor office visits fell 31 percent since the imposition of copayments on physician care in the five clinics surveyed. The financial condition of the clinics worsened measurably. Where utilization fell, receivables rose — by as much as 190 percent in

[83]Research Triangle Institute, *An Assessment of the Effects of Coal Movement on the Highways in the Appalachian Region* (Washington, D.C.: Appalachian Regional Commission, November 1977).
[84]Ibid., p. vi.

[85]*A Medical Survey of the Bituminous Coal Industry* (Washington, D.C.: U.S. Department of the Interior, Coal Mine Administration, 1947).

Photo credit: Douglas Yarrow

Shopping center construction, Beckley, W. Va., 1977

one clinic. Operating deficits were recorded in each quarter studied. Physician staffing declined by 42 percent from July 1977 through September 1978. Nonphysician staff declined by 25 percent. Special preventive health programs have been eliminated. Facilities and equipment are not being adequately maintained. Pharmacy and medical supply inventories have plummeted.[86]

The Funds population had a higher-than-normal hospitalization rate, which may be related to the shortage of ambulatory physican care. Some operators and physicians interpret

high hospitalization rates as being a reflection of a kind of "cultural hypochondria" that predisposes Appalachian miners and their families to overuse medical facilities. Hospital utilization will be encouraged by the physician drain that began with the Funds cutbacks in the summer of 1977 and by implementation of the 1978 UMWA contract, which imposes coinsurance payments on physician care but not on hospitalization.

Coalfield health care issues revolve around questions of availability and quality. In most coal counties—with certain prominent exceptions—there are too few quality doctors and health care extenders, facilities, and prevention programs. Much of this shortage is due to the remoteness and isolation of these counties, and the lack of certain kinds of cultural and recreational facilities for health professionals. But these limitations were overcome by the UMWA Fund in the 1950's and do not present an overwhelming obstacle to recruitment to-

---

[86]William Kissick and American Health Management and Consulting Corp., "West Virginia Primary Care Study Group: Problems of Reimbursement," Department of Community Medicine, West Virginia University, update, November 1978. See also Virginia Gemmell and Jane Ray, *Physician Loss in Central Appalachian Coalfield Hospitals and Clinics*, draft report by the Appalachian Regional Commission, 1978; and Charles Holland, et al., *West Virginia Primary Care Clinics*, 1978.

day. Federal and State programs designed to encourage general physician care in rural areas have had some success. But even where new doctors with rural-practice specialization go into coal towns, communities have trouble retaining them. Where significant expansion of the coal population occurs rapidly, currently overloaded treatment systems will function even less adequately.

The system of group-practice clinics and regional hospitals initiated by the UMWA Fund may still serve as a model for coalfield health care. The Fund recruited health personnel carefully. Quality control on services was maintained on non-Fund providers. Outpatient and community outreach programs were emphasized. The entire system was supported by an operator-paid tonnage royalty. When demand was steady or rising, the system worked well. Only when demand fell steadily over a period of years, did the Fund falter. If future demand increases as most estimates predict, a coal-supported health system of this kind may again prove to be the most practical solution.

Six Case Studies

Six high-growth Eastern coal counties were examined to determine their socioeconomic problems related to coal development and to assess their capacity to manage additional mining (volume II, appendix XII). Most Eastern coal counties have had to cope with some level of mining-related population growth even though, in many cases, production did not increase at all. These six counties represent the spectrum of capabilities and problems that exist in the Eastern fields.

The picture that emerges suggests that—to various degrees—short-term coal expansion has already overloaded Raleigh and Mingo Counties, W. Va., and Breathitt County, Ky. Greene County, Pa., Tuscaloosa County, Ala., and Perry County, Ill., have been less coal-dependent historically and appear to be able to manage increased coal production with little disorientation.

Although coal production has not increased significantly in Raleigh County, its county seat, Beckley, has been turned into a regional growth center for southern West Virginia. The price rise of metallurgical coal during the mid-

1970's turned Beckley into a boomtown. The city is trying desperately to manage a host of housing, water, sewage, and transportation problems. Although Raleigh has been a fair-sized coal producer for decades, the fact that Beckley has been turned into a center of regional commerce and finance has broadened the county's economic base. Yet prosperity continues to be keyed to coal. Because of Beckley's regional importance, its resources are greater than most coalfield county seats. Raleigh County is now, ironically, experiencing a severe economic slump that came with slack demand for local metallurgical coal. With little noncoal industry in the county, its economic fortunes rise and fall with demand for local coal. Little attention appears to have been given to economic diversification during the short-lived boom.

Mingo County, W. Va., has not expanded production, although more miners have been hired in the last 6 years. Some new mines have opened, but others have cut back. A flood swept through Williamson, the county seat, in 1977, devastating the community. The economy was slow to recover. A report to President Carter in March 1978 found that:

It has been almost impossible to acquire land to accommodate coal mining families whose homes were ruined in early 1977 floods or who are coming in for additional coal employment. The land not on steep slopes or on flood plains is largely held by landowners unwilling to sell their land for housing, preferring to hold it for speculation or resource development,[87]

As nearly 80 percent of Mingo County's land is owned by nonlocal, coal-owning corporations, tight land policies present major roadlocks to efficent growth management. Many public services—education, water, sewage, recreation, and health—are inadequate. More flooding has occurred since 1977. The viability of the coal industry in the Tug River Basin rests partly on better flood-prevention measures.

Breathitt County, Ky., experienced a 109-percent increase in coal production between 1970 and 1976. Almost all of its output is sur-

[87]Report to the President on Energy Impacts Assistance (Washington, D.C.: U.S. Department of Energy, Energy Impact Assistance Steering Group, March 1978), app. A., p. 3.

faced mined. Mining accounts for 40 percent of all public and private employment there. Production is expected to increase by 50 to 75 percent by 1986. Breathitt County was one of America's poorest counties; coal growth is welcomed by its residents. Still because of earlier underdevelopment, Breathitt is experiencing adverse impacts. (Because the expansion is in surface mining, the population impacts are expected to be less severe than if significant underground production were scheduled.) Breathitt County lacks many services. It has no hospital. Sewage treatment is found only in the county seat. No public transportation system exists. The County has a comparatively high tax rate, but its base is very small. Unmined coal was assessed at only $11.3 million in 1976, or about $0.025 per ton of recoverable coal (441 million tons). The State has increased assessments in recent years, but revenue is unlikely to cover needed services.

Greene County, Pa., is a deep-mining, coal county where almost 25,000 coal-related additional persons are estimated to be living by 1996—a number equal to the county's current population. A Westinghouse study of Greene County projects a total population increase of 118,300 by 1996.[88] The county is not well prepared to handle the impacts of massive and rapid growth.

In contrast, Perry County, Ill., and Tuscaloosa County, Ala., appear generally to be able to benefit from anticipated coal expansion. Both counties have diversified coal economies, with mining representing 1 percent of Tuscaloosa's employment and 17 percent of Perry's. The rate of coal expansion and its extent are more modest than those in Greene or Breathitt Counties, which enable growth to be managed more effectively. Individual communities in each county may be seriously affected by the opening of a mine, but this appears to be little cause for concern countywide.

Any rendition of the historical impacts of mining on Appalachia obscures the benefits that coal has brought and the region's growth

potential. Several factors will work to facilitate eastern growth. First, there exists a tradition and orientation among the population toward coal mining. Miners tend to be the children of miners. The lore of the work is handed down from generation to generation. The problems of the craft are understood by wives and families. Local schools are geared toward mining-related training programs. Local colleges usually offer mine engineering degrees. A second strong point is the production infrastructure that already exists. Despite the problems of coal-haul roads and rail and bargeline bottlenecks, a mine-to-market transportation system is in place. Further, there is an adequate support system consisting of equipment manufacturers, repair shops, and supply retailers, who can respond to increased mining. Local lawyers and brokers understand the coal business. The operators know the lay of the land both literally and figuratively. Coal resources are well known. The components of a production system are familiar. Problems of expanded development are not new. What is not greatly in evidence is the recognition that the ability to expand coal production is linked to the ability of mining communities to benefit from that expansion.

Since the early 1970's many eastern communities have experienced expanding mining populations and/or increased production. The seriousness of community impacts varies according to the variables discussed above. Most coal counties are aware that coal impacts will increase, but few have begun to do the necessary planning and resource accumulation that impact-management requires because of the dollar costs and time involved. To be effective, planning must occur before coal impacts begin. Existing Federal programs have not met the planning and front-end financing needs of coal-affected communities. Planning and resource management push communities to evaluate their tax systems, revenues, service-delivery systems, and likely sources of non-local assistance. As coal impacts have fallen on coalfield governments, communities are becoming aware that local taxation systems are inadequate. First, they rarely generate sufficient income to provide adequate public services. Second, they do not adequately tax the

[88]Westinghouse Environmental Systems Department, *Greene County and Environs Energy Impact Study* for Greene Hill Coal Co. (Pittsburgh, Pa.: Westinghouse Corp., 1977).

community's main assets: undeveloped mineral wealth and current coal production.[89] In Mingo County, for example, coal and land companies—which own almost 80 percent of the land—had an aggregated property tax levy in 1978 of $231,000 compared with $960,000 for private residents and local businesses.[90] Many coal counties assessed undeveloped coal at only a penny or two per ton, and tax revenues amounted to only a few dollars per acre each year. Individual counties and States have been reluctant to impose burdensome coal-related taxes. They feared the increased cost of local coal would disadvantage local operators in the marketplace. Nevertheless, many counties needing to raise substantial additional revenue are now eyeing local coal resources. Coal companies are challenging higher assessments in several counties. From a national perspective, differential tax burdens (from State to State) may create inefficient incentives for wise resource development.

Perhaps the most severe coal impact that communities have been unable to resolve is the issue of land availability and housing. Coal operators have generally chosen to stay out of the housing business and are reluctant to assist private developers. The private housing market is inadequate to build middle-income housing. Most coal counties have had little experience in public housing. Multiunit townhouses or high-rise developments do not fit the housing expectations of local residents. The stopgap solution of mobile homes is not likely to produce community stability and growth or to benefit the industry in terms of employee morale and absenteeism.

The level of current and future costs and benefits partly depends on the extent of existing underdevelopment. As plans are carried out to increase Eastern coal production, community development strategies will be required to achieve a positive ratio of benefits to

[89]Some Eastern States such as West Virginia have undertaken a comprehensive reevaluation and assessment program that promises to bring more coal-related tax revenues to coalfield governments. But West Virginia tax officials admit that the higher valuations of undeveloped coal property will not produce sufficient income to finance the range of needed services.

[90]See *Sandy New Era*, Feb. 1, 1979.

costs. Strategies of this sort are likely to call for economic diversification in coalfield communities, citizen participation, new land and taxation policies, regulated growth, and fewer cost externalizations. Strategies of this sort may slow the rate of coal development initially, but advance the East's ability to produce coal over the long term.

## Impacts on Western Communities

### Introduction

Coal development is occurring in eight States of the northern Great Plains, the Rocky Mountains, and the Southwest—North Dakota, Montana, Wyoming, Colorado, Utah, Texas, New Mexico, and Arizona. (Limited development is also found in the State of Washington.) Coal production in these States amounted to some 166 million tons, or nearly 24 percent of the Nation's coal output in 1977. By 1985, some 630 million additional tons of annual capacity may be available there. Currently, about 17,000 miners are employed in Western coal mining. Between 34,000 and 42,000 miners will work in this region in 1985. Coal mining is only one cause of coal-related impacts in the West. The construction and operation of coal-combustion facilities, such as coal-fired powerplants and coal-gasification plants, also affect isolated communities.

The West shares some characteristics with the Appalachian coalfields, but significant differences stand out. Both regions export raw materials, although the variety of Western resources extracted is far greater. Both regions have depended on imported capital to develop their resources. As the economic return to communities where minerals are extracted is often less than most other industrial activities, both areas have experienced relatively limited growth rates. Finally, each region has evolved an identity and cohesiveness that come, in part, from the collective perception of economic dependency or subservience to the rest of the Nation.

The West is not a homogeneous region geographically or socially. Its coal-producing areas are characterized by widely scattered towns, with typical populations of 5,000 persons or fewer. The counties are much larger in area than counties in the East and exhibit a varied topography. Much of the area is subject to Federal jurisdiction. Roughly 44 percent of the land area in the major Western coal States is owned either by the U.S. Government or by various Indian tribes.[91] Actual land control, owing to Federal coal-leasing policy, is even higher.

In contrast to the relative cultural homogeneity of the Appalachian coal region, the Western fields encompass a majority white population and three major cultural subgroups: Indians, Hispanic Americans, and members of the Church of Jesus Christ of Latter Day Saints (Mormons). Regionwide generalizations about attitudes toward growth are risky given this cultural mosaic. For historical and cultural reasons, attitudes toward coal development by minorities may differ from those of coal producers. The problems faced by Indian communities in managing coal-related development and the governing structure of the reservation system are unique. They are considered separately in this chapter.

Water is a prized commodity in the arid West. The ability of communities and coal operators to obtain sufficient water will shape the extent of coal development. There is little or no western water that is not already allocated to existing water-rights holders, primarily agricultural users. Obtaining water rights for coal development may occur smoothly through negotiated purchase. But in some areas of high water demand, or where State policies promote conservation of the agricultural sector, conflicts may develop between private and public entities. In market competition for scarce water, energy developers will almost always be able to offer higher prices than agricultural and ranching interests. Special protection for certain groups of water consumers may be warranted in this regard.

## Identification of Impacted Communities

With one exception, the distribution of new production throughout the West will be relatively uniform. Six States—Colorado, Montana, New Mexico, North Dakota, Texas, and Utah—are each expected to add 55 million to 61 million tons of new annual capacity between 1977 and 1987, according to estimates presented by *Coal Age*.[92] (These estimates are referred to as the "Keystone Case" throughout the remainder of this discussion.) The exception is Wyoming, which *Keystone* estimates to be able to produce more than 270 million additional tons annually by 1987. Campbell County, Wyo., alone is scheduled to have an annual production capacity of nearly 203 million tons per year over its 1976 level. Arizona and Washington are expected to have minor increases in capacity of 3 million and 1 million tons per year, respectively. Table 51 presents a county-by-county breakdown of estimated additional western capacity by 1987, by both surface and underground mining methods, developed from the *Keystone* data.

These estimates are generally considered to be optimistic. They are derived from a survey of coal producers on projected new mines and planned expansions of existing mines. Many of the proposed mines have not begun to acquire the necessary permits and approvals. Some capacity will never materialize because of unfavorable market conditions or unanticipated problems.

In order to evaluate more accurately the likely range of impacts on western communities, a low-production "case" (the OTA case) was calculated by State and county. Likely slippages in each State's production, owing to a slowdown in coal gasification efforts, reluctance to resume large-scale Federal coal leasing, and other factors were identified. (See volume II, appendix XIII for complete discussion of statistical methodology). Total new miner requirements for both production estimates, along with the 1976 population estimate for each county, are presented in table 52.

---

[91]*Public Land Statistics, 1975* (Washington, D.C.: U.S. Department of the Interior, Bureau of Land Management, 1976).

[92]George F. Nielsen, "Keystone Forecasts 765 Million Tons of New Capacity by 1987," *Coal Age*, February 1978, pp. 113-114.

## Table 51.—New Western Coal Production Capacity by County, 1977-86
### (in millions of tons)

| State and county | Keystone case | | | OTA low-production case | | |
|---|---|---|---|---|---|---|
| | Surface | Underground | Total | Surface | Underground | Total |
| **Arizona** | | | | | | |
| Navajo | 3.30 | — | 3.30 | 0.00 | — | 0.00 |
| **Colorado** | | | | | | |
| Adams | 12.50 | — | 12.50 | 2.44 | — | 2.44 |
| Archuleta | 0.10 | — | 0.10 | 0.10 | — | 0.10 |
| Delta | 0.25 | 1.80 | 2.05 | 0.25 | 1.02 | 1.27 |
| Elbert | 1.80 | — | 1.80 | 1.80 | — | 1.80 |
| Garfield | 3.00 | — | 3.00 | 0.40 | — | 0.40 |
| Gunnison | — | 3.70 | 3.70 | — | 0.27 | 0.27 |
| Huerfano | 0.10 | — | 0.10 | 0.10 | — | 0.10 |
| Jackson | 3.00 | — | 3.00 | 3.00 | — | 3.00 |
| Las Animas | — | 3.25 | 3.25 | — | 0.96 | 0.96 |
| Mesa | — | 8.85 | 8.85 | — | 3.76 | 3.76 |
| Moffat | 6.30 | 1.85 | 8.15 | 6.30 | 1.85 | 8.15 |
| Pitkin | — | 1.50 | 1.50 | — | 1.29 | 1.29 |
| Rio Blanco | — | 4.20 | 4.20 | — | 1.81 | 1.81 |
| Routt | 3.60 | 4.30 | 7.90 | 3.60 | 2.74 | 6.34 |
| Weld | — | 0.30 | 0.30 | — | 0.30 | 0.30 |
| Total | 30.65 | 29.75 | 60.40 | 17.99 | 14.00 | 31.99 |
| **Montana** | | | | | | |
| Big Horn | 35.50 | — | 35.50 | 14.28 | — | 14.28 |
| Crow Reservation | 13.00 | — | 13.00 | — | — | — |
| McCone | 5.00 | — | 5.00 | — | — | — |
| Musselshell | 1.20 | — | 1.20 | 0.17 | — | 0.17 |
| Richland | 0.10 | — | 0.10 | 0.10 | — | 0.10 |
| Rosebud | 8.16 | — | 8.16 | 4.09 | — | 4.09 |
| Total | 62.96 | — | 62.96 | 18.64 | — | 18.64 |
| **New Mexico** | | | | | | |
| Colfax | 0.50 | 1.50 | 2.00 | 0.50 | — | 0.50 |
| McKinley | 20.70 | — | 20.70 | 11.13 | — | 11.13 |
| San Juan | 37.70 | — | 37.70 | 15.37 | — | 15.37 |
| Total | 58.90 | 1.50 | 60.40 | 27.00 | — | 27.00 |
| **North Dakota** | | | | | | |
| Bowman | 0.60 | — | 0.60 | 0.60 | — | 0.60 |
| Dunn | 14.00 | — | 14.00 | 4.07 | — | 4.07 |
| Grant | 1.50 | — | 1.50 | 0.44 | — | 0.44 |
| McLean | 22.00 | — | 22.00 | 11.34 | — | 11.34 |
| Mercer | 11.80 | 4.00 | 15.80 | 5.63 | 2.00 | 7.63 |
| Oliver | 2.60 | — | 2.60 | 2.60 | — | 2.60 |
| Stark | 0.46 | — | 0.46 | 0.32 | — | 0.32 |
| Total | 52.96 | 4.00 | 56.96 | 25.00 | 2.00 | 27.00 |
| **Texas** | | | | | | |
| Atascosa | 6.00 | — | 6.00 | 2.98 | — | 2.98 |
| Bastro | 2.50 | — | 2.50 | — | — | — |
| Freestone | 0.60 | — | 0.60 | 0.60 | — | 0.60 |
| Grimes | 4.00 | — | 4.00 | 0.18 | — | 0.18 |
| Henderson | 8.80 | — | 8.80 | — | — | — |
| Hopkins | 0.18 | — | 0.18 | 0.18 | — | 0.18 |
| Milam | 9.60 | — | 9.60 | 3.57 | — | 3.57 |

Table 51.—New Western Coal Production Capacity by County, 1977-86 (continued)
(in millions of tons)

| State and county | Keystone case | | | OTA low-production case | | |
|---|---|---|---|---|---|---|
| | Surface | Underground | Total | Surface | Underground | Total |
| Panoca | 16.00 | —— | 16.00 | 11.98 | —— | 11.98 |
| Robertson | 6.00 | —— | 6.00 | 1.98 | —— | 1.98 |
| Titus | 4.60 | —— | 4.60 | 4.60 | —— | 4.60 |
| Total | 58.28 | —— | 58.28 | 26.07 | —— | 26.07 |
| **Utah** | | | | | | |
| Carbon | —— | 10.20 | 10.20 | —— | 7.00 | 7.00 |
| Emery | 4.00 | 15.00 | 19.00 | 2.19 | 8.86 | 11.05 |
| Kane | 11.00 | 6.00 | 17.00 | 6.03 | —— | 6.03 |
| Sevier | —— | 8.25 | 8.25 | —— | 4.33 | 4.33 |
| Wayne | 1.00 | —— | 1.00 | 0.77 | —— | .77 |
| Total | 16.00 | 39.45 | 55.45 | 8.99 | 19.99 | 29.18 |
| **Washington** | | | | | | |
| Lewis | 1.00 | —— | 1.00 | 2.00 | | 2.00 |
| **Wyoming** | | | | | | |
| Campbell | 202.50 | —— | 202.50 | 112.73 | —— | 112.73 |
| Carbon | 17.10 | 7.90 | 25.00 | 6.80 | —— | 6.80 |
| Converse | 13.00 | —— | 13.00 | 3.43 | —— | 3.43 |
| Hot Springs | 1.00 | —— | 1.00 | 0.78 | —— | 0.78 |
| Lincoln | 7.60 | —— | 7.60 | 5.69 | —— | 5.69 |
| Sheridan | 7.00 | —— | 7.00 | 6.04 | —— | 6.04 |
| Sweetwater | 10.25 | 0.95 | 11.20 | 7.89 | 0.95 | 8.84 |
| Uinta | 3.00 | —— | 3.00 | 2.63 | —— | 2.63 |
| Total | 261.45 | 8.85 | 270.30 | 145.99 | 0.95 | 146.94 |

SOURCE: George F. Neilsen, "Keystone Forecasts 765 Million Tons of New Coal Capacity by 1987," *Coal Age*, February 1978, pp. 113-134.

Table 52.—Additional Manpower Requirements for Western Coal Development, 1977-86
(*Keystone* case and OTA low case)

| State and county | 1976 est. pop. | Manpower Requirements | | Annual population growth rate, %/year | |
|---|---|---|---|---|---|
| | | *Keystone* case | OTA low case | *Keystone* case | OTA low case |
| **Arizona** | | | | | |
| Navajo | 61,200 | 184 | 0 | 0.18 | 0.00 |
| **Colorado** | | | | | |
| Delta | 19,100 | 624 | 353 | 1.81 | 1.06 |
| Gunnison | 8,800 | 1,068 | 78 | 5.62 | 0.52 |
| Las Animas | 16,200 | 1,217 | 360 | 3.79 | 1.26 |
| Mesa | 65,400 | 3,051 | 1,297 | 2.50 | 1.11 |
| Moffat | 8,900 | 576 | 576 | 3.34 | 3.34 |
| Pitkin | 9,600 | 544 | 413 | 2.97 | 2.32 |
| Rio Blanco | 5,300 | 762 | 328 | 6.42 | 3.21 |
| Routt | 10,200 | 987 | 769 | 4.68 | 3.80 |
| Total | | 8,829 | 4,174 | | |

**Table 52.—Additional Manpower Requirements for Western Coal Development, 1977-86 (continued)**
(*Keystone* case and OTA low case)

| State and county | 1976 est. pop. | Manpower Requirements | | Annual population growth rate, %/year | |
|---|---|---|---|---|---|
| | | *Keystone* case | OTA low case | *Keystone* case | OTA low case |
| **Montana** | | | | | |
| Big Horn | 10,600 | 1,804 | 727 | 7.29 | 3.51 |
| Crow Reservation | 2,383 | 661 | 0 | 10.30 | 0.00 |
| McCone | 2,700 | 333 | 0 | 5.70 | 0.00 |
| Rosebud | 9,900 | 538 | 270 | 2.86 | 1.53 |
| Total | | 3,336 | 997 | | |
| **New Mexico** | | | | | |
| Colfax | 13,300 | 430 | 383 | 1.79 | 1.61 |
| McKinley | 56,000 | 1,478 | 794 | 1.48 | 0.82 |
| San Juan | 67,700 | 1,336 | 545 | 1.13 | 0.47 |
| Total | | 3,244 | 1,722 | | |
| **North Dakota** | | | | | |
| Bowman | 4,200 | 15 | 15 | 0.21 | 0.21 |
| Dunn | 4,800 | 300 | 87 | 3.24 | 1.04 |
| McLean | 11,800 | 1,100 | 567 | 4.54 | 2.57 |
| Mercer | 6,700 | 1,764 | 874 | 9.94 | 5.95 |
| Oliver | 2,400 | 85 | 85 | 1.95 | 1.95 |
| Total | | 3,264 | 1,628 | | |
| **Texas** | | | | | |
| Atascosa | 19,800 | 396 | 197 | 1.14 | 0.58 |
| Freestone | 12,100 | 40 | 40 | 0.20 | 0.20 |
| Grimes | 12,200 | 264 | 12 | 1.23 | 0.06 |
| Henderson | 30,600 | 582 | — | 1.09 | 0.00 |
| Milam | 19,900 | 634 | 236 | 1.76 | 0.69 |
| Panoca | 16,400 | 926 | 692 | 2.96 | 2.28 |
| Robertson | 14,300 | 396 | 131 | 1.55 | 0.54 |
| Titus | 18,000 | 304 | 304 | 0.97 | 0.97 |
| Total | | 3,542 | 1,612 | | |
| **Utah** | | | | | |
| Carbon | 19,100 | 3,617 | 2,482 | 7.89 | 5.93 |
| Emery | 7,600 | 5,437 | 3,200 | 18.13 | 13.43 |
| Kane | 3,400 | 2,796 | 399 | 19.49 | 5.48 |
| Sevier | 12,400 | 2,855 | 1,498 | 9.06 | 5.60 |
| Total | | 14,705 | 7,579 | | |
| **Washington** | | | | | |
| Lewis | 48,900 | 66 | 132 | 0.08 | 0.16 |
| **Wyoming** | | | | | |
| Campbell | 14,500 | 6,237 | 3,459 | 13.61 | 9.29 |
| Carbon | 17,200 | 1,410 | 372 | 4.08 | 1.23 |
| Converse | 9,400 | 858 | 226 | 4.46 | 1.36 |
| Lincoln | 10,500 | 473 | 355 | 2.42 | 1.86 |
| Sheridan | 21,100 | 371 | 321 | 1.01 | 0.88 |
| Sweetwater | 31,300 | 522 | 448 | 0.96 | 0.83 |
| Total | | 9,871 | 5,181 | | |
| Total U.S. | | 47,041 | 23,025 | | |

The real population growth associated with coal development encompasses not only the miners themselves, but their spouses and children. Secondary employment is also generated. An average family size of three for both primary and secondary workers, coupled with an assumption of one local service job per primary job, yields an estimate of five additional persons per mining job. Projected annual growth rate in total population for each of the affected counties is presented in table 44. Some officials on the scene in Western States say the available estimates of county population severely understate reality. Where this is true, it hinders their efforts to secure outside aid based on population. It also would have the effect of lowering estimated growth rates because the base population is higher than the number used in these calculations).

Population growth that exceeds a rate of 5 to 10 percent annually is considered to generate significant socioeconomic consequences. Table 52 shows that 11 counties in the West will achieve growth rates exceeding 5 percent per year in the Keystone case. Four of these counties will have growth rates exceeding 10 percent per year over the 10-year period. (High-growth county data are in bold face.) Under the Keystone assumptions, the Utah counties of Emery and Kane, where underground mines are planned, are the most impacted, with growth rates of 18 and 19 percent, respectively. Coal development in Campbell County, Wyo., yields a population growth rate of nearly 14 percent per year. (It is assumed that population impacts of coal development are felt only in the county in which development occurs. When miners do not reside in the county where they work, this assumption needs adjustment.)

When the OTA low-production case is used, only six western counties appear to be vulnerable to rapid growth. Four of these counties— Carbon, Emery, Kane, and Sevier—are in Utah. The others—Campbell and Mercer—are in Wyoming and North Dakota respectively. The OTA case removes both Colorado counties and the three Montana counties from the highly vulnerable list.

The list of vulnerable counties grows slightly if the time frame shifts from 1977-86 to 1977-81. With Keystone's assumptions, three more Colorado counties—Moffat, Pitkin, and Routt—are now experiencing rapid population growth. The shorter time frame does not add counties to the OTA estimates.

Two other factors should be considered. Mine construction workers appear 2 to 5 years earlier than the mine-operating work force. Construction crews are about the same size as mine-operating crews, so actual population effects are moved ahead in time but not increased in magnitude. Second, some utilities have located mine-mouth powerplants near coal mines. Powerplant construction crews are from 5 to 10 times larger than operating crews. Some communities have had trouble adjusting to the short-term impacts of these crews, (e.g., Rock Springs, Colstrip, Huntington, Craig, and Gillette) and if coal development continues, so will many of their problems. Only eight counties will experience both new coal production and operational powerplants after 1981 (table 53), according to these forecasts. When individual plant situations are investigated, these eight can be pruned because of time-phased unit construction (which maintains a semipermanent construction work force in a general area) and time lags between startup and operation. Mercer County, N. Dak., and Emery County, Utah, have already been identified in both cases as high-growth areas. The powerplant construction variable adds several Texas counties to the list of impacted communities. However, since all of these Texas counties have relatively high populations (ranging from 12,000 to 30,000) and are within 100 miles of Dallas or Houston, local population growth will probably be spread out and absorbed without great difficulty.

In sum, this analysis shows that from 6 to 11 western counties will show rapid population growth from coal development in 1977-86. The population effects will stem from coal mine construction and operation rather than from powerplant development. In both the Keystone and OTA cases, four counties in Utah are identified. Mercer County, N. Dak., and Campbell

**Table 53.—Western Counties With Both New Coal Production and Powerplants to Come Online After 1981**

| State | County | Online date |
|---|---|---|
| North Dakota | Mercer | 1981, 1984 |
| Texas | Freestone | 1983, 1984 |
| | Grimes | 1982 |
| | Henderson | 1982 |
| | Morris | 1980, 1982 |
| | Robertson | 1984, 1985 |
| | Titus | 1982 |
| Utah | Emery | 1978, 1980, 1983, 1985 |

County, Wyo., are also singled out in both cases. Most of these counties have small populations, the range being between 2,383 on the Crow Reservation to 19,100 in Carbon County, Utah. The four Utah counties and Campbell County, Wyo., will experience the fastest rates of growth. Several dozen western towns are also likely to grow rapidly from coal development even though their respective counties do not appear in the high-growth lists.

Boomtown Effects

How any particular community is affected by rapid coal development depends on the size of the mining enterprise, mining method, size and quality of existing public and private infrastructures, rapidity of development, tax structure, and ratio of transient to permanent employees among other variables. The duration of coal impacts is likely to run for 10 to 30 years at one or another level of intensity.

Coal impacts can be grouped according to broad subject classifications (economic, political, and social). Both public and private sectors are affected. But it is impossible for two reasons to calculate a definitive, cost-benefit bottom line. First, all of the costs and benefits are not translatable into dollars. Second, costs and benefits are distributed unevenly among private-sector groups (workers, owners, and others) and public-sector interests (local communities, national energy needs, etc.) Therefore, although policymakers can become aware of the implications of various coal policies, there is no neat formula for weighing one kind of cost against another kind of benefit. The ensuing discussion talks about the costs

and benefits of Western coal production, but does not make generalizations about when costs equal benefits, when costs exceed benefits, and when they are less.

Private Sector: Costs and Benefits

It is possible to make very rough calculations of some of the private-sector, economic impacts of projected coal development.

Annual coal-related wage income and benefits are the principal economic gain from Western coal development in local communities. Assuming that each new miner will earn $25,000 annually ($100 per day times 250 work days), the 34,000 to 42,000 additional miners in 1985 will earn between $850 million and $1 billion per year in pretax income. To this must be added the wages paid to workers involved in secondary coal employment.

Coal development does not bring large inputs of local capital investment or large profit returns to coal areas. Almost all of the major mines in the West are being opened by nonlocal investors, usually utilities, energy companies, or conglomerates. Apart from costs of acquiring the minable coal resources, the major capital expense involved in opening a new mine is equipment, almost all of which comes from nonlocal manufacturers. Once a mine is operating, net income is either reinvested or distributed as profits to stockholders. Little of the mine's economic surplus can be used by local banks or local entrepreneurs and residents. The surplus that is likely to remain in local communities is hard to quantify, but is likely to be insignificant in most cases. However, where a mine-mouth generating plant is part of the coal development, the local return increases even though the profit distribution pattern is not likely to change.

Coal development will affect existing business sectors in western communities. Many will be able to expand their sales and profits. But for some it will be a mixed blessing, and for others no blessing at all. Noncoal wage income does not rise as quickly as miners' wage rates. Local businesses will have difficulty retaining current employees and recruiting new ones. Secondly, local entrepreneurs may want to expand their business, but local loan capital may be tight. There is also the prospect of increased competition from new business started by local or, more likely, out-of-State entrepreneurs or national chains. Some local business may be wiped out.

In the past, the short-term costs of rapid coal development to the private sector have been high. The experience of Rock Springs, Wyo., is a dramatic example. When coal development increased the population of Rock Springs rapidly, its quality of life deteriorated.[93] The preboom ratio of two service jobs to every one primary job declined leaving local business without sufficient personnel to maintain services. Workforce turnover was high. Mining and construction caused inflation, bringing financial hardships to noncoal employees and residents on fixed incomes. After the initial disruption the ratio of local service jobs to primary jobs tended to normalize as they did in Gillette. Coal-induced inflation frequently pinches the wallets of noncoal employees and fixed-income residents.[94] High turnover rates among both coal and noncoal workers is often seen. People leave the community, although a net increase may be registered in the census. Much social stress occurs. In the Rock Springs case, the costs of boomlike coal development to many new miners and noncoal local residents may have outweighed the benefits.

The most significant private-sector problem in boomtowns is housing. Usually, there are few local builders capable of undertaking large developments. Often local banks cannot finance major housing projects. Many small banks are reluctant to tie up their capital in

---

[93]John S. Gilmore and Mary K. Duff, *Boom Town Growth Management: Rock Springs-Green River, Wyoming* (Boulder: Westview Press, 1975).

[94]Interviews with bankers in northwestern Colorado found that fixed-income residents were borrowing to meet increased living costs by using the appreciating value of their homes as collateral. Such short-term loans are rolled over each year to repay previous loans and obtain additional money for rising living expenses. This pattern, the bankers predicted, will certainly end in economic disaster for some of those involved. Don Kash, Mike Devine, and Allyn Borsz, *Impacts on Western Coal-Producing Communities*, OTA contractor report, April 1978.

housing loans, as higher interest rates can be charged on personal and consumer loans. Consequently, housing costs skyrocket. In western boomtowns the average $50,000 price of a new home is about 30 percent higher than the national average for nonmetropolitan areas. Apartment rents are nearly double the median national figure of $120 per month. With the housing supply inadequate and costly, new residents are forced into mobile homes. Little property tax revenue is generated by these units in some States because they are bought and owned as personal rather than real property. State legislatures can remedy this. The shortage and high price of housing are also caused by the natural reluctance of local builders to risk major investment on the basis of promised coal expansion; the West has seen its share of booms burst.

The uncertainty of demand for western boomtown housing is a major part of the housing supply problem. Lenders, housing developers, and builders are reluctant to rely on the announced schedules for new energy projects. They fear being stuck with unsold homes if projects are delayed or canceled. The private, primary mortgage market is composed principally of savings and loan associations and the secondary mortgage market is insurance companies and other investors. These capital lenders may be unwilling to back investments that are not in low-risk categories.

The reluctance of local lenders is attributable partly to the funds available. Local financial markets in the West are unable to accommodate the demand for new mortgages in rapidly growing small communities. Outside financing is constrained by the absence of financial relationships with out-of-State institutions and the inability of local financial institutions to accumulate large enough blocks of mortgages to sell in secondary mortgage markets. These shortcomings of the mortgage market in the rural West may become a greater problem in the future. FmHA recently announced guidelines for providing assistance to energy-impacted areas where energy employment increases 8 percent or more in a year, a housing shortage exists and local and State financial resources are inadequate. Only $20 million is available under this program which applies to

*Photo credit: Earl Dotter*

**Coalfield housing near Peabody Coal's Kayenta mine on the Navajo Reservation, 1977**

both coal and uranium production, processing, and transportation.

The potential demand for new homes is somewhat more certain in larger towns, such as Gillette and Farmington, where large numbers of energy executives or administrative personnel live. However, even in these larger towns the temporary nature of powerplant construction and uncertainty about future energy development result in a fear of overbuilding. This is particularly true of towns that have had previous booms collapse.

Several coal developers have provided housing for construction worker and mineworker families. Much of this has been planned mobile-home developments. Arco is building a planned town (Wright, W. Va.) primarily for employees of its Black Thunder MineWright's single-family housing is expensive—in the $50,000-and-up range—and most miners seem to be opting for the cheaper, more readily

available mobile home. By living closer to the mine than they would have in Gillette, the only town in Campbell County, workers will commute shorter distances. Wright will cost ARCO about $18 million.

The company-town approach is one alternative to chronic housing shortages. But the cost of building and operating such towns discourages most companies from trying it. More common is a catch-as-catch-can approach to housing. Coal developers will offer some assistance to towns or to their employees, but principal responsibility is left to local authorities and individuals. Several examples illustrate the mixed record of western towns in coping with housing needs.

In Emery County, Utah, housing is desperately needed. One cause of the shortage is a deficient public infrastructure. Without water and sewer hookups, housing of any kind cannot be built. Another problem has been Emery County's lack of licensed contractors. Inflated costs, housing shortage, and the limited supply of contractors has significantly changed the composition of the housing inventory. In 1977 mobile homes were 41 percent of Emery County's housing supply. In Huntington, where in 1970 there were only two mobile homes, 375 were counted in 1976. The Southeastern Utah Association of Governments sees mobile homes as the only housing choice available to new households. A somewhat different pattern is found in Gillette, Wyo., where there were no large-scale builders or developers before the coal boom. Market uncertainties and the possibility of environmental litigation over Federal coal leasing presented an unattractive level of risk for large-scale land developers and creditors. These initial problems have been alleviated. Many developers now see a single-family housing market in Gillette. The risks involved in large subdivisions have been largely absorbed by the coal companies. In some instances, the companies have acted as land developers; in others they have either guaranteed the sale of lots and/or houses or guaranteed the credit of a housing subdivider. Yet single-family housing production has not increased substantially. The primary barriers are the delays and high costs of producing new units of acceptable quality and the relationship between the cost of producing the types of dwellings desired by the newcomers and their ability to pay for the desired units.[95] Many residents are forced to live in mobile homes, which have become the fastest-growing housing alternative in Campbell County. Sixty percent of the increase in housing since 1970 has been in mobile homes. Currently 67 percent of the housing in the county outside the Gillette city limits is mobile homes.[96]

[95]In a recent Wyoming survey, 67 percent of the respondents preferred a single-family unit and 56 percent indicated both the willingness and ability to purchase one, but only 45 percent had achieved the goal of owning a single-family unit. (Dale Pernula, *1977 Citizen Policy Survey* (Gillette/Campbell County Department of Planning and Development, 1977).)

[96]Keith D. Moore and Carrie Loomis, *Housing Market Dynamics in Campbell County, Wyoming and the Impacts of the Proposed Mobil-Consol Pronghorn Mine,* a Denver Research Institute report prepared for the U.S. Geological Survey, 1977.

Housing on Indian reservations raises several problems that are different from those discussed above. A boomtown syndrome probably won't occur on most reservations, although the social stress created may be similar. The number of new miners that will be involved in Indian coal production will not be more than several thousand. Most of these will be Indians already living on reservation land if the tribes succeed in making the coal companies give hiring preference to local Indians. The housing problem on the reservations is less one of shortage and more one of adequacy. The serious deficiencies in Indian housing are well-known—to Indians, at least. Table 54 presents a snapshot summary of Indian housing quality—including the Navajo Reservation—where housing needs are most severe.

In sum, the distribution of private costs and benefits of increasing Western coal production vary. Coal workers and mineowners will obviously benefit from new employment opportunities. Coal-related services will establish themselves in regional centers, and locally owned commercial enterprise may expand. Spendable income in local communities will increase rapidly. The material quality of life will rise for some.

On the other hand, these private benefits are not distributed evenly. Local capital liquidity is unlikely to increase significantly, apart from the savings deposited by wage earners. Local businessmen may not be able to expand or modernize their services, and national chains may take over their markets. Opportunity costs may be high in Western coal development. Ranching, farming, and tourism may be curtailed in coal areas. Local inflation penalizes those on fixed incomes and those whose wages are independent of the coal business. Data are not available comparing boomtown inflation rates with national or regional rates, although anecdotal evidence suggests that the former is abnormally high. Coal profits will not generally be distributed locally. Much investment will not be spent locally, either.

One important and relatively unexplored issue is the dollar cost of impact internaliza-

**Table 54.—Housing Conditions of Indians in Three Western Regional BIA Areas—1973**

| | BIA area office | | |
| --- | --- | --- | --- |
| | Billings[a] | Navajo | Albuquerque |
| Total number of families | 6,071 | 23,801 | 8,349 |
| Number of standard units | 3,335 | 3,126 | 3,180 |
| Number of substandard units | 2,269 | 19,242 | 2,841 |
| Families doubled up | 467 | 1,433 | 2,328 |
| Total need as percent of families | 45.1 | 86.9 | 61.9 |
| New units needed .... | 1,861 | 7,324 | 3,332 |
| Rehabilitations needed | 875 | 13,351 | 1,837 |

[a]Includes Crow and Northern Cheyenne Indians in Montana.
SOURCE: U.S. Congress, 1975.

tions on the dollar cost of producing Western coal. Coal companies in the West generally offer extremely high wages to attract labor. To the degree that rapid coal expansion decreases quality of life, operators may be forced either to offer increased monetary incentives to retain workers or to become directly involved in the costly business of providing community services.[97] If coal operators absorb the capital costs of, for example, a water and sewage system, their product prices will rise. It is debatable whether price-escalator provisions in utility contracts can be invoked. If the labor force perceives coal development as the source of quality-of-life decline, workers may seek unionization and even higher monetary rewards or community development clauses in collective bargaining. State and local governments may impose taxes on coal developers to finance needed services. To the degree that coal-related public costs are paid by tax revenues on Western coal, higher coal taxes may lower company profits and raise coal prices.

Western communities have had much experience coping with the short-term stresses created by booms and the long-term depressions left by the "busts." From today's perspective, it appears that Western coal growth, based as

[97]Richard Nehring and Benjamin Zycher, *Coal Development and Government Regulation in the Northern Great Plains* (Santa Monica, Calif.: RAND Corp., 1976), p. 127.

it is on long-term utility contracts, will smooth out the boom-bust cycle. However, it is likely that a considerable lag will occur in the perceptions of local residents as they weigh this boom against those of the past. Their caution is justifiable because economic diversification and planning for the postcoal years are not occurring. A long boom followed by an even longer bust may be in store. In any case, boom growth is the least preferable path of economic development. Slower and more steady growth is less disruptive and more beneficial to more people over a longer period of time. However, because coal projects are usually matters of corporate decisionmaking, many western communities may have no choice. If this perception is shared by westerners it would not be surprising to observe a political backlash against current coal development by citizens who see their long-term interests depreciated by rapid coal developments.

**Public Infrastructures: Costs and Benefits**

The balance between public costs and benefits of Western coal development changes over time. In the 5 years needed for a mine to be developed, public services in small communities will be strained by the rapid increase in population. Demand for more and better services will usually exceed a community's tax revenue and ability to deliver services. Where county or State tax revenue increases but heavily impacted towns can't get a fair share, the financial situation is aggravated. Antitax feeling, expressed in Proposition 13-type movements, further binds local governments.

Coal development also strains local physical infrastructures—roads and bridges, water supplies and purification plants, sewage facilities, schools, and fire and police protection. Existing service systems are likely to be severely stretched and weakened. Individual pieces of these systems may collapse entirely. Short-term public costs may also be expressed in terms of community destabilization, cultural strain, social disorganization, and a general perception of decline in the quality of life.

As the mine reaches full capacity some public costs will become less serious. Tax revenue will generally increase; services should expand. The pace of growth will slow. The rate at which any given community begins to normalize demand for services with its ability to deliver them depends on a number of local factors. The most important of these are: the degree of predevelopment need, tax policy, rate of development, priorities and attitudes of the coal producer, reliability of demand for local coal, skill of local officials, extent of economic diversification, and ability of State and county governments to adapt their policies to meet the needs of localities. In some communities front-end problems, such as lack of revenue, may extend for years. The deficit may never be eliminated with available resources. These problems will be solved in some communities, to be sure, but western history shows a spectrum of success and failure concerning the ability of communities to assemble public capital and meet boomtown needs.

Western mineral development also starkly portrays the common "back-end" problem of resource depletion and community decay. When mines play out or market conditions force closure, local communities are "busted." When communities were tied to a single company or industry in the past, back-end problems were often disastrous. Since western mines have operational lives of 20 to 60 years, planning and diversified development of Western coal areas may be needed to avoid the worst public cost of earlier mineral development: ghost towns.

Most impacted western communities are now struggling to cope with the front-end problems of coal development. Such communities face the necessity of either expanding existing public services or building them from scratch. In either case, a period of 2 to 5 years exists before those services appear. Most western communities do not have adequate funds in the early stages of coal development to plan their growth and seek Federal assistance. Where coal-processing or electric-generating plants are being constructed, front-end financial problems are magnified by the overnight appearance of the temporary plant-construction crew. State-administered severance-tax income does not return to local communities on

a dollar-for-dollar basis and does not begin at all until coal is actually sold.

Aside from education, public services that usually need expansion in boomtowns include fire and police protection, roads, water and sewage treatment, solid-waste collection and disposal, health-care facilities and services, detention facilities, county and municipal facilities, and recreation. For each new resident, between $1,500 to $2,300 (1975 dollars) is estimated to be required for just the capital costs of these facilities. Of this amount, roughly 75 percent is devoted to water and sewage treatment. A hypothetical town of 2,500 persons with an annual municipal budget of $20,000 would require between $1.5 million and $2.3 million capital investment if its population were to increase 40 percent (1,000 persons). Clearly, the front-end costs are huge in comparison to local tax revenues and bonding capacity.

Existing transportation facilities may be suddenly overloaded when coal mining grows fast. Highways break down. Coal haulage disrupts small towns. Unit trains bisect communities. A recent Colorado Department of Highways study,[98] estimated that more than $250 million would be required for primary and secondary highway construction by 1985. In six energy-impacted counties, more than four times the annual level of construction for primary and secondary highways in all 63 counties of Colorado, is needed. It has also been estimated that at least 30 new grade separations are required to maintain intracommunity access in communities suddenly beset by unit trains. Current funding in Colorado provides for one or two new grade separations per year for all purposes.

Boomtown problems are aggravated where the coal project creates revenues for one jurisdiction while the social impacts of the new mine occur in another. Jurisdictional mismatches are one of the most intractable problems. The most common mismatch occurs when coal mines or powerplants are located outside a town's limits. The county collects the tax dollars while the municipality struggles to

[98]Briefing to commissioners by staff of Colorado Highway Dept.

meet the needs of new miners. But the jurisdictional mismatches within a county are more easily handled than those occurring across county or State lines. There are no examples of where this inequity has been addressed except through usually meager State impact assistance programs.

The evolving balance between public benefits and public costs depends largely on local and State coal tax policies. The level and equity of public revenues depend on the mix of taxes imposed, the level of each tax, and the efficiency of the collection system. These matters often involve two or three levels of government and are always matters of political debate. Local authorities rarely impose a corporate income tax on local business. Property tax usually generates most local revenue. This burden often falls more heavily on residential and agricultural property owners than on mineral-resource owners. Leasers of publicly owned coal pay no property tax although they pay a tax on leasehold value based on production. Some counties impose specific taxes on coal. For example, most coal counties in Illinois have imposed a 1-percent sales tax at the mine mouth; local and county taxes in Montana average about 5 percent; Wyoming counties average about 6.3-percent ad valorem tax based on production; counties in other Western States impose no coal-specific taxes. States differ significantly in their coal taxation schemes. These Western States impose coal taxes: Wyoming (10.1 percent in 1978); North Dakota (65 cents/ton with an inflation escalator); New Mexico (38 cents/ton steam coal and 18 cents/ton metallurgical coal); Montana (30 percent on the value of the coal at the mine mouth); Colorado (30 cents/ton for deep-mined coal and 60 cents/ton surface-mined coal).[99] Utah, South Dakota, Texas, and Oklahoma have no coal taxes. Western States also use different formulas to redistribute State-collected severance taxes to counties of origin. State and local tax policies do not seem to be a criteria for mine siting. Depending on the level of predevelopment need in each community, future tax income may or may not be sufficient to

provide adequate public services following initial coal development.

The net public gain or loss from coal growth also depends on how much system deficit is present when development begins, the extent of the overload, and the level of services perceived necessary to meet the perceived need. Communities that are created from scratch have extremely high startup costs and may need subsidies for many years. But they may be ultimately more successful in providing services and achieving a net gain The more common western pattern appears to be coal-growth overloads of existing systems, which may already be below par. In such cases, arithmetic gains in population may require geometric increases and investments in services. (For example, a 50-percent increase in population may require a doubling of the local police budget.) The greater the existing deficit and the greater the development impacts, it is more probable that there will be a small net public gain. Conversely, the smaller the existing service deficit and the smaller the development impact, the more probable that public benefits will be greater. The level of public services differs from town to town. A few examples suggest the range of experience.

Craig, Colo., is a typical rural community on Colorado's western slope. Preboom services provided by the community cost $208 per capita and were considered adequate by most residents. The 1979 estimated per capita expenditure is $344 (in constant 1973 dollars). With coal development the water system was expanded to serve a population of 15,000 but is almost at capacity with 11,000. The sewer system will be expanded to a 20,000-population capacity. Craig was not able to finance these capital projects totally with local revenue. Most of the money came from Federal and State sources, along with some industry contributions. The city has not yet resolved the issue of how to pay for a new reservoir (estimated to cost $10 million).

The public infrastructure in Emery County, Utah, communities was not ready to absorb the coal-based growth that came. In some cases, services were not even adequate for the

[99]Coal Week, Apr. 17 and 24, 1978.

existing population. Many of the small towns in the county were not required by the State to prepare annual budgets. Those that did often had budgets only 1-page long. Little planning or forecasting of capital expenditures occurred. A 1978 report by the Southeastern Association of Utah Governments said that almost every town in the county needs a new municipal building. Many need to upgrade their water and sewer facilities; others need sanitary landfill operations. Every town is in need of a fire station and new equipment, and many need a massive street improvement program.[100]

Gillette, Wyo., had energy-related growth in the oil and gas boom of the 1960's and the coal development of the 1970's. Even with this growth, the city's tax base has fallen further and further behind the county's. Gillette has not always had the capital needed to improve its public services. In the 1960's Gillette's bonds were issued piecemeal, with no provision for long-range planning of financial strategies. The result was poor use of the city's existing tax base in creating bonding capacity. For a time, this severely limited what the city could do. However, through the efforts of a bond counsel and a new professional town manager, these debts were largely refinanced in 1977.[101]

Gillette and Campbell County have been fortunate in receiving assistance from industry in the form of in-kind contributions such as loans of equipment to clean streets or help with digging the foundation for the hospital. The members of the industrial subcommittee of the chamber of commerce have also contributed money that has been used for studies and plans for long-term city projects.[102]

Education

Public education is usually financed by State and county governments, with special categories of assistance provided through Federal programs. The local school system's ability to manage coal-generated enrollment depends on the preboom use of existing facilities, the rate of expansion and size of the new student population, and the financial ability of the community to provide additional facilities and services.

School costs are usually the largest portion of local public expenditures. Per capita expenditures for education in the West are already high. For example, the capital costs for each additional student in Gillette, Wyo., are projected to be approximately $2,500. An increase of 500 in the school population would amount to $1.25 million more in capital expenditures. Operating expenditures are about $2,000 per student, so that 500 new students would require an additional yearly operating expenditure of $1 million.[103]

The long-term financial demands for education in Western coal communities are substantial. Whether this situation poses serious financial problems will be determined by how soon the school-age population begins to expand rapidly. Where school-age populations do not grow until coal facilities are operating, school districts may have surplus revenue. An example of this favorable situation is found in Rosebud County, Mont., where school districts are expected to enjoy substantial financial surplus at current tax rates. However, this is the exception.

Typically, the increase in school enrollment from coal-related work occurs before any increase in ad valorem tax receipts. This may cause operating expenditures per pupil to decrease. However, most States "guarantee" that these expenditures will not fall below a set amount fixed by the legislature. Therefore, the operating portion of the budget is not the main cause for concern; rather, it is capital expenditures. For example, if additional permanent classrooms are to be built or temporary ones purchased, the existing population pays, not the newcomers or the coal companies. These initial problems may be corrected if mines and

---

[100]*Monograph Series: Local Government* (Price, Utah: outheastern Association of Utah Governments and Development District, 1978).
[101]From 1976 and 1977 interviews with Mike Enzi, Mayor of Gillette and a 1977 interview with Flip McConnaughey, City Manager.
[102]From 1978 interview with Dave Bell, Director, Gillette Chamber of Commerce.

[103]1976 and 1977 interviews with J. O. Reed, Superintendent, Gillette, Wyo.

powerplants are placed on the tax rolls and no jurisdictional mismatches are involved.

In Moffat County, Colo., for example, school enrollment increased 29 percent from 1976-78. In the same period assessed property valuation increased 144 percent. While these statistics may suggest optimism, the fact is that the county's schools are overcrowded. Increases in enrollment have put the total student body at about 300 students over capacity. In October 1978, a $9 million bond issue was finally passed (after having failed twice) for construction of a new high school. In addition to overcrowded conditions, the school district lost 35 teachers in the summer of 1978 for a variety of reasons. A number went to work in higher paying coal projects. Although the positions were filled, there were problems in attracting new teachers, mainly owing to the high cost of living in Craig.

Much of the locally perceived impact on education involves subjective judgments about what constitutes good education. "Quality" education is an intuitively meaningful concept to most parents, but it is hard to quantify. Different communities may have different collective perceptions as to what "it" is and whether their children are currently getting "it." To the difficult question of educational quality must be added the question of who pays for it. Long-time residents sometimes object to higher property taxes and more indebtedness to pay for the education of the children of "newcomers." Moreover, coal development pressures may inflame residents over what is perceived as a lessening of educational quality. This may diminish over time as new facilities are built. But the absence of front-end monies and the high costs of permanent school facilities make the process of resolving this problem difficult. Adaption to boom conditions may take the form of increasing student-teacher ratios, shifting to half-day sessions and installation of temporary school trailers. This is likely to keep the issue of coal and education alive indefinitely. Other problems may be perceived. New teachers will have to be hired, and some, at least, may have nonlocal educational values and methods. The education system may be perceived as becoming less personal and more bureaucratic, hence less accountable. The destabilizing effects of rapid economic growth may be played out in school disruptions and student stress.

Boomtown educational needs have both a quantitative and qualitative dimension. Quantitatively, the problem is finding capital and personnel to sustain and expand existing facilities and programs. Qualitatively, the problem is matching perceptions of preboom and postboom reality with the changing nature of local education. The concern of the community about the "quality" of their boomtime education will vary according to the precise mix of short-term adaptation mechanisms and longer term expansion of services that authorities are able to orchestrate. Both newcomers and oldtimers may continue to be dissatisfied with their educational system for different reasons. Both groups may have legitimate grievances. It may be unlikely that quantitative improvement in the local educational system will produce a one-for-one, stepped increase in the community's perception that education has improved.

## Health Care

Both the public and private sectors are involved in health care delivery and financing. Coal development will enlarge the health-treatment needs of local communities in various ways: more people need more care; more people need a wider range of health care; coal miners have special health needs.

Health services, which are often inadequate in rural communities before development starts, are among the most difficult to maintain in boomtowns. Medical needs typically expand at least as rapidly as population, and faster when a large proportion of the new residents are young children. If health services are limited, a smaller proportion of children may be immunized against preventable diseases. Fewer screening tests may occur. Inability to obtain early treatment may permit disease to become more serious.

The health care delivery system in western communities is limited and, by national standards, less than adequate. One eight-State study of 185 potentially impacted western towns found that fewer than half (87) had a resident doctor and about one-third (60) had a hospital.[104] Generally, the doctor/population ratio is much lower in western towns than in the country overall. Like other rural areas, the West has difficulty attracting and holding physicians. Apart from the availability of physicians and health care extenders, health care norms depend on the quality of care provided and the rate of use, both of which depend to a greater or lesser part on the individual's ability to pay. It is easy to understand the frustration coal miners and others feel when "good" health care is unavailable even though adequate insurance plans cover them.

Recent survey research in Sweetwater County, Wyo., sheds light on the importance of adequate health services. Fully 82 percent of the 157 newcomers interviewed ranked medical and mental health services in their top five community priorities. Sixty percent of the sample said they would leave town if these services were not improved. These opinions were acted on: "The labor turnover rate was running up to 150 percent as mining employees were leaving the area because of dissatisfaction with the quality of life."[105]

Illness-prevention systems are even less developed than treatment systems. Prevention programs are provided by both the public and private sectors. In small, rural communities, public health services are minimal, where they exist at all. Immunizations, water and sewage treatment, prenatal and postnatal care, and nutrition education are not well developed. Adequate systems are even less likely to be found on many Indian reservations. Much preventive care depends on an individual's awareness of its importance and his ability to pay. Absence of prevention measures often results in higher treatment costs, some of which are borne by public agencies. Generally, the

public and private prevention systems in Western coal towns are haphazard and not well synchronized.

Coal miners have special health needs stemming from their occupation. Although western surface mining is statistically safer and healthier overall than underground mining, surface miners still experience a significant level of industrial injury and disease. Respiratory and pulmonary disease, hearing loss, and stress are found more frequently in miners than in the population generally. Coal's accident rate requires adequate emergency medical services at the mines and fast transportation to treatment centers. Provision of emergency services—including the training of miners as emergency medical technicians—is not required under Federal law or by most States. Some coal operators provide such training and service; others don't. Specialists in miners' occupational health problems are not found in the newer, Western coal development areas. Some practice in the older coalfields of Utah and Colorado, where UMWA and the UMWA Fund established clinics after 1950.

Other health-related services are required in stressful boom conditions, including treatment and counseling for alcoholism, child abuse, marital strain, juvenile delinquency, and mental health The rapid infusion of newcomers into well-established communities creates stressful situations among natives and newcomers alike. The smaller the community, the less likely that such services are available. However, high absenteeism and turnover rates in Western coal towns suggest that services are needed to treat the symptoms and consequences of growth-related stress.

Political Impacts

The political impacts of Western coal development may be extensive, both from the perspective of boomtown residents and in terms of the politics of further coal expansion.

Rapid population growth often generates conflict between antigrowth and progrowth factions, newcomers and oldtimers, labor and management, environmentalists and prodevel-

[104]Regional Profile: Energy-Impacted Communities (Lakewood, Colo.: Federal Energy Administration, Socioeconomic Program Data Collection Office, 1977).
[105]John S. Gilmore, et al., Analysis of Financing Problems in Coal and Oil Shale Boom Towns (Denver Colo.: University of Denver Research Institute and Bickert, Browne, Coddington & Assoc., Inc., 1976), p. 72.

opment factions, high-wage jobholders and fixed-income or low-wage persons, transients and permanents, and coal-related business and local business. These conflicts may benefit local communities in the long run if they mobilize political interest constructively and create pressure for better public services. Yet more public services almost always involve more Government regulation, which may itself become a political issue. Political cleavages resulting from coal development may shatter community life. In company towns, coal developers may dominate local politics, for better or worse. In other places, two-party competition may result from the introduction of new interest groups. One-party systems may be factionalized. Tightly run political machines may develop. The conflict that change produces may interact with economic and social conditions to produce widespread community disintegration and dissatisfaction. But, on the other hand, growth can create conditions where political cooperation is required to meet new demands.

Coal-impacted western communities vary tremendously with respect to attitudes toward politics and coal development, so that regional generalizations are especially risky. Nonetheless, it is reasonable to assume that some communities will eventually group themselves into two polarized alliances: progrowth and less growth. As the rate and scope of development increase, polarization is probably more likely to occur and be irreconcilable. But exceptions will be found. Some Mormons of Utah, for example, favor coal development without reservation. This prodevelopment view comes from the desire for economic development and from a set of socioreligious values arising from the Mormon experience.

Although procoal factions among Indians and Mormons may favor procoal Federal policy, they may also oppose the Federal regulation that often accompanies such policies. Western hostility toward Federal policy may be intensified by the lack of applicability of Federal aid to western boomtowns. Most relevant Federal programs make funds available to States and local governments only to remedy problems of economic stagnation or natural

disasters. These conditions do not exist in Western coal communities. Hence, most Federal impact assistance funds are unavailable to Western coal towns.

Western rural county governments are largely dominated by farmer/rancher interests. County commissioners deal with few major conflict-producing issues. Those they do handle usually concern the level and distribution of taxes and public expenditures. Because of the western populist tradition and the prevalent skepticism of government in general, rural officials tend to be ambivalent toward land use planning and zoning. Their ranching and landowning constitutents are strongly opposed to any intervention in their control of their property.[106]

In Gillette, Wyo., it was not until 1976 that the Gillette/Campbell County planning office really became active when the need to handle boom conditions became obvious. When planning and zoning codes were established, a 3-by 5-mile box was drawn around Gillette. The planning department's control is much greater within this area than in the rest of the county. Outside the box, the planning commission can only make recommendations to the county commissioners, who need not (and do not always) follow them.[107]

Coal growth is apt to cause drastic changes in existing political structures. Whether these changes are seen as good or bad depends on the observer's point of view. Eventually, ranching interests may lose their role as the dominant political force in local politics. Newcomers may share power with older residents, or they may supplant them entirely. Newcomers are likely to bring with them different

[106]"Our findings reveal that a significant feature of the study area is its traditional rural abhorrence of planning. The area's residents consistently and strongly resist the idea of telling people what they can and cannot do with/on their land. Almost 60 percent state that no one but the owner should have any say about how privately owned resources are used." Community Service Program, University of Montana. *A Study of Social Impact of Coal Development in the Decker-Birney Ashland Area* (Helena, Mont.: Montana Energy Advisory Council, Office of the Lieutenant Governor, 1975), p. 20.

[107]Based on a 1978 interview with Dave Ebertz, former part-time City Manager, Gillette.

values and expectations that may be at odds with those of the existing residents. Political change in coal communities is driven by the influx of new residents. One study of southeastern Montana found that newcomers and old-timers tend to stay apart. For the most part, newcomers have not been accepted into established social structures.

Often, changes in boomtown political culture will not come easily. Conflicts may occur between newcomers and oldtimers. There may be more urban vs. rural strain, as well as an increase in industry vs. environment issues. There may be more concern over and need for Federal intervention as urbanization takes place. Unions may become more of an issue and possibly a political force.

It is difficult to predict whether such changes will be beneficial to the community in the long run. Many preboom residents will not favor them. Conflict may not be viewed favorably by either oldtimer or newcomer. Yet as assimilation and adaptation occur, conflicts may be resolved.

### Social Impacts

In preboom western towns, informal social controls shape community social structure.[108] Doors are rarely locked because everyone knows everyone else. Family violence occurs less often. Teenage drug usage may not be a serious problem because of stable and familiar social norms and the community's isolation. Alcoholism is considered more an individual than a community problem. This kind of social structure is vulnerable in boomtowns. Some of the changes that occur are not beneficial. High-growth communities often have increased crime and mental health problems. Juvenile delinquency grows. Divorce and family disruptions are more common.

---

[108]Sociologist William Freudenberg has concluded that the primary reason for stability in small communities is because there is a kind of "social buffering" at a very localized small group/primary acquaintance level, which provides a source of continuity for feelings of personal worth and social integration. William Freudenberg, *The Social Impact of Energy Boom Development on Rural Communities: A Review of the Literature and Some Predictions,* paper presented for the American Sociological Association, session 77, August 1976.

In response, communities usually expand their official control mechanisms to compensate for the loosening of informal social controls. Police forces are enlarged. Mental health clinics either increase in number or add to their caseloads. The emergency room of the hospital and social service agencies are busier than before.

Craig, Colo., provides a good example of what can happen to a community impacted by a large construction work force. Rapid population growth (a 47-percent increase between 1973-76) coincided with an increase in negative social behavior, as suggested by the following:[109]

| Problem area | Percent increase 1973-76 |
|---|---|
| Substance abuse . . . . . . . . . . | 623 |
| Family disturbance . . . . . . . . | 352 |
| Self-respected emotional disorder . . . . . . . . . . . . . . . | 45 |
| Child abuse/neglect . . . . . . . . | 130 |
| Child behavior problems . . . . | 1,000 |
| Crimes against property . . . . | 222 |
| Crimes against persons . . . . . | 900 |

While negative social impacts are frequently reported, they represent only part of the picture. The other part involves many examples of citizens organizing to provide social, recreational, and cultural activities. Many residents of coal-impacted communities speak with pride and satisfaction of the informal initiatives they have taken to improve the quality of life in their communities. Mayor Enzi of Gillette reflects this pride. "Gillette is virtually in the process of building a new town, and there is room for everyone's participation. In Gillette, the only newcomer is the spectator who is unwilling to pitch in."

The farmer/rancher interests also have ambivalent attitudes toward coal. Development may make marginal farmland valuable for future housing sites, thus giving its owners the opportunity to sell at a great profit. The farmer/rancher who does not want to sell may work his place part time and seek full-time employment in the mines. On the other hand, ranchers

---

[109]Includes cases reported to law enforcement, social services, hospital and mental health clinic in Craig. The number of cases was averaged from November/December 1973 and November/December 1976. Denver Research Institute, *Impacts on Western Communities* (OTA contractor Report, 1978), p. 63.

and farmers may not be able to replace their workers who switch over to high-paying jobs. Also, they may resent using good agricultural land for mining. Fruit growers around Paonia, Colo., were unhappy about orchards being taken out of production. Finally, they may be skeptical about different kinds of people moving into a homogeneous community.

Utah's policy favors energy development almost without qualification because of the State's desire for economic expansion. As a whole, Mormon communities share this desire, but some have reservations. Rural, somewhat isolated Utah towns that have high percentages of Mormons may not welcome outsiders. Newcomers are not easily integrated. Residents of Kane County, Utah, wanted the Kaiparowits powerplant built but did not want plantworkers to live in their communities. They persuaded coal developers to plan a new community to house the outsiders.

People in places like Gillette, Wyo., which can now be classified as midboom, see coal development as inevitable.[110] They know their county contains vast coal resources and that new mines will inevitably open. According to Mayor Enzi, the people in Gillette no longer challenge growth but have accepted it. This attitude has taken 10 years to evolve.

Some westerners see no real benefits from coal development. Those who have left an urban environment and have settled in beautiful, small western communities are often against development. They have chosen their new home, typically on Colorado's western slope, because of its environment. They do not want it changed. In a few places on the western slope, miners with Brooklyn accents can be found working in the mines and arguing for no more growth. Their attitudes are not contradictory. Their feelings are often shared by some long-time residents and their children, who are concerned about the influx of outsiders and environmental protection. They often form or ally themselves with environmental organizations on issues like air and water quality and degradation of the quality of life. Sportsmen

[110]From a 1978 interview with Flip McConnaughey, City Manager, Gillete, Wyo.

and outdoorsmen are often against development because they do not want to lose their hunting, fishing, and backpacking areas to the coal industry.

The tourist industry shares these concerns. It may face competition for employees and loss of land for recreational purposes, as well as a lessening of the area's attractiveness to tourists. Tourism may suffer from the increased demand for temporary accommodations caused by transient construction workers.

The elderly may or may not have definite feelings about coal development, but they are usually hit hard. Some rural communities have a relatively high proportion of retired people living on fixed incomes. As the town booms, prices (particularly rents) are inflated, squeezing those with low or fixed incomes.

The degree to which prodevelopment, go-slow development, or antidevelopment attitudes exist in a given area depends on the makeup of the community and the characteristics of the coal project. Articulate, politically active newcomers can play an important role in deciding whether or how much coal development will take place.

The one point that can be drawn from the sparse information on the social impacts of Western coal development is that instability is created. Sometimes this appears to be as much a result of the uncertainty attending coal development as from any actual changes that have occurred. Social change will come with coal development, but it is doubtful that it will affect western towns uniformly. Some towns will ultimately benefit from this change; others probably will not.

## Coal Development on Indian Lands

### Present and Potential Production

It is estimated that some 100 to 200 billion tons of recoverable coal lie beneath Indian lands in the West. There are currently 10 coal leases on 239,402 acres of Indian land. Although they are only a fraction of the total coal under lease in the West, Indian leases have a special importance because of their size and coherence. The average Indian lease covers almost 24,000 acres—16 times larger than the average Federal and 25 times larger than the average Western State lease. Each Indian lease is large enough to support large-scale mining projects.

Five mines now operate on Indian land. Four of these mines rank in the top 10 largest in the country. They produced approximately 22.9 million tons of coal in 1977, almost 14 percent of the coal mined in the West.

Three reservations—Navajo, Hopi, and Crow—now host coal mines. Several others— Southern Ute, Uintah and Ouray, Fort Berthold, Northern Cheyenne, and Zuni—have excellent potential for coal mining. The Department of the Interior lists 25 Indian reservations as coal owning. Estimates of surface-minable capacity on these reservations ranges from 217 million to 326 million tons per year by the mid- to late-1980's. Actual production, however, may be considerably less.

The extent to which Indian coal is developed will depend on decisions made by each tribe.

A key to understanding how—and under what conditions—coal development will take place may be found in the history of coal leasing and mining on Indian lands. The legal framework within which coal development has been conducted is shaped by the unique trust relationship between the Federal Government and tribes.

The basis of the Federal-Indian relationship may be found in the treaties, congressional acts, and executive orders that began in colonial days. Treaties were used primarily to secure lands for white settlement and to maintain peace with the tribes. In return, some lands—a small portion of the land the tribes once owned—were reserved for Indian tribes, with guarantees of protection and certain services from the Federal Government.

Among the services provided by the Federal Government is advising the tribes on the management of their resources. In the past, Indian coal was developed under standard-form leases arranged for and managed by the Bureau of Indian Affairs (BIA) in the Department of the Interior. BIA advised the tribes as to whether, to what extent, and under what terms their coal would be developed. Tribes are no longer satisfied to rely on BIA for decisions regarding their resources.

In March 1973, one tribe—the Northern Cheyenne—became so dissatisfied with leases it had signed with BIA advice, that it petitioned the Secretary of the Interior to cancel all outstanding coal leases and permits on its reservation. A year later, the Secretary ruled in favor of the Northern Cheyenne and declared the leases "technically invalid."

Since this ruling, other lease-holding tribes have tried to renegotiate or cancel contracts, accusing Interior of violating its trust responsibilities to protect tribal interests.

Despite the controversy surrounding these contracts, many tribes see coal development as an opportunity to bring badly needed revenues to tribal governments as well as income and jobs to tribal members. Most feel, however, that they will not realize the potential benefits of coal development unless they develop the capability to manage development themselves. Tribes are building their own staff capabilities so that they can determine the quantity and quality of their resources, analyze the impacts of potential development, make informed decisions concerning their resources, and manage and regulate coal production activities and impacts.

Coal Development Impact

Rapid or extensive coal development often causes adverse social, economic, and environmental impacts. This is especially true on Indian reservations.

Most reservations are sparsely populated and have few towns. In a few cases, coal mining may spawn the growth of instant towns to provide support services for new mines. Even if future tribal contracts provide for Indian hiring preferences, non-Indians will be drawn to the mines as technicians and other management personnel. Non-Indian businessmen may open stores on or near reservations. Although only 5,000 to 10,000 miners are likely to be involved in Indian coal development over the next 10 years, the ripple effect of their new-found income, work habits, and social relations are likely to be disproportionately large on established Indian societies.

Coal development would also affect tribes more than other rural communities because the reservations have a dearth of services and facilities, such as sewers, housing, schools, etc. Any population increase—even a small one—would strain already-overburdened systems. Cultural differences between the tribal members and outsiders who migrate to the area to work on the project can augment the problem.

In addition, coal development impacts Indians more than other rural residents because of the importance of the land to Indian tribes. For some tribes land has religious significance; for others, certain areas are sacred. For all of the tribes, land is an important tie to their history and future.

Finally, coal development will affect tribes more than most small, isolated communities because tribal governments have great difficulty generating revenues needed to offset demands created by development. Conventional methods of raising revenues, such as taxation, issuing bonds, obtaining private financing, or receiving Federal funds directly, are either not available or difficult for tribes.

All of these impacts can serve as deterrents to the development of Indian coal unless tribes can effectively manage them.

Conclusion

Coal development on Indian reservations is likely to increase as tribes gain greater control over their own resources.

Tribes are building up their staff capabilities so they can determine the quality and quantity of their coal reserves, negotiate with energy companies, and regulate the companies' operations. Tribes are looking at alternatives to the standard form lease used in the past. Tribal coal holdings are attractive to energy companies because of their large size and the quality of the minerals. And coal development—if carefully managed by the tribe—is attractive

to many Indian leaders as a solution to their dire poverty—average per capita income is between $1,000 and $1,500 annually—and unemployment—40 to 70 percent depending on the reservation. However, many energy projects on the reservation have been met with bitter grass-root protests.

Coal development will have major socioeconomic and cultural impacts on Indians. The reservations lack public infrastructures capable of absorbing mining-related population. The tribes have difficulty raising revenues to manage the impacts of coal development.

Tribal leaders interviewed agree on several conditions for coal development. First, that the majority of coal-related jobs be filled by Indians. Second, that fair market value is obtained for the coal. Third, that coal development will enhance tribal sovereignty by recognizing tribal jurisdiction over environmental standards and by preserving the integrity of Indian land and culture. Tribes want to ensure that coal development is not a one-crop economic development that leaves them several decades from now without any resources, with scarred lands and shattered cultures. Thus, tribes are expected to demand higher coal royalties, more control over the conditions of mining, and more social benefits from coal development.

## Social Impacts of Transportation

Much of the increase in mine-to-market coal traffic will probably make use of existing routes. Expanded operation, rather than new construction, involves additional safety and environmental costs to the public.

All forms of moving coal—or coal-fired electricity—have social costs. Trains interrupt motor vehicle traffic and contribute to accidents at grade crossings. More than 1,900 persons were killed and almost 21,000 injured in railroad accidents in 1974.[111] Most of these accidents were at grade crossings. Perhaps, 15 percent or more of all train accidents involved coal-hauling trains.[112] Truck haulage, which is

[111]Compiled by OTA from Federal Railway Administration data.
[112]Coal represents about 20 percent of all rail traffic, but for reasons such as frequency of grade crossings and routes, coal traffic probably accounts for less than 20 percent of rail accidents compiled.

extensive in Appalachia, adds to highway hazards and congestion. The big trucks accelerate the deterioration of the roadways, especially when weight limits are not enforced. Trucks and trains are noisy. They disturb large numbers of people where they pass through heavily populated areas. Barge haulage is safer to the public than either truck or rail carriers, but it adversely affects recreational boating, fishing, and swimming to some degree. Barge wakes may erode river shores and islands. High-voltage transmission lines may produce harmful electric-field effects.

Railroads now carry about 65 percent of all coal from mine to consumer. This pattern is likely to continue. Aside from the needs to improve rail-haulage capability, safety and environmental costs may require attention. Two problems, in particular, stand out. First, disruption to community life caused by rail traffic through population centers will increase as large mines are developed in the West for customers several hundred miles distant. Haulage through a given point may approach the equivalent of 18 hours a day in some heavily used areas. Rerouting may be desirable in some communities to minimize this inconvenience.

Second, grade-crossing accidents will increase as haulage increases unless remedial steps are taken. Underpasses and overpasses are standard ways of eliminating rail-haulage risks in population centers. Automatic, lighted barrier gates are safety features at crossings that are often not found at rural crossings. Because of the cost and inconvenience, railroads are often reluctant to reconstruct their lines or add safety features, such as lights and mechanical crossing guards. Table 55 illustrates four costs—noise, vehicle delay, injuries, and deaths—associated with a hypothetical 36-percent increase in coal trains per day (measured in ton miles) between a Montana mine and a Wisconsin utility. At 26.6 trains per day, 53,000 persons would be exposed to excessive noise; 983 hours of vehicle delay per year would occur; and 9 fatalities and 37 injuries would happen at grade crossings. Yet it should be noted that the increase in noise and accidents is not proportional to the

increase in volume although vehicular delay is greater.

Deaths and injuries associated with the total transport component of the coal cycle were calculated in a recent study.[113] For a 1,000-MW coal-fired powerplant with a 0.75 load factor, MITRE estimates 2.3 fatalities and 23.4 injuries occur annually. That translates nationally to about 460 deaths and 4,680 injuries each year from coal transport. As coal production increases, the scale of human costs will grow unless countermeasures are initiated.

The alternative to coal haulage is high-voltage electric transmission, which carries its own costs. Intense electric fields surrounding these lines disturb people living along their rights-of-way. Noise and radio interference have been noted.

Concern has developed about health and safety near the lines. Tests are now underway to measure the effect of high electric fields on plants and animals. For voltages up to 500 kV, no evidence exists to indicate health hazards, but very little is known about the long-term effects of low-voltage electric fields on human health. At higher voltages—say at 765 kV—adverse reactions have been reported that may have been caused by the fields. As the voltages increase, potential health effects may increase. Much work is needed to determine the possible health costs of high-voltage transmission lines carrying more than 765 kV. DC presents fewer potential problems because it carries an equivalent amount of AC power at lower voltages. But because the character of DC electric fields differs from AC fields, other human health effects may occur. Objects near DC lines may be charged with voltages of 20 to 40 kV, enough to cause annoying shock.

## Impacts on Coal-Utilization Communities

Analyses regarding the socioeconomic and environmental impacts of an increased national reliance on coal have heretofore focused on

[113]MITRE Corp./Metrek Division, *Accidents and Unscheduled Events Associated With Non-Nuclear Energy Resources and Technology*, February 1977, p. 51.

Table 55.—Impacts of Increased Coal Train Traffic Carrying an Assumed Added
Volume of 13.5 Million Tons Per Year from Colstrip, Montana to the Vicinity of
Becker, Wis.

| | Present | Future | Percent increase |
|---|---|---|---|
| Trains per day (average) | 19.6 | 26.6 | 36 |
| Population exposed to noise levels exceeding EPA community Guidelines[a] | 45,000 | 53,000 | 18 |
| Vehicle hours delay per year at grade crossings[b] | 602 | 983 | 63 |
| Injuries per year at grade crossings | 33.1 | 36.6 | 11 |
| Deaths per year at grade crossings | 8.6 | 9.4 | 9 |

[a] Day night average sound level, a noise exposure measure weighted more heavily for night then for daytime, of 55 DBA has been judged by EPA to be the maximum allowable to protect public health and welfare with an adequate margin of safety in residential communities.
[b] 695 crossings were analyzed.

SOURCE: Science Applications, Inc., *Environmental Impacts of Coal Slurry Pipelines and Unit Trains*, Office of Technology Assessment, 1977.

the extraction phase, i.e., on coal mine communities where new facilities for the preparation and combustion of coal are located.

A general knowledge of the coal industry suggests a number of categories of potential impact. These can be divided roughly into those that should be on balance beneficial to the host community and those that will be predominantly harmful. The former consists mainly of economic impacts. The establishment of a coal preparation plant or powerplant can be expected to provide employment opportunities in the construction, operation, and maintenance of the plant and in supporting community services. The plant may also contribute substantially to the county tax base and benefit other local taxpayers.

The major potential negative impacts would appear under the headings of environment, public safety, and convenience. The first category includes air and water pollution, noise, esthetic (i.e., visual) impacts, and possible concerns relating to the health effects of high-voltage transmission lines from powerplants. Safety and convenience refers to concerns such as coal-carrying truck or rail traffic and the potential disruption of recreational and other community activities.

Dehue, Logan County, W. Va., 1977

Photo credit: Earl Dotter

Surprisingly, virtually no attempt has been made to ascertain public attitudes toward existing coal powerplants. At a minimum, such information should be useful in developing siting strategies for future installations. Consequently, OTA undertook a survey of residents living near three coal-fired powerplants in the Middle Atlantic region: the Chalk Point plant in Maryland, the Chesterfield plant in Virginia, and the Bruce Mansfield plant in Pennsylvania. The results of that survey indicate that the public regards increased employment opportunities as the most significant advantage of living near such a plant (table 56). Other advantages cited by the public included the perception that coal-generated electric power is more economical than other sources of energy such as oil and gas. The public also perceived benefits to the Nation in shifting from imported oil to domestic coal. A relatively small percentage of respondents cited tax benefits and the perception that coal-fired plants were safer than nuclear. On the negative side, a large percentage of the public (52.3 percent) cited air pollution as the most significant disadvantage associated with living near a coal-fired powerplant, regardless of plant location or distance. Other disadvantages cited were water pollution, noise, adverse visual impacts, and recreational impacts (fishing, boating, and swimming). In an attempt to measure the air and water pollution impact, respondents were asked to rate the level of pollution in their communities and indicate the impact the plant

had on that level. Almost half of the respondents indicated the air and water pollution levels were severe but only a small percentage thought the plant was largely responsible (6.4 percent for air; 4.8 percent for water). The disparity between the level of impact and its perception is due in part to the fact that respondents felt that industries located in the same area contributed to the pollution problem. A relatively small percentage of respondents indicated that traffic to and from the plant had a negative impact. In an effort to elicit an overall summary judgment concerning the local impact of such a plant, respondents were asked their attitude regarding construction of an additional coal-fired powerplant in or near their community. In all three cases, a large percentage of respondents living within 3 miles of the plants opposed the construction of a new plant. In addition, when given a choice between coal and nuclear powerplants, the public did not express a strong preference for either (39.3 percent preferred coal; 36.8 percent preferred nuclear). In many cases the preference for coal was attributed to perceptions that nuclear powerplants were unsafe and fear of radioactive fallout. In general, the balance of positive and negative attitudes toward coal plants correlated more closely with the distance respondents lived from the plant than with any other variable, including the degree of plant compliance with State environmental standards.

**Table 56.—Survey Results**

| Chalk Pt. | Chester. | B. Mansfield | Percentages |
|---|---|---|---|
| | | | **3. Air pollution level** |
| 28.2 | 17.5 | 8.1 | ___ Not a problem. |
| 37.2 | 34.1 | 24.3 | ___ Some, but does not disturb anyone. |
| 23.8 | 42.7 | 46.8 | ___ Occasionally reaches severe levels. |
| 1.1 | 3.3 | 12.7 | ___ Usually severe and interferes with some community activities. |
| 9.7 | 2.5 | 8.1 | ___ Cannot rate level. |
| | | | **4. Plant impact on air pollution** |
| 4.8 | 2.0 | 13.8 | ___ Severe with frequent odor, dust, or damage. |
| 29.7 | 18.7 | 34.3 | ___ Moderate with occasional odor, dust, or damage. |
| 31.5 | 28.0 | 27.0 | ___ Slight with little odor, dust, or damage. |
| 34.1 | 51.3 | 24.9 | ___ No noticeable impact. |

## Table 56.—Survey Results (continued)

| Chalk Pt. | Chester. | B. Mansfield | |
|---|---|---|---|
| | **Percentages** | | |
| | | | **5. Water pollution level** |
| 8.6 | 33.9 | 8.6 | _____ Usually severe and interferes with some community activities. |
| 33.0 | 27.0 | 25.5 | _____ Occasionally reaches severe levels and sometimes interferes with community activities. |
| 24.4 | 15.5 | 27.6 | _____ Some, but does not disturb anyone. |
| 16.8 | 11.5 | 14.8 | _____ Not a problem. |
| 17.2 | 12.1 | 23.4 | _____ Cannot rate level. |
| | | | **6a. Plant impact on water pollution** |
| 7.9 | 2.9 | 5.1 | _____ Severe with major changes in temperature, level, or chemicals in the water. |
| 25.7 | 13.1 | 24.3 | _____ Moderate with some changes in temperature, level, or chemicals in the water. |
| 29.4 | 31.1 | 27.5 | _____ Slight with minor changes in temperature, level, or chemicals in the water. |
| 37.0 | 52.9 | 43.1 | _____ No noticeable impact. |
| | | | **6b. If plant has impact on water pollution, which of the following apply?** |
| 19.3 | 6.0 | 6.1 | _____ Fishing is better. |
| 35.0 | 21.9 | 33.7 | _____ Fishing is not as good. |
| 2.5 | 4.1 | 7.5 | _____ Boating and swimming are better. |
| 41.4 | 23.8 | 31.6 | _____ Boating and swimming are not as good. |
| 7.5 | 6.8 | 22.1 | _____ Drinking water is polluted. |
| 13.2 | 18.9 | 10.5 | _____ Other, please specify. |
| | | | **7. Plant impact on public health** |
| 0 | 0 | .7 | _____ Public health very much better. |
| 4.1 | 1.5 | 3.0 | _____ Public health somewhat better. |
| 79.0 | 84.0 | 61.1 | _____ No difference in public health. |
| 15.7 | 13.3 | 29.3 | _____ Public health is somewhat worse. |
| 1.1 | 1.2 | 5.9 | _____ Public health very much worse. |
| | | | **8. Plant impact on public safety or convenience** |
| 3.0 | 4.7 | 3.6 | _____ Public safety or convenience very much better. |
| 5.2 | 9.4 | 8.0 | _____ Public safety or convenience somewhat better. |
| 79.9 | 79.8 | 50.0 | _____ No difference in public safety or convenience. |
| 10.4 | 5.3 | 29.6 | _____ Public safety or convenience somewhat worse. |
| 1.5 | .9 | 8.8 | _____ Public safety or convenience very much worse. |
| | | | **9. Plant impact on area attractiveness** |
| .7 | .6 | 3.9 | _____ Very much more attractive. |
| 6.2 | 1.7 | 11.0 | _____ Somewhat more attractive. |
| 35.4 | 61.2 | 24.9 | _____ No difference in attractiveness. |
| 38.0 | 30.1 | 32.4 | _____ Somewhat less attractive. |
| 19.7 | 6.4 | 27.8 | _____ Very much less attractive. |
| | | | **10. Impact of power transmission lines on area** |
| .4 | .6 | 1.8 | _____ Very desirable effect. |
| 2.2 | 2.3 | 5.1 | _____ Somewhat desirable effect. |
| 47.6 | 52.9 | 46.9 | _____ No noticeable effect. |
| 35.4 | 38.8 | 30.7 | _____ Somewhat undesirable effect. |
| 14.4 | 5.5 | 15.5 | _____ Very undesirable effect. |
| | | | **11. Plant impact on area activities** |
| 1.1 | 3.8 | .7 | _____ Plant grounds and facilities available for community use. |
| 17.8 | 16.8 | 19.2 | _____ Plant tax payments prevent taxes on home from being higher. |
| 14.4 | 11.2 | 30.8 | _____ Noise or air pollution or water pollution discourage community activities. |
| 61.0 | 63.5 | 42.0 | _____ No noticeable impact. |
| | | | **12a. Mode of transportation** |
| 56.6 | 64.6 | 0 | _____ Train. |
| 2.9 | 1.7 | 49.4 | _____ Barge. |
| 5.4 | .3 | .4 | _____ Truck. |
| 23.3 | 13.6 | 8.2 | _____ Don't know. |

## Table 56.—Survey Results (continued)

| Percentages | | | |
|---|---|---|---|
| Chalk Pt. | Chester. | B. Mansfield | |
| | | | **12b. Transportation impact on community** |
| 6.1 | 2.8 | 22.9 | _____ Yes. |
| 93.9 | 97.2 | 77.1 | _____ No. |
| | | | **13. Favor/oppose construction of new plant** |
| 28.2 | 11.8 | 23.6 | _____ Strongly oppose. |
| 21.1 | 16.7 | 21.6 | _____ Somewhat oppose. |
| 24.6 | 33.4 | 27.7 | _____ Neither favor nor oppose. |
| 14.6 | 23.0 | 12.7 | _____ Somewhat favor. |
| 11.4 | 15.1 | 13.4 | _____ Stongly favor. |
| | | | **14. Preference for coal or nuclear plant** |
| 34.4 | 46.7 | 38.7 | _____ Coal-fired. |
| 41.1 | 39.0 | 33.9 | _____ Nuclear. |
| 24.4 | 14.3 | 27.4 | _____ Don't know. |
| | | | **15. Distance from home to plant** |
| 31.4 | 9.3 | 61.9 | _____ Less than 3 miles. |
| 27.1 | 38.5 | 27.9 | _____ Between 3 and 6 miles. |
| 22.5 | 29.2 | 4.8 | _____ Between 6 and 10 miles. |
| 14.3 | 13.9 | 1.0 | _____ More than 10 miles. |
| 4.6 | 9.0 | 4.4 | _____ Don't know. |

Photo credit: Pennsylvania Power Company

Bruce Mansfield plant, Shippingport, Pa.

# ESTHETIC IMPACTS

## Introduction

The esthetic impacts of increased coal utilization are difficult to quantify and evaluate because procedures for measuring esthetic impacts have not been established. In part this is a result of the subjective nature of esthetics. For example, Webster defines esthetics as "beauty or sensitivity to beauty." The response to an esthetic impact, positive or negative, is largely dependent on value judgments.

Consequently, a great deal of uncertainty exists regarding the role esthetics should play in affecting public policy.

Section 102(2)(B) of the National Environmental Policy Act calls for the consideration of esthetic impacts in decisionmaking but does not provide assistance in determining what methods and procedures should be developed to facilitate that consideration. Some attention has been given to the formulation of quantifiable indicators but these are still in the very early stages of development. Table 57 illustrates work in this area.

The most obvious problem is the fact that esthetic impacts are perceived differently by different individuals and groups. Although no one finds strip mines esthetically pleasing, there are individuals who feel that the construction and operation of coal utilization facilities will have a positive impact. For example, some feel reclamation is an opportunity for improving the landscape. Others argue, however, that the construction and operation of coal utilization facilities will have adverse esthetic impacts; changing the natural characteristics of an area and its ecosystems is viewed as detrimental. Furthermore, in some instances esthetic impacts can be used as a surrogate for other concerns, such as opposition to government and industry, and resistance to outsiders. An example of this is the Kaiparowits controversy where fear of esthetic impacts were used as a basis to mobilize support from a broad-based national constituency. Although strong local support was given to the development of the Kaiparowits plant, a national constituency that views air pollution impacts on nearby pristine areas as unacceptable actively opposed the plant.

The variable most often used to measure esthetic impact is change. Present EISs generally reflect the view that changes in the environment are negative. In addition, esthetic impacts identified in EISs appear to be of an immediate or first-order kind; EISs do not identify beneficial second-order esthetic impact associated with social and economical benefits (libraries, theaters, etc.). Legislation also appears to embody the assumption that change is negative. This is evident in the language of the prevention of significant deterioration standards for air quality. There does not appear to be any firm basis for defining all change from the natural state to be bad, however.

#### Table 57.—Esthetic Judgment Categories

| Esthetic judgment | | | | | |
|---|---|---|---|---|---|
| Air esthetics | Water esthetics | Landscape esthetics | Biota esthetics | Sound esthetics | Equality of esthetic opportunity |
| Visibility | Clarity | Urban dominated | Population | Background sound | |
| Odor | Floaters | Mountain dominated | Variety | Intermittent sound | |
| Irritants | Odors | Desert dominated | Health | | |
| | | Agriculture dominated | Location | | |
| | | Forest dominated | | | |
| | | Water dominated | | | |

SOURCE: Adapted from Gum, et al., 1974: 34.

Photo credit: Earl Dotter

**Ellsworth, Pa. Slag pile in the background**

## Mining

A major concern in the West is whether reclaimed land can be revegetated in an area with limited rainfall. Although similar concerns are expressed in the Central and Eastern regions, revegetation is easier there because of the higher levels of rainfall and thicker soil profiles. Because of thin coal seams, however, as much as 10 times more land in the East must be stripped in order to provide the same quantity of coal as in the West. Contour mining in mountainous areas produces still another serious esthetic impact. Often the highwall is visible for miles. Reclamation of contour mining is costly and difficult and may appear slightly artificial. But if properly reclaimed, nonmountainous land can be used for many purposes, including agriculture, recreation, forestry, housing, and industrial development. Another impact associated with strip mining is a reduction in visibility due to dust. This impact is

probably of greater concern in the West where rainfall is limited. Noise is another impact from mining. Although blasting noise is not identified as a serious problem, it is an annoyance to individuals who live near the mine. The most significant visual impact resulting from underground mining is subsidence (uneven sinking of land surfaces). This impact is a significant concern in the Central and Eastern coal regions.

## Transportation

Several esthetic impacts result from the transportation of coal. Unit trains produce a noise level of about 95 decibels; these levels may remain as high as 55 decibels within one-half mile of the track. At present, noise from transportation affects trackside residents and as rail haulage increases with expanded production, noise, and dust will become more frequent and more serious annoyances. Highways also contribute to noise pollution, and both highways and railroads may be considered unsightly. However, the most obvious visual transportation impact is from transmission lines. The lines are especially unsightly in scenic and residential areas and are visible for miles. Tower design as well as proper siting could possibly reduce the impact.

## Combustion

Visually, coal-fired electric powerplants are rarely pleasing to the eye. Often stacks are visible for miles. In addition, coal piles surrounding the plant are unsightly and may produce dust. Large-scale coal utilization facilities can significantly reduce visibility and alter the natural coloration of the sky. During an inversion, visibility may be reduced even more. Gray skies associated with heavily industrialized areas may suggest what is to come if coal utilization is accelerated. Individuals living in close proximity to the facility may experience noise and odor impacts, but these do not appear to be a serious concern at this time.

Chapter VII
# PRESENT FEDERAL COAL POLICY

# Chapter VII.—PRESENT FEDERAL COAL POLICY

## TABLES

# PRESENT FEDERAL COAL POLICY

Neither Congress nor the President has yet articulated a comprehensive and consistent policy for coal. Nevertheless, the major elements of such a policy are in place. The National Energy Act of 1978 and the earlier Energy Supply and Environmental Coordination Act of 1974 promote the use of coal to increase the Nation's capability to use domestic energy sources. The National Environmental Policy Act, the Clean Air Act and Amendments, the Clean Water Act, the Surface Mine Control and Reclamation Act, and other Federal policy actions minimize environmental degradation resulting from, among other causes, the production and use of coal. The Mine Safety and Health Act is intended to reduce the hazards to miners, while black lung benefits compensate for past abuses. Various provisions also have been made to assist communities experiencing rapid growth as a result of coal development. Thus coal policy might be summarized briefly as: When fossil fuel is to be used, coal shall be the choice wherever practical, but it must be mined and burned in ways that minimize the negative environmental and human impacts. These goals are obviously contradictory, but they need not be mutually exclusive.

Because Congress has never legislated a national coal policy, the current Federal role has evolved incrementally as part of a continuing political process where different interests — both inside and outside of Government — compete for influence. This competition involves lobbying Congress, media campaigns, gaining access to policymakers, researching issues, bargaining with the opposition, and mobilizing and targeting pressure. The consequence of this process is that Federal policy is often contradictory, frequently delayed by legal challenges, occasionally jarred by legislative initiatives, and rarely settled once and for all.

This chapter explores the present structure of Federal coal policy and how it is being implemented. The next chapter analyzes policy strategies that may be considered in the future.

## POLICY EVOLUTION

At the end of World War II, coal supplied about half the Nation's energy needs. By 1950, it had been eclipsed by oil, and natural gas surpassed it in 1958. Such a rapid decline in absolute as well as relative importance indicates that more was occurring than an industry succumbing to stronger competition. Both market prices and Federal policies discouraged the use of coal relative to oil and gas. Similarly, the negative impacts of coal were not of paramount concern to the Federal Government except in so far as they contributed to the tilt toward oil and gas. Federal coal on western lands was made easily available, but little development was expected to take place.

In the last decade, the realization of the resource limitations of oil and gas has renewed interest in coal as the logical substitute, result-ing in a variety of Federal policies related to coal. Federal legislation first addressed the occupational health and safety risks of coal mining. Subsequent Federal policy directives addressed the air and water quality effects of mining and combustion on human health and the environment. Recently, the disposal of mine and combustion wastes and the reclamation of surface mines have come under Federal jurisdiction. However, there are many points in the entire pattern of coal production and use where the different goals — increased production, environmental protection, and maximum social benefit — conflict. Both Congress and the executive branch have sought to fashion policies and administrative frameworks that can resolve these conflicts, but they ofte are frustrated by overlapping and conflicting mandates that result in inconsistent or fragmented policies. For example, an adequate water supply is crucial to mining operations but the lack of a comprehensive Federal water policy offers little direct support to a policy of increased coal development.

The Federal legislation that constitutes the major elements of a national coal policy are analyzed below, categorized according to their goals: environmental and social impact management, and the promotion of coal use.

## National Environmental Policy Act

The National Environmental Policy Act of 1969[1] (NEPA) was intended to restructure Federal agency decisionmaking in favor of an interdisciplinary approach to ensure that environmental amenities and values would receive appropriate consideration along with the traditional economic and technical factors. NEPA was the first major environmental legislation passed by Congress and it has remained the most far-reaching in scope.

In general, NEPA has a threefold purpose: to declare a national policy to create and maintain conditions under which man and nature can exist in productive harmony and can fulfill the social, economic, and other requirements of present and future generations; to increase the understanding of ecological systems and natural resources; and to promote efforts that will prevent or eliminate damage to the environment. As one means of achieving these purposes, NEPA requires all Federal agencies to include a detailed environmental impact statement (EIS) in every recommendation or report on proposals for legislation and other major Federal actions significantly affecting the quality of the human environment. The EIS requirement has been held to be applicable to a wide variety of Federal actions, ranging from the construction of a Federal building to the issuance of a construction permit for a nuclear powerplant.

An EIS is required to include detailed information about:

[1] 42 U.S.C. 4321 et seq.

• the environmental impact of the proposed action,
• any adverse environmental effects that cannot be avoided should the proposal be implemented,
• alternatives to the proposed action,
• the relationship between local short-term uses of man's environment and the maintenance and enhancement of long-term productivity, and
• any irreversible and irretrievable commitments of resources that would be involved in the proposed action should it be implemented.

In addition, the Council on Environmental Quality (CEQ) regulations for the preparation of EISs require discussions of indirect effects, such as population and growth; energy requirements and the conservation potential of various alternatives and mitigation measures; and possible conflicts between the proposed action and the objectives of other Federal, regional, State, tribal, or local policies.

All coal-related activities that have a significant impact on the environment and that need Federal authorization require an EIS. This includes coal leases on Federal lands and large coal combustion facilities. Although permits issued by the Environmental Protection Agency (EPA) under the Clean Air and Water Acts are exempt from the EIS requirement, those Acts require separate analyses of a project's impact on the environment.

The original CEQ regulations implementing NEPA were limited to the preparation of EISs and did not address NEPA's other provisions intended to improve agency planning and decisionmaking. In addition, CEQ's guidelines for preparing EISs were only advisory; more than 70 different sets of agency regulations were promulgated to implement the guidelines. As a result, agency practices varied widely and the EIS tended to become an end in itself rather than a means to making better decisions. The NEPA process was criticized by environmentalists for its attention to procedure rather than substance, by permittees for delays it caused in project authorization, and by Government officials for the amount of paperwork it required.

In 1977, CEQ was given the authority to issue binding regulations to replace its advisory guidelines and the various agency regulations. The new CEQ requirements, which were published in November 1978, are designed to produce better agency decisions and to reduce delays and paperwork.[2]

The new CEQ regulations are intended to comply with the original intent of NEPA that the EIS be an action-forcing procedure to implement the substantive requirements of the Act and thus to produce better agency decisionmaking. They require agencies to publish a concise public record that indicates how the EIS was used in making a decision. This record must indicate which alternative is preferable on environmental grounds. If that alternative is not the one chosen, the record must identify the essential considerations of national policy that were balanced in making the decision, including factors not related to environmental quality such as economic and technical considerations or legislative mandates, and must explain why the environmental considerations were outweighed by these other factors. A second provision of the new regulations, which is intended to produce better decisions, requires agencies to monitor projects to ensure that mitigation procedures and other conditions established in the EIS are implemented. Finally, the new regulations require that a list of the people who helped prepare an EIS and their professional qualifications be included in the statement to encourage accuracy and professional responsibility and to ensure that an interdisciplinary approach was followed.

To reduce paperwork, the regulations require agencies to reduce the length of EISs (CEQ suggests 150 pages as the limit for a normal EIS), prepare analytic rather than encyclopedic statements, use plain language, follow a clear format, and reduce emphasis on background material. In addition, Federal agencies can eliminate duplication by preparing EISs jointly with State and local agencies that require similar assessments. The regulations also institute a new "scoping" procedure for deciding which issues should be emphasized in

[2]43 F.R. 55978 (Nov. 29, 1978).

the EIS and how the responsibility for the EIS should be apportioned among the agencies involved. This scoping procedure is to begin as early as possible in the NEPA review and must be integrated with other planning.

Provisions of the regulations designed to reduce delays in project authorizations include emphasizing interagency cooperation before the EIS is prepared rather than submission of adversary comments on a completed draft, establishing time limits for EIS preparation, and using categorical exclusions to define actions that are exempt from the EIS requirement. In addition, the regulations call for a "tiered" approach in which material common to a broad program or a related series of actions is included in broader EISs, such as national program or policy statements, and then summarized and incorporated by reference in successively narrower EISs, such as regional or basinwide program assessments and, ultimately, site-specific statements. Finally, many of the regulations intended to reduce paperwork also will reduce delays, such as the scoping process and the elimination of duplicate statements.

It is too soon to tell whether these regulations will restructure agency decisionmaking in order to implement fully NEPA's environmental policy goals. However, the regulations should significantly reduce the resources required to comply with the Act and thus prevent NEPA from becoming a limiting factor in decisions related to coal development.

## Federal Coal Leasing

The National Energy Plan calls for expanded reliance on coal. Some commentators have suggested that Federal coal reserves will be a key factor in determining whether the Nation will meet its energy goals. Western coal comprises roughly half of the total U.S. coal reserves; the Federal Government owns 65 percent of Western coal and indirectly controls another 20 percent. On the other hand, the 1977 amendments to the Clean Air Act may considerably reduce the advantages that have accrued to low-sulfur Western coal in the past if those amendments require pollution controls regardless of the sulfur content of the fuel.

Even if Western coal is required to meet national energy goals, it is not clear how many new Federal coal leases would be needed. Sixteen billion tons of Federal coal already are under lease and another 9 billion are subject to existing applications for preference-right leases. In addition, there are an estimated 93.4 billion tons of recoverable coal reserves on private lands in the West.

Thus it is unclear how important Federal coal will be to the Nation in the future or whether additional Federal leasing is needed. Yet no comprehensive Federal coal leasing policy currently exists, and even if more Federal coal is required to meet the Nation's energy goals, it probably will not be able to play a central role until the early 1980's. This section reviews the history of Federal coal leasing, examines recent congressional action, and outlines the major obstacles to a rational leasing policy.

### Past Leasing Law and Practice

Originally, the Federal Government sold or gave away its coal-bearing lands. Then, under the Mineral Leasing Act of 1920,[3] it became Federal policy to issue leases for prospecting and mining. Where there were known coal deposits, leasing tracts were awarded by competitive bidding. In other areas, the Department of the Interior issues 2-year prospecting permits. If commercial quantities of coal were discovered during those 2 years, the permittee was entitled to a preference-right lease for all or part of the land.

Although the 1920 Act contained numerous provisions designed to protect the public interest and to ensure a fair return to the Government, most leases were issued on an ad hoc basis without regard to energy needs or environmental factors. Much of the leased land fell into the hands of a few large companies that were interested only in speculation. The Federal Government often charged lower royalties than private lessors and Government leases

could be held without development at little cost and transferred at will.

Leasing generally was based on a site-by-site response to applications, usually at the State level with no National policy guidance. Consequently, what mining did occur was hampered by irrational landownership patterns in the Western coalfields. Areas that logically should have been mined as single units were divided among private owners and Federal lessees. Similarly, there was often a split in ownership of surface and mineral rights. In addition, data necessary to develop a coherent National policy, such as information on the location and extent of coal deposits, were not available.

In November 1970, the Bureau of Land Management (BLM) released a study that summarized the results of these abuses.[4] The study found that while the number of acres being leased was rising, production from leased lands was dropping; 91 percent of all coal leases were not producing anything. BLM concluded that "existing policies and procedures with respect to the development of federally managed coal resources are inadequate to encourage their development." As a result of this study, the Department of the Interior imposed a moratorium on coal leasing while it reassessed its own policies. In February 1973, Interior announced that a new long-term policy was under development and in the interim no new prospecting permits would be issued and coal would be leased only where certain "short-term" criteria were met.

The short-term criteria were designed to prevent undue hardship to companies as a result of the moratorium. Initially, they permitted coal leasing when necessary to maintain an existing operation or to serve as a reserve for production in the near future. Unfortunately, these criteria were abused and received scathing criticism from the General Accounting Office.[5] The criteria were revised in July 1977; 2 months later they were included in an in-

---

[3]30 U.S.C. 181 et seq.

[4]*Holdings and Development of Federal Coal Leases,* (U.S. Department of the Interior, Bureau of Land Management, unpublished, November 1970).

[5]*Further Action Needed on Recommendations for Improving the Administration of Federal Coal-Leasing Program* (Washington, D.C.: General Accounting Office, U.S. Comptroller General, April 1975).

junction issued by a Federal court against the entire Federal leasing program.[6] At present, short-term leasing is permitted only when necessary to maintain the level of production at an existing mine or to meet existing contracts and when the extent of the lease is not greater than required to meet these conditions for 3 years.

Short-term leasing has produced little coal. Only 12 leases covering 30,459 acres have been issued since 1973. Industry is reluctant to commit itself to short-term coal leases and the court's criteria are difficult to meet.

Over the same period, the Federal Government's long-term leasing program has gone through two stages, known as EMARS I and EMARS II. In May 1974, the Department of the Interior issued a draft programmatic EIS, outlining its plans for a long-term leasing program, or Energy Minerals Allocation Recommendation System (EMARS I), which required the Government to set leasing targets based on national energy needs. Under EMARS I, leasing tracts were to be nominated by the Government based on land use plans. Environmental impact assessments were to be performed on the nominated tracts and those that were suitable for leasing put up for competitive bidding.

In September 1975, after a comment period on the impact statement and EMARS I, the Department of the Interior issued the final programmatic EIS. The impact statement adopted an Energy Minerals Activity Recommendation System (EMARS II) instead of the Allocation System contained in the draft statement. Under EMARS II, industry would nominate leasing tracts, indicating the type, amount, and location.

Both the final impact statement and the EMARS II program were criticized severely. The General Accounting Office noted, among other deficiencies, that the Department of the Interior was assuming its traditional role of reacting to industry proposals rather than managing national energy resources.[7] In September 1977, a Federal court enjoined Interior from taking any steps whatsover, directly or indirectly, to implement the new coal leasing program, and ordered the Department to correct the deficiencies in the final EIS and to circulate the revised statement for review and comment.[8] Until the new EIS is completed—which Interior indicates will be not earlier than mid-1979—no long-term Federal coal leasing will occur.

### Development of a New Federal Coal Leasing Policy

In 1976, Congress passed the Federal Coal Leasing Amendments Act[9] to address the deficiencies of the Mineral Leasing Act of 1920. The 1976 amendments discourage speculation by making leases more difficult to obtain and to hold cheaply. Federal leases now must be developed within 10 years or they terminate automatically, they cannot be issued for less than their fair market value, and mimum royalties are set. In addition, all leasing must be by competitive bidding with 50 percent leased under a deferred bonus bidding system that makes it easier for small companies to compete. Preference-right leases are abolished, and no person or corporation may hold more than 100,000 acres of leases.

The amendments also contain various provisions to encourage land use and environmental planning, including the creation of "logical mining units." In addition, one-half of all royalties must be turned over to the States to provide public services and mitigate community impacts. Finally, the Department of the Interior is directed to conduct a comprehensive exploratory program to determine the extent of Federal coal reserves.

In 1977, BLM promulgated regulations to implement the Federal Coal Leasing Amendments Act. Because they are so recent and because

---

[6]*Natural Resources Defense Council* v. *Hughes,* 7 E.L.R. 20785 (D.D.C., 1977).

[7]*Role of Federal Coal Resources in Meeting National Energy Goals Needs To Be Determined and the Leasing Process Improved* (Washington, D.C.: General Accounting Office, U.S. Comptroller General, April 1976).
[8]*NRDC* v. *Hughes.*
[9]30 U.S.C. 201-209.

new leasing has been almost nonexistent since 1970, the degree to which they will remedy past abuses is unclear. However, it already is obvious that a number of the regulations are confusing and may have to be clarified or revised.

In his May 1977 environmental message, President Carter directed the Secretary of the Interior to undertake a major review of Federal coal leasing policy.[10] In particular, the Secretary was directed to ensure that the Department of the Interior could respond to reasonable production goals while leasing only those areas where mining is environmentally acceptable and is compatible with other land uses. Special attention is to be given to existing leases that are nonproducing or environmentally unacceptable.

In response to this directive and to a request by the court that enjoined EMARS II, the Department of the Interior created a Federal Coal Management Review Policy Committee.[11] The committee has undertaken 13 management tasks in response to the Presidential directive. One of its first products is a paper that outlines three options for an overall approach to long-term leasing: industry nomination of both areas and tracts for leasing, Government identification of areas with industry nomination of specific tracts, and Government selection of both areas and tracts.[12]

Until the committee's work is complete, it is unclear what direction the new Federal coal leasing policy will take, or which, if any, of these three options will be adopted. However, the Secretary of the Interior has stated that the new Federal policy will restrict coal leasing to those areas where the Federal Government owns the surface rights.[13]

## Problem Areas

Many of the problems that still must be resolved by the Federal Coal Management Review Policy Committee, the Department of the Interior, and Congress are carryovers from earlier leasing policy. They include logical mining units, preference-right lease applications, the requirements of diligent development and continued operation, estimated recoverable reserves, advance royalty payments, and the exchange of environmentally sensitive leased lands for other unleased Federal land.

A logical mining unit (LMU) is a contiguous area of coal land under the control of one operator that can be developed effectively and mined within a defined period of time. The theory of LMU makes sense; it asserts that national leasing policy should consider the nature of the coal resource in each case and then assemble the physical resources necessary to produce the coal economically while preserving environmental safeguards. However, the regulations promulgated by BLM to implement the Federal Coal Leasing Amendments Act depart from this theory in two respects. First, under the regulations, boundaries of an LMU are set according to the legal boundaries of leases rather than according to natural formations of the coal seam. Second, an LMU is required to produce a fixed amount of coal each year without regard to the individual characteristics of each coal mining operation.

Technical problems also surround the formation of LMUs. The Secretary of the Interior does not appear to have the authority to order private, State, or Indian lands to be combined with Federal leases to form an LMU nor does the Secretary have the authority to divide existing leases into several LMUs. In addition, practical and legal problems may arise from the requirements that all lands in an LMU be contiguous and that the holders of existing leases must consent to their inclusion in an LMU. Finally, industry representatives have argued that the congressional mandate that all reserves in an LMU be mined within a 40-year period may result in inefficiencies in some circumstances.

[10]The President's Environmental Program (Washington, D.C.: Council on Environmental Quality, 1977).
[11]Federal Coal Management Review Policy Plan (U.S. Department of the Interior, unpublished, July 25, 1977).
[12]Memorandum to the Secretary of the Interior From the Deputy Executive Secretary on Departmental Approach for the Long-Term Coal Leasing Program: Decisionmaking (unpublished, Oct. 17, 1977).
[13]220 Energy Users Report 24 (Oct. 22, 1977).

Much uncertainty and controversy also surrounds the preference-right lease applications (PRLAs). At present, there are approximately 180 outstanding PRLAs covering an estimated 9 billion tons of coal on 446,000 acres. The Department of the Interior contends that it must issue a PRLA when a permittee has established that the land contains commercial quantities of coal. "Commercial quantities" was interpreted as coal "of such character and quantity that a prudent person would be justified in the further expenditure of his labor and means with a reasonable prospect of success in developing a valuable mine . . ."[14] Some critics have questioned the appropriateness of the prudent person test and suggested the substitution of a "workability" or "paying quantities" concept. Others have asserted that, in addition to the commercial quantities test, Interior must consider environmental impacts before issuing a PRLA. Finally, even if all the foregoing issues are resolved quickly, it will take years to process the pending applications. It appears that a minimum of 5 to 7 years will pass before any coal can be produced from PRLA tracts.

A number of problems also exists in the regulations that implement the requirements for diligent development and continued operation of leases. First, the regulations define those requirements in terms of LMUs, yet old leases cannot be designated as LMUs without the consent of the lessees. The Department of the Interior has taken the position that it can designate an old lease as an LMU when it is readjusted despite statutory language that suggests that the consent of the lessee is required. Second, it is not clear that the present definition of diligent development, which does not require periodic reports, will prevent lessees from holding land for long periods of time without substantial development.

Additional problems exist in implementing the concept of estimated recoverable reserves. Many of the key terms in the coal leasing regulations are defined in terms of this concept, but there is no commonly accepted definition of "estimated recoverable reserves" and various organizations compute reserves in different ways at different times. As a result, industry and Government estimates can vary by as much as 100 percent. Moreover, the present regulatory scheme provides Federal leaseholders with strong incentives to underestimate their recoverable reserves.

Finally, problems exist in the regulations designed to implement those portions of the 1976 amendments that relate to advance royalty payments and exchanges of land or leasing rights. The protection afforded by advance royalty provisions of the amendments appears to have been diluted by the Department of the Interior regulations that grant the Mining Supervisor blanket authority to permit advance royalty payments in lieu of continued operation without requiring a specific finding that such a substitution would be in the public interest. In addition, Interior currently is developing regulations that would authorize the exchange of Federal land under lease or subject to a PRLA for future bidding rights. The Department would prefer the broader authority to exchange leases in environmentally sensitive areas for other unleased land as well as for future bidding rights.

## Surface Mining Control and Reclamation Act of 1977

Many surface mining operations result in disturbances that adversely affect commerce and the public welfare by destroying or diminishing the utility of land for commercial, industrial, residential, recreational, agricultural, and forestry purposes. Many operations have caused erosion and landslides, contributed to floods, polluted the water, destroyed fish and wildlife habitats, impaired natural beauty, damaged private property, and undercut Government policies and programs to conserve soil, water, and other natural resources. The existing State regulatory programs intended to deal with these impacts varied widely and often were not enforced adequately. These findings, and others, formed the basis for congressional action leading to passage of the Surface Mining Control and Reclamation Act (SMCRA) of 1977.[15] A detailed analysis is presented in volume II, appendix XVII.

---

[14]41 F.R. 18847 (May 7, 1976).

[15]30 U.S.C. 1201 et seq.

Based on the assumption that mining should be a temporary activity, the Act was intended to change coal mining practices that generate severe social and environmental costs and to prohibit mining operations in areas that cannot be reclaimed. To accomplish these goals, the Act mandates State permit programs for surface mines and for the surface operations of underground mines and State procedures for designating areas unsuitable for mining. In the interim, or in the event of a State's failure to establish an adequate program, the Federal Government will retain regulatory authority. In addition, the Act establishes a fund for the reclamation of abandoned mines.

This section discusses the permit programs for surface and underground mining. Mandated procedures for designating areas as unsuitable for mining are discussed in chapter IV.

Each application for a surface coal mining and reclamation permit must include detailed information about the type and method of coal mining operation, engineering techniques, and equipment to be used; the probable hydrologic consequences of the mining and reclamation, both on and off the minesite; any manmade features or significant archeological sites that may be affected by mining; the geological and physical characteristics of the coal, including a chemical analysis of potentially acid- or toxic-forming strata; a soil survey of potential prime farmland; and the reclamation plan.

The probable hydrologic consequences of mining and reclamation must be determined relative to the hydrologic regime and the quantity and quality of surface and ground water systems including dissolved and suspended solids under seasonal flow conditions. Sufficient data must be collected to enable the regulatory agency to assess the probable cumulative impacts of all mining in the area on hydrology and water availability.

The reclamation plan must describe the condition of the land prior to mining including its existing and potential land uses and its productivity as well as its average yield of food, fiber, forage, or wood products under optimum management. The plan also must specify the proposed postmining land use and describe in detail how this use will be achieved including the engineering techniques and equipment to be used, the cost per acre of reclamation, and a detailed timetable for accomplishing reclamation. In addition, the plan must describe the means of compliance with applicable air and water quality and health and safety regulations.

All surface mining permits issued under the Act must require that the coal mining operations meet all applicable environmental protection performance standards. These standards govern the maximum recovery of fuel; restoration of the land to its original contour; use of explosives; waste disposal, including the use of waste piles as dams or embankments; construction of access roads; and revegetation. Additional, more stringent standards apply to environmentally sensitive areas such as prime farmland, steep slopes, alluvial valleys, and timber lands.

Permits for underground mining must also require the mine operator to prevent subsidence to the extent possible, seal all openings to the surface, and prevent acid or other toxic drainage.

Enforcement of SMCRA's performance standards and reclamation plans will play a critical role in determining the impact of the Act on coal production and environmental quality. As with administration of other SMCRA provisions, enforcement can be delegated to States if their mining standards and reclamation requirements are at least as stringent as the national standards. If this strategy of delegating the primary enforcement role to the States is successful, the Federal Office of Surface Mining (OSM) would not require large-scale enforcement resources. If the strategy does not succeed, the ability of OSM to enforce the Act adequately is uncertain. During interim enforcement of the Act by OSM, about 160 inspectors were responsible for ensuring compliance at more than 3,900 surface mines and approximately 2,550 underground mines (or approximately 1 inspector per 35 surface and per 56 underground mines).

An inadequate inspection force will result in an inability to make the required number of inspections and therefore an emphasis on "problem mines" and flagrant violations. It is doubtful that all problem mines will be identified; those mines that are not inspected as frequently as required could escape strict compliance with the Act.

Other issues related to enforcement of SMCRA include the availability and adequacy of performance bonds, which are intended to secure compliance with the Act. The amount of a bond depends on the characteristics of the site, but must be sufficient to cover completion of the reclamation plan if the work had to be performed by the regulatory agency. Where reclamation is difficult and therefore expensive, performance bonds may be difficult to secure. The Small Business Administration, which guarantees bonds from private sureties in other industries, reportedly does not plan to do so for coal mine reclamation.[16] In addition, where the adverse impacts of mining may not become apparent until long after mining has ceased (for example, acid drainage from underground mining), the period of liability may be insufficient.

The mining industry has been highly critical of SMCRA. The principal complaints include the complexity and detail of the regulations and the resulting increased costs of mining and reclamation. The industry argues that the Act and OSM's draft regulations are too detailed and do not permit a mine operator to tailor the engineering designs to site specific factors. The result, according to industry, will be substantial increases in the cost of mining and reclamation, and therefore in the cost of coal. Many of these requirements will be made more flexible in OSM's final regulations, issued spring of 1979. However, much of the specificity is written into the Act itself and must be changed legislatively. Once these changes have been made, the cost increases attributable to SMCRA probably will not be substantial enough to limit either the supply of or demand for coal. For example, OSM's cost analysis of the discretionary portions of the Act (that is, excluding the requirements that were

specified by Congress) indicates that their final regulations would result in only a 0.25-percent increase in the cost of electricity to the average residential customer in 1985.[17]

## The Federal Mine Safety and Health Act of 1977

The hazards in mining coal and other materials, and the need to provide for the health and safety of the Nation's miners, have long been matters of Federal law. The Federal Mine Safety and Health Act of 1977[18] is the most recent expression of congressional intent to remedy unsafe conditions and practices and to reduce the number of mining fatalities and injuries. The 1977 Act is based on the Federal Coal Mine Health and Safety Act of 1969;[19] it incorporates many of the provisions of the 1969 Act but increases the level of protection for miners.

The 1969 Act complemented the earlier Metal Act,[20] which regulated the occupational health and safety of all miners except coal miners, and represented a direct response to the number of deaths and serious injuries from unsafe and unhealthy conditions and practices in coal mines. To remedy these abuses, the Act established interim mandatory health and safety standards and directed the Departments of the Interior and of Health, Education, and Welfare to develop permanent standards; to ensure that mine operators comply with those standards; to provide benefits for victims of black lung; and to assist the States in developing and enforcing effective health and safety programs.

Implementation and enforcement of the mandatory standards established under the 1969 Act resulted in dramatic decreases in the

---

[16]"Excessive Regulation Drowns the Industry," *Coal Age*, January 1979, pp. 11-13.

[17]*Permanent Regulatory Program Implementing Section 501(b) of the Surface Mining Control and Reclamation Act of 1977* (U.S. Department of the Interior, Office of Surface Mining Reclamation and Enforcement, Final Environmental Statement OSM-EIS-1, January 1979).

[18]30 U.S.C. 801 et seq. (Public Law 95-164, Nov. 9, 1977).

[19]30 U.S.C. 801 et seq. (Public Law 91-173, Dec. 30, 1969).

[20]30 U.S.C. 721 et seq. Repealed by Public Law 95-164 (Nov. 9, 1977).

number of coal mine fatalities and injuries. However, a 1974 report on occupational safety and health[21] found that the incidence of work-related injuries and illnesses for miners still exceeded the "all industry" rate by about 14 percent. Work-related deaths showed, even more forcefully, the continuing inadequacy of mine safety and health laws and their enforcement. According to this report, about 1 of every 1,500 mine workers was killed on the job or died from work-related injuries or illnesses in 1973. This compared with 1 of every 2,800 railroad workers, 1 of every 4,000 construction workers, and only 1 of every 12,400 for all workers covered by the Occupational Safety and Health Administration.

In 1977 the Senate Committee on Human Resources drew a number of conclusions from its oversight of the Metal Act and the 1969 Coal Act:

First, the Metal Act does not provide effective protection for miners from health and safety hazards and enforcement sanctions under that Act are insufficient to encourage compliance by operators.

Second, enforcement of safety and health laws should be the responsibility of agencies which are generally responsible for the needs of workers.

Third, both the Coal and the Metal Acts do not provide means to react quickly enough to newly manifested health hazards.

Fourth, the procedures by which safety and health standards are made under both the Metal and the Coal Act are much too slow and cumbersome for standards promulgated under those Acts to keep pace with developments in a dynamic and expanding industry.

Fifth, the assessment and collective civil penalties under the Coal Act have resulted in penalties which are much too low, and paid much too long after the underlying violations to effectively induce meaningful operator compliance.

Sixth, enforcement sanctions under the current laws are insufficient to deal with chronic violators.[22]

---

[21]*Annual Report on Occupational Safety and Health* (Washington, D.C.: U.S. Department of Labor, Occupational Health and Safety Administration, 1974).

[22]Senate Report No. 95-181, 95th Cong., 1st sess., Federal Mine Safety and Health Act of 1977, p. 7.

By enacting the Federal Mine Safety and Health Act of 1977, Congress combined protection of all miners under a single comprehensive law that adopts the best features of both earlier statutes relative to health and safety. Insofar as the 1977 Act's objectives affect coal mine productivity, the Act adopts the consensus of labor and industry witnesses that a safe coal mine is also a productive mine. The resulting extensions of regulatory power in coal mining are more the product of technical improvements in the standard-setting and enforcement process than substantial departures from prior law.

The 1977 Mine Safety and Health Act adopted the provisions of the 1969 Act that prescribed mandatory health and safety standards and provided black lung benefits. That these parts of the Act were not appreciably altered is consistent with the legislative intent that standard-setting and enforcement procedures be made uniform throughout the mining industry while the standards themselves remain responsive to the characteristics of different segments of the industry.

Although much of the 1969 Coal Act remains as it was, several important changes are expected to increase the level of safety and health protection in the Nation's coal mines: standard-setting and enforcement provisions were made more effective, admininstration was transferred from the Department of the Interior to the Department of Labor, the Federal Mine Safety and Health Review Commission was established, mandatory health and safety training of miners was instituted, and the exercise of safety and health rights by miners was given added protection and support.

Past procedures for promulgating and revising standards in both the Metal and the Coal Acts had resulted in long delays between the perception of needed improvements and the implementation of new or revised standards. Each step in the process now requires compliance within a specified period. Enforcement procedures now also must comply with a more rigorous timetable.

A key element of the 1977 Act is the shift of administration from the Department of the Interior to the Department of Labor. Interior's responsibility for maximizing energy resource development was found to be incompatible with concurrent responsibilities for enforcing mine safety and health regulations; the Labor Department was perceived as having but one purpose: the welfare of the workers. The responsibilities of Interior's Mining Enforcement and Safety Administration also have been transferred to the Department of Labor, where an Assistant Secretary for Mine Safety and Health will preside over the new Mine Safety and Health Admininstration.

Under the 1969 Coal Act, review of contested matters was an internal function of the Secretary of the Interior, who established a Board of Mine Operations Appeals to separate prosecutorial and investigative functions from his adjudicatory functions. With the transfer of administration to the Department of Labor, a similar system was considered but ultimately rejected. While recognizing organizational and administrative drawbacks, Congress was persuaded to establish a completely independent adjudicatory authority, the Mine Safety and Health Review Commission. An independent commission was considered essential to provide administrative adjudication that preserves due process and instills more confidence in the program. Affected miners or their representatives have an opportunity to participate in the Commission's proceedings, and it is the intent of Congress that the Commission develop procedures to facilitate the participation of parties appearing pro se or not represented by counsel. This attention to adjudicative detail and purpose is better suited to serving both the interests of the parties and the underlying purposes of the Act.

One of the historic problems in the American coal industry has been the inadequate training afforded coal miners. Many miners still go underground with little or no training and, until very recently, the Federal requirements for training were weak. The 1977 Act provides for at least 40 hours of training for new underground miners, 24 hours for new surface miners, and 8 hours per year of refresher training for all miners. Miners must be paid

their normal rate of pay and any costs incurred while attending this training. If an inspector determines that any miner has not received the requisite training, then the miner must be withdrawn from the mine until his training is complete. Any miner so withdrawn may not be discriminated against by the operator and is entitled to full compensation during the training.

Congress expressed its displeasure with the Interior Department's repeated attempts to restrict miners' protection from retaliation for engaging in safety activity by changing not only the substantive aspects of the antidiscrimination provisions but also the procedural aspects. Under the new provisions, once a complaint is filed, the Secretary must conduct an investigation. If as a result of the investigation the Secretary believes that a violation has occurred, the Secretary must file to intervene in the proceedings and attempt to prove to the Commission that a violation has occurred. If the Secretary concludes that no violation occurred, the miner can still prosecute a case before the Commission. Even where the Secretary participates in the proceedings, the miner has the right to offer evidence, cross-examine the respondent's witnesses, and generally participate as a party. In both the 1969 Act and the 1977 Act, the miner is entitled to an award of attorney's fees if a violation of the antidiscrimination provision is proved. The 1977 Act, in clearly setting out the procedural steps in an antidiscrimination proceeding and clearly delineating the role of the Secretary, has greatly strengthened the antidiscrimination provision.

One of the most valuable provisions in the 1977 Act is the possibility of temporary reinstatement. Under the 1969 Act, miners often were severely disadvantaged if they chose to prosecute complaints, including significant economic hardship during the 1 or 2 years it might take to litigate their claims. The 1977 Act solves this problem by requiring the Secretary to make an initial determination, after a factual investigation of a discrimination complaint, as to whether the miner's complaint is frivolous. If the Secretary determines that the complaint is not frivolous, the Secretary must petition the Commission for temporary reinstatement of the miner, and the Commission must order reinstatement absent a showing of bad faith on the part of the Secretary. This simple measure should go a long way in providing practical protection for the miners who exercise their safety rights and are discharged as a result.

Both the 1969 and the 1977 Acts provide for reinstatement with back pay for discharged miners who prevail on their discrimination claims. The 1977 Act also provides for interest on an award of back pay to compensate miners for the loss of wages during litigation.

The loss of income that occurs when a miner is fired presents serious problems that back pay and interest alone may fail to remedy. Neither the 1969 Act nor its legislative history explicitly provided for the award of special damages to a miner who had been discriminated against for attempting to enforce safety rules. Not surprisingly, the Interior Department never exercised its discretionary authority to award special damages, even where clear economic damages resulted from a discriminatory act. The legislative history of the 1977 Act rejects the Interior Department position and specifically authorizes the award of special damages. In addition to compensatory relief, the 1977 Act anticipates the use of affirmative relief, such as a cease and desist order, where appropriate.

The Federal Mine Safety and Health Act of 1977 is not a major departure from prior law governing the operators of the Nation's coal mines. Those most affected will be miners employed in other mining sectors and mine operators. The present law does represent an incremental improvement, however, in what is an ongoing congressional effort to reduce mining deaths and injuries to the lowest practicable levels. In addition to the major changes already discussed, several smaller amendments in the law can be expected to further these objectives.

For example, the definition of a mine "oper-ator" is expanded to include any independent contractor performing construction services at a mine. Thus, employees of mine construction contractors now are considered miners and the Secretary should be able to issue citations, notices, and orders, and the Commission should be able to assess civil penalties, against these contractors. Statistics on mine-related deaths and injuries also will be more accurate because these contractors previously were not required to comply with recordkeeping and reporting obligations.

In addition, some penalties have been made more harsh, the collection of fines has been facilitated by an 8-percent interest charge, and an additional enforcement mechanism, injunctive-type relief, has been adopted to provide a flexible method of dealing with habitual or chronic violators of the Act.

Because the 1977 Act is modeled on its predecessor, the Mine Health and Safety Act of 1969, no substantial adaptations by the coal industry appear necessary. While noncoal mining operations previously subject to the Metal Act may face adjustment difficulties, coal industry and labor familiarity with the prior law should permit a relatively smooth transition. Increased operating costs may result from mandatory safety and health training, but whether the total additional costs directly associated with the legislation represent a net loss to the operators remains unclear.

Costs and benefits associated with safety and health regulations have been difficult to define accurately. If productivity can be improved as a result of vigorous Federal enforcement and operator compliance, all parties will benefit. If productivity remains unchanged or declines, however, coal consumers can be expected to share the burdens of improving the safety and health of the Nation's miners. Not unmindful of these costs, Congress has nevertheless declared mine safety to be an overriding concern, and the several changes described can be expected to lessen the annual rate of mining deaths and injuries.

## The Clean Air Act

In drafting the Clean Air Act Amendments of 1977,[23] Congress was aware of the related but often conflicting demands of environmental and energy policies. However, the amendments do not always reflect a consistent unified approach to the fundamental problems involved, and the overall effect of the Act is difficult to assess. To the extent that coal-fired facilities have a greater potential for emitting air pollutants than do facilities using other energy sources, they may require greater expenditures for pollution control equipment under the Act or be subject to stricter siting and other preconstruction review procedures and thus may be at a competitive disadvantage. On the other hand, some provisions of the Act exempt coal-burning sources from regulatory restrictions applicable to other fuels. Still other provisions limit the growth and development of all stationary sources in certain areas.

It is not possible to quantify the impact of any of these provisions on increased coal use, much less to assess their combined impact. Such an analysis would require a major technical and economic inquiry. Rather, this section identifies and briefly discusses the nature of the Clean Air Act provisions that affect the use of coal by new fossil-fuel-fired sources as well as the conversion from oil or gas to coal by existing sources. Some of these provisions have been in effect since 1970; others are included in the 1977 amendments.

### Background and History

Before 1970, air pollution control essentially was left to the States, with the Federal Government providing technical and financial assistance for planning and for R&D. The Clean Air Act Amendments of 1970[24] inaugurated direct Federal regulation, mandated specific State implementation plans (SIPs), and required Federal intervention in the absence of State action.

The 1970 amendments represented a balance between uniform national requirements

and the preservation of States rights. Those factors that were deemed to require national uniformity in order to prevent regional subversion of the Act's goals or regional economic advantages included numerical standards for how clean the air must be and for emissions of airborne pollutants from future sources as well as guidelines for air quality control plans. The States were required to devise and implement plans in accordance with these guidelines for achieving and maintaining the specified levels of air quality. The Act also allowed the States to set more stringent standards. This fundamental role division continues under the 1977 amendments, which were intended to remedy problems that had developed under the 1970 Act and to strengthen some regulatory programs.

### National Ambient Air Quality Standards

The central feature of the 1970 Clean Air Act Amendments was the requirement that EPA promulgate National Ambient Air Quality Standards (NAAQS). NAAQS define air quality in terms of ambient concentrations of pollutants. These standards represent target levels for air quality; they do not regulate emissions from individual sources. The amendments required two sets of standards that reflected the latest scientific knowledge about the effects of various air pollutants. Primary standards were intended to protect public health; secondary standards were designed to protect the public welfare from pollution damage to soils, vegetation, animals, materials, and other environmental factors not related to human health.

Pursuant to the 1970 amendments, the EPA Administrator listed six pollutants as having potentially adverse effects on public health and welfare and established primary and secondary NAAQS for each. Standards have been established for sulfur oxides ($SO_x$) (measured as sulfur dioxide ($SO_2$)), particulate matter, nitrogen dioxide ($NO_2$), hydrocarbons, photochemical oxidants, and carbon monoxide (CO); standards for lead were proposed in 1978. Under the 1977 amendments, these standards are to be reviewed and revised if necessary every 5 years beginning in 1980. The standards for $SO_x$, particulate matter, and $NO_2$, the pri-

---

[23]Public Law 95-95 (Aug. 7, 1977), 42 U.S.C. 1857 et seq.
[24]Public Law 91-604 (Dec. 31, 1970).

mary byproducts of coal combustion, are listed in table 58.

**Table 58.—Standards for Particulate Matter, Sulfur Dioxide, and Nitrogen Dioxide**

| Pollutant | Averaging time | Primary standard | Secondary standard |
|---|---|---|---|
| Particulate matter | Annual (geometric mean) | 75 $\mu$g/m³ | 60 $\mu$g/m³ |
| | 24-hour | 260 $\mu$g/m³ | 150 $\mu$g/m³ |
| | 3-hour | — | — |
| Sulfur dioxide.... | Annual (arithmetic mean) | 80 $\mu$g/m³ | — |
| | 24-hour | 365 $\mu$g/m³ | — |
| | 3-hour | — | 1,300 $\mu$g/m³ |
| Nitrogen dioxide . | Annual (arithmetic mean) | 100 $\mu$g/m³ | 100 $\mu$g/m³ |
| | 24-hour | | |
| | 3-hour | | |

These ambient air quality standards are implemented at the national level through standards of performance for new stationary sources and guidelines for State control strategies for existing sources as well as through guidelines for regulatory programs designed to improve air quality in areas that have not attained the national standards, to prevent degradation of air quality in areas cleaner than the national standards, and to protect visibility in important scenic areas.

The State role centers on the preparation and implementation of a plan, consistent with EPA guidelines, that sets out control strategies for meeting and maintaining NAAQS in various parts of the State (known as air quality control regions (ACQRs)). States have considerable discretion in deciding what emission limitations and other controls on individual sources to use in cleaning up their air, as long as their SIPs are shown to be capable of achieving the national standards.

The 1970 Act required the States to attain the primary standards as expeditiously as practicable but not later than 3 years after the date the SIP became efective. The SIPs also were required to specify a reasonable time at which the secondary standards would be achieved; "reasonable time" was defined by EPA to depend on the degree of emission reduction

needed to attain the standards and on the social, economic, and technological problems involved in doing so. However, as of 1977, 116 of the 247 ACQRs reported violations of the primary annual particulate standard while 108 reported violations of the 24-hour standard. Similarly, 12 ACQRs reported violations of the primary annual SO₂ standard while 37 reported violations of the 24-hour standard.[25] The 1977 amendments require the States to revise their SIPs by 1979 in order to provide for attainment of the primary standards by 1982. However, the amendments did not change the "reasonable time" requirement for achieving the secondary standards.

### New Source Performance Standards

The Clean Air Act Amendments of 1970 required the EPA Administrator to establish standards of performance for large new or substantially modified stationary sources to ensure that they would not exacerbate existing air quality problems or contribute to new ones. Under the 1970 Act, New Source Performance Standards (NSPS) included federally determined allowable rates of emissions from 19 categories of sources. SIPs were required to include a procedure for preconstruction review of new sources to ensure that these standards were met.

The 1977 amendments significantly tightened these requirements. In order to meet NSPS, 28 categories of sources now must apply the best technological system of continuous emission reduction that has been demonstrated adequately. In addition, fossil-fuel-fired sources are subject to an enforceable percentage reduction in emissions that would have resulted from the use of untreated fuels.

In determining which technological systems of continuous emission reduction have been demonstrated adequately, the EPA Administrator is required to consider the energy requirements of a technology as well as its cost and any nonair quality health and environmental impacts. In calculating the percentage reduction requirements, the Administrator may give

[25]43 F.R. 8962 (Mar. 3, 1978).

credit for mine-mouth and other precombustion fuel-cleaning processes. NSPS have been promulgated for fossil-fuel-fired steam generators and for coal preparation plants. EPA plans to announce new standards for industrial boilers in 1980.

As discussed in chapter IV, the NSPS provisions in the 1977 amendments have become controversial. If the final regulations require continuously operating flue-gas desulfurization (FGD) systems, it probably will promote the use of locally available high-sulfur coals, especially in the Midwest, while removing the advantage that accrued to low-sulfur coal under the previous regulations. In addition, a continuous control requirement could delay the construction of new coal-fired powerplants, causing a greater reliance on exisiting plants than would have occurred under the previous regulations.[26] Regardless of whether the final regulations require full, partial, or a sliding-scale control, they will increase the pollution control costs of new coal-fired plants.

Nonattainment Areas

The 1970 Clean Air Act Amendments did not specify the consequences of a State's failure to attain NAAQS by the deadline. While EPA regulations promulgated in 1976[27] filled this gap to some extent, it was clear that congressional guidance was necessary. Consequently 1977 amendments add new requirements that must be incorporated into all SIPs by July 1979.

The 1977 amendments basically adopt EPA's 1976 offset policy for nonattainment areas until July 1979. In general, this policy imposes the following conditions for the issuance of construction permits for new or modified sources in nonattainment areas: the source meets the lowest achievable emission rate (LAER), the permit applicant certifies that all its other facilities are in compliance with all applicable Clean Air Act control requirements, the permit applicant has secured emissions reductions from existing sources that more than offset the emissions from the proposed source, and a

positive net air quality benefit results. LAER must reflect the most stringent emission rate required by any State or the lowest rate achieved in practice, if the latter is more stringent. In no event, however, may emissions from a source subject to LAER be allowed to exceed the applicable NSPS.

This offset policy applies to any source capable of emitting 100 tons per year of a pollutant that would exacerbate an existing NAAQS violation. EPA has announced that it will implement the policy through the use of significance levels patterned after the prevention of significant deterioration (PSD) increments (see below). A source will not be considered to exacerbate air quality problems if its emissions are below the specified significance level.[28]

Whether a source is subject to the offset policy is determined on a case-by-case basis through air quality modeling. Thus a source still could be located in an AQCR with localized violations so long as its emissions will not exacerbate those local violations.

After June 30, 1979, construction of new or modified stationary sources that would adversely affect air quality in nonattainment areas must be prohibited unless the applicable SIP meets the requirements of the 1977 amendments. SIP revisions must provide for attainment of the primary standards not later than 1982, and in the interim must require annual incremental emissions reductions from existing sources through the implementation of reasonably available control measures.

As with the new NSPS requirements, EPA's offset policy is controversial. It places significant constraints on siting in many areas because the offsetting reductions are difficult to obtain and LAER is expensive to meet; siting constraints could become even more severe if SIP revisions are not accomplished in a timely manner. In addition, most industries feel it is inappropriate to place the burden of securing emission reductions on the private sector

---

[26] 43 F.R. 42154 (Sept. 19, 1978).
[27] 41 F.R. 55524 (Dec. 21, 1976).

[28] 44 F.R. 3274 (Jan. 16, 1979).

rather than on Government.[29] Although many States are expected to adopt EPA's offset policy in their SIP revisions, they are being encouraged by EPA to experiment with alternative programs for cleaning up "dirty air" areas, such as emission taxes, or banking of emission reductions to allow for future growth.[30]

### Prevention of Significant Deterioration/ Visibility Protection

The 1970 Clean Air Act Amendments merely required EPA and the States to achieve and maintain NAAQS; they did not address the question of air quality in areas already cleaner than NAAQS require. In 1972 environmental groups brought suit against EPA to prohibit the Agency's approval of SIPs that failed to prevent significant deterioration of air quality. Relying on the Act's stated purpose of protecting and enhancing the Nation's air quality, the court ordered EPA to develop a program to prevent the degradation of air quality in clean air areas.[31] EPA's PSD regulations were promulgated and incorporated into all SIPs in 1974, and were adopted with some changes in the 1977 amendments.

In general, the PSD program divides clean air areas into three classes and specifies the maximum allowable increases in ambient concentrations of pollutants, or PSD increments, for each class.

As discussed in chapter IV, construction of a major emitting facility (defined to include most large fossil-fuel-fired steam electric plants and other coal-burning sources) that will affect air quality in a clean air area is subject to extensive preconstruction review and permit requirements. Review of a permit application must be preceded by an analysis of the ambient air quality at the proposed site and in areas within 50 km downwind that may be affected by emissions from the proposed facility.

The PSD regulations could constrain the development of coal combustion facilities in two main situations: where the difference between the baseline concentration and the maximum allowable increase already is lower than the applicable increment, and where sources that are exempt from PSD review (because of their size or date of construction) will use up available increments. However, a 1975 Federal Energy Administration (FEA)/EPA study of the effects of the then-pending PSD legislation on the electric utility industry concluded that the requirements would not significantly hamper siting of even the largest sources of air pollution.[32] These conclusions are summarized in appendix XVI of volume II. In addition, the study analyzed the economic impacts of PSD requirements on the utility industry; it concluded that the industry's total capital requirements and annual operating costs as well as the costs to consumers would increase less than 3 percent to the year 1990.

Siting considerations also could be affected by regulations intended to protect visibility in Class I areas (national parks and wilderness areas subject to the lowest PSD increments) primarily valuable for scenic factors. The 1977 amendments established a national goal of preventing any future—as well as remedying any existing—impairment of visibility in these areas. By August 1979, EPA is required to promulgate regulations to assure reasonable progress toward meeting the national goal, taking into account the economic, energy, and environmental costs of compliance. In addition, States must revise their SIPs to require all existing major stationary sources constructed in the last 15 years to install the best available retrofit technology for controlling emissions that may impair visibility. A principal contributor to reduced visibility is fine particulate matter; these particles are not captured by the current technology-of-choice, electrostatic precipitators. Their control may require the use of baghouses, a costly alternative. However, the Act does provide exemptions for sources that

[29]Sixteen Air and Water Pollution Issues Facing the Nation (Washington, D.C.: General Accounting Office, U.S. Comptroller General, October 1978).

[30]44 F.R. 3274 (Jan. 16, 1979).

[31]Sierra Club v. Ruckelshaus, 344 F.Supp. 253 (D.D.C., 1972), affd. 4 E.R.C. 1815 (D.C. Cir. 1972), affd. Fri v. Sierra Club, 412 U.S. 541 (1973).

[32]An Analysis of the Impact on the Electric Utility Industry of Alternative Approaches to Significant Deterioration (Washington, D.C.: Federal Energy Administration and Environmental Protection Agency, October 1975).

can demonstrate that they will neither cause nor contribute to a significant impairment of visibility in mandatory Class I areas, or where the costs of compliance would be too high (for example, if the source is scheduled for retirement soon). Until the final regulations are published, it is not possible to estimate the extent to which visibility protection requirements will constrain increased coal consumption.

### EPA Studies and Potential New Standards

In adopting the 1977 amendments, Congress directed EPA to undertake an extensive review of existing standards and to study potential new standards. To the extent that EPA's efforts result in tightening existing standards or in promulgating new standards, these provisions may impose new and potentially costly constraints on the use of coal. At this time, however, it is possible only to identify the areas of potential concern.

The amendments require EPA to review air quality criteria and NAAQS by the end of 1980, and every 5 years thereafter, and to revise the criteria and standards as appropriate. Any revision of a NAAQS for a pollutant may result in major changes in SIPs and control requirements applicable to sources of that pollutant. In addition, the amendments direct EPA to promulgate a 3-hour primary standard for $NO_2$, unless EPA determines that there is no significant evidence that this standard is necessary. A stringent short-term $NO_2$ standard may require flue-gas denitrification, a technology that is not expected to be available until after 1985.

Another provision of the amendments requires EPA, in conjunction with the National Academy of Sciences, to complete a study of the health effects of fine particulates; preliminary findings will be published by June 1979. The amendments also require EPA to determine whether emissions of radioactive pollutants, cadmium, arsenic, and polycyclic organic matter may endanger public health. If EPA finds that these emissions do pose a hazard to human health, then EPA must establish air quality standards, NSPS, or hazardous pollutant standards for them. For radioactive pollutants, EPA must make this determination by

August 1979; background documents on the sources, health effects, population exposures, and risks associated with the other pollutants were available for public comment in spring 1978. All of these pollutants are present in trace amounts in the fly ash of coal emissions; moreover they are concentrated preferentially on the fine particulate matter that conventional control technologies do not remove effectively. Baghouses entailing large costs and space requirements would be required to capture these materials if the standards mandated highly efficient removal.

Finally, EPA currently is developing a standard for the control of sulfates, a transformation product of $SO_x$. Sulfates also adhere to fine particulates and contribute to visibility impairment and acid precipitation, primarily in the Northeast. A stringent sulfate standard may impose severe constraints on new coal-fired sources in the Eastern United States.

### Interface With Energy Legislation

The Energy Supply and Environmental Coordination Act (ESECA) of 1974[33] (discussed below) amended the Clean Air Act to provide for coordination between national energy and air quality goals. EPA's primary responsibilities under ESECA were to determine the earliest date that a source converting to coal could meet applicable air pollution control requirements and, if necessary, to grant an order to the source extending its date of compliance with the Clean Air Act. The 1977 Clean Air Act Amendments repealed this requirement under ESECA and placed EPA's coal conversion oversight within its general authority to grant delayed compliance orders.

In 1978, ESECA was replaced by the Powerplant and Industrial Fuel Use Act[34] (see below), which prohibits the use of oil or natural gas as a primary energy source in new fuel-burning installations and the use of natural gas in existing facilities after 1990. Temporary and permanent exemptions from these prohibitions are

[33]Public Law 93-319 (June 22, 1974), as amended by Public Law 94-163 (Dec. 22, 1975) and Public Law 95-70 (July 21, 1977).
[34]Public Law 95-620 (Nov. 9, 1978).

provided for facility operators who demonstrate that they cannot meet the prohibition without violating environmental requirements such as emissions limitations; the operator must demonstrate that he has made a good faith effort to comply with environmental requirements before he is entitled to use gas or oil instead of coal. The Act specifically states that it is not intended to permit any existing or new facility to delay or avoid compliance with applicable environmental requirements.

The 1977 Clean Air Act Amendments encourage increased coal use through measures designed to prevent significant local or regional economic disruption or unemployment. Where a source or class of sources intends to use petroleum products, natural gas, or non-local coal in order to comply with the Clean Air Act requirements, and it is determined that the result would be local economic disruption or unemployment, the President can prohibit the source from using any fuel other than locally or regionally available coal. A source may be ordered to enter into long-term contracts of at least 10 years for supplies of local coal as well as contracts to acquire additional pollution controls. This provision does not exempt a source from the requirements of the Act, but prevents it from relying on fuels with a sulfur content lower than that of locally available fuels.

Implementation of the Amendments

This review of the 1977 amendments indicates that, if effectively implemented, the Act may increase the costs of coal utilization and may impose siting constraints in both dirty and clean air regions, primarily in the vicinity of Eastern and Western coal resources. However, the extent to which the environmental and health objectives of the amendments are achieved and, conversely, the extent to which substantial constraints on increased coal use are created, depend on the implementation of the Act. Major implementation factors to be considered include monitoring requirements, air quality modeling, and the level of enforcement.

The Act contains several provisions designed to ensure that major stationary sources

such as fossil-fuel-fired powerplants monitor their emissions. A condition for EPA approval of SIPs is that they include requirements for the installation of monitoring equipment and for the submission of periodic reports on the nature and amount of emissions. In addition, EPA has independent authority to require a source to monitor and report its emissions when EPA determines that this information is required to assess compliance with the Act.

Monitoring data are used in conjunction with dispersion modeling techniques in order to determine a source's impact on air quality under a variety of Clean Air Act provisions. Current regulations limit the applicability of air quality models to a downwind distance of not more than 50 km.[35] However, recent research suggests that under certain meteorological conditions, such as after prolonged periods of stagnation or during extremely persistent winds, air pollutants and their transformation products may be transported over distances greater than 50 km. To the extent that this longer range transport of pollutants occurs and is not regulated directly, it may interfere with enforcement of measures directed at visibility protection and it may consume PSD increments in clean air areas or contribute to NAAQS violations in nonattainment areas downwind. Thus the limits imposed on air quality modeling may further constrain siting downwind.

Finally, the 1977 amendments enhance EPA's enforcement authority in several important respects. First, EPA now has the authority to seek civil penalties to a maximum of $25,000 per day of violation. Second, criminal actions now may be brought against any responsible corporate officer. Third, and most important, the amendments provide for the imposition of noncompliance penalties that operate in addition to the other penalty provisions. These noncompliance penalties essentially remove the economic advantages accruing to a noncomplying source. The Act sets forth elab-

[35]43 F.R. 26380 (June 19, 1978). See also *Guideline on Air Quality Models* (Research Triangle Park, N.C.: U.S. Environmental Protection Agency, April 1978), OAQPS 1.2-080.

orate procedures and standards for this penalty provision, which, in general, requires a source not in compliance as of July 1979 to pay a quarterly penalty equal to the money saved during that quarter as a result of non-compliance.

However, a variety of problems still exist in enforcing the Clean Air Act and in air quality management in general. Gaps may be created in the ability to enforce the Act when new regulations are challenged in court (as they almost always are). New regulations supplant old ones, yet the courts will grant a stay of enforcement of the new ones, leaving EPA with nothing to enforce. In addition, enforcement problems may occur from a lack of communication among the various agencies and governments involved. Thus action taken by the Department of Justice, the EPA Office of General Counsel, or the State may undercut enforcement actions in progress at the EPA regional offices. Similarly, there is a general lack of integration within EPA for overall environmental management strategies. Personnel responsible for developing air programs may be unaware of the implications of their actions for solid waste disposal or water pollution and vice versa.

General air quality management also is constrained by States rights in that administration of the Clean Air Act must be turned over to the States as soon as their SIPs meet the minimum requirements, yet the States may not be ready or able to administer and enforce the Act adequately. Conversely, whenever a State program falls below the minimum requirements, administration and enforcement revert to the EPA regional office regardless of whether that office has sufficient funds or personnel.

In addition, most of the Clean Air Act requirements are dependent on the technological and economic availability of control technology, yet the Act places most of the burden for developing that technology on Federal R&D programs that are not funded adequately. A shift in regulatory philosophy that would force industry to take the initiative, such as an emissions tax, could alleviate some of EPA's management problems. Finally, because EPA must use its available R&D funding to attempt to remedy existing pollution control problems the Agency is unable to allocate funds to the anticipation of future control problems.

## Conclusions

The Clean Air Act of 1970 failed to achieve the primary NAAQS nationwide by the target date of 1975. In 1977 Congress responded with thorough revisions of the Act in order to achieve those standards in areas that had not done so and to protect the quality of the air in regions that already are cleaner than NAAQS. Whether in light of continued industrial growth the new amendments will achieve the desired air quality is uncertain. In an effort to do so, however, the amendments may impose new constraints on stationary sources.

Whether the 1977 amendments will impede increased coal use is largely a question of economics. Within certain limits, increased costs resulting from clean air requirements, such as NSPS, will not significantly affect the amount of coal used; fuel demand has been shown to be rather unresponsive to moderate price increases. However, regulations that may delay the siting of facilities, such as those for nonattainment areas and for PSD, or that may substantially increase the cost of new facilities, could encourage greater reliance on existing capacity in lieu of new plants. In addition, provisions of the Act that require stringent controls regardless of the sulfur content of the fuel may tend to favor production of locally available high-sulfur coals over higher priced low-sulfur coal.

## The Clean Water Act

Water pollution associated with the direct use of coal stems from three major sources: surface and deep mining operations, preparation plants (including ancillary storage areas and washing facilities), and combustion facilities. Effluents from these sources are regulated under the Clean Water Act[36] (formerly known as the Federal Water Pollution Control Act) through ambient water quality standards, effluent limitations for new and existing sources, limitations on thermal discharges, permit programs, and areawide planning.

---

[36]33 U.S.C. 466 et seq.

## Water Quality Standards

Section 101 of the Clean Water Act establishes two national water quality goals. The first, to be achieved by 1983, is an interim goal that provides for the protection and propagation of fish, shellfish, and wildlife and for recreation in and on the water. The second national goal is the elimination of all pollutant discharges. The States have the primary responsibility for achieving these goals and for planning the development and use of land and water resources consistent with them. In 1977, Congress recognized that significant progress toward these goals was not being made and that the discharge elimination goal probably is unrealistic. Accordingly, the Clean Water Act Amendments of 1977 extended the deadline for compliance with the stricter limitations under the Act by 1 to 3 years.

Each State is required to develop and implement, subject to the approval of the Administrator of EPA, a comprehensive water quality management plan that includes water quality standards. These standards consist of the designated uses of the waters involved, including their use and value for public water supplies; propagation of fish and wildlife; recreational, agricultural, industrial, and other purposes; and navigation. In addition, the standards include water quality criteria for the waters based on these uses.

In general, the water quality standards are to be achieved through effluent limitations on discharges from point sources (see below). However, for those waters for which the effluent limitations are not stringent enough to implement the applicable water quality standard, the State must establish a total maximum daily load for the relevant pollutants. This load must be established at the level necessary to implement the applicable water quality standards with seasonal variations and a margin of safety that takes into account any lack of knowledge concerning the relationship between effluent limitations and water quality. In addition, the State must estimate the total maximum daily thermal load required to assure protection and propagation of a balanced indigenous population of shellfish, fish, and wildlife. This estimate must take into account the normal water temperatures, flow rates, seasonal variations, existing sources of heat input, and the dissipative capacity of the waters, and also must include a margin of safety that takes into account any lack of knowledge about the development of thermal water quality criteria.

## Effluent Limitations

Effluent limitations are restrictions established by a State or the EPA Administrator on quantities, rates, and concentrations of chemical, physical, biological, and other constituents that are discharged from point sources. Effluent limitations may be categorized by: 1) the sources for which they have been established, 2) whether those sources discharge directly into receiving waters or into a publicly owned treatment works, and 3) the degrees of control required for each category of sources or pollutants and the dates those controls become mandatory.

In general, the 1977 amendments require all categories of point sources to apply the best practicable control technology currently available (BPCTCA) not later than July 1, 1977. Those point sources that discharge conventional pollutants (including, but not limited to, pollutants classified as biological oxygen demanding, suspended solids, and hydrogen-ion concentration (pH)) must apply the best conventional pollution control technology (BCPCT) not later than July 1, 1984. Finally, all categories of point sources must apply the best available technology economically achievable (BATEA) that will result in reasonable further progress toward the stringent discharge elimination goal if the Administrator finds that the goal is technologically and economically achievable. BATEA is required not later than 3 years after the date the effluent limitations for a pollutant have been established or by July 1, 1984, whichever is later, but in no case later than July 1, 1987. In determining the control measures and practices to be applicable to point sources, the EPA Administrator must take into account the age of equipment and facilities involved, the process employed, the engineering aspects of the application of various types of control technologies, process changes, and nonwater quality environmental impacts (including energy requirements) as well as the total cost of achieving the limitation in relation to the effluent reduction benefits to be achieved.

Where pollutants are introduced into publicly owned treatment works they are subject to pretreatment standards to ensure that the effluent limitations applicable to the treatment works will not be violated. In addition, as mentioned above, where the general effluent limitations are not strict enough to contribute to the attainment or maintenance of the water quality goals for a particular stream, the EPA Administrator may establish stricter standards for point sources located along that stream. These stricter, water quality related effluent limitations must take into account: 1) the tradeoff between the economic and social costs of achieving the limitation, including any economic or social dislocation in the affected communities, and the social, economic, and water quality benefits to be obtained; and 2) whether the limitation can be implemented with available technology or alternative control strategies.

The principal coal-based activities for which effluent limitations have been established are steam electric-power generation and coal mining; these are summarized in tables 59 and 60. The steam electric-power generating point source category applies to all units that produce electricity for distribution and sale. It is broken down further by the size and age of the unit. In practice, however, neither the distinctions based on size and age of the facility nor those based on degree of control required

**Table 59.—Effluent Limitations: Steam Electric Power Generating Units**

| Source category | Technology requirement | Type of discharge | Pollutant | Limitation[a] | Measurement time |
|---|---|---|---|---|---|
| Generating unit[b] and small unit[c] subcategories: existing and new[d] sources; | BPCTCA and BATEA | All except once-through cooling water; all | pH | 6.0-9.0 | At all times |
| | | | PCBs | No discharge | At all times |
| old unit[e] subcategory | | low-volume waste sources; ash transport water; | TSS | 100 mg/l | Maximum/day |
| | | | | 30 mg/l | 30-day average |
| | | metal cleaning wastes; boiler blowdown | Oil and grease | 20 mg/l | Maximum/day |
| | | | | 15 mg/l | 30-day average |
| | | Metal cleaning wastes; boiler blowdown | Copper | 1.0 mg/l | Maximum/day |
| | | | | 1.0 mg/l | 30-day average |
| | | | Iron | 1.0 mg/l | Maximum/day |
| | | | | 1.0 mg/l | 30-day average |
| | | Once-through cooling water; cooling tower blowdown | Free available chlorine | 0.5 mg/l | Maximum concen. |
| | | | | 0.2 mg/l | Average concen. |
| Generating unit and small unit subcategories: existing sources; old unit subcategory | BATEA | Cooling tower blowdown | Zinc | 1.0 mg/l | Maximum/day |
| | | | | 1.0 mg/l | 30-day average |
| | | | Chromium | 0.2 mg/l | Maximum/day |
| | | | | 0.2 mg/l | 30-day average |
| | | | Phosphorus | 5.0 mg/l | Maximum/day |
| | | | | 5.0 mg/l | 30-day average |
| Generating unit and small unit subcategories: new sources | | Cooling tower blowdown | Materials added for corrosion inhibition, including but not limited to zinc, chromium, phosphorus | No detectable amount | Maximum/day |
| | | | | No detectable amount | 30-day average |
| Area runoff[f] subcategory: existing and new sources | BPCTCA and BATEA | Material storage and construction runoff | TSS | 50 mg/l | Maximum/day |
| | | | pH | 6.0-9.0 | at all times |

[a]All limitations expressed in mg/l are to be multiplied by the volume of the waste flow.
[b]"Generating unit" means any unit except those defined as small or old.
[c]"Small unit" means any unit (except one defined as old) of less than 25 MWe rated capacity or any unit which is part of an electric utilities system with a total net generating capacity of less than 150 MWe.
[d]New sources are those on which construction is commenced after the applicable limitations are published.
[e]"Old unit" means any unit of 500 MWe or greater capacity first placed in service on or before January 1, 1970, and any unit of less than 500 MWe capacity first placed in service on or before January 1, 1974.
[f]Applicable to discharges resulting from material storage and construction runoff except untreated overflow associated with a 10-year, 24-hour rainfall.

**Table 60.—Effluent Limitations: Coal Mining Point Source Category**

| Source category | Technology requirement | Quality of discharge | Pollutant | Limitation | Measurement time |
|---|---|---|---|---|---|
| Coal preparation plants and associated areas; coal mines | BPCTCA | Acid[a] and alkaline[b] | Iron | 7.0 mg/l | Maximum/day |
| | | | | 3.5 mg/l | 30-day average |
| | | | TSS | 70 mg/l | Maximum/day |
| | | Alkaline | | 35 mg/l | 30 day average |
| | | | pH | 6.0-9.0 | At all times |
| | | | Manganese | 4.0 mg/l | Maximum/day |
| | | | | 2.0 mg/l | 30-day average |

[a]"Acid or ferruginous mine drainage" means mine drainage that before any treatment either has a pH of less than 6.0 or a total iron concentration of more than 10 mg/l.
[b]"Alkaline mine drainage" means mine drainage that before any treatment has a pH of more than 6.0 and a total iron concentration of les than 10 mg/l.

(BPCTCA or BATEA) is meaningful. As can be seen in table 59, except for pollutants from corrosion-inhibiting materials, the effluent limitations are the same across the board. The coal mining point source category applies to all active mining areas (surface and deep) including secondary recovery facilities and preparation plants but excluding surface mines in which grading has been completed and reclamation work has begun. The limitations are broken down into those applicable to acid drainage and to alkaline drainage.

The EPA Administrator may modify any of the limitations for a point source if the owner of the source demonstrates that the modified requirement will represent the maximum use of technology within his economic capability and will result in reasonable further progress toward the discharge elimination goal.[37] The 1977 amendments provide that such a modification is mandatory if the owner also demonstrates that it will not interfere with attainment of a water quality standard or the 1983 water quality goal, and it will not result in additional requirements on any other point source. As mentioned above, this provision reflects congressional doubt about the reasonableness and practicability of the 1985 discharge elimination goal.

Thermal Discharges

Limits on thermal discharges from steam electric-generating plants are included in the effluent limitations for those sources. These include the BATEA requirements for existing

generating units and NSPS for generating units and for small units.

Existing generating units are required to eliminate the discharge of heat from the main condensers by July 1, 1981, through the application of BATEA. Exceptions to this general limitation include:

• Blowdown from recirculated cooling water systems, provided the temperature of the discharge does not exceed the lowest temperature of recirculating cooling water prior to the addition of the makeup water. (Systems technologically incapable of meeting this exception are exempt provided they begin construction prior to July 1, 1981.);
• Blowdown (overflow) from a cooling pond under construction or in operation prior to July 1, 1981, and used to cool water before it is recirculated to the main condensers;
• Where sufficient land for mechanical draft evaporative cooling towers is not available (on property owned before March 4, 1974) and where no alternate recirculating cooling system is practicable;
• Where the total dissolved solids concentration in blowdown exceeds 30,000 mg/l and land not owned by the owner of the source is located within 150 m in the prevailing downwind direction of every practicable location for mechanical draft cooling towers and no alternate recirculating system is practicable; and
• Where the cooling tower plume would, in the opinion of the Federal Aviation Administration, cause a substantial hazard to commercial aviation in the vicinity of a

---

[37]40 CFR pts. 402, 423, 434.

major commercial airport and no alternate recirculating cooling water system is practicable.[38]

In addition, the effective date of the retrofit requirement may be extended for from 1 to 2 years where reliability would be jeopardized by timely compliance.

The New Source Performance Standards for generating units and small units provide that there shall be no discharge of heat from the main condensers except in the case of blowdown from recirculated cooling water systems or from cooling ponds where the temperature of the blowdown does not exceed the lowest temperature of recirculated cooling water prior to the addition of the makeup water. In addition, the Act requires that the location, design, construction, and capacity of cooling water intake structures reflect the best technology available for minimizing adverse environmental impacts (e.g., impingement and entrainment).

Where these limitations on thermal discharges are deemed to be more stringent than necessary for the protection and propagation of a balanced, indigenous population of shellfish, fish, and wildlife, the EPA Administrator may modify the thermal discharge limits for a source. In addition, sources upon which modifications are begun after 1972 and which, as modified, meet all applicable effluent limitations, are exempt from more stringent limitations for a period of 10 years following completion of the modifications.

### Permit Systems

Effluent limitations and water quality standards are implemented through State certification programs and through the National Pollutant Discharge Elimination System (NPDES).

An applicant for a Federal license or permit to conduct any activity, including the construction or operation of facilities, that may result in a discharge into navigable waters, must obtain State certification that the discharge will not violate any effluent limitations, water quality standards, or NSPS. Where the discharge will affect more than one State, the

[38]Id.

Federal licensing or permitting agency must condition the permit to ensure that all water quality requirements will be met. In addition, when Federal regulations require only a construction permit, the certifying State must be given an opportunity to review the manner in which the facility will be operated in order to ensure that water quality requirements will not be violated. If the State finds that the operation of the facility will result in violations, the Federal agency may suspend the license or permit.

NPDES is designed to ensure the orderly and timely achievement of water quality goals without sacrificing economic or energy growth. Under NPDES, a facility may be issued a permit for a discharge on the condition that the discharge will meet all applicable water quality requirements. NPDES permits are issued under State programs approved by EPA, or, where a State program has not been approved, by the EPA Administrator. The permits are for fixed terms not to exceed 5 years and can be terminated or modified for violations. Compliance with the conditions under which an NPDES permit is issued is deemed compliance with the effluent limitations and water quality standards promulgated under the Clean Water Act.

### Areawide Planning

Section 208 of the Clean Water Act encourages areawide land and water management planning for regions with substantial water quality problems because of urban-industrial concentrations or other factors. The features of section 208 plans that could affect coal development include programs:

1. to regulate the location, modification, and construction of facilities that may result in a discharge;
2. to control mine-related sources of pollution including new, current, and abandoned surface and deep mine runoff;
3. to control construction activity-related sources of pollution; and
4. to control the disposal of residual waste material and of pollutants on land or in subsurface excavations.

Section 208 programs were intended to provide the long-range planning basis for the implementation of other Clean Water Act programs. However, the other programs addressed immediate pollution problems and received funding priority over section 208. Consequently, implementation of areawide land and water management planning has been slow; to date only 9 of 216 plans have received final approval. For reasons primarily related to local politics, when the section 208 plans are implemented they are not expected to affect the siting of new coal combustion facilities significantly.

## Resource Conservation and Recovery Act of 1976

Prior to passage of the Resource Conservation and Recovery Act (RCRA) of 1976,[39] the Federal Government faced a policy vacuum in regard to the control of solid wastes from coal-producing and coal-consuming facilities. Because there were no major Federal programs related to the land disposal of solid wastes, disposal practices could not even be regulated indirectly through the requirements of NEPA. Yet scrubbers eventually will produce large quantities of sludge, and coal mines are notorious for their waste piles.

In general, RCRA seeks to control open dumping under a system of State plans and permits for solid waste disposal. All forms of solid waste are covered, both hazardous and nonhazardous, with more stringent regulations for the former. In neither case, however, are there prohibitions on waste disposal.

The implementation of this Act by EPA could have far-reaching consequences for the handling of solid wastes from coal production and use. For example, section 1004(5) of the Act defines "hazardous wastes" to include those which, because of their quantity, concentration, or physical, chemical, or infectious characteristics may pose a substantial present or potential hazard to human health or to the environment. Under section 3001, EPA is required to: 1) establish criteria for identifying

[39]42 U.S.C. 6901 et seq.

the characteristics of hazardous wastes, taking into account toxicity, persistence, and degradability in nature, potential for accumulation in tissue, and related factors such as flammability, corrosiveness, and other hazardous characteristics; and 2) based on these characteristics, list particular hazardous wastes that will be regulated.

The generator of a substance that is listed as a hazardous waste is responsible for its disposal in accordance with the applicable State plan. In addition, generators of hazardous wastes are subject to extensive recordkeeping provisions that require them to identify the quantities of wastes generated, the constituents of the wastes that are significant in quantity or in potential harm to human health or the environment, and the eventual disposition of the wastes. Finally, the generator must furnish information on the general chemical composition of hazardous wastes to anyone transporting, storing, treating, or disposing the wastes.

The Act also imposes Government-wide responsibilities that affect coal production and use. It requires Federal agencies to conform to EPA solid waste management guidelines if their activities, such as leasing or permitting, embrace production or disposal of solid wastes, particularly hazardous solid wastes. The impact of this provision is unclear until the Act's implementation by EPA is understood more fully within Federal agencies. However, it can be expected to involve operating agencies such as the Tennessee Valley Authority and permitting agencies such as the Departments of the Interior and Agriculture, which grant mining leases and approve powerplants on Federal land.

The overall effect of RCRA probably will be to increase the cost of waste disposal; the Act also could increase the amount of waste to be disposed. If neither ash nor sludge is determined to be hazardous, the costs of waste disposal could increase as much as 45 percent (see table 61). However, a preliminary analysis by the utility industry of various waste streams from coal-fired powerplants indicates that bottom ash, fly ash, and scrubber sludge all could

**Table 61.—Potential Sludge and Ash Disposal Costs Under RCRA**

| | Ash (mills/kWh) | Sludge (mills/kWh | Total (mills/kWh) | Percent increase over base |
|---|---|---|---|---|
| Base case (current average treatment)................... | 0.45 | 0.41 - 0.69 | 0.86 - 1.14 | — |
| Case 1: Ash: Nonhazardous Sludge: Nonhazardous......................... | 0.57 - 0.65 | 0.65 - 1.03 | 1.22 - 1.68 | 45 |
| Case 2: Ash: Hazardous Sludge: Nonhazardous......................... | 0.77 - 0.84 | 0.65 - 1.03 | 1.42 - 1.87 | 65 |
| Case 3: Ash: Hazardous Sludge: Hazardous ............................ | 0.77 - 0.84 | 0.90 - 1.17 | 1.67 - 2.01 | 84 |

(Case 4, where ash is nonhazardous but sludge is hazardous, is considered extremely unlikely)

SOURCE: Energy Resources Co. Inc.

be classified as hazardous.[40] All three wastes yield substances (primarily toxic trace elements) in concentrations that exceed the allowable limits for drinking water. In addition, ash contains concentrations of radionuclides and corrosive substances that approach the allowable limits. If either or both of these wastes is listed as hazardous, disposal costs could increase as much as 84 percent (see table 61).

Listing ash and/or sludge as hazardous also would change utility disposal practices significantly. Sludge ponds could be prohibited as "open dumps" (an area in which there is a reasonable probability of adverse effects on health or the environment) and utilities forced to find alternative disposal methods. In addition, the sale or use of wastes probably would be discouraged. The National Ash Association estimates that more than 80 percent of the utilities that burn coal sell or use a portion of their ash and sludge wastes. The provisions of RCRA that govern transportation of these wastes could make them noncompetitive. This could increase disposable ash wastes by approximately 18 million tons annually at current levels; no comparable estimates were available for sludge. This effect directly contradicts RCRA's stated objective of promoting the recovery and recycling of solid wastes.

Cost increases also could be expected in the disposal of coal mine wastes, which contain significant quantities of potentially toxic trace metals, if they are listed as hazardous. Potential overlaps between RCRA and SMCRA could add to these cost increases.

[40]Electric Power Research Institute, *The Impact of RCRA (Public Law 94-580) on Utility Solid Wastes* (FP-878, TPS 78-779, August 1978).

## Coal Conversion Authority

ESECA[41] was enacted following the 1973 oil embargo in order to reduce imports of natural gas and oil by increasing the use of coal in their place. Under ESECA, the FEA Administrator was required to prohibit powerplants, and possibly other major fuel-burning installations, from burning natural gas or oil as their primary energy source if certain conditions were met. These conditions are:

• that the facility has the capacity and necessary plant equipment to burn coal;
• that the burning of coal by the facility is practicable, that adequate coal supplies and transportation facilities will be available, and that the order would not impair a powerplant's reliability of service; and
• that EPA either certifies that the facility will be able to comply with applicable portions of the Clean Air Act by the effective date of the FEA order or grants a compliance date extension.

In addition, ESECA authorized the FEA Administrator to require a powerplant or other major fuel-burning installation in the early planning process to be designed and constructed to use coal as its primary energy source. As with the prohibitions, design or construction requirements were conditioned on an adequate supply of coal and preservation of reliability of service, as well as the ability of the owner of the facility to meet existing contractual commitments and to recover capital investments made as a result of the FEA order.

[41]Public Law 93-319 (June 22, 1974), as amended by Public Law 94-163 (Dec. 22, 1975) and Public Law 95-70 (July 21, 1977).

Finally, the FEA Administrator was authorized to allocate coal, while EPA was empowered to issue temporary suspensions of some SIP provisions, in order to facilitate coal conversion. FEA authority to issue conversion orders under ESECA expired on December 31, 1978.

For a variety of reasons, including ineffective management and inadequate funding and personnel, ESECA did not result in many coal conversions. On June 30, 1975, FEA issued prohibition orders to 74 generating units at 32 powerplants and construction orders on 143 units at 97 powerplants. By early 1977, EPA had reviewed 65 of the 74 prohibitions, finding that only 11 of the 65 could burn coal immediately and remain in compliance with the Clean Air Act. Compliance date extensions were given to 20 more, while 34 were found to require additional pollution control equipment.[42] Other stumbling blocks to the success of ESECA included the lack of financial incentives (other than market price) to stimulate coal conversions and the lack of statutory prohibitions for new sources.

The first National Energy Plan proposed by the Carter administration anticipated that two-thirds of the reduction in projected 1985 oil imports (3.3 million bbls/d out of 4.5 million bbls/d) would be achieved by coal conversions. The primary means by which these conversions were to be accomplished was a tax on industry use of oil and gas. (The utility sector was expected to switch to coal because of market incentives and was not considered the chief target of the proposed legislation.) However, Congress failed to approve the industrial use tax, and the final energy package is designed to achieve coal conversions through a regulatory program with slightly more authority than that granted under ESECA.

The primary purposes of the Powerplant and Industrial Fuel Use Act of 1978[43] (part of the National Energy Act) are: 1) to reduce petroleum imports and increase the capability to use indigenous energy resources, 2) to conserve natural gas and petroleum for uses other than electrical generation for which there are no feasible substitutes, and 3) to encourage the greater use of coal, synthetic gas derived from coal, and other alternate fuels in lieu of natural gas and petroleum. Supporting purposes of the Act include the rehabilitation and upgrading of railroad service and equipment necessary to transport coal, compliance with all applicable environmental requirements, and assurance of adequate supplies of natural gas for agricultural uses.

The Powerplant and Industrial Fuel Use Act (PIFUA) strengthens the regulatory program under ESECA in two primary ways. First, PIFUA prohibits, with certain exemptions, the use of natural gas or petroleum as a primary energy source in new electric powerplants and new major fuel-burning installations[44] and provides that no new electric powerplants may be constructed without the capability to use coal or any other alternate fuel as a primary energy source. In addition, PIFUA prohibits existing powerplants from using natural gas as their primary energy source after 1990 and, in the meantime, from switching from any other fuel to natural gas and from increasing the proportion of natural gas used as the primary energy source. The Secretary of Energy is granted additional authority to prohibit the use of petroleum and natural gas where certain conditions related to technical and economic feasibility are met.

Second, these prohibitions are reinforced with a shift in the burden of proof. That is, under ESECA, the choice of fuel was left up to the owner of a facility, and the Federal Government was required to prove that a particular facility could and should use coal. PIFUA, however, begins with a blanket prohibition against the use of oil and natural gas, and the owner of a facility must make a good faith effort to comply with the prohibition and show that despite these efforts he will be unable to

---

[42]Progress in the Prevention and Control of Air Pollution in 1976: Annual Report of the Administrator of the Environmental Protection Agency to Congress, S. Doc. No. 95-75, 95th Cong., 1st sess. (November 1977).

[43]Public Law 95-620 (Nov. 9, 1978).

[44]The provisions of the Powerplant and Industrial Fuel Use Act are applicable to powerplants and other stationary units that have the design capability to consume any fuel at a heat input rate of at least 100 million Btu per hour or to a unit at a site that has an aggregate heat input rate of at least 250 million Btu per hour.

comply and that he is entitled to an exemption from that prohibition.

The prohibitions also are reinforced with a variety of financial assistance provisions. PIFUA provides an additional 10-percent tax credit for industrial investment in alternative energy property such as boilers, pollution control technology, and equipment for producing synthetic fuels from coal. In addition, investment tax credits and accelerated depreciation were denied for new gas and oil burners. PIFUA also provides loans for up to two-thirds of the cost of pollution control equipment for powerplants. Finally, funds were made available for the rehabilitation and maintenance of rail lines used to transport coal.

These statutory prohibitions are, however, subject to numerous temporary and permanent exemptions. In addition, the Department of Energy (DOE) is given great latitude in interpreting the Act. Therefore, PIFUA's success in achieving coal conversions will not be possible to predict until DOE promulgates the final regulations under which exemptions will be granted. For example, PIFUA permits an exemption if the cost of coal "substantially exceeds" the cost of oil or natural gas. The definition of "substantially exceeds," the costs to be included in the determination and the methods for arriving at those costs will be set out in the regulations and will determine the availability of exemptions under this provision.

Despite these uncertainties, DOE estimated in June 1977 that a regulatory program alone (that is, without an industrial use tax) would increase industrial coal consumption by 66 million tons in 1985 (as compared to no further legislation) and would yield 700,000 bbls/d in oil savings. However, the impact of PIFUA may be difficult to ascertain because most utilities and major industries are not planning new oil- or gas-burning facilities. Utilities have reported plans to bring 250 new coal-fired units on line by 1985; no new large industrial oil- or gas-fired boilers have been ordered since March 1977 and industry projections indicate none are expected. However, these projections could be undercut by environmental regulations. EPA plans to announce New Source Performance Standards for industrial boilers

under the Clean Air Act soon after PIFUA takes effect. If those standards are stringent, industrial coal use could be constrained severely in the Eastern United States. Because large powerplants and industrial boilers are the chief target of PIFUA, and because those facilities already plan to use coal to the extent possible, the main effect of PIFUA could be to provide financial assistance to ensure those plans do not change. Where PIFUA could have a major impact—on smaller industries—the amount of coal involved is not as great and exemptions are more easily obtained.

## Other Federal Policy Actions

In addition to the protection from adverse impacts of coal use afforded to health and the environment by the above legislation, a variety of other Federal policy actions may affect coal production and use. These include the Endangered Species Act, the National Historic Preservation Act, the Fish and Wildlife Coordination Act, and a variety of measures related to transportation and transmission.

To the extent that coal production and use disrupt the ecology, it may be constrained by the Endangered Species Act of 1973.[45] The Act is designed to protect all forms of wildlife (including mammals, birds, reptiles, amphibians, fish, shellfish, and other crustaceans, and insects) and plants through conservation programs for endangered and threatened species and for the ecosystems upon which these species depend. The provisions of the Act that are most relevant to coal production and use are the prohibitions against "taking" any species that the Secretary of the Interior has determined to be endangered or threatened and against violating any regulation promulgated under the Act, and the requirement that all Federal departments and agencies consult with the Secretary to ensure that actions authorized, funded, or carried out by them do not jeopardize the continued existence of these species or result in the destruction or adverse modification of their habitat.

Actions related to the production or consumption of coal, such as mining or facility construction, that would result in the death of

---

[45]16 U.S.C. 1531 et seq.

endangered or threatened species, are constrained by the prohibition against taking. In addition, the regulations promulgated under the Act designate habitats that are critical to the survival of some species. Any action affecting a designated critical habitat is an offense under the Act if it might be expected to result in a reduction in the number or distribution of the species of sufficient magnitude to place the species in further jeopardy or to restrict the potential expansion or recovery of the species. Several of these designated critical habitats are in major coal-producing counties.

The requirement that all Federal agencies consult with the Secretary of the Interior to ensure that actions authorized, funded, or carried out by them will not result in ecological displacement or otherwise adversely affect endangered species directly affects almost all coal-related activities. For example, most surface mines require a permit under the Surface Mining Control and Reclamation Act. Similarly, most coal facilities, such as preparation and generating plants, require permits and authorizations under a variety of Federal laws such as the Clean Air and Water Acts. In determining whether actions will adversely affect an endangered or threatened species, agencies must consider not only the particular action to be authorized or funded, but also the probable secondary effects, such as induced private development.

Once an agency has consulted with the Secretary of the Interior, the final decision on whether or not to proceed with an action lies with the agency itself. That is, the Secretary of the Interior does not have veto power over actions of other agencies that might adversely affect critical ecosystems. However, the Act makes liberal provision for citizen suits to enjoin actions that violate the Act or regulations promulgated under it. In these suits the court will give great weight to the Secretary of the Interior's opinion about the effect of an action, and will defer to the Secretary to determine what modifications are necessary to ensure that an action does not adversely affect endangered or threatened species.

The National Historic Preservation Act of 1970[46] (NHPA) requires all Federal agencies to

obtain comments from the Advisory Council on Historic Preservation whenever any action may affect a site or structure listed, or found eligible for listing, in the National Register. In addition, regulations promulgated under NHPA require the agencies to determine whether there are historic, acheological, architectural, or cultural resources that may be eligible for listing in the National Register. If there are, the proposed project must be referred to the agency charged with protection of those resources for review.

As with the Endangered Species Act, NHPA's requirements affect almost all coal-related activities. Significant historic and cultural resources already have become the focus for concern over the impacts of energy development in the Eastern United States, while the preservation of acheological sites is a key issue in the Indian lands of the West. However, in very few instances may a proposed project be blocked by NHPA. Rather, the Act requires that an adequate opportunity to study the historic or other resource be provided before the project proceeds. Therefore, the primary impact of NHPA would be to delay proposed coal-related activities that might endanger these resources.

The Fish and Wildlife Coordination Act of 1970[47] offers some protection against the modification of any water body as the result of a Federal or federally permitted project. Under the Act, Federal agencies must consult with the Department of the Interior's Fish and Wildlife Service and with the State agency having jurisdiction over fish and wildlife, prior to taking any action on a proposed project. Serious consideration must be given to mitigating adverse impacts to fish and wildlife, and a project may be enjoined until the mandated coordination and consultation have occurred.

The principal coal-related activities that may be affected by this legislation include the construction of cooling water intake structures and of barge and other transportation facilities. However, these activities also are subject to the requirement that all projects affecting navigable waters obtain a permit from the

[46]16 U.S.C. 470 et seq.

[47]16 U.S.C. 661 et seq.

Army Corps of Engineers. In practice, therefore, the consultation required under the Fish and Wildlife Coordination Act usually occurs at the time of application for the Corps permit. If water resources will be taken by the project and wildlife values destroyed, mitigating measures may be required to protect the water flow or even to replace lands lost to construction.

Finally, the Transportation Act of 1966[48] forbids the taking of publicly owned wildlife refuges and parks as well as public or private historic sites for highway construction unless there is no feasible and prudent alternative and all steps are taken to minimize harm. This provision probably will not pose significant constraints to coal use because adequate transportation facilities already exist in most

parts of the country. Its greatest potential impact would occur if greatly increased production of Western coal required new highway construction.

Similarly, where transportation or transmission facilities are routed across public lands, or where the site for a combustion facility is on public land, the Federal agency having jurisdiction over the land must issue a permit for use of the right-of-way. There usually are statutory limits on the width of the corridor or on the size of the site, and other limitations may be imposed by the agency in order to prevent adverse environmental impacts. Again, these requirements probably would have a greater effect in the West where there is the largest concentration of Federal land.

# IMPLEMENTATON

How these major Federal policy actions are implemented plays a central role in determining whether they contribute to or obstruct a coherent national coal policy. The primary factors relevant to their effective implementation include the number of agencies with responsibility for regulating coal production and use, the extent to which those agencies' mandates may overlap and conflict, and the number of gaps in regulatory programs created by unanticipated problems.

## Institutional Factors

The responsibility for implementation of the various laws that affect coal production and use and their impacts on health and the environment is divided among a variety of administrative departments and agencies. The major areas of responsibility for environmental and social impacts are environmental impact assessment, air and water quality management, resource extraction, occupational health and safety, solid waste disposal, and siting and land use. For the promotion of coal use, agency responsibility covers resource exploration and acquisition, and resource use. Although

the focus of this discussion is on Federal agencies and their responsibilities, it must be kept in mind that State and local laws and their implementation also have a substantial impact on the direct use of coal.

### Environmental and Social Impacts

Since the passage of the National Environmental Policy Act (NEPA), all Federal agencies must prepare a detailed EIS on all major actions significantly affecting the quality of the human environment. Two Federal agencies have EIS oversight responsibilities: CEQ and EPA. CEQ was established by NEPA to advise the President on environmental affairs, to provide the public with information about environmental issues, and to monitor other agencies' compliance with NEPA. CEQ regulates the preparation of EISs, serves as the Federal repository for EISs, reviews and comments on draft EISs, and develops comparative analyses on the EIS process. However, CEQ's role is primarily advisory; it can neither compel compliance with NEPA nor block actions it feels would have unacceptable impacts.

EPA was established in 1970 to administer a variety of environmental laws and to police the environmental activities of other agencies

---

[48]49 U.S.C. 1650.

by reviewing and publicly commenting on "environmentally impacting actions." As with CEQ, EPA's role in EIS review is advisory; it can make recommendations but cannot block other agencies' actions unless those actions violate other environmental legislation.

A number of State and local governments have enacted environmental policy acts loosely modeled on NEPA. Their requirements range from detailed EISs to advisory planning mechanisms.

The Federal agency primarily responsible for air quality management is EPA. As discussed in previous sections, EPA's role under the Clean Air Act includes publishing information about the effects of airborne pollutants on human health and the environment and about means of monitoring and controlling those pollutants, establishing numerical standards for ambient air quality and for pollutant emissions from various sources, promulgating guidelines for State plans to implement those standards, and implementing and enforcing the standards in the absence of State action. HEW sponsors some research into the health effects of airborne pollutants, and the Department of Justice assists EPA in enforcing the Clean Air Act.

The States have the primary responsibility for implementing and enforcing the Federal air quality standards in accordance with EPA guidelines. In addition, implementation and enforcement authority may be delegated to local government where State law permits and where the locality is deemed to be capable of handling this responsibility.

As is the case with air quality, EPA has the primary Federal oversight responsibility for water quality management. EPA publishes information about water pollutants and methods of control, sets water quality standards and effluent limitations for sources in order to meet those standards, establishes guidelines for State regulatory and permit programs, and implements and enforces the Clean Water Act in the case of State failure to do so. A variety of Federal departments and agencies assist EPA in fulfilling this responsibility including the Department of Agriculture, which is responsible for watershed management and runoff ero-

sion prevention; the Geological Survey, National Oceanic and Atmospheric Administration, National Aeronautic and Space Administration, and the Coast Guard, which assist EPA in water quality surveillance; HEW, which assists EPA in researching the harmful effects of water pollutants; the Department of Justice, which aids EPA and the States in enforcing the Clean Water Act; and the Department of the Interior, which assists EPA in determining the standards of water quality necessary for wildlife protection.

The State role under the Clean Water Act centers on developing and implementing plans and permit programs consistent with EPA guidelines.

As mentioned above, there is no comprehensive Federal policy for water resource management. The allocation of water rights traditionally has been the responsibility of the States, and means of allocation vary widely. However, many Federal agencies have been assigned duties that affect State allocations. First, the Water Resources Council[49] (WRC) assesses water supplies and, through its River Basin Commissions, coordinates regional water resource planning. In addition, WRC evaluates water resource requirements and availability for nonnuclear energy technologies. The Department of the Interior sells or leases water supplies from its reclamation projects and, in cooperation with the Army Corps of Engineers, evaluates reservoir projects for the development of domestic, municipal, or industrial water supplies. The Corps also is responsible for inland waterways in cooperation with the Department of Commerce. The Department of Housing and Urban Development assists in water supply and distribution planning. The Departments of the Interior and Agriculture cooperate on water conservation and utilization projects. These and a variety of other agencies have additional minor and advisory responsibilities for the management of water resources as they relate to irrigation, aquatic life, recreation, flood control, and other purposes.

---

[49]42 U.S.C. 1962a et seq.

The Department of the Interior has primary Federal responsibility for the environmental and social impacts of resource extraction. Within the Department of the Interior, the Office of Surface Mining (OSM) implements the Surface Mining Control and Reclamation Act (SMCRA). Under SMCRA, OSM sets environmental performance standards for all aspects of surface mining and for the surface operations of underground mines and establishes guidelines for State permit programs designed to implement and enforce those standards.

A variety of other offices within the Department of the Interior are responsible for environmental management of Federal coal leases. The Geological Survey has overall regulatory authority over extraction operations after leasing decisions have been made, including the site-specific conditions to be incorporated in leases as well as the resource conservation regulations for Federal lands. The Bureau of Land Management develops the reclamation requirements to be included in mining plans, while the Bureau of Indian Affairs regulates mining activities on Indian lands.

Prior to the passage of SMCRA, State regulation of the environmental impacts of mining varied widely. States that wish to retain regulatory authority over surface coal mining and reclamation operations now must develop comprehensive plans and permit systems in accordance with SMCRA. OSM will regulate the mining activities in those States that fail to develop or enforce a regulatory program.

Until the 1960's, regulation of the occupational health and safety impacts of mining was characterized by conflicts between State and Federal jurisdiction. These conflicts were resolved in 1969 with passage of the Coal Mine Health and Safety Act, which gave the Department of the Interior primary regulatory responsibility for miner's health and safety. Within the Department of the Interior, the Mining Enforcement and Safety Administration (MESA) set minimum standards, outlined penalties for violations, and established mine closure criteria. However, the Department of the Interior's concurrent responsibility for maximizing energy resource development was found to be incompatible with its duty to enforce mine health and safety regulations, and in 1977 the Mine Safety and Health Act shifted the latter duty to the Department of Labor. The Department of Labor's Mine Safety and Health Administration (MSHA) will take over MESA's responsibilities. Both HEW's National Institute for Occupational Safety and Health and the Department of the Interior's Bureau of Mines conduct mine health and safety research.

Federal responsibility for solid waste disposal is divided among EPA, the Department of the Interior, and the Army Corps of Engineers. Under the Resource Conservation and Recovery Act (RCRA), EPA has general oversight authority for solid waste disposal. EPA is required to promulgate guidelines for the transportation and disposal of hazardous and nonhazardous wastes of all types (except radioactive wastes). The Department of the Interior's OSM sets minimum environmental performance standards for the disposal of spoil and coal-processing wastes. If either of these wastes is listed by EPA as hazardous or if EPA's environmental protection standards are more stringent than OSM's, the mine operator will be required to meet the stricter standards. Finally, the Army Corps of Engineers issues permits for the disposal of solid wastes or of dredge and fill material in navigable waters.

Implementation and enforcement of EPA's requirements under RCRA and of OSM's regulations under SMCRA may be turned over to States that establish approved regulatory programs. To the extent allowed by these State programs, local governments may control the location of waste disposal through land use planning and zoning.

As with water resource management, no comprehensive Federal policy exists for land use and facility siting. Although the Nation's energy goals call for increased reliance on coal, the recent escalation in parties-at-interest to energy development makes it difficult to find acceptable sites for new coal combustion facilities. In addition, most of the Federal regulatory programs discussed above indirectly control facility siting to prevent unacceptable site-specific environmental and social impacts, but there is little coordination among these programs.

Only on Federal lands do Federal agencies have any direct control over land use and energy facility siting. The Departments of the Interior and Agriculture supervise most Federal coal lands and have the authority to forbid mining where it would be environmentally unacceptable. For all Federal lands, the agency having jurisdiction over the land must issue a permit for use of the land as the site for a coal-related facility or for the right-of-way for transportation or transmission purposes.

Indirect controls—primarily constraints—on facility siting may be exercised by CEQ and EPA through their role in reviewing EISs. In addition, a variety of EPA regulations for air and water quality management limits the number of sites available to utilities and industry.

The States have the greatest amount of control over facility siting and land use, either through their general police power to protect the public health and welfare or through the implementation of federally mandated programs. Under their police power, States can control energy resource development through land use planning and zoning, permit requirements, and regulation of public utilities. A few States have enacted comprehensive statewide land use planning legislation; some only provide statewide planning for energy facility siting; in others all land use and siting remains under piecemeal legislation, much of it implemented at the local level.

### The Promotion of Coal Use

Federal responsibility for resource exploration and acquisition rests with the Department of the Interior, which administers the Federal coal-leasing program. Within the Department of the Interior, the Geological Survey evaluates coal resource data while BLM records lease applications and collects various fees, rents, and royalties. In addition, other agencies with jurisdiction over the surface of public lands, such as the Department of Defense, may block coal leases on their lands.

The responsibility for coal extraction and use is divided between the Departments of the Interior and Energy. As discussed above, the Department of the Interior oversees Federal coal leases. But the 1977 Department of Ener-

gy Organization Act gives DOE control over economic leasing terms and conditions. DOE's duties under the 1977 Act include establishing long-term production goals for federally owned energy resources, developing standards for rates of production from Federal leases, specifying economic terms and conditions of individual leases (for example, eligibility of joint ventures), and setting guidelines for postlease conditions (such as recommending forfeiture of a lease that does not meet production rates). To facilitate cooperation between the Departments of the Interior and Energy in administering the leasing program, the 1977 Act created a Leasing Liaison Committee within DOE but composed of equal numbers of members from both departments.

The use of coal is governed mainly by DOE through conversion of existing facilities to coal, the prohibition of new large facilities from burning gas or oil, R&D on new technology, and regulatory price setting.

## Evaluation

Many critics of present Federal coal policy and of the agencies that implement it argue that energy development is overregulated. Certainly the scope of Federal intervention has grown dramatically in the last decade. The number of permits, certifications, and authorizations required to operate a coal mine or a coal combustion facility has increased substantially. Often several agencies share the responsibility for regulating a particular activity, such as leasing federally owned coal. Where those responsibilities overlap, conflicts may occur between the goals of the agencies involved. For example, limitations imposed on Federal coal leases by the Department of the Interior to prevent unacceptable environmental or social impacts may be incompatible with the economic terms and conditions imposed by DOE to achieve national energy goals. Whether the present Federal coal policy is perceived as counterproductive usually is a function of the interests being represented. Impartial analysis is rare and conventional cost-benefit analysis often cannot adequately weigh the tradeoffs between the dollar costs of regulation and the resulting unquantifiable environmental and health benefits.

On the other hand, some commentators argue that more regulation of coal-related activities is required, either because of the manner in which agencies have interpreted their Federal mandates, or because gaps or inconsistencies in those mandates preclude the existence of either a coherent national coal policy or a coherent national environmental policy. The most significant obstacles include the lack of comprehensive Federal programs for coal, water, and land resource management; the conflict between States rights and the need for uniform Federal legislation; the lack of workable mechanisms for solving interstate or interregional problems; and the lack of mechanisms for long-range planning.

No comprehensive Federal policies currently exist for coal leasing, water resource management, or land use and facility siting. A Federal coal-leasing policy should be developed and implemented by the early 1980's, long before any significant coal supply constraints are expected to arise. Water availability and land use, however, may present obstacles to increased coal use.

As discussed above, water supplies traditionally have been allocated by the State with some Federal oversight. For the most part, State control of water resources is logical because of the wide variation in water availability. Thus a regulatory scheme that may be workable in the East, where surface and ground water resources are relatively abundant, would not be appropriate for the arid and semiarid regions of the West. Yet as the competition for water for agricultural, industrial, residential, and energy uses increases, a national system of priorities may become necessary.

Similarly, while land traditionally has been considered the Nation's most abundant resource, past abuses and the increasing concerns of parties-at-interest have begun to limit the land available for energy resource development. As with water, the concerns over land use and siting have regional variations. In the more industrialized East the concerns center on facility siting patterns; concentrations of coal-fired powerplants in particular areas may lead to cumulative and interactive impacts that are not fully understood. In the Western States, the primary concern is the preservation of environmentally valuable scenic areas and the prevention of the adverse social and economic impacts of rapid development. To date, proposals for Federal land use legislation have been designed to encourage statewide planning. But many of the existing siting problems, such as air quality management and the long-range transport of pollutants, do not respect State boundaries. Without comprehensive nationwide land use planning that directly addresses facility siting problems it may not be possible to meet national energy goals.

Problems also have arisen in implementation of those regulatory schemes that mandate comprehensive State programs in accordance with Federal guidelines. Although nationwide legislative uniformity in these areas is in the public interest, States are under increasing pressure to reduce spending and are reluctant to accept the responsibility for major new regulatory programs. Yet if the States wish to preserve their rights to regulate energy development within their borders, they must accept that responsibility. Similarly, where federally mandated State regulatory programs are found to be inadequate, responsibility for their implementation and enforcement reverts to Federal agencies that have neither adequate personnel nor funding to perform those duties.

As discussed in chapter V, a variety of environmental impacts of energy resource development have become regional problems that do not lend themselve to management on a State-by-State basis. Yet the mechanisms for solving interstate pollution problems are cumbersome and ineffective, and the result will be an increase in the number of suits between States. For example, the State of Kentucky has passed a law that requires powerplant operators to obtain a permit from Kentucky if they take water from or discharge waste into the Ohio River along the State's border. This law effectively gives Kentucky control over powerplant siting along the Ohio River in the States of Ohio, Indiana, and Illinois. Kentucky has announced that those States must work together to solve their common air and water pollution and siting problems or face legal action by

Kentucky. To some extent, the existing Federal EPA regions could be used to manage interstate pollution problems, but the same issues of coordination and cooperation exist at the regional level; there has been very little interaction among the EPA regional offices even though environmental management strategies developed for one region may significantly affect another. For example, Kentucky's programs are under the jurisdiction of EPA region IV while Ohio, Indiana, and Illinois are in region V. The neighboring States of Pennsylvania and West Virginia, which share the same pollution problems, are under the jurisdiction of region III. In September 1978, these three EPA regions established an interregional task force to coordinate pollution control in the Ohio River Valley. Such interstate and interregional cooperation must become the rule rather than the exception if environmental problems are to be solved. Existing mechanisms for this cooperation, such as interstate compacts, should be adequate if used effectively.

Finally, most of the Federal policy actions discussed represent a legislative response to an existing problem, such as lack of development of Federal coal leases or the already polluted condition of the Nation's air and water. Consequently most of the programs and the R&D funding is aimed at solving these problems and little attention is given to long-range planning or to researching potential future problems.

In summary, the piecemeal legislative approach to energy resource development and environmental management has resulted in a variety of implementation problems. Some of these result in additional dollar costs to the developer. Others result from a lack of coordination within and among regulatory agencies, such as the solid waste impacts of air pollution control or interstate and interregional energy development impacts. Still other problems, such as the obstacles to energy facility siting, may require additional regulation if the Nation's energy goals are to be met. For the most part, however, these problems could be solved if existing legislation were implemented in an effective, coordinated manner.

Chapter VIII
# POLICY OPTIONS

# Chapter VIII.—POLICY OPTIONS

## TABLES

# Chapter VIII
# POLICY OPTIONS

Each of the national energy supply and demand scenarios in this assessment involves a very substantial increase in coal use over the next two decades. There is no doubt that the resource to sustain a high level of use over that period is physically present and accessible. It is also clear that from an engineering standpoint coal can be extracted, processed, and burned at an economic cost that will make it very competitive with alternative fuels. What is not so clear is how the external costs, institutional and social constraints, and other nonmarket factors associated with coal use will affect the validity of the economic and technological analysis. At the extreme, increased coal use could pose such serious external costs to the environment and public health that would make it unacceptable. At a minimum, the process of reducing external costs (e.g., by imposing pollution controls) and coping with internal constraints (e.g., labor-management problems) will moderately increase the economic costs of coal utilization. Given the central place of coal in future U.S. energy planning, it makes a great deal of difference where we ultimately come out on the continuum between minimum and maximum constraints on coal use. In short, the stakes involved in formulating a national coal policy are very high.

The tasks of policy analysis in this area are to identify the potential problems and constraints and to examine the range of governmental policies that offer some promise of ameliorating them. This study does not recommend specific policies, but it identifies policy options, the sorting criteria for choosing among them, and the implications of available choices.

The sorting criteria are of three basic types: 1) national objectives concerning the level of coal production and use, 2) political and normative values, and 3) pragmatic calculations concerning the relative efficacy of policies and technologies in stimulating production and use and/or minimizing adverse impacts. These are analyzed in sequence.

## NATIONAL ENERGY OBJECTIVES

National objectives concerning the magnitude and timing of coal use set the context for formulating coal policy. For example, acceptable policies for the leasing of Federal coal reserves, workplace health and safety, and clean air legislation may be different in kind or degree depending on whether the Nation seeks 100, 125, or 150 quadrillion Btu (Quads) of energy supply. Similarly, policies designed to compel the conversion of existing industrial boilers from gas and oil to coal may or may not be necessary depending on the Government's timetable for increased coal use.

In actual fact a sufficient supply of coal should be available to meet the three coal use scenarios cited above while satisfying existing and pending environmental, health and safety, leasing, and related legislative and regulatory requirements. Nevertheless, there are actions that will provide an additional margin of safety against the possibility that these supply projections are overly optimistic or that it becomes necessary to raise coal's fraction of U.S. total energy supply above the levels posited in this report. Many of these measures have merit independent of their potential effect on coal supply. The list includes efforts to: 1) mitigate the adverse community impacts that might constrain coal development, 2) remedy the sources of labor-management disputes and promptly settle strikes that do occur, 3) anticipate and avert potential coal transportation bottlenecks by upgrading existing modes (e.g., railroads) and facilitating the creation of new ones (e.g., slurry pipelines), 4) expedite the formation of a leasing policy and the designation of eligible tracts, 5) streamline the permitting process for new mines, and 6) develop procedures for anticipating and accommodating potential objections to new coal facilities in order to avoid extensive litigation and delay.

Demand is more likely to be a constraint on coal development over the next two decades than is supply. While demand will probably be adequate to sustain all but very high energy scenarios, this is far from certain. Several broad policy options are available to strengthen the future market for coal. These include: 1) tax pressures and incentives to induce utility and industrial conversion to coal; 2) R&D support for technologies, e.g., fluidized-bed combustion (FBC) and solvent-refined coal (SRC) that can help make coal an acceptable fuel for small users; 3) RD&D support for improved, less expensive emission control technologies; 4) RD&D support for coal gasification and liquefaction technologies; and 5) higher prices for natural gas and fuel oil.

In general, however, the different plausible targets regarding coal production and use for the remainder of the century do not emerge as a critical basis for sorting among legislative and regulatory policy options.

## POLITICAL AND NORMATIVE VALUES

Values play a critical role in the policymaking process, yet policy analysis often proceeds under the assumption that policymaking is or can be a clinically objective, value-free process. In fact, the choice between conflicting courses of action will often and inescapably reflect subjective judgments concerning what is desirable. With regard to coal policy, the most important value conflict involves the relative priority assigned to increasing production as opposed to reducing adverse impacts. Taken together, existing legislation and regulations define a rough but discernible balance between these two value sets. In broad terms future policy must either accept that balance or shift it in favor of production or impacts amelioration. This tradeoff, perhaps more than any other, lies at the heart of national coal policy. In one sense, however, the dichotomy is a false one. Adverse impacts lead to constraints on coal use, and a major reason for controlling them is to facilitate coal development. The environmental goal conflicts with the production goal only when protection measures increase the cost of coal sufficiently to dampen demand. Other potential value conflicts involve the proper allocation of decisionmaking authority between the public and private sectors and among Federal, State, local, and tribal governments. Conflicts may also occur between various impacts-related values; e.g., operating a new coal mine may help solve a number of community problems, including unemployment, but may have serious adverse impacts on the physical environment.

What follows is a more detailed examination of three conflicting value sets: 1) production maximization or impacts amelioration, 2) the allocation of decisionmaking responsibility concerning coal development between Federal or State and local authorities, and 3) the allocation of decisionmaking between the public or private sectors. The first choice between production and impacts values can be analyzed in terms of specific tradeoffs.

**The first tradeoff** is between coal extraction and environmental quality. Mining has a number of inevitable adverse environmental impacts. To some extent these are specific to either surface or underground, but both forms of mining have a range of effects. Although complete control over these effects is impossible, all can be greatly reduced with existing technologies and procedures. But these technologies have both costs and limitations. They will certainly make coal more expensive. Higher prices for coal can have secondary impacts on much of the economy, including inflation and unemployment levels. The limitations of reclamation and control technologies mean that mining certain coal reserves (e.g., under prime farmland) may be precluded altogether.

On the other hand, failure to employ available controls has its own costs. These can include scarred landscapes (often permanent), subsidence (some of it continuing for a century or more), pollution and siltation of surface waters, pollution and disruption of aquifers, and in some cases, flooding. Secondary costs can include a reduction in tourism, damage to agriculture, diminished opportunities for recreation, esthetic impairment, and the need for control and reclamation. Whereas costs of employing controls in the form of higher energy prices are borne by the beneficiaries of coal use, the costs of not using controls tend to fall disproportionately on the coalfield communities and their inhabitants.

**The second tradeoff** is between coal combustion on the one hand and environmental quality and public health on the other. Burning coal to produce energy will unavoidably generate emissions with potential adverse impacts on health and the environment. Some emissions cannot be practicably controlled with known technologies, while the others can be partly, but not entirely, eliminated. Carbon dioxide ($CO_2$) emissions belong in the first category and sulfur oxides ($SO_x$) and nitrogen oxides ($NO_x$) in the second. Nevertheless, most immediately threatening emissions can be minimized with present or pending technologies. There are costs. Coal becomes more expensive to burn; environmental problems are created by the need for disposal of flue-gas desulfurization (FGD) sludge; and the process of coal conversion by plants is made more expensive and time consuming.

The costs of not controlling emissions include agricultural crop losses, damage to forests and freshwater fisheries, adverse impacts on esthetic and property values, possible alteration of global climate over time, and possible increased incidence of human illness and mortality due to lung and other diseases.

To the extent to which powerplants are located in urban areas, there is a basic symmetry between the costs and benefits of emission control. The urban consumer population bears the primary burden of increased energy costs and benefits from the reduction in pollution from nearby plants. The increasing tendency, however, is to site new plants in rural areas or on the outer fringes of urban centers. Under

these circumstances the symmetry is overturned with potential implications for the balance of political pressures concerning the tradeoff between energy price and pollution control.

Where the policymaker or analyst ultimately draws the balance in the tradeoff between energy availability and cost on one side and environmental and health considerations on the other depends on four considerations:

1. The severity of the specific impacts being analyzed and the equity of their distribution.
2. Personal values, e.g., economic growth as opposed to conservation.
3. Attitudes toward risk. The economic costs of imposing various controls are reasonably predictable, but the environmental and health costs of foregoing those controls are not, and the range of possible consequences is very wide. Consequently, policy choices often involve tradeoffs between the known and unknown or partially known.
4. Relative value assigned to present and future costs and benefits. The economic consequences of higher coal prices will be felt in the short term, while many of the most important and environmental effects (e.g., from carcinogens) will not be felt for years or even decades.

**The third tradeoff** is between coal extraction and community well-being. Mining can have substantial community benefits relating to economic growth and employment. The costs can be high as well, including overloaded community services, economic dislocation, and social disruption. Which effects will tend to predominate depend on the circumstances of the particular community. The costs of alleviating community distress related to coal development tend to take the form of transfers of State and Federal resources to the locality— e.g., through loans, grants, and bond guarantees. A bill submitted to the 95th Congress would have allotted $1 billion over the next decade for this purpose.

The costs of doing nothing are also real, though not readily quantifiable. These costs include worker dissatisfaction (with consequent high turnover and unrest), alcoholism and related social ills, rapid inflation in the coalfields, disruption of the local economy, opportunity costs due to losses in potential income from noncoal economic activities (e.g., declines in tourism), and congestion and disruption of settled community life. The decision of where to place the burden of dealing with these community problems hinges partly on whether future coal development revenues are seen as ultimately sufficient to cover the community costs or not. If so, the choice may be sidestepped by providing loans to cover the immediate "front-end" costs, which can be repaid out of future revenues. If not, a choice must be made where the burden will fall—on the society as a whole, which presumably benefits from the increased availability of energy; on the individual coalfield communities; or on the utilities and mining companies and, through them, on the direct consumers of the energy.

**The fourth tradeoff** is between coal extraction and workplace health and safety. A certain number of casualties in the form of injuries, deaths, and occupational disease among miners is an inevitable result of coal production. Mining, particularly underground, is an inherently dangerous occupation. Nevertheless, with current technologies and procedures (e.g., dust sampling and control, safety training, inspections) the risks can be lowered. The costs of doing so include higher priced coal, and possibly a dampening of productivity and output. The latter relationship has yet to be clearly demonstrated. The costs of not acting to reduce the risks to miners are a higher incidence of accident and disease, which could result in increased labor unrest. Once again policymakers face a choice between allocating the costs of ameliorating impacts to society as a whole through increased energy prices or allowing the costs to fall upon a single group—the miners.

The policy implications of these tradeoffs are simply illustrated by positing two opposite ideal types of policymaker—an "impacts minimizer" and an "energy maximizer." In addressing the major issues involved in coal policy the *impacts minimizer* will favor such measures as stringent standards (requiring the best available control technology (BACT)) under the Clean Air Act for removal of $SO_x$ and $NO_x$, application of Federal point source air emission standards to existing as well as new facilities, the imposition of stringent and detailed regulations implementing the Surface Mining Control and Reclamation Act (SMCRA), the classification of sludge as a hazardous substance subject to rigorous disposal standards, and the enforcement of strict criteria (e.g., regarding the impact on water availability in arid regions) in selecting sites for coal facilities.

With regard to community effects of coal development, the impacts minimizer will favor policies that cause the coal companies to internalize an increased portion of the public costs (e.g., concerning roads and housing) consequent to their activity. He or she will tend to favor the use of State eminent domain powers to obtain coal company lands for housing in Appalachia and in both the East and West will support increased community participation and control concerning the decisions of coal companies that significantly affect the community. A go-slow, careful approach to further leasing of Federal coal lands will be preferred. In the tension between production/output objectives on the one hand and health/safety goals on the other, the impacts minimizer will give preference to the latter. Examples include the application of more rigorous dust control standards and procedures and the use of more Federal safety inspectors in the mines. With regard to labor/management disputes over such questions as mine safety and health care centers, the tendency will be to support the miners.

The *energy maximizer* is defined as the mirror image of the above and, as such, would embrace an opposite set of policies. The above list of policy issues is merely illustrative and is far from complete. Moreover, not all policies sort in terms of production and impacts values.

A second set of competing values relevant to coal policy relates to Federal Government regulation of the private sector—a classic issue of political ideology. For the sake of convenience, the two competitive perspectives

Minemouth powerplants represent one option in an intricate tradeoff of costs and benefits

Photo credit: EPA Documerica

can be called "proregulatory" and "antiregulatory"—recognizing that these are oversimplified caricatures. In recent years the *proregulatory* perspective has been the predominant influence on national policymaking concerning coal. Rooted in the liberal political tradition, proponents of this view tend to give less priority to the production of coal than to the mitigation of its adverse impacts. They see major tensions between the interests of the coal companies and the public and are generally skeptical of the will or ability of these corporations to avoid actions detrimental to the public interest without strong Government pressure to do so. They have a high regard for the value of public, grassroots participation in decisions affecting coal development. Regulations are viewed positively as an indispensable means of protecting the environment and public health and safety. They provide a uniform, detailed, obligatory code of conduct on the coal companies that is enforceable in the courts.

In contrast, adherents of the *antiregulatory* perspective are associated with the conservative part of the political spectrum, tend to value energy production over impacts mitigation, and see the public interest as being best served by according maximum freedom of action to the productive genius of the private sector. Government attempts to regulate economic activity are seen as a recipe for higher costs and reduced productivity to the detriment of all. Some regulation of coal development may be unavoidable, but it should take the form of general performance criteria that leave to the private company freedom to determine how best to achieve the standard. Wherever possible, broad guidelines and reliance on voluntary compliance should be the rule.

These two contrasting philosophies lead to very different approaches to specific problems. Three illustrative examples are noise abatement, air pollution control, and coal facility siting. With regard to noise, advocates of the first perspective tend to favor detailed requirements concerning the devices and procedures that must be used to achieve a reduction in noise levels in a mine. Scorning this "cookbook" approach, the antiregulators favor only performance standards, e.g., miners

must not be exposed to sounds above a certain decibel level. How the company achieves this objective is not the Government's concern. Similarly, with regard to air pollution, the proregulators argue for BACT requirements whereas their counterparts favor simple emission standards. As for siting, the first perspective sees merit in detailed uniform site selection criteria whereas the second viewpoint tends to favor case-by-case negotiation of sites by interested parties within very general guidelines.

A final value choice concerns how decisionmaking authority over coal development should be allocated among the various levels of government: Federal, State, local, and tribal. Considerations that argue for a growing Federal role include: 1) there is a clear need for a national policy concerning coal development and use, 2) a substantial portion of known coal deposits underlie Federal lands, 3) Federal money plays a key role in R&D concerning coal, and 4) scientific evidence increasingly documents the regional/interstate nature of the pollution problem. Coal burned in Ohio produces acid rain in New Hampshire. Pesticides entering the Mississippi River in Illinois eventually pollute the drinking water of New Orleans, La. The failure to impose minimum national standards would permit one State to attract industrial investment by lowering its environmental standards, thereby becoming a pollution haven to the economic and perhaps environmental detriment of its neighbors. Moreover, it is argued that the scenic quality of a State like Utah is a national treasure and, as a consequence, the Federal Government has a responsibility to protect it.

On the other hand, the potential costs of an increased Federal role are: 1) reduced State and local initiative, 2) a diminished ability to fine-tune policy to fit local and regional conditions, 3) the possibility of a halfhearted State and local commitment to the successful implementation of national coal policy, 4) a risk of creating a precedent for increased Federal intervention in regional and local affairs generally, and 5) a belief that Federal requirements concerning air quality, surface mining, water quality, and potentially, land use unjustly interfere with the right of States to develop their resources, attract economic investment, and draw their own balance between energy and environmental goals. For example, the prevention of significant deterioration (PSD) requirements under the Clean Air Act limit the scope for industrial development in Utah and Montana. Provisions of SMCRA constrain the ability of Kentucky to exploit in-State coal reserves. Rising pressures on State budgets for fiscal austerity exacerbate the problem be-cause much Federal environmental legislation sets standards and then mandates implementation by the States—at the latters' expense. Despite this difficulty the potential tradeoff between national and subnational decision-making can be at least partly sidestepped by setting broad Federal standards of performance to be elaborated and applied by the States. This strategy has been followed in recent legislation concerning coal development.

# EFFICACY CRITERIA

The remaining type of criteria for selecting among available policies and associated technologies involves an assessment of their utility in solving the specific production, utilization, and impact problems associated with coal. Unlike the previous two sets of criteria, which are to a significant degree discretionary, the third class of criteria involves a determination of fact, i.e., an assessment of the feasibility and effectiveness of various policy options. The unevenness of the data base means that the gaps in relevant knowledge must be clearly identified during the analysis.

Five major areas of policy concern have been identified, each with a potential for significant influence on efforts to expand the production and combustion of coal. They include environmental impacts, community and social impacts, labor-management relations, workplace health and safety, and leasing of Federal coal reserves. The policy options relevant to these areas of concern are analyzed in terms of the efficacy criteria cited above.

## ENVIRONMENTAL IMPACTS

Environmental considerations are an important potential constraint on any substantial increase in coal production and combustion. A number of special characteristics tend to distinguish environmental from other concerns. One is the range and scale of potential impacts—from minor aggravations to global catastrophe. Perhaps the best known example in the latter category is the possible climatic ef-fects resulting from increased concentrations of atmospheric $CO_2$. The potential time scale of environmental damage covers a similarly broad range from virtually instantaneous and short-lived phenomena to impacts that are slow in developing but that will endure for centuries. Environmental issues relate directly to matters of great importance to people. What for the environmentalist is a question of public health and the quality of life is for the developer a matter of personal livelihood.

Governmental concern for the environment is evidenced by an imposing body of legislation and regulation administered by a substantial bureaucracy. To a remarkable degree the legislation has kept pace with advances in scientific knowledge. In terms of control technology, legislation has more than kept pace, i.e., it has in some instances assumed a technology-forcing function.

Achieving a policy consensus on environmental questions is always difficult. This reflects, in part, the value conflicts noted above. It also reflects important gaps and ambiguities in the scientific evidence, for example, with regard to the health effects of chronic long-term exposure to relatively low levels of certain pollutants. The sheer magnitude and unprecedented nature of some of the potential impacts tend to induce skepticism regarding the available data. The combination of strong emotional commitments and uncertain data is a sure recipe for political conflict. This tendency is exacerbated in the case of coal by the time, effort, and cost involved in developing and implementing environmental controls.

Many of the participants in the policy process approach environmental issues from quite different perspectives. One viewpoint assumes the worst regarding potential adverse impacts on the environment from coal development and use. The burden of proof concerning the environmental acceptability of coal is placed on the industry. Adherents to this perspective would sharply limit coal development until control technologies have proven their effectiveness. As such technologies become available their universal application would be a condition of coal development. The alternative perspective is the mirror image: impacts are assumed to be acceptable and manageable pending clear evidence to the contrary, the costs of environmental controls are emphasized, and control technologies must meet a test of eocnomic acceptability as well as effectiveness. The threshold tolerance of environmental disruption is predictably higher among supporters of the second perspective.

A final factor complicating the task of environmental policy analysis is the sheer complexity and comprehensiveness of the field. Policy options designed to minimize the adverse environmental impacts of increased coal use range from very specific technological or managerial actions (e.g., lining of ponds used to hold toxic wastes) to regulatory modifications (e.g., regulation of stack plume opacity), to broad questions of strategy and philosophy (e.g., the use of effluent charges and other market-oriented mechanisms in lieu of regulated controls).

The era of unregulated environmental impacts is clearly past for coal, as for other fuels. There is in place an elaborate, though still incomplete, framework of legislation, regulation, and implementing institutions that constitute a national policy system for managing the environmental impacts of increased coal use. The relevant control technologies are at various stages in their evolution from conception to maturity—but most have at least reached the point where a first generation technology can actually be applied in coal industry operations. Control technologies for combustion emissions are particularly impor-

tant and, while existing technologies may not be optimal, they are workable and effective and the outlook for improvement is reasonably promising. In short, after the investment of substantial economic, technological, and human resources over recent years, the foundation for a viable environmental policy for coal now exists.

Under these circumstances the priority task of policy analysis is to identify ways the existing policy system can be upgraded. Five major tasks appropriate to this effort are:

1. Identify gaps in present knowledge regarding the nature and magnitude of the risks to the environment associated with coal utilization and the data required to fill those gaps.
2. Indicate the prospects and priority needs for the development of specific control technologies as a guide to possible Federal support.
3. Examine the existing body of law and regulation for omissions, inconsistencies, and unproductive or counterproductive requirements.
4. Analyze the prospects for effective implementation.
5. Identify those specific major issues that warrant the priority attention of policymakers.

A discussion of preliminary findings relevant to each of these tasks follows.

## Data Gaps and Needs

Present scientific understanding of the environmental impacts associated with coal is deficient in a number of areas. The relevant policy response by the Federal Government would be support for research designed to supply the required data. The following is a preliminary list of areas where such additional research is needed:

- rate of accumulation of atmospheric $CO_2$ and its impact on climate;
- atmospheric transport of pollutants, the chemical transformations they undergo, and the paths they travel;

- correlation between particular pollutants or levels of pollutants and human health (dose-response relationships);
- relationship between coal combustion, acid rain, and the productivity of croplands, forests, and freshwater fisheries;
- patterns and consequences of plume touchdowns;
- impacts of $SO_x$ and photochemical oxidants on human health and agricultural productivity;
- impacts of fine particulates on human health;
- role of coal combustion in the formation of hydrocarbons and their impact on human health; and
- physical and chemical interaction of different pollutants, soils, and hydrological configurations (i.e., the role of site characteristics in determining the impact of pollutants and pollutant disposal methods on ground and surface waters).

## Development Priorities for Control Technologies

A variety of control technologies and techniques is presently or prospectively available for dealing with the multiple assaults on the environment from coal development. These vary widely in effectiveness, stage of development, and future promise. The task from a policy standpoint is to identify those areas where problems remain, where technological innovation and development are needed, and where Federal actions should make a difference.

With regard to control of the combustion products of coal the priority needs are for improved $NO_x$ control technologies, electrostatic precipitator designs that are effective against small particulates and low-sulfur coal, lower cost baghouses, techniques for minimizing hydrocarbon emissions from small boilers, and the improvement of new technologies for combustion (FBC) and fuel cleaning (SRC). The prospects in each case are sufficiently promising to warrant the commitment of substantial R&D funds. There is also need for continued upgrading of FGD scrubbers with regard to their reliability, maintenance requirements, and costs. One major emission from coal combustion, $CO_2$, is not susceptible to practical control by any known technology.

A similar agenda of priority needs can be identified with regard to the control of adverse environmental impacts of mining. They include ways of making constructive use of land that is subject to uncontrolled subsidence, materials and methods for backfilling and/or sealing abandoned mines, improved techniques for the safe burial of mine wastes, new methods for constructive use or recycle of mine wastes, improved techniques for controlling acid mine drainage, and methods for reclaiming particularly sensitive land forms (e.g., steep slopes, prime farmland, etc.) after surface mining.

Other control technologies that warrant attention and the commitment of resources are regenerable scrubbers, impermeable land fills for FGD sludge, and water-conserving designs for energy facilities including dry and wet/dry cooling systems.

## Adequacy of Existing Law and Regulations

A substantial body of law and regulation relevant to the control of environmental impacts from coal development is presently in place. There remain, however, areas of omission, inconsistency, and weakness that warrant congressional attention. In part, the problems transcend coal and relate to a lack of Federal policy in such major related areas as water resource management, conservation, land use, and energy facility siting. Mechanisms for long-range planning and for resolving interstate and interregional problems are generally weak. There is, moreover, still no comprehensive, consistent national policy toward energy generally and coal specifically. For example, there is no Federal policy to have the prices of all fuels reflect their true relative costs or to provide a consistent regulatory framework for all modes of transporting coal. Federal legislation and regulation on coal reflect an unresolved tension between the goals of environmental protection and energy development, with the task of drawing the balance left to the courts.

Agency mandates and jurisdictions overlap, as with the Department of Energy (DOE) and the Department of the Interior responsibilities for leasing and the Department of the Interior and the Environmental Protection Agency (EPA) jurisdictions over the implementation of SMCRA. In some cases legislation has resulted in environmental programs seemingly at cross purposes. In the past, environmental legislation tended to focus on a specific media and be blind to the effects of control efforts on other media. More recent legislation has attempted to address this problem, e.g., in formulating New Source Performance Standards (NSPS) regulations, their impact on nonair environmental quality must be taken into account. The problem is in a sense irresolvable because it is an immutable physical law that matter cannot be destroyed. Thus the Clean Air Act amendments that in effect mandate FGD scrubbers create a substantial land and water pollution problem in the form of large quantities of sludge.

Existing law deals creatively with another tension—that between the need for uniform national standards and States rights—by mandating State implementation of Federal environmental guidelines. This solution may be threatened, however, by an increasing inability or unwillingness of States to bear the costs of such programs in a time of financial stringency.

There are some specific areas of omission in environmental legislation and regulations concerning coal. Current regulations under the Clean Air Act do not deal with the long-range transport and transformation of combustion products. Thus the regulations control sulfur dioxide ($SO_2$) but not the more dangerous transformation product—sulfates. They also leave small (respirable) particulates and trace hydrocarbons inadequately regulated. The small boilers that would be the principal source of hydrocarbons are not federally regulated. Other areas where the present legal-regulatory framework is incomplete include leasing policy and whether sludge will require disposal as a hazardous substance.

Table 62 presents a summary of key problems and possible new policy initiatives organized around specific pollutants in each medium.

### Table 62.—Environmental Impacts (Land)

| Problem | Impacts | Solutions/policies |
|---|---|---|
| Subsidence | Disrupts surface use, damages structures, lowers property values. | • Prohibition of underground mining beneath densely populated areas pending demonstrated ability to prevent subsidence.<br>• Alter law to make mine owner directly responsible to surface owner.<br>• Government-financed insurance to compensate surface owners for damage resulting from subsidence.<br>• Tax on underground mining to compensate for future damages.<br>• Research regarding controlled subsidence: Planned subsidence for active mines; induced subsidence for inactive mines.<br>• Increased research regarding preventing subsidence by using FGD sludge, FBC sorbent, or other materials, to backfill abandoned mines.<br>• Research regarding productive uses for land subject to uncontrolled subsidence. |
| Disposal of mine wastes (gob, preparation plant wastes, sludge from treatment of acid mine drainage) and combustion products (ash, slag, FGD sludge, FBC sorbent) | Surface disposal source of declining esthetic and property values, erosion, landslides, and gob pile fires.<br><br>All disposal methods risk pollution of surface, ground water by leaching. | • R&D regarding reuse of waste materials (e.g., in highway pavement, as a mineral source, and as backfill for underground mines. Major objective is to find methods to prevent leaching of toxic materials, to allow disposal under RCRA. |

## Table 62.—Environmental Impacts (Land)—Continued

| Problem | Impacts | Solutions/policies |
|---|---|---|
| Reclamation of surface-mined lands | Inadequately reclaimed land is less productive and may be subject to erosion, landslides, and acid mine runoff.<br><br>Have had enforcement problems with smaller mines. | • Monitor implementation of Surface Mine Control and Reclamation Act (SMCRA).<br>• RD&D regarding feasibility of fully reclaiming and arid Western forest lands, steep slopes, prime farmland and alluvial valleys.<br>• Limit surface mining to lands where reclamation to former higher use is virtually certain. Develop criteria by which all coal lands can be classified in terms of their eligibility for surface mining. SMCRA would accomplish this if State programs are strong.<br>• Adopt measures that favor underground mining (e.g., requirement that FGD scrubbers be placed on existing as well as new plants). |

### Environmental Impacts (Water Quality and Quantity)

| Problem | Impacts | Solutions/policies |
|---|---|---|
| Acid mine drainage | Pollution of surface and ground water to the detriment of both flora, fauna, and community drinking water.<br>Problem most acute regarding abandoned underground mines in pyrite rich strata.<br><br>Have had enforcement problems with smaller mines. | • **Active mines:** Strengthen enforcement effort to insure full coverage of small mines, detection of illegal mining.<br>• **Abandoned mines:** Lengthen bonding period to ensure permanent acid control.<br>• Establish a fund to pay for control failure, paid for by tax on mining.<br>• RD&D regarding improved control methods and technologies. |
| Land disposal of ash and FGD sludge | Leaching of salts and toxic trace elements into ground water and salts, trace elements, and small particles into surface flows. | • (See disposal of mine wastes and combustion products under land impacts above.)<br>• Research regarding pollutants, soils, and hydrology to obtain detailed understanding of site factors that influence impacts.<br>• Incentives to use regenerable scrubbers or to recycle waste from nonregenerable systems.<br>• Monitor employment of existing technologies to determine if present incentives are adequate. |
| Water consumption requirements of plant cooling systems | Stress on limited water supplies in arid regions (e.g., Upper Colorado River Basin). | • Require water conservation in the design and operation of energy facilities (e.g., dry and wet/dry cooling) in water-short areas.<br>• Water conservation in arid lands agriculture as a means of freeing additional water supplies. Methods include:<br>• More efficient irrigation.<br>• Switching to less water intensive crops.<br>• Adding irrigation efficiency and/or water use requirements to rules governing Federal water projects.<br>• Reexamination of new impoundment and irrigation projects designed to supply low cost water to farmers. |

## Table 62.—Environmental Impacts (Air Quality)—Continued

| Pollutant | Impacts | Solutions/Policies |
|---|---|---|
| $CO_2$ | Possible atmospheric heating with consequent climatic shifts. | • Pursue energy strategies that preserve non-fossil fuel options. |
| $SO_x$ | Stunts agricultural crops.<br>Possible adverse health effects.<br>Acid rain.<br>Visibility degradation. | • Increased research regarding $SO_x$ chemistry and transport.<br>• Increased research regarding environmental effects.<br>• Increased research regarding health effects (demographic studies).<br>• Upgrade air quality monitoring network.<br>• Upgrade capability for modeling long-range transport.<br>• Designation of "regional problem areas" forcing upgrading of SIPs including stricter emission standards for existing plants and more stringent siting criteria. |
| $NO_x$ | Photochemical oxidants that have adverse effects on health and agricultural productivity. | • Use of Japanese scrubbing technology.<br>• Increased R&D regarding $NO_x$ controls; maximum priority given to demonstrating low $NO_x$ combustion.<br>• Increased research regarding health and environmental effects of photochemical oxidants.<br>• Increased research regarding $NO_x$ chemistry and transport.<br>• Continue R&D regarding new combustion technologies and fuels.<br>• Increased research effort regarding environmental effects of acid rain. |
| Particulates | Adverse health effects of fine particulates.<br>Visibility degradation. | • Institute particulate emission standards that distinguishes by particle size.<br>• Require baghouses on new plants.<br>• R&D regarding new designs for electrostatic precipitators. |
| Hydrocarbons | Possible high levels of hydrocarbon emissions from small boilers with consequent adverse impacts on health. | • Increased research regarding chemistry of formation and health impacts.<br>• Increased research regarding levels of hydrocarbon emissions from small boilers.<br>• Establish design standards for small units.<br>• R&D regarding emission controls for small units.<br>• Devise system for monitoring and inspecting small units.<br>• Avoid promoting coal use in small boilers. |

## Implementation

Data may be adequate and regulations appropriate, but if the law is not implemented little will be achieved. The process of implementing environmental rules will require careful attention because implementation questions will loom increasingly large as the basic framework of environmental legislation and regulation is put in place. In the simplest case implementation involves the installation of a particular piece of equipment—with Government imposed sanctions the price of failure or recalcitrance. Increasingly, however, implementation rests on more complex and less tangible actions—planning, management, and general procedures. This is due to the site-specific nature of activities related to coal development (e.g., reclamation) and the interaction of environmental impacts with social, political, and economic concerns. To the extent that implementation (e.g., of combustion controls) can be based on technical hardware, it will be easier than if it depends on procedures. Implementation questions will focus on the various major items of recent legislation and regulation, notably the SMCRA, the Clean Air Act amendments, the revised requirements and procedures for environmental impact statements (EIS), and the Toxic Substances Control Act. The stringency and vigor, including timetables, with which these are interpreted and enforced will have a major influence on the way the balance is drawn between coal development and environmental protection. This in turn will depend in part on how effectively the responsible executive agencies (EPA, DOE, the Department of the Interior, and the Council on Environmental Quality (CEQ)) can coordinate their activities. More specific implementation questions relate to the content of forthcoming regulatory decisions regarding leasing and whether sludge will be classified as hazardous. Also, the thorny issue of how to deal with a State that is out of compliance with Federal clean air standards as a consequence of pollution transported from out of State will have to be addressed.

## Major Policy Problems

Among the many policy issues relating to the control of the environmental consequences of coal development, three stand out:

1. the utility of cost/benefit analysis,
2. the role of Government relative to the private sector in the development of control technologies, and
3. whether national point source emission standards should be imposed on existing as well as new facilities.

These are addressed in turn.

## THE ROLE OF COST/BENEFIT ANALYSIS

All environmental decisionmaking involves at least an implicit weighing of costs and benefits. Even a conscious decision not to measure costs—as in the Clean Air Act's requirements for establishing ambient air quality standards to protect public health—can be interpreted as a decision that the benefits of health are so high that they must outweigh any costs incurred in its protection.

Much environmental legislation contains language requiring a consideration of "costs" in standard-setting. For example, the Clean Air Act requires EPA to consider, in specifying a NSPS for controlling air pollution, "the cost of achieving . . . emission reduction(s), any nonair quality health and environmental impact, and energy requirements." However, in virtually every case the costs of control are described as a constraint rather than as a factor to be balanced against benefits. In all ai and water legislation spelling out the terms for selecting emission limitations, EPA is asked to select the "best technology available," not the most cost-effective.

In most public opinion surveys, Americans have supported spending for environmental improvement, even when such spending causes some economic hardship. However, this support is based on the public's perception that the benefits of environmental standards outweigh their costs. In the wake of a variety of regulatory decisions (e.g., the Clean Air Act amendments and the new proposed NSPS for control of $SO_x$, the proposed mining regulations issued by the Office of Surface Mining) designed to strengthen environmental controls, the industries subject to them have attempted to change public perception about the balance of costs and benefits. In addition, there have been many calls for requiring regulatory agencies to balance costs and benefits and explicitly defend their selected regulatory strategies in these terms.

Arguments raised in favor of requiring cost/benefit analysis include the following:

1. While general quantitative understanding of environmental impacts of coal development may not be well developed, many

individual areas of impact can be measured. (E.g., some of the costs of reclamation failure, pollution crop damages, costs to municipalities of pollution of drinking water sources.) Thus, cost/benefit type analyses may be appropriate for some standards, if not for all.

2. The problems currently associated with identifying and appropriately quantifying benefits may never be resolved unless regulatory agencies are forced to take benefits into account in their decision-making. With a limited research budget, the environmental agencies are not likely to pursue vigorously research that is not directly required for their regulatory functions. For example, EPA virtually abandoned its research program on environmental benefits and disbanded the responsible organization (the Washington Environmental Research Center) in a 1975 reorganization. (Some funding has recently been restored to this research area.)

3. Analytical techniques for dealing with risk and uncertainty can address many of the problems associated with cost/benefit analysis.

4. Public acceptance of expensive environmental controls may be jeopardized unless the benefits that these controls provide are clearly identified.

There are a number of arguments against requiring such a formal weighing of costs and benefits in environmental standard-setting:

1. The cost/benefit calculations performed in the past by public works agencies such as the Bureau of Reclamation and the Army Corps of Engineers have produced a widespread aversion to this form of analysis. Environmental benefits, being inherently difficult to quantify, have tended to be neglected in quantitative evaluations to the detriment of environmental values.

2. Knowledge about some of the most critical environmental effects of coal development is inadequate. Major controversies rage as to the magnitude and even the existence of specific environmental and health impacts. A prominent example is the controversy surrounding the associa-

tion that has been claimed between community death rates and sulfate pollution levels. Respectable scientific support can be found for estimates of annual deaths caused by today's air pollution ranging from zero to tens of thousands.

3. Even when environmental impacts can be quantified in physical terms, it is difficult to translate these impacts into a "language" that allows comparison with monetary costs of control. The state of the art of such translation is not well advanced.

4. The level of uncertainty involved in identifying and quantifying the benefits of pollution control could considerably increase the incidence of judicial rejection of environmental standards. For example, EPA has had considerable difficulty in promulgating enforceable effluent guidelines for water pollution control, which require mainly technical and economic analysis. A requirement for a careful balancing of costs and benefits could make matters far worse.

5. Any requirement to determine costs and benefits and/or to balance them in arriving at regulatory decisions may substantially delay the standard-setting process. Aside from the environmental damage that may occur, delays could hamper energy development by adding to the uncertainty currently faced by entrepreneurs.

At least two options are available for dealing with these questions short of requiring explicit cost/benefit analysis for all standards. Congress could:

## Option A.

Require regulatory agencies to state the expected benefits of their proposals. This would avoid the need for the agencies to conduct the difficult analysis involved in balancing economic costs and poorly quantified benefits (such as esthetic improvements, statistical risks of health injuries, etc.), but would force an explicit public discussion of benefits and probably would provide incentives for benefits research.

## Option B.

Establish an independent commission that will decide which, if any, forthcoming environmental standards must be set by an explicit balancing of costs and benefits. The basis for the commission's decisions would be the state of the art of impact assessment for the pollutants in question. This would take into account the sharp variations in the state of environmental research, although it may not provide an incentive for more vigorous research (actually, it could provide a negative incentive if the agencies perceive cost/benefit analysis as an undesirable requirement). The commission could be given the authority to review the environmental research programs of the regulatory agencies to ensure that fruitful areas of research are appropriately pursued.

### WHO SHOULD DEVELOP
### POLLUTION CONTROLS?

Interest in the control of $SO_x$ and $NO_x$ emissions has led to the development of flue-gas $SO_x$ and $NO_x$ scrubbers, new combustion technologies such as fluidized-bed combustion, and new fuels such as SRC. Many of these technologies are in early stages of development, and all will undergo continued refinement so long as they are considered desirable. The continued rapid development of such control technologies is critical to maintaining an effective environmental protection strategy, but the most efficient means to achieve such development has been a subject of considerable debate within the Government. The major issue is the role Government should play in either carrying out the required development itself or supporting comparable efforts by industry. This debate is equally applicable to water and land pollution control. Positions range from advocacy of Government "in house" development to the commercial stage, to total reliance on industry. (For example, EPA and the Office of Management and Budget (OMB) have argued for years over EPA's active role in developing scrubbers.)

The major arguments in favor of active Government participation are:

- NSPS must be based on demonstrated technologies; therefore, industry has an incentive to avoid demonstrating new controls or improvements in existing controls.
- An equipment manufacturer cannot "demonstrate" a control technology unless he places it on a commercial-scale plant. To do this he needs the cooperation of the polluting industry, which might not be forthcoming without Government intervention for the reason just indicated.
- EPA has been generally quite successful in its control development and demonstration program. Examples of successful controls developed or improved through this program include combustion modification for $NO_x$ control and flue-gas scrubbers for $SO_x$ control.
- An active R&D role by EPA and DOE is necessary in order to maintain the in-house expertise to allow competent, informed policymaking about control levels and technologies required under the Clean Air and Water Acts.
- In those areas where the polluting industries are public utilities, the behavior of their regulatory commissions can skew their behavior away from searching for the most efficient, least expensive controls. For instance, widespread allowance of fuel transportation cost "pass-throughs" and delays in granting rate increases for capital expenditures tend to push utilities away from capital-intensive control solutions even if these are the least costly and most effective options in the long run.

The major argument in favor of allowing industry to be the prime mover in developing control technologies is that Government should not attempt to do what private industry can do as well or better:

- The polluting industries have the largest reserve of personnel who are intimately familiar with the processes to be controlled and in an ideal position to develop the most efficient technologies possible. In addition, the industries have the strong incentive to develop controls that are inexpensive and conserving of their resources; this incentive may be weak in Government-research activities.

- A private pollution control industry exists that will pursue control technology development even if the polluting industry does not wish to. Also, there are cases of industries generating considerable profits from marketing controls they have developed for their own plants; this phenomenon adds considerably to their incentive for successful and efficient operation of control innovations.
- Private entities such as the Electric Power Research Institute (EPRI) are heavily involved in pollution control R&D, apparently at a high level of professional competence.
- The need to get facility siting approval in difficult areas—with bad meteorology, pre-existence of high levels of air and water pollution, or proximity to sensitive ecological areas—provides a continuing incentive for industry to develop improved pollution controls.
- The Clean Air Act allows exemptions to NSPS and BACT requirements for firms installing promising experimental equipment, providing additional incentive for private development.

The options available for dealing with this question are:

## Option A.

Maintain the status quo, i.e., continue Government spending in control development at its current level while continuing to cooperate with private development of controls (for instance, EPA and EPRI have a cooperative agreement for sharing research results). This option recognizes the value of encouraging a "dual track" of control development and accepts the existing incentives for industry development of controls as requiring Government participation.

## Option B.

Substantially increase Government spending in pollution control technologies, especially in areas where control development does not appear to match the seriousness of pollution problems. Although spending for controls on energy-producing industry is high, the en-

vironmental impacts of energy development depend strongly on the extent of control of nonenergy industry. Thus the environmental hazard posed by $SO_2$ emissions from a power-plant may be great or slight depending whether there are other sources of sulfur emission (e.g., smelters) nearby. EPA's control program for industrial processes has been at a low level of support for several years; increased spending in this area could conceivably have an eventual payoff in removing some constraints from energy development (e.g., in present nonattainment areas). A corollary of this option could be to restore EPA's authority to pursue large-scale demonstration of energy-related pollution controls; at the moment, DOE has this responsibility. Although DOE has recently indicated an interest in pursuing the development of improved controls for existing conversion technologies, its major emphasis in the past has been on new technologies such as FBC, gasification, etc.

This option clearly dismisses the argument that Government should not be heavily involved in activities that private industry is capable of doing (however, it must be recognized that most of the actual work sponsored by the Federal Government is contracted out rather than being conducted in-house).

## Option C.

Substantially decrease Government spending in pollution control development, while taking steps to provide increased incentive for industry to expand its efforts. These incentives might include: 1) changing the language of the Clean Air Act to make control requirements more "technology forcing" (as has been done with requirements for automobiles); 2) using economic inducements such as accelerated rates of depreciation for experimental technology, liberal Federal support for capital and operating costs of industry demonstration projects, Federal assistance in persuading local rate commissions to allow utilities to immediately incorporate the expense of new technologies into their rate base, and tax credits for testing of new technologies.

This option clearly is derived from the philosophy that private industry is the ideal developer of controls for its own technologies, while recognizing that the nature of the marketplace requires special incentives to encourage this development. This option accepts the risk that industry could choose to avoid development of new controls and presents the Federal Government with the choice of either taking harsh measures (fines or plant closings) or delaying or loosening control requirements.

## NATIONAL CONTROL STANDARDS FOR EXISTING FACILITIES

Under the present regulatory structure for control of air pollution jurisdiction over new sources is shared by the States and EPA but control over existing sources is the exclusive responsibility of the States. This has created concern about control of long-range, interstate transport of pollution and our ability to construct a rational, cost-effective strategy for emissions reductions. The divided regulatory structure may be adequate for protecting local air quality from locally produced pollution, but it contains no effective mechanism for protecting against air quality degradation caused by the interstate transport of pollution from existing sources. This happens because the characteristics that tend to lead to long-distance transport—tall stacks and persistent winds—also tend to minimize impacts on local air quality. Pollution sources with these characteristics usually would be loosely controlled by their SIPs. For example, large coal-fired powerplants in the Ohio River Basin have been associated with elevated sulfate levels hundreds of miles downwind, while their SIPs allow them to burn high-sulfur coal with no controls.

Because of EPA's lack of direct control of existing sources; national strategy must concentrate mainly on restricting emissions from new sources. The Clean Air Act's requirement for continuous technological controls leads to a high cost of $SO_2$ emission reduction for new sources; estimates of the cost of the EPA and DOE proposals range up to $1,000/ton of emissions reduced, and total costs of the NSPS controls will be several tens of billions of dollars by 1990. Although these controls will slow the rise of $SO_x$ emissions in the face of expanding coal use, the large quantity of emissions from existing powerplants will remain unaffected by NSPS and is not expected to decrease substantially in this time period. The questionable wisdom, from a national perspective, of simultaneously requiring maximum controls on new powerplants and lenient or no controls on a number of large existing plants is highlighted when the potential costs of controlling some of the existing plants are examined. For example, if a utility currently using 4 percent sulfur coal could obtain 2 percent sulfur coal at a $10/ton premium, it would achieve a 50-percent reduction in $SO_2$ emissions at a cost of $250/ton of emissions reduced.

Because long-term stabilization and reduction of $SO_x$ emissions depend on the gradual movement to reduced operations and eventual retirement of the older uncontrolled plants, changes in their expected operational patterns and retirement schedules can have severe effects on national emissions levels and air quality. The sharp differences between the operating costs of old and new plants due to emission control requirements or the latter may make the intensive operation of older plants more attractive to the utilities. NSPS requirements could have a perverse effect of increasing the emissions in some areas by encouraging a shift of baseload operations to the older, poorly controlled plants and by discouraging their retirement. Computer modeling has tended to confirm the potential for this effect.

Individual States are unlikely to change voluntarily their SIPs to eliminate these potential problems. An increase in control requirements would trade an economic cost to a State's constituents for a benefit to the residents of other States. In some cases, increased control of $SO_x$ could require a shift to out-of-State sources of coal, hurting in-State producers and miners. Finally, there is substantial controversy surrounding the impacts of long-range transport (acid rain, sulfate, health effects, etc.) so that the known benefits of emission reductions are less tangible than the costs. The options available to Congress for dealing with the issue of further controls on existing facilities are:

## Option A.

Maintain the present divided regulatory structure and forego additional reductions in emissions from existing sources until either further research results are obtained or less expensive or disruptive control alternatives are developed. This option can be combined with increased funding for research on the impacts and mechanisms of long-range pollutant transport and for development of controls. Control development would focus on low-cost alternatives suitable for retrofit. Examples are competitive low-sulfur/low-ash fuels like SRC, low-$NO_x$ burners, and modifications to electrostatic precipitators to provide better fine-particle control. This option avoids additional regulatory costs unless they are absolutely necessary or unless new controls can sharply reduce costs. The option is essentially a delaying tactic until more information is obtained and more options are available. It can be supported by recognizing that emissions are not expected to rise rapidly in the next decade or so. However, it involves the implicit acceptance of the risk that present emission levels are causing significant levels of health and environmental damage. These risks are discussed in detail in chapter V.

## Option B.

Amend the Clean Air Act to provide for Federal control of existing pollution sources. Require existing facilities to satisfy the same emissions standards as new facilities. Exempt plants that are close to retirement from full compliance, but require fuel switching to clean or cleaned coals when this alternative is available. Successful implementation of this option would drastically cut emissions of $SO_x$ and particulates. But because retrofit of controls is more expensive than incorporating the controls in the plant design, and the existing coal-fired capacity will represent the majority of total coal-fired capacity for several decades, the cost of this option should be in excess of that predicted for the proposed NSPS for steam electric utility boilers through 1990, or several tens of billions of dollars. For some plants, electricity production costs would increase by 20 percent or more. This option em-

bodies the idea that risks of the type associated with coal pollutants are unacceptable. However, implementation would be difficult. Besides the cost, the option would present substantial resource problems: the need to quickly construct large numbers of scrubber installations, the requirement to safely dispose of large quantities of scrubber sludge, and the need to train large numbers of operating personnel for new control systems. Given the difficulties that American utilities have had in operating scrubbers, this resource problem could substantially delay implementation.

## Option C.

Amend the Clean Air Act to provide for Federal control of existing pollution sources. Selectively increase controls on large facilities where the SIPs have been determined to be lenient or when the facilities have been determined to be a major source of problems associated with long-range transport of pollutants. Also, increase pressure to enforce SIP deadlines with tight enforcement resources and/or economic incentives. The $SO_x$ controls envisioned in this option would be primarily low-sulfur coals, cleaned coals, and eventually, SRC. Selective use of scrubbers might be justified for newer plants. Also, the R&D program described in Option A would be adopted. As in Option B, this course of action places a high premium on risk avoidance. It considers the potential dangers attributed to acid rain, fine particulates, and sulfates as significant enough to warrant considerable expenditure of control funds, but attempts to keep expenditures significantly below the costs of total control. It recognizes the variations among SIP regulations and the role that location plays in determining the seriousness of impacts. Nevertheless, it remains (like scenario B) an expenditure of considerable funds to combat impacts that for the most part have yet to be definitely proved; it can be expected to produce considerable opposition from utilities and industry on these grounds. Also, it can be expected to be opposed by Midwestern States whose powerplants, now burning local coal, might be asked to shift to low-sulfur coals from out-of-State sources. This latter problem can be overcome by speeding up the demonstration of advanced coal-cleaning processes and allowing the States to require their use in favor of out-of-State coals.

## COMMUNITY IMPACTS

The impacts of coal development on local communities and the response of those communities will be a significant factor influencing the future of coal as a national energy source. Some of the impacts—notably the generation of investment and employment in economically depressed areas—are almost universally viewed as positive. The adverse impacts, however, are of greatest concern to policymakers because they may jeopardize the projected conversion to coal and require ameliorative action by Government. The dislocations and social ills associated with boomtowns may lead to high worker turnover and low productivity in western mining areas. The utter inadequacy of such basic social infrastructure as housing, roads, and sewers may severely inhibit efforts to rapidly expand Appalachian coal production. Community resistance to siting of coal-fired powerplants may induce further slippage in already lagging construction timetables. Consequently, a viable national energy program would logically include policies designed to alleviate the negative community impacts associated with coal development.

Such policies, if they are to be effective, must recognize two basic characteristics of the present situation. First, there is considerable uncertainty regarding the nature of coal development impacts and the balance of benefits and costs that will accompany them. Policy perspectives vary accordingly. Proponents of community aid programs contend that future tax revenues and other economic benefits of coal development will never cover all the costs to the locality, thus necessitating a real net transfer of Government funds. An alternative view sees energy growth as similar in its impacts to other sources of growth and more likely than most to pay its own way in terms of community costs and benefits. From this perspective, the Federal Government role should be limited, except in special circum-stances, to supplementing existing policy mechanisms. Second, value disagreements concerning what are positive and what are negative impacts result even where the nature of these impacts is understood. Whether economic growth should be viewed as a positive or negative phenomenon is itself the subject of dispute. Value conflicts extend to other areas. For example, should community impact costs be borne by energy consumers through forcing the coal companies to internalize such costs and pass them along in the form of higher prices? Or should such costs be paid out of general tax revenue in the form of Government loans and grants to impacted communities? What is the extent, if any, of Government obligation to assist communities that will ultimately benefit economically from Federal policies to stimulate coal use but are unable to adequately cover the "front-end" costs of infrastructure and community service development in the short term?

Clearly, policies designed to ameliorate adverse impacts must recognize these uncertainties and value disagreements. This can be done with a policy approach that is basically accommodational, that seeks to anticipate the concerns of interested parties and deal with them in a way that encourages compromise. Such an approach has two principal characteristics.

First, all parties affected by increased coal use are able to participate in decisions concerning the location, timing, and scale of coal developments that directly concern them. Second, Federal policies are designed to distribute the risks, costs, and benefits equitably among all parties affected by increased coal development. The process of participation and consultation has two drawbacks; it can be expensive and time consuming. However, to the extent that adverse impacts can be anticipated and forestalled through consultation with the affected communities, the economic costs of ameliorating those impacts will be lessened. Moreover, the result can be a general upgrading of community capabilities.

Adverse community impacts associated with coal development take three major forms: 1) community services and infrastructure are overloaded, 2) local economies become over-heated and distorted, and 3) social instability accompanies a decline in some aspects of the quality of life. These are briefly summarized as a prelude to the analysis of possible policy responses.

## Overload of Existing Community Services

In many parts of the Eastern coalfields existing services are incapable of coping with the effects of a rapid expansion in coal development. The weakness has multiple dimensions. Local government services and infrastructure (utilities, police, roads, flood control, etc.) are often woefully inadequate. Public utilities are frequently incapable of meeting any significant additional demands without an overhaul of the entire system. Consequently the marginal costs of increased utility services are often very high. The commercial services and the supporting infrastructure in the private sector are also weak. Professional and skilled personnel are difficult to attract, given the living conditions in the coal towns. Private investment is discouraged by the risks associated with the historic boom and bust cycle of the coal industry. Local financial institutions have very limited resources. The tax system, particularly as it relates to the coal industry, is inadequate, reflecting a general weakness in local political institutions. Coal lands and enterprises are characteristically taxed at low, sometimes spectacularly low, rates. In many locales housing is in desperately short supply due to rugged topography, large landholdings by coal companies, a shortage of mortgage money, high utility costs, a low volume capacity of local builders, the absence of public housing programs, and other causes. Thus, in coping with the impacts of increased coal production the Eastern coalfield communities begin in a serious deficit situation with already inadequate services and supporting infrastructure. This situation rests, in turn on basic economic, social, and political underdevelopment rooted in a lack of economic diversification and absentee ownership of coal-related resources.

The Western coalfields present a different, although still troubled, picture. Community services and infrastructure are often extremely limited because of the characteristically small size and isolated location of Western coal towns. The explosive increase in demand for community services associated with large-scale coal development will occur before coal revenues become available. This front-end financing problem is exacerbated by the present small tax base and lack of bonding history or authority of the communities in question and the long leadtimes required to initiate coal production. The problem is in some instances aggravated by jurisdictional mismatches where the productive enterprise (and consequently the tax revenue) is located in a different political jurisdiction from the locality that must bear the brunt of the social costs. As in the East, supplying sufficient housing is a serious problem. The principal causes are inadequate sources of financing and the limited volume capacities of local builders. In sum, western efforts to provide community support to coal development must start largely from scratch. Compared to the East there is virtually no capability in place, but often there is also no service deficit to make up and no antiquated infrastructure that must be dismantled. Further, there is nothing comparable to the pervasive constraint in the East imposed by the high and virtually irreducible ratio of population to habitable land.

## Economic Dislocation

For the coal towns the benefits of a rapidly expanding coal industry are principally economic, but there are economic costs to the community as well. In the East as coal-related incomes and demand for goods and services rise, the commercial sector is often unable to keep pace, with rapid inflation the result. For those in the community on fixed incomes or otherwise not positioned to benefit from the income effect of an expanded coal industry, the consequences can be devastating. Other costs of coal growth can include a reinforcement of the existing single primary product economy, with its associated vulnerabilities. There are opportunity costs as coal development inhibits or forecloses other beneficial economic activities, notably recreation and tourism.

In the coal boomtowns of the West severe inflation is also a problem. Opportunity costs are present here as well, particularly disruption of the rural ranching economy because of coal development.

## Social Instability and Quality of Life

For many citizens of the community the social and psychological costs of rapid coal development may be the most difficult to bear. Heavy coal truck traffic on eastern rural roads not designed for such use can pose a growing threat to public safety and convenience. Overloading of local services plus sheer congestion can mean increasing emotional stress and even mark a perceived decline in the quality of life of established eastern communities. The boom and bust history of the industry can induce a basic mood of uncertainty about the future that contributes to the social and psychological malaise. In the West, boomtown conditions can lead to crowding, emotional stress, family problems, juvenile delinquency, crime, alcoholism, and other social ills rooted in a pervasive sense of rootlessness and impermanence. The situation is exacerbated by the tendency of coal development to erode the settled ranching culture of the area. Here too, uncertainties about the future of the industry can reinforce other problems.

These factors can adversely influence coal output in a number of ways, but particularly in terms of their impact on the work force. Miner dissatisfaction with community living conditions can translate into high worker turnover, absenteeism, low productivity, and strikes.

The policy options available for dealing with these problems fall into two categories: generic measures designed to improve the process of coping with coal development impacts nationally and specific initiatives designed to solve particular problems, some of which are peculiar to one geographical area. These two categories are discussed in turn. ·  .

## General Policy Options

Five major policy initiatives of this type can be identified:

First, improve the access of communities to information concerning coal development impacts. One way would be to establish a single national energy facility siting schedule, perhaps as part of a national energy information service. The objective would be to provide advance information about energy development plans to potentially affected communities. In this regard, DOE might be empowered to gather and disseminate energy-impact-related data available within the Federal Government, including OCS and BLM leasing plans and industry projects submitted for Federal licensing or approval. (Some regional efforts of this nature exist. For instance, EPA's Region 8 office in Denver publishes a continually updated list of planned facilities). On request the Federal Government might provide any locality scheduled to be impacted by a major new energy facility with an assessment of the effect of that development on the community. Such an assessment might be conducted as a joint Federal-State-local effort. This could be accomplished through existing EIS procedures if they were modified to give added weight to community impacts along with the existing attention to impacts on flora and fauna.

Second, increase opportunities for local authorities and citizens to participate in decisions concerning the siting, construction, or expansion of coal facilities. This might involve support, in the form of financial assistance and access to information, of citizen interest groups acting as intervenors in Government proceedings concerning facility siting, and related decisions. Similarly, mechanisms might be established for giving State, local, and tribal authorities earlier access to relevant Government deliberations than is now commonly available. Consideration might also be given to supplementing the modified EIS procedures suggested above with a requirement that companies planning to construct energy facilities propose actions they will take to alleviate those adverse community impacts that have been identified.

Third, improve the capability of localities to manage coal development impacts. This might take the form of Federal technical assistance for development, planning, and impact assessment. The Government could also ease the financial burden on localities seeking to cope with adverse impacts. This might involve grant and loan programs to finance public sector personnel, construction, and other costs. The eligibility requirements of existing Federal community assistance programs might be modified to give priority to coal-impacted communities. The Federal tax code could be amended to encourage prepayment of State and local taxes by industries planning coal facilities. Also Federal programs can be used to encourage State initiatives to deal with community impacts, e.g., by according priority to States that have established mechanisms for State assistance to coal communities. Finally, Federal assistance could facilitate direct communication between responsible officials from different coal communities, thereby enabling community governments to share experiences in dealing with energy development impacts.

Fourth, improve coordination of Federal assistance programs. A variety of steps might be taken in this regard. DOE, for example, could be designated as the lead Federal agency with responsibility for coordinating Federal impact assistance programs through an interagency board at the regional level. An office within DOE could be established to coordinate community impact assistance planning with other energy-related planning. The Federal Regional Commissions could be asked to sponsor regular assessments of coal-related community impacts on a regional basis. The Commissions might also be accorded a substantive voice in the allocation of Federal impact assistance. Federal-State coordination might be improved by a requirement that Federal decisions affecting the siting or expansion of energy facilities be compatible with federally approved State impact mitigation strategies.

Fifth, modify the point criteria by which the eligibility of communities for Federal assistance is determined. These criteria are presently designed to permit assistance to communities in economic decline by basing eligibility on such measures as unemployment and the percentage of substandard housing. Programs under the Department of Housing and Urban Development and the Economic Development Administration of the Department of Commerce are prominent examples. A community suffering the dislocations of a coal boom cannot qualify under these criteria.

## Specific Policy Options

Listed below are specific Federal actions that might alleviate some of the potential adverse impacts associated with coal development. These are in addition to the broader generic actions outlined above.

### COMMUNITY SERVICES AND INFRASTRUCTURE

- Target Federal highway funds toward road construction (including overpasses), improvements, and repairs necessitated by coal development.
- Increase Federal funds available through the Farmers Home Administration for water and sewage systems in coal-impacted localities and broaden eligibility for such assistance to include communities with over 10,000 population. Larger communities do not presently qualify for FmHA funding.
- Establish a new Federal loan program to finance the construction of public works needed to meet demands resulting from coal development. An alternative would be to provide a Federal guarantee of local bonds to finance such construction.
- Compel coal development companies to internalize some of the risk associated with front-end financing by prepayment of taxes, provision of certain community services, or guarantees of municipal bonds as a condition for obtaining needed Government licenses and approvals. Alternatively, an attempt could be made to induce such actions by coal companies in return for tax breaks or other inducements.
- Expand Federal subsidy programs for low- and moderate-income housing such as the FmHA homeownership program. Other possible Government initiatives to alleviate the shortage of housing include the use of Federal and State eminent domain authority to obtain land, encouragement of community nonprofit housing corporations with subsidies, modifications of the

tax laws to encourage coal development companies to provide employee housing, and Federal subsidies or guarantees designed to make mortgage loans with low downpayments available to young couples.

• Impose a Federal severance tax on coal with revenues going to States or localities to provide needed community infrastructure and services. Use funds generated by leases of Federal coal lands for the same purpose.

### ECONOMIC DISLOCATION

• Create a stable and predictable demand for coal by means of a national energy policy clearly committed to substantially increased coal use through the end of the century.

• Encourage non-coal-related private investment in the coalfield communities by use of tax incentives, preference in awards of Federal contracts, etc.

• Give preference to coal-producing areas in siting of Federal installations.

• Levy a national severance tax or royalty on coal production and use the revenues to establish development banks or trusts for investment in coalfield communities.

• Provide compensatory payments to communities that bear the social costs of mining federally owned coal or of coal development in adjacent Federal lands.

• Upgrade the capabilities of the Appalachia Regional Commission and its western counterparts to support a broad range of development programs.

• Help coalfield communities identify and develop alternative sources of economic growth (e.g., reconstruction of a historic narrow gauge railway as a tourist attraction).

### SOCIAL INSTABILITY AND QUALITY OF LIFE

• Provide loans, grants, bond guarantees, etc., to assist coal communities in establishing recreational and social service programs (e.g., family counseling, day care centers, adult education, parks).

• Assist community self-help projects (e.g., civic beautification).

• Fund study by Federal Regional Commissions concerning possible regional approaches to preserving existing farms and ranches.

• Fund joint study by Federal and State governments and Indian tribal governments regarding ways of limiting the corrosive impact of energy development on Indian culture and social systems.

Any comprehensive program to deal with the community consequences of coal development will have at least two objectives: the mitigation of specific adverse impacts and the diversification of the local economy. Clearly, such an effort will strain the resources of local jurisdictions and may require supplemental Federal funding—in the form of loans, grants, or both.

If it is assumed that revenues generated by coal development ultimately will prove adequate to cover the costs of impacts amelioration, the money can be raised by local taxes on the activities of the coal companies supplemented, if necessary, by Federal loans to cover front-end costs of public works and other infrastructure. The coal companies will presumably pass along the cost of taxes to consumers in the form of higher coal (i.e., higher energy) prices. This, in effect, puts coal development on a pay-as-you-go basis with the ultimate beneficiaries of coal production, the energy consumer, paying most of the bill. The feasibility of the pay-as-you-go option will depend on:

1. how heavily coal development activities can be taxed without pushing the price of coal to uncompetitive levels,
2. whether coal communities are willing and able to impose those levies that will capture the potential revenue, and
3. the availability of Federal loans.

If it is assumed instead that coal development revenues will not cover the external costs involved, the choice is between turning to the Federal or State governments for some sort of net transfer of funds or continuing the past practice of allowing the costs to fall on the local communities in the form of environmental and social deterioration. The limiting factors will be the availability of Federal or State funding and the absorptive capacity, in terms of environmental and social pressures, of individual coal communities.

Policymakers will also have to consider how the needed funds will be acquired and distributed. In practical terms, the communities have two methods of raising revenue from coal development, a property tax on company-owned coal lands and facilities within the local jurisdiction, and a severance tax on the value or quantity of coal actually mined. The difficulty with any local tax is that it may place the community at a competitive disadvantage compared to other localities in attracting coal investment.

The Federal Government has a much wider choice of sources for funds, including general Federal revenues based primarily on the income tax and a variety of levies specifically on coal production and use—including fees, rents, and royalties for leases on Federal land, and a national coal severance tax. Federal net transfers can be made available to coal communities in the form of direct payments (e.g., revenue sharing, grants, trust funds) according to some established formula (e.g., amount of Btu equivalents mined) or through a development bank that would provide assistance on a project-by-project basis.

## LABOR-MANAGEMENT RELATIONS

The history of labor-management relations in the coal industry has been a tumultuous one. Any analysis of the contribution of coal to U.S. energy requirements must take into account the possiblity of supply disruptions due to work stoppages. This was dramatically demonstrated in 1977-78 when the United Mine Workers of America (UMWA) and the Bituminous Coal Operators of America (BCOA) locked horns in a prolonged 109-day strike that effectively shut down about half of U.S. coal output. The reasons for this troubled history are multiple and complex. They include characteristics of the coal market, the ownership structure of the industry, the nature of the

workplace and work process, the social and political environment of the coalfields, and the accumulated ill-will and mistrust built up over decades of conflict between the operators and miners.

Policymakers face two basic tasks: first, how to ameliorate the sources of destructive labor-management relations and lay the groundwork for a more constructive long-term relationship; and second, how to deal with another lengthy strike should it occur.

Under the present legislation, the Federal Government can do little to directly alter the terms or context of labor-management relations. The principal exceptions are measures designed to ensure a growing or stable market for coal by mandating or inducing the use of coal instead of gas or oil. The recently passed National Energy Act contains a number of such provisions, including a prohibition against the use of oil and gas in new utility or industrial boilers, DOE authority to require capable facilities to use coal, and restrictions on the use of gas in existing utility powerplants. A stable market should ease the historic insecurity of both operators and miners that has been such a large factor in the industry's labor problems. Other measures are more indirect and relate to the basic social, environmental, and other ills of the coalfields that contribute to the miners' discontent. These options are outlined in earlier portions of this chapter.

In the event of another major coal strike the Federal Government will have an interest in achieving a settlement that is prompt, noninflationary, and that establishes the basis for long-term labor-management stability. If the latter condition is to be met, the settlement must be supported by a substantial majority of rank-and-file miners and must address some of the underlying problems noted previously.

In pursuit of these objectives, five major strategies or approaches will be available, as they were during the 1978 strike:

1. reliance on collective bargaining with limited Government intervention,
2. collective bargaining with strong Government involvement,

3. use of Taft-Hartley with limited efforts at enforcement,
4. Taft-Hartley with vigorous enforcement, and
5. Government seizure of the mines.

Each has its own set of opportunities and liabilities. Of the five, only the last will require legislation. It should be emphasized that these options are purely instrumental; they are independent of any judgment on the substantive merit of the labor/management issues in dispute.

## Collective Bargaining With Limited Government Intervention

Under this approach resolution of the strike would be essentially left to bargaining between negotiators for UMWA and BCOA. The Federal Government would limit itself to encouraging the negotiators by "jawboning," meditation, and public appeals to the industry rank-and-file miners.

The major argument in favor of this policy is that it enables the Government to avoid a direct confrontation with either the operators or the miners. If strike-related damage to the economy can be kept to manageable levels (due perhaps to mild weather and increasing production from nonunion mines), this strategy promises to leave BCOA and UMWA increasingly isolated as their mutual leverage over the rest of society declines. Under these conditions the strike may be seen by both labor and management as increasingly self-defeating, thereby hastening a negotiated settlement.

The principal arguments against this approach are that it offers no assurance of a quick end to the strike or that a settlement, when achieved, will be anything more than a temporary truce borne of mutual exhaustion that leaves the underlying causes of labor unrest in the coalfields intact. Moreover, if a prompt settlement is not achieved, the credibility of the Government's energy policy may be seriously undermined with consequent long-term damage to the economy and the effort to reduce reliance on imported fuels. Finally, a hands-off policy may contribute to

an image of Government ineffectiveness in dealing with the problems of an industry that has been identified as the key to America's energy future.

## Collective Bargaining With Strong Government Intervention

This approach would involve preserving the framework of collective bargaining while bringing substantial Government pressure to bear on one or both parties to modify their negotiating positions. The instrumentalities of pressure will differ depending on whether the operators or the union is the target. Means of influencing the operators could include:

1. manipulation of Government contracts,
2. threat of Government seizure of the mines,
3. vigorous implementation of antitrust laws regarding horizontal and vertical divestiture,
4. proposed changes in the tax laws (e.g., the coal depletion allowance, the write off for black lung benefits, rapidity of amortization, investment credits),
5. modifications in leasing regulations,
6. a tightening or loosening of regulations concerning coal imports and exports,
7. changes in the frequency of Federal health and safety inspections of mines, and
8. a general increasing or lessening of the Government regulatory and permitting burden on the industry.

Pressure on the union could take the form of:

1. withholding food stamps from strikers,
2. threats to investigate union finances,
3. modifications in National Labor Relations Board regulations to make it either easier or harder to organize new mines,
4. threats of preferential Government purchasing from nonunion mines,
5. greater or lesser Government willingness to make coal miners exceptions to national wage guidelines, and
6. an offer to explore means of using public money to strengthen union health care programs.

In addition, statements by Government spokesmen can be used in an effort to pressure or persuade both miners and operators.

## Taft-Hartley With Limited Enforcement

This strategy is similar to that apparently adopted by the administration following the second rejection of a tentative contract by the rank and file in early 1978. It involves the use of Taft-Hartley to obtain police protection for those miners who want to return to work. The principal focus would be nonunion mines shut down due to threats of violence by strikers at nearby union mines. No effort would be made to coerce unwilling miners back into the mines. As an inducement, miners who went back to work under a Taft-Hartley injunction could be paid at a new higher wage scale (within Government wage-price guidelines) pending a final settlement. The basic advantage of this approach is that it permits the Government to facilitate a return to production of nonunion mines without a major confrontation with striking miners. To the extent that production is increased, the pressure on UMWA and BCOA will mount. Warmer weather, the growing economic strain on miners and coalfield communities, and an increase in coal prices caused by supply shortages can all contribute to that pressure.

There are several arguments against this strategy. First, it does not assure a quick end to the strike and thus leaves open the possibility of all the negative economic effects identified in the limited intervention option. Second, by tacitly accepting defiance of Taft-Hartley by UMWA miners this policy tolerates disregard for the law and may contribute to a general impression of Government ineffectiveness in energy matters. Third, this approach will probably be widely perceived as favoring the operators at the expense of the miners. Fourth, the underlying assumption is that there is significant nonunion coal production that is not forthcoming because of a union strike. This may be a false premise. Finally, as with the first policy option, there is little reason to think that a settlement reached under this strategy will successfully address the root causes of labor unrest.

## Taft-Hartley With Vigorous Enforcement

With this strategy, every effort would be made to use Taft-Hartley to persuade and, if necessary, force miners to return to work. Tools available to the Government include all those listed under the second option plus fines levied on recalcitrant union locals and arrests of pickets. If necessary, separate agreements between UMWA locals and individual coal companies would be encouraged. Other aspects of the strategy might include the offer of a provisional wage increase to miners returning under Taft-Hartley and creation of a White House Commission to recommend terms of a new contract. The basic effect of this policy is to increase pressure on the union by making a strike illegal for 80 days. Any efforts toward encouraging local settlements and thereby fragmenting the industry are a threat to both the union and the operators. The result may induce renewed and productive BCOA-UMWA negotiations under threat of such fragmentation.

The principal argument in favor of this option is that it is designed to achieve an immediate restoration of coal production. At present both UMWA and BCOA are seriously divided internally and consequently vulnerable to the threat of fragmentation. Given that vulnerability, a possible outcome of this strategy would be to induce the miners and operators to resume serious negotiation in order to forestall the mutual danger posed by Government intervention. Also, an activist posture by the administration should have some political benefit in terms of providing an image of decisive national leadership. There may even be an important benefit if it becomes necessary to resort to individual agreements between specific union locals and coal companies because fragmentation may create the possibility of breaking the historic pattern of labor relations in the industry. It could create the opportunity for new leadership, new ideas, and a new structure for union-industry bargaining. This in turn may make it possible to address some of the root causes of labor unrest in the Eastern coalfields.

Arguments against the invocation and vigorous enforcement of Taft-Hartley center on the danger that the whole effort could be counterproductive. Previous attempts by the Government to use Taft-Hartley to force coal miners back to work have been uniformly unsuccessful. Even a successful attempt could be politically damaging by antagonizing other unions and blue-collar workers generally. On close examination, some of the tools available to Government are of dubious utility. Because many union locals may already be bankrupt, fines could be ineffective. A cutoff of welfare benefits and food stamps may be successfully challenged in the courts with the result that new legislation will be required. Reliance on Taft-Hartley will clearly antagonize UMWA, and to the extent fragmentation is threatened, BCOA as well. Fragmentation might well result in anarchy rather than a new industry structure, with competitive inflationary wage increases and labor instability the result. Localized settlements or an industrywide settlement reached under the threat of fragmentation are unlikely to systematically address the underlying causes of labor instability. Also, a strong and cohesive union will be required if the pervasive community and environmental problems of Appalachia are ever to be solved. It will not be easy to reconstitute an industrywide union; dismemberment of UMWA might prove to be effectively permanent. Moreover, a basic tenet of national labor policy is to facilitate, not undermine, collective bargaining.

## Government Seizure of the Mines

If invocation of the Taft-Hartley Act fails to bring settlement of a strike, the remaining option of last resort is seizure of the mines by the Federal Government. Such action would require the passage of enabling legislation by Congress. The case for legislation would presumably be made in terms of the strike's short-term impact on the economy and the long-term injury done to the Government's effort to meet the Nation's energy requirements through increased use of coal. It should be noted, however, that retrospective analysis of the 1978 strike indicates that the economic impacts, with localized exceptions, were quite manageable. Because of a projected increase in nonunion western production, a future strike would have even less damaging consequences for the national economy.

History suggests that if seizure is to be effective, the authorizing legislation must provide the President with the power to control the conditions under which seizure is carried out. This means authority to change the terms and conditions of employment, to decide when the property should be returned to private ownership, and to seek injunctions in Federal court if there is resistance to this control. The President also has at his command a number of sanctions—moral, economic, military, judicial, and legislative—the use of which Congress must be prepared to support if they are to be effective.

On taking control of the mines, the Government might try to resolve the impasse that led to the strike by mediating between the parties. More severe Government actions would include removal of the existing management personnel, alteration of the conditions and terms of employment, and wage increases (or decreases). Strikes would be forbidden. A decision concerning the conditions of takeover might follow an inquiry by a White House Commission and consultations with miners and operators. The Government could choose to negotiate a new contract with the union and make its acceptance by management a condition for returning the mines to their former ownership. By directly negotiating a new labor contract the Government might be in a position to break the historic cycle of mistrust and hostility in coal labor relations. For example, a Government-negotiated contract might go far to meet the miners' demands concerning such key noneconomic issues as medical care and safety. At the same time a systematic effort to improve the quality of life in the coalfield communities could alter the environment that nourished the miners' discontent. To the extent these objectives are achieved, the climate for a successful implementation of a national energy policy will be markedly improved.

Arguments for this approach center on the contention that it is a way of acheiving both a quick and durable settlement. It is seen as an extreme remedy necessitated by the severity and intractibility of the problem. This strategy is also attractive on the grounds that it will probably be welcomed by UMWA and organized labor generally and should convey an image of governmental decisiveness and vigor to the public as a whole.

Arguments against this option begin with the observation that the historical record of the use of seizure in labor disputes has not been altogether happy. Of the 71 instances in which it has been used since the Civil War, only 40 resulted in agreements between labor and management before the property was returned to its owners. In the 31 remaining instances negotiations took place after the seizure was terminated and in 20 cases strikes occurred.

Moreover, in 38 of the seizure cases either labor or management obstructed production in some way while the Government was running the business. President Truman's seizure of the steel mills in 1952 was welcomed by the unions but eventually resulted in a 53 day strike. The fact that the seizure of the coal mines would require congressional legislation raises the possibility of congressional delay or veto. Significant congressional reservations. seem likely given the predictable opposition of the coal companies and a generalized uneasiness with seizure as socialist. To the extent that seizure undermines business confidence in the administration, the political costs could be high. Finally, the seizure or an attempted seizure of the mines will preoccupy congressional and executive energies and attention at the expense of other high priority concerns.

# WORKPLACE HEALTH AND SAFETY

Underground mining is a hazardous occupation as measured by the incidence of work-related accidents and disease. Health and safety issues have been a major factor behind labor unrest in the coal industry—including the 109-day UMWA strike of 1977-78. Any substantial increase in coal production will inevitably be purchased at the price of thousands of ill and injured miners.

## Health

A number of measures designed to upgrade the existing dust control system can be identified. They include:

- Assignment of higher priority to efforts to control nonrespirable dust.
- Development of new methods of mitigating the health impacts of dust from longwall mining, including improved respirator designs, fans, and special ventilation systems.
- Development of area sampler technologies (with appropriate standards) that would supplement personal samplers. Unlike the latter, these area devices should

be capable of providing an immediate, on the spot reading of dust levels.
- Reorganization of the respirable dust-sampling program around the new sampler technology and around the concept of miner control or joint control of the program. Shared management should minimize the opportunities for falsification of results that exist in the present program.
- Monitoring of the incidence of trace elements in coal dust.
- Support for research to resolve some of the present uncertainty concerning the adequacy of the 2-mg standard.

The Mine Safety and Health Administration (MSHA) is now considering alternative sampling systems to improve measuring and reporting reliability. Two promising approaches are in-mine dust measurement (allowing immediate correction of excessive dust) and continuous, machine-mounted monitoring that would automatically cut power to operating machinery when dust levels are too high. Both of these approaches would require new capital investment and could impede production during the year or two they were installed. These costs would probably be offset over time by a lower prevalence of coal workers' pneumoconiosis (CWP) among workers and lower operator-financed compensation expenses. The present public costs of black lung compensation are substantial.

Other hazards—"nonrespirable" dust, trace elements, emissions from machinery fires and diesel engines, and the like—contribute to respiratory illness to one degree or another. These pollutants may work synergistically with coal dust, thereby increasing the hazard to the worker. Since the individual miner experiences these hazards cumulatively, research and prevention programs should be structured in holistic, multifactor fashion rather than in terms of single-factor hazards. Federal regulation may be necessary to cover the respiratory hazards the 1969 Act did not address specifically (e.g., trace elements). Improved monitoring and control technologies for these pollutants may be needed.

Noise and stress are other significant mine-health concerns. The National Institute of Occupational Safety and Health (NIOSH) has recommended a tightening of the current noise standards, but neither MSHA nor the Occupational Safety and Health Administration (OSHA) has proposed implementing regulations. Miners experience significantly more hearing loss than nonminers, and some studies have linked noise to accidents. A more stringent noise standard would undoubtedly require additional investment by mine operators. After this initial expense, however, production costs should not be increased and productivity may be enhanced by lowering workplace noise levels. Medical costs and compensation are likely to be reduced if a tougher standard is adopted.

Job stress to which noise contributes, has been implicated as an important causal agent in coronary heart disease, gastrintestinal malfunctions, severe nervous conditions, and other disorders.

Recent studies have found more anxiety, depression, irritation, and somatic complaints among miners than other blue-collar workers. Underground coal miners reported a high level

of emotional strain in a recent survey. Stress and strain are probably associated with absenteeism, workplace hostility, and lower productivity. Shift rotation may be an important source of job stress and disruption of family life. It may also be related to adverse health and safety effects. A Federal policy response to this situation could include banning or modifying certain work practices that research identifies as key sources of job-stress, and establishing a program for monitoring workers for signs of stress.

## Safety

As coal production increases, more workers are likely to be hired. If accident rates do not improve, the number of fatalities and disabling injuries will double and triple as production and employment doubles and triples. It is unlikely that any major labor-saving technology will be commercialized over the next 15 to 20 years that would sharply raise productivity and reduce the number of workers exposed to safety hazards. Therefore, it is necessary to assess accident prevention strategies in the context of existing mining systems.

The 1969 Coal Act sought to prevent coal mine disasters. It has succeeded. The number and severity of disasters have been cut substantially in the last 8 years. Other legislated safety measures—involving roof support, electrical hazards, blasting—have probably reduced fatalities and injuries, but the effects are often difficult to separate and measure.

The principal need in the area of safety is to develop a program to reduce the frequency of disabling injuries (there were 15,000 such injuries last year, each of which cost more than 2 months of lost time).

Several measures designed to mitigate the problem can be identified:

- Comprehensive, mandatory safety standards and design features for machinery.
- More safety conscious work procedures in the mine such as preventing the accumulation of debris along haulage ways. A related measure would be to package supplies (e.g., cement) in small enough units

to be handled easily in confined quarters thereby avoiding back and other muscle injuries.

- Specially designed prevention programs for high-accident mines devised jointly by labor, management, and Federal officials.
- Improved safety education and job training of workers and supervisors. This should have measurable safety benefits since so much of the work force is young and inexperienced.
- Greater frequency of visits by Federal safety inspectors. Studies have shown an inverse correlation between the incidence of inspections and accidents.
- Research into the relationship among the pace of work, production quotas, and accidents.

- A greater voice for miners in safety decisions.

Safety does not come easily or cheaply. Management and workers must practice it continually. Prevention-consciousness must be incorporated into the work attitudes of both management and labor. The costs of inadequate attention to safety are measured in lower productivity, compensation payments, lost production time, lower morale, absenteeism, and medical bills. Upgrading safety often involves capital investment, and may lower productivity and output. On the other hand, a conscientious approach to safety can help production and productivity, and often brings dollar benefits in terms of lower insurance premiums and medical bills.

# LEASING

The most basic policy question concerns whether additional leasing of Federal coal lands will be required to meet projected increases in coal demand. Western coal comprises roughly half of the Nation's coal reserves and the Federal Government owns 60 percent and indirectly controls another 20 percent of Western coal. The extent to which Western coal will be used to meet national energy demand is presently unclear due to the 1977 amendments to the Clean Air Act, which may have reduced the attractiveness of Western coal to utilities. Even if it turns out that substantial quantities of Western coal are required, it is not clear how much (if any) new Federal coal reserves would have to be leased. There are already 18 billion tons of Federal coal under lease and another 9 billion tons on land for which preference right lease applications (PRLAs) are pending. However, the status of PRLAs is very much uncertain as a consequence of the "commercialization quantities" test to be applied to all applications. Coal from PRLAs could meet national demand for a decade or, alternatively, many PRLAs could be found illegal or unable to pass the commercial quantities test. Given the uncertainties it is impossible to predict when any new Federal lands will have to be made available for leasing. In

any case, available sources of Western coal will be adequate for the next several years, regardless of the disposition of PRLAs. Moreover, recent court cases mean that no new leasing will be possible until the early 1980's at the earliest.

By that time it is possible that a tight coal market could create the conditions for a coal land rush in the West. The clearest way for the Federal Government to defuse such an event is to prepare contingency plans for the orderly resumption of coal leasing at that time. This will require the Government to select among various options for an overall long-term approach. Recently a high-level review committee within the Department of the Interior addressed this problem and produced an option paper.

The stated goal of the paper is to suggest alternative approaches for managing coal resources that emphasize environmental and land use planning and enable the Department of the Interior to adjust the amount of coal leased to production goals set by DOE.

For each suggested option, land use planning and environmental standards are employed to assess which "coal resource areas" are suitable for mining. Within those areas specific tracts would be identified for leasing. Production goals set by DOE would determine how many mines would be opened. The options differ as to whether industry or Government identifies the areas and tracts for leasing and designates the areas to be subject to environmental and land use assessment.

## Option I:
### Industry Nominates Areas and Tracts for Leasing

Under the first option proposed by the Review Committee, industry would nominate those areas it desired to mine. The Department of the Interior would then perform various land use and environmental studies to determine which of the nominated resource areas was suitable for mining. Interior would estimate the production potential of the suitable areas relation to the production goals set by DOE. A decision on which areas actually to lease would be made taking into account "by trading off the projected resource needs and environmental consequences of development." If the areas nominated by industry and found suitable for development failed to provide sufficient coal to meet production needs, then the Department of the Interior would designate additional areas on its own initiative. Finally, industry would nominate specific tracts that it desired to lease within each selected leasing area. The frequency and size of leases would be adjusted to meet DOE production goals.

## Option II:
### Government Identifies Areas for Leasing and Industry Nominates Specific Tracts

Under the proposed second option the selection of potential coal leasing areas would first be made by the Department of the Interior. Coal areas appearing to have "significant national potential" would be identified and subjected to environmental and land use analysis. Those areas found suitable for leasing would be "selected and prioritized by comparing resource, socioeconomic, and environmental

values." Using DOE production targets as a goal, leasing targets would be set for those areas found to have the highest priority. Industry would then be asked to nominate tracts within the selected areas that it wished to lease. The frequency and size of leases would be timed to meet the leasing targets for each area.

## Option III:
### Government Selects Areas and Tracts for Leasing

Under the third option suggested by the Review Committee, the Government would follow the procedure specified under Option II for selecting areas for mining and setting leasing targets for each area. In addition, however, the Government would select those specific tracts to be offered for lease without requesting nominations from industry. As with Option II the frequency and size of leases would be keyed to the area leasing targets.

### Comparison of the Options

Table 63 summarizes the Department of the Interior's comparison of these three options. As can be seen, in the Department's opinion, the more involved the Government becomes in the planning process the higher the cost to the Government and the greater the chance of production shortages produced by reason of Government errors in selecting areas that are not economically suitable for mining. On the other hand, the greater the Government role the more assurance there will be that adverse social, economic, and environmental impacts can be minimized.

Whichever approach is ultimately selected, a number of specific issues will have to be clarified. They include logical mining units, preference right lease applications, the requirements of diligent development and continued operation, estimated recoverable reserves, advance royalty payments, and the exchange of environmentally sensitive leased lands for other unleased Federal land. These topics are examined in chapter VI. There are also some important institutional issues concerning the division of responsibility between the Department of the Interior and DOE regarding leasing. Given the inherent complexity of the leasing process, considerable effort will be required to prevent it from becoming hopelessly lengthy and cumbersome.

**Table 63.—Comparison of Policy Options Under Consideration by the Department of the Interior**

|  | Option I | Option II | Option III |
|---|---|---|---|
| 1. Determination of production goals | Government | Government | Government |
| 2. Identifies areas for leasing | Industry | Government | Government |
| 3. Identifies tracts for leasing | Industry | Industry | Government |
| 4. Defines areas for environmental planning | Industry | Government | Government |
| 5. Cost of planning and administration | — | Increasing | ⟶ |
| 6. Chances for environmental mistakes | — | Decreasing | ⟶ |
| 7. Chances for production shortages | — | Increasing | ⟶ |
| 8. Likelihood of litigation | — | Decreasing | ⟶ |
| 9. Consideration of socioeconomic concerns | — | Increasing | ⟶ |

SOURCE: U.S. Department of the Interior.

# CONCLUSION

With the possible exception of carbon dioxide pollution, all the significant problems associated with substantially increased coal use appear to be solvable. That is, policy remedies exist to make coal a viable fuel option for the United States through 2000. But the fact that effective policies can be identified does not mean they will be adopted and implemented. If coal use is to be facilitated and the potential adverse impacts of such use controlled, a complex network of law and regulation will be required to deal with environmental, community, health, safety, and other impacts and to a lesser extent, coal supply and demand. Most of that framework already exists but some does not. Under the best of circumstances there will be difficult problems of coordination, administration, enforcement, technological improvement, and cost. In short, the ingredients for an effective national coal policy exist, but that achievement will not come easily or inexpensively.

# Glossary

# ACRONYMS

| | | | |
|---|---|---|---|
| AC | —alternating current | MSHA | —Mine Safety and Health Administration |
| AQCR | —air quality control region | NAAQS | —National Ambient Air Quality Standards |
| BACT | —best available control technology | | |
| BATEA | —best available technology economy achievable | NBS | —National Bureau of Standards |
| | | NEF #1 | —National Environmental Forecast No. 1 |
| BCPCT | —best conventional pollution control technology | NEPA | —National Environmental Policy Act |
| BCOA | —Bituminous Coal Operators Association | NHPA | —National Historic Preservation Act |
| | | NIH | —National Institutes of Health |
| BIA | —Bureau of Indian Affairs | NIOSH | —National Institute for Occupational Safety and Health |
| BLA | —Black Lung Association | | |
| BLM | —Bureau of Land Management | NIRA | —National Industrial Recovery Act |
| BNL | —Brookhaven National Laboratory | NO | —nitric oxide |
| BPCTCA | —best practicable control technology currently achievable | $NO_2$ | —nitrogen dioxide |
| | | $NO_x$ | —nitrogen oxide |
| CEQ | —Council on Environmental Quality | NPDES | —National Pollutant Discharge Elimination System |
| CHESS | —Community Health and Environmental Surveillance System | | |
| | | NRDC | —Natural Resources Defense Council |
| CO | —carbon monoxide | NSPS | —New Source Performance Standards |
| $CO_2$ | —carbon dioxide | OSHA | —Occupational Safety and Health Administration |
| CWP | —coal worker's pneumoconiosis | | |
| DC | —direct current | OSM | —Office of Surface Mining |
| DOE | —Department of Energy | PAN | —peroxyacyl nitrate |
| EIS | —environmental impact statement | PBN | —peroxybenzoyl nitrate |
| EPA | —Environmental Protection Agency | PIFUA | —Powerplant and Industrial Fuel Use Act |
| EPRI | —Electric Power Research Institute | | |
| ESECA | —Energy Supply and Environmental Coordination Act | POM | —polycyclic organic matter |
| | | PRLA | —preference right lease application |
| ESP | —electrostatic precipitator | PSD | —prevention of significant deterioration |
| FBC | —fluidized-bed combustion | RCRA | —Resource Conservation and Recovery Act |
| FGC | —flue-gas cleaning | | |
| FGD | —flue-gas desulfurization | SCS | —Soil Conservation Service |
| GAO | —General Accounting Office | SIP | —State implementation plan |
| GNP | —gross national product | SMCRA | —Surface Mining Control and Reclamation Act |
| HEW | —Department of Health, Education, and Welfare | | |
| | | $SO_2$ | —sulfur dioxide |
| HGMS | —high-gradient magnetic separation | $SO_3$ | —sulfur trioxide |
| IEA | —International Energy Agency | $SO_x$ | —sulfur oxide |
| LAER | —lowest achievable emission rate | SRC | —solvent-refined coal |
| LMU | —logical mining unit | TDS | —total dissolved solid |
| MESA | —Mining Enforcement and Safety Administration | TSP | —total suspended particulates |
| | | TVA | —Tennessee Valley Authority |
| MFD | —Miners for Democracy | UARG | —Utility Air Regulation Group |
| MHD | —magnetohydrodynamics | UMWA | —United Mine Workers of America |
| MOPPS | —Market-Oriented Program Planning Study | WHO | —World Health Organization |
| | | WRC | —Water Resources Council |
| mpg | —miles per gallon | | |

# DEFINITIONS

*Acre-foot:* A measure of water 1 foot deep by an acre in area, or 43,560 ft³.

*Ambient Air Quality Standards:* According to the Clean Air Act of 1970, the air quality level which must be met to protect the public health (primary) and welfare (secondary). Secondary standards are more stringent than Primary Ambient Air Quality Standards.

*Anthracite Coal:* A hard, high rank coal with high fixed carbon.

*Aquifer:* A subsurface zone that yields economically important amounts of water to wells; a water-bearing stratum or permeable rock, sand, or gravel.

*Ash (fly ash):* Light-weight solid particles that are carried into the atmosphere by stack gases.

*Base Load:* The minimum load of a utility, electric or gas, over a given period of time.

*Best Available Control Technology (BACT):* A technology or technique that represents the most effective pollution control that has been demonstrated, used to establish emission or effluent control requirements for a polluting industry.

*Bituminous Coal:* The coal ranked below anthracite. It generally has a high heat content and is soft enough to be readily ground for easy combustion. It accounts for the bulk of all coal mined in this country.

*Black Lung Disease:* A group of pulmonary diseases that are common among coal miners.

*BTU:* British thermal unit, a measure of the energy required to raise one pound of water one degree Fahrenheit.

*Coal Gasification:* The process that produces synthetic gas from coal.

*Coal Liquefaction:* Conversion of coal to a liquid for use as synthetic petroleum.

*Commercial Sector:* A subsector of service industries that includes wholesale and retail trade, schools and other government nonmanufacturing facilities, hospitals and nursing homes, and hotels. As defined, this sector does not include transportation and household services.

*Criteria Pollutants:* Six pollutants identified prior to passage of the Clean Air Act Amendments which now have established Ambient Air Quality Standards, i.e., sulfur dioxide, particulate matter, carbon monoxide, photochemical oxidants, nonmethane hydrocarbons, and nitrogen oxides.

*Effluent:* Any water flowing out of an enclosure or source to a surface water or groundwater flow network.

*Elasticity:* The fractional change in a variable that is caused by a unit change in a second variable. Income elasticities are important in energy estimates, since these estimate the changes in quantities of energy demanded as incomes change.

*Electrostatic Precipitator (ESP):* A device for cleaning stack gas of particulates. The device first charges particles in the gas stream and then collects them on an oppositely charged surface.

*Enhanced Oil Recovery (EOR):* A variety of techniques (other than conventional pumping) for extracting additional quantities of oil from a well.

*Fertility Rate:* Average number of lifetime births per woman.

*Flue-Gas Desulfurization:* The use of a stack scrubber to reduce emissions of Sulfur oxides. See stack scrubber.

*Fluidized Bed:* A fluidized bed results when gas is blown upward through finely crushed particles. The gas separates the particles so that the mixture behaves like a turbulent liquid. Being developed for coal burning for greater efficiency and environmental control.

*Greenhouse Effect:* The potential rise in global atmospheric temperatures due to an increasing concentration of $CO_2$ in the atmosphere. $CO_2$ absorbs some of the heat radiation given off by the Earth, some of which is then reradiated back to the Earth.

*Gross Energy Demand:* The total amount of energy consumed by direct burning and indirect burning by utilities to generate electricity. Net energy demand includes direct burning of fuels and the energy content of consumed electricity. The difference between gross and net energy demand is a measure of the energy losses by utility conversion to electricity. The difference between gross and net energy demand is a measure of the energy losses by utility convers to electricity. About two-thirds of the energy input at the utility is lost in generation and transmission.

Gross National Product (GNP): The value of all goods and services produced in a given year. GNP is a "value added" concept. It is stated in either current or constant (real) dollars.

Groundwater: Subsurface water occupying the saturation zone from which wells and springs are fed; in a strict sense, this term applies only to water below the water table.

Heat Pump: A device that moves heat from one environment to another. In the winter it moves heat from the outside of a building to the inside, and in the summer it moves heat from the inside to the outside.

Hydroelectric: Electricity generated by water power.

Industry: Industry is an aggregate of three sectors—manufacturing, mining, and construction.

In Situ Processing: In-place processing of fuel by combustion without mining applies to oil shale and coal.

Joule: A unit of energy which is equivalent to 1 watt for 1 second. 1 Btu = 1,055 Joules.

Labor Force: The number of persons 16 years of age or older who are either employed or actively looking for work.

Lignite: The lowest rank coal from a heat content and fixed carbon standpoint.

Metallurgical Coal: Coal used in the steelmaking process. Its special properties and difficulty of extraction make it more expensive than steam coal.

Methane: $CH_4$, carburated hydrogen or marsh gas formed by the decomposition of organic matter. It is the most common gas found in coal mines.

NEP: National Energy Plan, the plan presented to Congress by President Carter in April 1977.

New Source Performance Standards: Standards set for new facilities to ensure that ambient standards are met and to limit the amount of a given pollutant a stationary source may emit over a given time.

Oil Shale: A finely grained sedimentary rock that contains an organic material, kerogen, which can be extracted and converted to the equivalent of petroleum.

Participation Rate: The percentage of persons 16 years of age or older who are either employed or actively looking for work. In 1976, the participation rate was 62.1 percent.

Particulate Matter: Solid airborne particles, such as ash.

Peak Power: The maximum amount of electrical energy consumed in any consecutive number of minutes, say 15 or 30 minutes, during a month.

Petrochemical Feedstocks: Petroleum used as an industrial raw material to manufacture goods, such as chemicals, rather than as a source of energy.

Prevention of Significant Deterioration (PSD): Pollution standards that have been set to protect air quality in regions that are already cleaner than the Ambient Air Quality Standards. Areas are divided into three categories determining the degree to which deterioration in the area will be allowed.

Prime Farmlands: Land defined by the Agriculture Department's Soil Conservation Service based on soil quality, growing season, and moisture supply needed to produce sustained high crop yields using modern farm methods.

Primitive Areas: Scenic and wild areas in the national forests that were set aside and preserved from timber cutting, mineral operations, etc., from 1930-39 by act of Congress; these areas can be added to the National Wilderness Preservation System established in 1964.

Process Steam and Heat: Steam and heat produced for industrial process uses, such as the activation of drive mechanisms and product processing.

*Productivity;* The value of goods or services produced by a worker in a given period of time, such as 1 hour. For the United States in 1975, this averaged $7.39. Increases in output over time are used to measure gains in productivity. A variety of time-periods are used, including output per worker per year. Also, productivity statements often refer to gains in private sector output per worker rather than output in the total economy.

*Quad:* One quadrillion ($10^{15}$) British thermal units (Btu).

*Reclamation:* Restoring mined land to productive use; includes replacement of topsoil, restoration of surface topography, waste disposal, and fertilization and revegetation.

*Reserves:* Resources of known location, quantity, and quality which are economically recoverable using currently available technologies.

*Residential Sector:* Includes all primary living units—houses, apartments, and mobile homes. Households are classified as follows: a) family households, which incorporate persons who are either married or blood related; b) primary individual households, which are made up either of single persons or incorporate two or more persons who are neither married or blood-related.

*Resources:* Mineral or ore estimates that include reserves, indentified deposits that cannot presently be extracted due to economical or technological reasons, and other deposits that have not been discovered but whose existence is inferred.

*Retrofit:* A modification of an existing structure, such as a house or its equipment to reduce energy requirements for heating or cooling. There are basic types of retrofit: equipment, such as a heat pump replacing less efficient equipment; and insulation, storm doors, calking, etc., designed to lower energy requirements.

*Seam:* A bed of coal or other valuable mineral of any thickness.

*Slurry Pipeline:* A pipeline that conveys a mixture of liquid and solid. The primary application proposed is to move coal long distances (over 300 miles) in a water mixture.

*Stack Scrubber:* An air pollution control device that usually uses a liquid spray to remove pollutants, such as sulfur dioxide or particulates, from a gas stream by absorption or chemical reaction. Scrubbers are also used to reduce the temperature of the emissions.

*Steam Coal:* Coal suitable for combustion in boilers. It is generally soft enough for easy grinding and less expensive than metallurgical coal or anthracite.

*Strip Mining:* A surface mining method that removes the overburden that covers the coal seam in a series of parallel strips.

*Subbituminous Coal:* A low rank coal with low fixed carbon and high percentages of volatile matter and moisture.

*Subsidence:* The sinking, descending, or lowering of the land surface; the surface depression over an underground mine that has been created by subsurface caving.

*Sulfates:* A class of secondary pollutants that includes acid-sulfates and neutral metallic sulfates.

*Sulfur:* An element that appears in many fossil fuels. In combustion of the fuel the sulfur combines with oxygen to form sulfur dioxide.

*Sulfur Dioxide:* One of several forms of sulfur in the air; an air pollutant generated principally from combustion of fuels that contain sulfur.

*Supply:* The functional connection between the price of a good and the quantity of that good that some agent is willing to sell at that price. The supply function is generally positive, or (geometrically speaking) up-sloping, meaning that as price goes up, the quantity supplied also goes up.

*Swing Fuel:* A fuel that plays a key role during the transition from exhaustible to inexhaustible fuels. Coal is viewed by many as the swing fuel during the transition.

*Synthetic Fuel:* A fuel produced by biologically, chemically, or thermally transforming other fuels or materials.

*Transportation Sector:* Includes five sub-sectors: 1) automobiles; 2) service trucks; 3) truck/bus/rail freight; 4) air transport; and 5) ship/barge/pipeline.

*Unit Train:* A system for delivering coal in which a string of cars, with distinctive markings and loaded to full visible capacity, is operated without service frills or stops along the way for cars to be cut in and out.

*Western Coal:* Can refer to all coal reserves west of the Mississippi. By Bureau of Mines definition, includes only those coal-fields west of straight line disecting Minnesota and running to the Western tip of Texas. Wyoming and Montana (sub-bituminous) and North Dakota (lignite) have the largest reserves.

# INDEX—THE DIRECT USE OF COAL

37, 112, 118, 173, 195, 218, 221, 224,
363, 368, 381-2, 385, 387, 389, 393,
394, 402-3
Department of Health, Education, and
Welfare, 345, 366, 367
Department of Housing and Urban Develop-
ment, 394
Department of Interior, 110
Department of Justice, 366
Department of Labor, 26, 260, 261, 346,
347
   Bureau of Labor Statistics, 140. See also
   Mine Safety and Health Administration
Department of the Interior, 26, 30, 261,
288, 342, 343, 345, 346-7, 366, 367,
368, 382, 385, 403, 404
Depression and job stress, 273
Desulfurization, flue-gas, 44
Devine, Mike, 313n.
Dewatering process, 77, 82
Diesels, health problems from, 272
Dingell, John, 140
Direct current (DC) transmission, 83
Directorates, interlocking, 118-19
Divorce, 323
Dochinger, L.S., 223n., 224n.
Double alkali, 96
Doyle, H.N., 263n.
Draglines, walking, 65
Drilling technology, 65
Drug use, 323
Dubofsky, Melvyn, 129n.
Duff, Mary K., 313n.
Dust, 24, 259
   Air pollution and, 187, 242
   Control of, 16
   Federal standards on, 266-9
   Health of workers, 262-3, 266-9,
   274
   Ions and, 94
   Particles of, 24
   Reduction of, 73
   Standards on, xi
Dyspnea, 264, 271

E

Eastern Associated, 133
Ebertz, Dave, 322n.
Economic dislocation, 392
Ecosystems, 9, 14, 187
Edgerton, E.S., 222n.
Education
   Opportunity for, 298-9
   School enrollments, 320
Electricity, 5, 11, 39
   Coal and, 11, 40-2, 53, 327
   Generation of decentralized, 42-4

Price of, 36
Production of, 53, 102
Transmission of, 83
Electric Power Research Institute
(EPRI), 37, 102, 199, 388
Electrostatic precipitators (ESPs),
9, 89, 93, 174-5, 196
El Paso Natural Gas, 117
Emery County
   Infrastructure in, 318-19
   Production in, 311, 314-15
Emphysema, 15, 208, 215, 259, 262
Endangered Species Act, 160, 164, 363
Energy
   Development of, 175
   Key factors in, 33-4
   Maximization of, 376
   Scenerios on, 34-8, 36, 39
   Supply alternatives in, 38-40, 39
Energy Minerals Allocation Recommendation
System (EMARS), 341, 342
Energy Supply and Environmental Coordin-
ation Act (ESECA), 337, 353, 361-3
Enterline, P.E., 263n.
Environmental impact statements (EIS),
25, 164, 332, 338, 339, 365, 366, 368,
385, 393
Environmental Protection Agency (EPA),
x, 11, 9, 26, 95, 99, 163, 166, 168-9,
171, 172, 173, 192, 195, 199, 200-1,
218, 223, 225, 231, 232, 235, 239,
243, 245, 249, 269, 338, 350, 351,
352, 353, 355, 356, 357, 358, 359,
368, 370
Environment and coal impacts, 183-6,
184-6
   Assumptions/predictions on, 186
   Impacts, 382-4
Enzi, Mayor, 323, 324
Epidemiological studies, xiii, 8, 203-4,
215, 219-20, 266
Equitable Life Assurance, 119
Erosion, control of, 81
Escavators, bucket wheel, 66
ESP, 12, 241
Esthetics and mining, 332-4, 332, 333
EXXON, 113, 117

F

Farmer, M.M., 45n.
Farmers Home Administration, 293, 394
Federal Aviation Administration, 358-9
Federal Coal Leasing Amendments, 341,
342
Federal Coal Management Review Policy
Committee, 342
Federal Coal Mine Health and Safety